Industrial Chemistry

Industrial Chemistry

Edited by **Gerald Cole**

Willford Press
New York

Published by Willford Press,
118-35 Queens Blvd., Suite 400,
Forest Hills, NY 11375, USA
www.willfordpress.com

Industrial Chemistry
Edited by Gerald Cole

International Standard Book Number: 978-1-68285-256-9 (Hardback)

Printed in China.

Contents

Preface

Industrial chemistry focuses on the processing of raw materials into end products. This book aims to understand such processes and their applications in the fields of nanochemistry, petrochemistry, etc. It delves into topics such as process control, coatings, lubricants, polymers, modeling, etc. Experimental and applied researches from across the world have been included in this extensive book.

This book is a result of research of several months to collate the most relevant data in the field.

When I was approached with the idea of this book and the proposal to edit it, I was overwhelmed. It gave me an opportunity to reach out to all those who share a common interest with me in this field. I had 3 main parameters for editing this text:

1. Accuracy - The data and information provided in this book should be up-to-date and valuable to the readers.

2. Structure - The data must be presented in a structured format for easy understanding and better grasping of the readers.

3. Universal Approach - This book not only targets students but also experts and innovators in the field, thus my aim was to present topics which are of use to all.

Thus, it took me a couple of months to finish the editing of this book.

I would like to make a special mention of my publisher who considered me worthy of this opportunity and also supported me throughout the editing process. I would also like to thank the editing team at the back-end who extended their help whenever required.

Editor

1

Investigation on the morphological characteristics of nanofiberous membrane as electrospun in the different processing parameters

T Mazoochi[1*], M Hamadanian[2,3], M Ahmadi[4] and V Jabbari[1,2*]

Abstract

In this study, polysulfonn nanofibers were electrospun and effects of electrospinning parameters including applied voltage and tip-to-collector distance on the morphology of electrospun PSf nanofibers were investigated by SEM. SEM images of electrospun nanofibers showed that morphology and diameter of the nanofibres were mainly affected by applied voltage. Fourier transform infrared spectrometer and thermo-gravimetric/differential scanning calorimeter analysis were used for investigate the chemical and thermal properties of PSf nanofibers, respectively. The applied voltage and tip-to-collector distance were all shown to have varying effects on fiber diameter and fiber uniformity. Increasing the applied voltage results in increases the surface charge of the jet and helps to reduce the frequency of occurrence of beads. It has shown that tip-to-collector distance directly affects spinnability and fiber diameter and at very low distance (5cm) droplets were formed due to the electro-spraying.

Keywords: Polysulfone, Electrospinning, Nanofiberous scaffold, Processing conditions

Background

Nanofibers have attracted interest and have been widely investigated for potential applications in electronics, mechanics, optoelectronics, and catalysis and for medical, biological, and environmental applications because of their several specific properties (e.g., high surface area to volume ratio, a very high aspect ratio, improved mechanical performance, relatively small pore size, and flexibility in surface functionalities) [1,2]. Nanofibers can be prepared by different processing techniques which include the following: (1) template synthesis [3], (2) self-assembly [4], (3) phase separation [5], (4) drawing [6], (5) melt-blowing [7], and (6) electrospinning [8]. Among them, electrospinning stands out as the most promising technique for fabrication of nanofibers. Although the electrostatic spinning processes have been discovered long time ago, electrospinning has gained much interest only by the end of the 20th century.

Electrospinning technique has attracted significant attention as a manufacturing process for producing nanofiber materials with diameters ranging from the micrometer to nanometer scales [9]. This technique provides a promising and straightforward way to fabricate infinite and continuous fibers applied as nanostructure and biomedical materials [10]. A wide range of materials such as engineered polymers, biological polymers, ceramics, and composites has been successfully electrospun into one-dimensional materials having many different microstructures [11]. Featuring the various outstanding properties such as very small fiber diameters, large surface area per mass ratio, high porosity along with small pore sizes [11], flexibility, and superior mechanical properties [12], nanofiber mats have found numerous applications in biomedicine (tissue engineering, drug delivery, and wound dressing), filtration, protective clothing, reinforcement in composite materials, and microelectronics (battery, transistors, supercapacitors, sensors, and display devices) [13-18].

The electrospinning process involves application of a high electric field to a polymer solution or polymer melt. A schematic of an electrospinning set-up is shown in

* Correspondence: mazoochi45@yahoo.com; vahid_jabbari.azeri@yahoo.com
[1]Anatomical Sciences Research Center, Kashan University of Medical Sciences, Kashan 87159-88141, Iran
[2]Institute of Nanoscience and Nanotechnology, University of Kashan, Kashan 87317-51167, Iran
Full list of author information is available at the end of the article

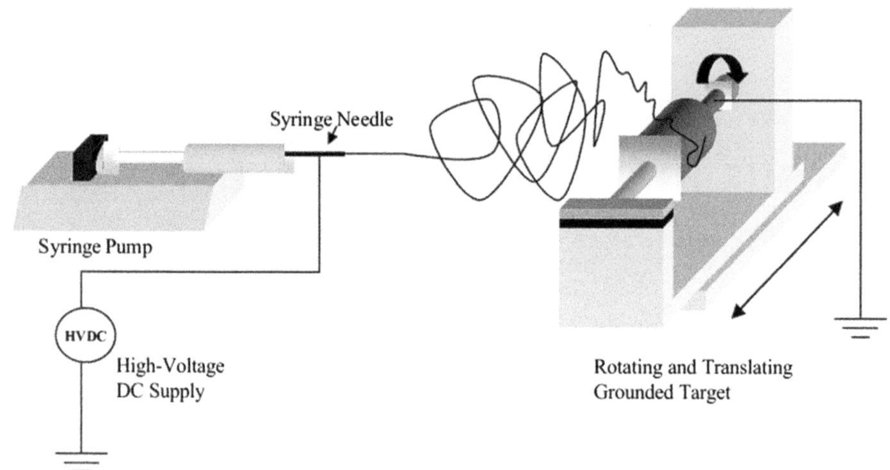

Figure 1 Schematic diagram of electrospinning setup.

Figure 1. There are three basic components including a high voltage supply device, a reservoir with a capillary tip for the spinning solution, and a metallic collector [19]. The polymer solution or melt is delivered through the capillary by means of an appropriate pump. One electrode lead of a high voltage power supply is immersed into the polymer solution or connected to the capillary tip of the reservoir, and the other one is connected to the collector. Applying high voltage (between 10 and 50 kV) on the solution induces electric charges. The mutual charge repulsion creates force acting oppositely to the solution surface tension. As the applied field strength is increased, the hemispherical solution surface at the tip of the capillary deforms into a conical shape (Taylor cone) [20]. When the applied field strength exceeds a threshold value, the repulsive electrostatic force overcomes the surface tension, and the charged jet is ejected from the tip of the Taylor cone. The small jet diameter permits rapid mass exchange, and the solvent usually evaporates during its travel from the capillary to the collector acting as a counter electrode. As a result, charged polymer fiber is deposited on the collector. The electrospun fiber charges are gradually neutralized in the environment. The end product of the process usually consists of randomly deposited fibers (mat) with diameters ranging from micrometers to nanometers.

Morphological characteristics of electrospun nanofibers, such as fiber diameter and uniformity, depend on many parameters which are mainly divided into the three categories: solution properties (solution concentration, solution viscosity, polymer molecular weight, and surface tension), processing conditions (applied voltage, volume flow rate, spinning distance, and needle diameter), and ambient conditions (temperature, humidity, and atmosphere pressure). It should be noted that among them, processing conditions are the most affecting parameter on the morphology of electrospun fibers [21]. In this study, electrospun polysulfone (PSf) nanofibers were fabricated by electrospinning 15% (*w*/*v*) PSf in *N*,*N*-dimethyl formamide (DMF) as solvent. Also, the effect of processing conditions (applied voltage and tip-to-collector distance (TCD)) on the fiber properties such as diameter and uniformity was analyzed, and the optimum conditions to make the best nanofiber mat were investigated.

Methods

Electrospinning of PSf nanofibers

PSf (Mw = 70,000, Sigma-Aldrich Corporation, St. Louis, MO, USA) and DMF as solvent (Merck Co., Germany) were used as received without further purification. It has been shown that electrospun PSf fibers can be produced using various solvent systems, and among them, DMF was found to be the most favorable solvent for producing uniform round fibers with smooth surfaces due to its high boiling point, high solution conductivity, and high dielectric constant compared to other solvents [22].

The electrospinning of the PSf solution was conducted using 15% (*w*/*v*) PSf (Scheme 1) solution in DMF. The complete electrospinning apparatus consisted of a syringe

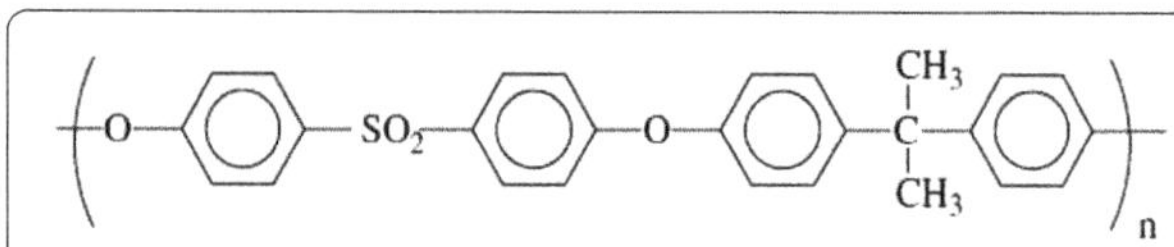

Scheme 1 Chemical structure of PSf.

Table 1 Processing parameters of electrospinning

Trial number	*V* (kv)	*D* (cm)	*C* (*w/v*)	*F* (ml/min)	*S* (rpm)	Average fiber diameter (nm)
1	10	5	15	0.5	1,000	320
2	10	10	15	0.5	1,000	300
3	10	15	15	0.5	1,000	270
4	15	5	15	0.5	1,000	500
5	15	10	15	0.5	1,000	430
6	15	15	15	0.5	1,000	350
7	25	5	15	0.5	1,000	370
8	25	10	15	0.5	1,000	330
9	25	15	15	0.5	1,000	305

C, concentration; *D*, spinning distance; *F*, flow rate of solution; *S*, speed of rotating collector; *V*, voltage.

and stainless needle, ground electrode, copper plate covered by aluminum foil as a collector, and adjustable high voltage supply (FNM Co., Tehran, Iran). The prepared solution was placed into the syringe, and a positive lead from the power supply was attached to the external surface of the metal needle. When high voltage was applied across the solution and the grounded collector, the solutions in the syringe would be ejected from the tip of the needle to generate fibers, which would be collected on the grounded collector. In this study, the resulting solutions were electrospun at 10- to 25-kV applied positive voltage, 5- to 15-cm working distance (the distance between the needle tip and the collector), and 1 ml/h solution flow rate controlled by a syringe pump (Table 1).

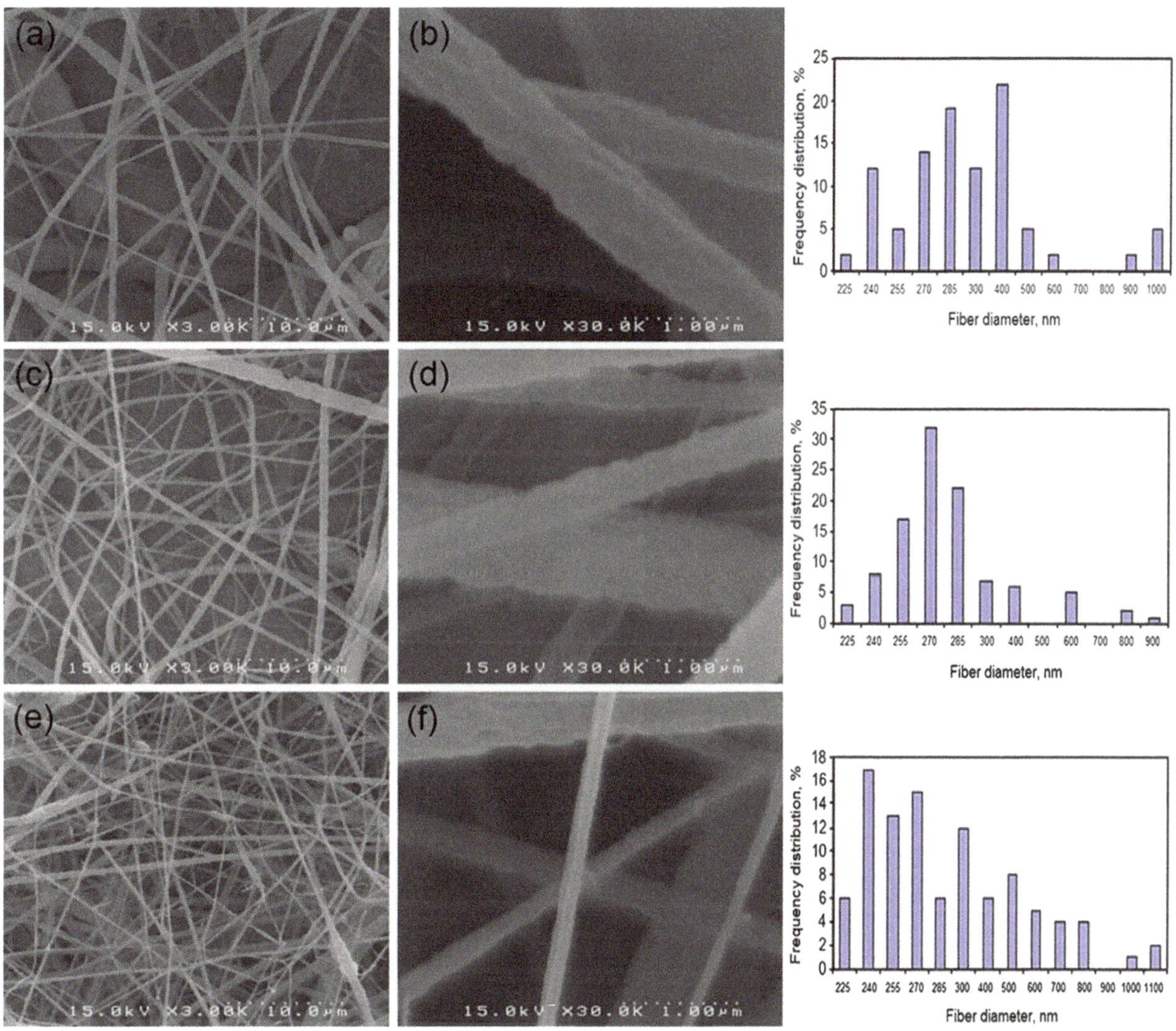

Figure 2 SEM images of PSf fibers at (a,b) 5, (c,d) 10, and (e,f) 15 cm TCD. In all of the experiments, $V = 15$ kV, $C = 8\%$ (*w/v*), $S = 1{,}000$, and $F = 0.5$ (mL/min).

Characterizations

The surface morphology of the electrospun fibers was observed on a JEOL JSM-5600LV scanning electron microscope (JEOL Ltd., Akishima, Tokyo, Japan) at an acceleration voltage of 10 to 25 kV. The samples for scanning electron microscopy (SEM) were dried under vacuum, mounted on metal stubs, and sputter-coated with gold. The mean PSf fiber diameters were estimated using image analysis J software and calculated by selecting 100 fibers randomly observed on the SEM image.

The samples used for Fourier transform infrared (FTIR) analysis were cast to be thin enough to ensure that the observed absorption was within the linearity range of the detector. FTIR spectra of the electrospun PSf mat were recorded using attenuated total reflection in an IR spectrometer (Nicolet Avatar 360, Madison, WI, USA.). The transmission infrared spectra of all samples exhibited broad peaks in a range from 4,000/cm to 400/cm.

Thermal stability of the electrospun PSf mats was examined by thermogravimetric/differential scanning calorimeter (TGA/DTA) experiments using SII model EXTAR TG 6200 (Seiko Instruments Inc., Chiba, Japan). All the samples were pre-weighed and allowed to undergo programmed heating in the temperature range of 50°C to 700°C at a rate of 10°C/min.

Results and discussion

Morphological properties: effect of processing conditions

Fiber formation using electrospinning technique is a complex process affected by a great number of

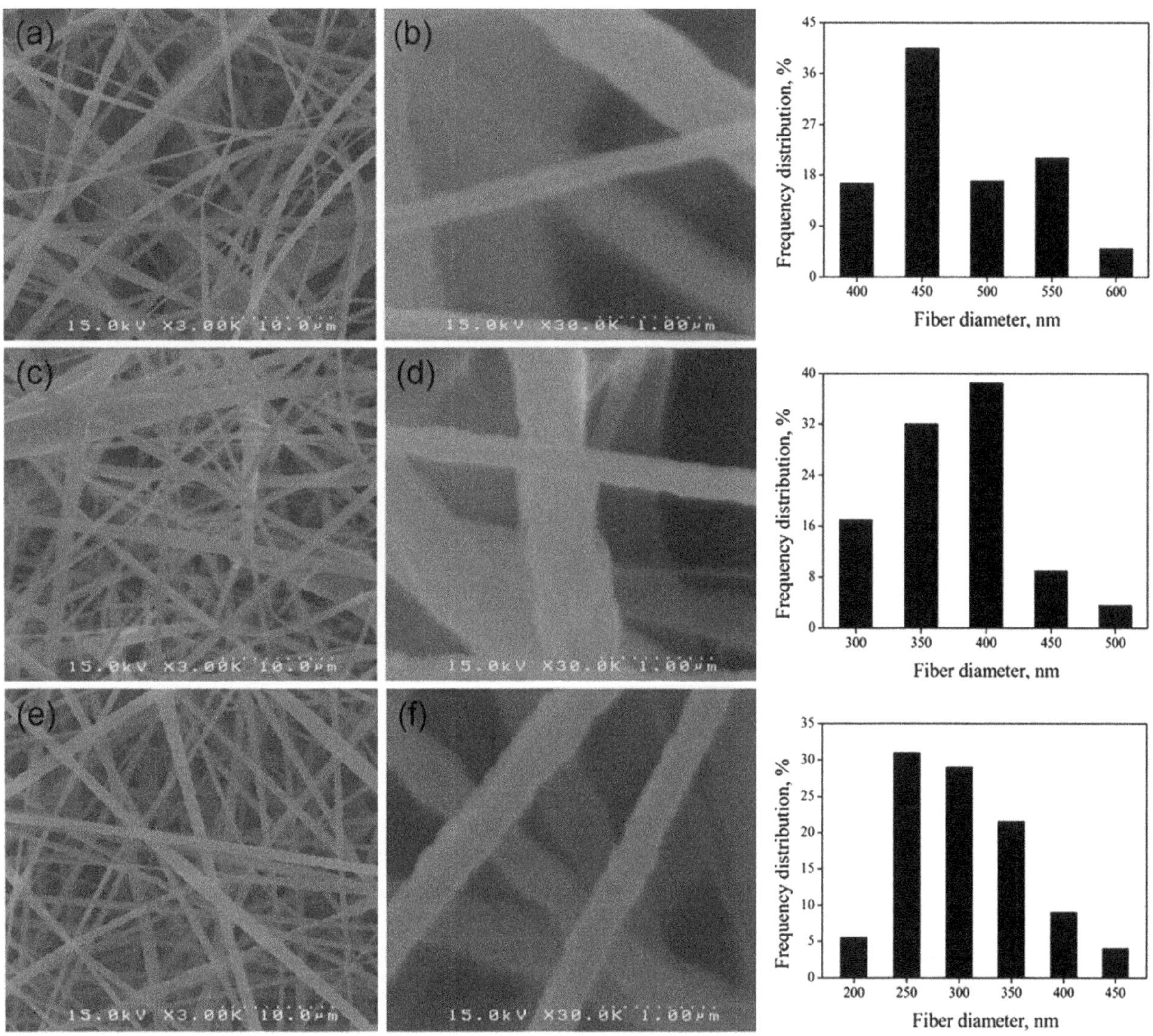

Figure 3 SEM images of PSf fibers at (a,b) 5, (c,d) 10, and (e,f) 15 cm TCD. In all of the experiments, $V = 20$ kV, $C = 8\%$ (w/v), $S = 1{,}000$, and $F = 0.5$ (ml/min).

parameters that generally include the following: (1) solution parameters, (2) process variables, and (3) ambient parameters.

Solution parameters include polymer type (natural or synthetic, organic or inorganic, monogenic or non-monogenic, linear or branched), polymer molar mass, solution characteristics (viscosity, volatility, dielectric constant), absence or presence of a low molecular weight organic or inorganic salt, conductivity of spinning solution, and solution surface tension. Process variables include electric potential at the capillary tip, the gap (distance between the tip and the collector (TCD)), and feed rate. Ambient parameters include air pressure, temperature, and humidity. Knowledge and control of these parameters are of great importance for successful preparation of micro- and nanofibers of polymers with a desired morphology. Among these parameters, the process variables have more effect on the morphological characteristics of fibers. In this study, effect of applied voltage (10 to 25 kV) and TCD (5 to 15 cm) on the fiber diameter and morphology were investigated and discussed with the results tabulated in Table 1.

Applied voltage

Figure 2 shows the morphology and average diameter of electrospun fibers fabricated at various levels of voltage applied to 15 wt.% PSf. Electrospinning was done at a solution flow rate of 100 ml/min and TCD of 5, 10, and 15 cm. It was shown that as the applied voltage was increased from 10 to 15 kV, the fiber diameter was increased, and after that with more increases of voltage, fiber diameters were decreased. At higher applied

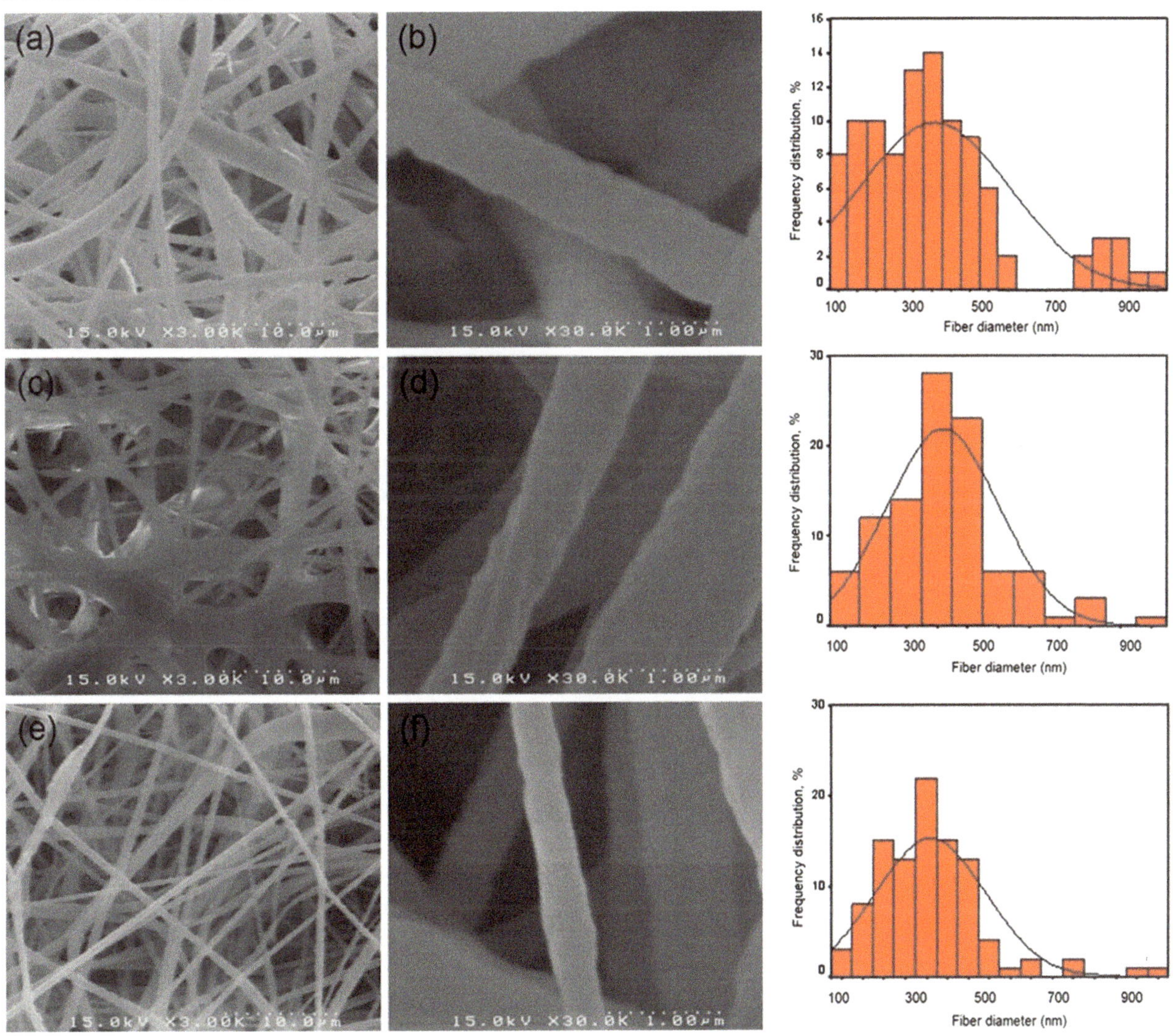

Figure 4 SEM images of PSf fibers at (a,b) 5, (c,d) 10, and (e,f) 15 cm TCD. In all of the experiments, $V = 25$ kV, $C = 8\%$ (*w/v*), $S = 1,000$, and $F = 0.5$ (ml/min).

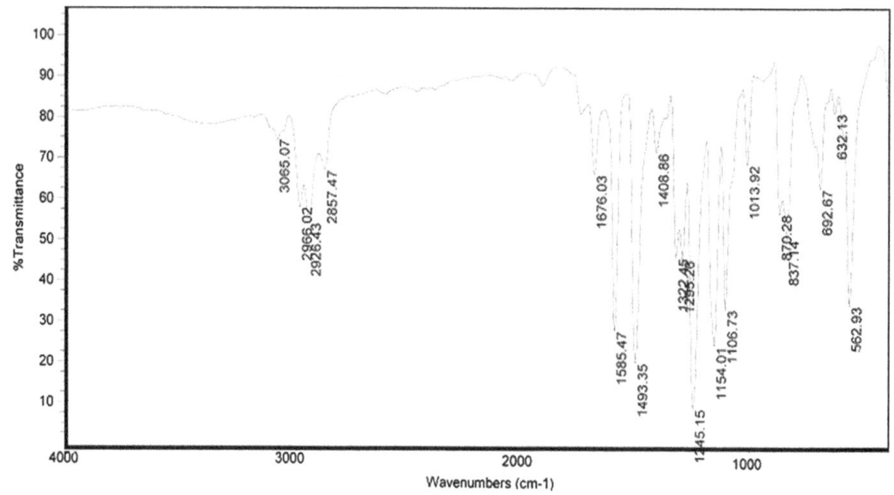

Figure 5 FTIR spectra of PSf.

voltages, the charge could be accelerated so that there is not enough time for the spinning solution to be developed, and thus, the fiber diameter increased as it was observed for the 20-kV applied voltage [23]. However, in most cases, a higher voltage will lead to greater stretching of the solution due to the greater columbic forces in the jet as well as the stronger electric field. These have the effect of reducing the diameter of the fibers as observed for 25-kV applied voltage [24].

In general, when high voltage is applied during the electrospinning process, Taylor cone formation becomes stable, and then, the columbic repulsive force within the jet of spinning solution makes the viscoelastic solution extended. If the applied voltage is higher than the critical point, wherein more charge will drop from the end of the needle due to the acceleration of charge, the Taylor cone becomes unstable [25].

Table 2 FTIR assignments of electrospun PSf nanofiber mat

Frequency (cm^{-1})	Assignments
3,065	O-H stretching vibrations
2,966; 2,926; 2,857	Asymmetric and symmetric C-H stretching vibrations involving entire methyl group
1,585; 1,493	Aromatic C=C stretching
1,408	Asymmetric C-H bending deformation of methyl group
1,365	Symmetric C-H bending deformation of methyl group
1,322; 1,295	Doublet resulting from asymmetric O = S = O stretching of sulfone group
1,245	Asymmetric C-O-C stretching of aryl ether group
1,154	Asymmetric O = S = O stretching of sulfonate group
1,106	Symmetric O = S = O stretching of sulfone group
1,013	Aromatic ring vibrations

Distance between tip and collector

Bead formation is the most common type of defect encountered in electrospun fibers and occurs primarily as a result of the instability of the jet under different process conditions [26]. Qualitatively, beads may be expected at times during the electrospinning whenever the surface tension forces tend to overcome the forces (including charge repulsion and viscoelastic forces) that favor the elongation of the continuous jet [27]. This process occurs intermittently, as fiber formation still remains a dominant process and consequently leads to the typical 'beads on a string' morphology described for a variety of different polymer-solvent systems (Figure 3) [28].

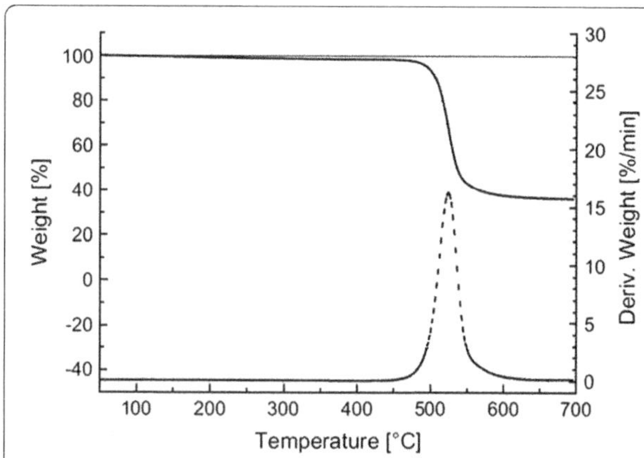

Figure 6 TGA/DTA diagram of PSf nanofibers. Solid line represents weight; dashed line represents derivative weight.

Theoretical analyses of the mechanism of the bead formation have been attempted by several groups. The rapidly elongating jet could undergo several different modes of instability. Analysis, for instance, predicted three modes of instability that can develop in extending the jet, of which two are axisymmetric. One of these is the Rayleigh instability, which is primarily governed by the surface tension, and the other is the conducting instability governed mainly by the electrical conductivity of the fluid. In axisymmetric instability, the axis of the electrospun fiber remains undisturbed, but its radius is modulated, yielding the wave-like deformations of the fiber that are the precursors of beads. Therefore, the processing conditions that favor axisymmetric instabilities also favor bead formation, whereas increased whipping instability discourages bead formation. Higher surface charge densities favor whipping instability over axisymmetric modes, therefore generally suppressing bead formation (Figure 4).

In general, depending on the solution property, the effect of varying the distance may or may not have a significant effect on the electrospun fiber morphology. In some cases, changing this distance has no significant effect on the fiber diameter. However, beads were observed to form when the distance was too low (Figure 3) [28]. The formation of beads may be the result of the increased field strength between the needle tip and the collector. Decreasing the distance has the same effect as increasing the applied voltage, and this will cause an increased in the field strength. As the field strength becomes too high, increased instability of the jet may encourage bead formation [29]. However, if the distance being such that the field strength is at an optimal value, there are less beads formed as the electrostatic field provides sufficient stretching force to the electrospun jet [30].

In other circumstances, increasing the distance results in a decrease in average fiber diameter (Figure 4) [31]. A longer distance means that there is a longer flight time for the polymer solution to be stretched before it is deposited on the collector [32]. However, there are some cases where at a longer distance, the fiber diameter increases. This is due to the decrease in the electrostatic field strength, resulting in less stretching of the resultant fibers [33]. When this distance is too large, no fibers are deposited on the collector. Therefore, it seems that there is an optimal electrostatic field strength; below which, the stretching of the PSf solution will decrease, resulting in increased fiber diameters.

Structural properties

FTIR spectroscopy

The FTIR spectra of the electrospun PSf nanofiber mat are shown in Figure 5, and the chemical assignments of PSf were illustrated in Table 2. The main peaks of PSf were characterized by the peaks at 1,585, 1,245, 1,322, 1,154, 1,106, and 1,013 cm^{-1} that correspond to the stretching of aromatic C = C, C-O-C (ether group), and O = S = O, respectively.

Thermal properties

The thermal decomposition of the PSf nanofiber mat has been studied using thermal analyses. Figure 6 shows the TGA/DTA diagram for the elecrospun fibers. TGA of PSf nanofibers showed two different stages of weight loss. The first stage ranges between 100°C and 520°C that may correspond to the loss of adsorbed and bound water and other organics. The second stage of weight loss starts at around 520°C, in which there was 90% weight loss, due to the degradation of PSf. The DTA of PSf shows a broad endothermic peak around 520°C. This peak is attributed to the decomposition of the PSf nanofiber mat.

Experimental

Polysulfone (PSf) (Mw = 70,000, Aldrich Co., St. Louis, 104Q6 MO, USA) and N,N-dimethyl formamide (DMF) as solvent (Merck Co., Germany) were used as received without further purification. It has been shown that electrospun PSf fibers can be produced by using various solvent systems, and among them, DMF was found to be the most favorable solvent for producing uniform round fibers with smooth surfaces due to its high boiling point, high solution conductivity and high dielectric constant compared to other solvents [22].

Conclusion

This study deals with the effects of processing variables including applied voltage and TCD on the morphological properties of electrospun PSf fibers that were investigated quantitatively as well as qualitatively. The appropriate range of parameters including dry, bead-free, and continuous fibers without breaking up to droplets was formed. It was observed that TCD has a direct influence on jet flight time and electric field strength. A decrease in this distance shortens flight times and solvent evaporation time and also increases the electric field strength, which results in more bead formation. Longer spinning distance resulted in more uniform PSf fibers. The effect of the spinning distances was more pronounced at higher applied voltages. Increasing the applied voltage increases the surface charge of the jet and helps to reduce the frequency of occurrence of beads.

Competing interests
The authors declare that they have no competing interests.

Acknowledgment
We acknowledge the scholarship, equipment and material support, and study fee waiver provided for T. Mazoochi by the Anatomical Sciences Research Center of Kashan University of Medical Sciences.

Author details
[1]Anatomical Sciences Research Center, Kashan University of Medical Sciences, Kashan 87159-88141, Iran. [2]Institute of Nanoscience and Nanotechnology, University of Kashan, Kashan 87317-51167, Iran. [3]Department of Physical Chemistry, Faculty of Chemistry, University of Kashan, Kashan 87317-51167, Iran. [4]Department of Chemistry, Faculty of Sciences, Shahid Beheshti University, Tehran 19835-389, Iran.

Authors' contributions
All authors read and approved the final manuscript.

References

1. Li D, Xia Y (2004) Adv Mater 16:1151
2. Yongyi Y, Puxin Z, Hai Y, Anjian N, Xushan G, Dacheng W (2006) Frontiers of Chemistry in China 1:334
3. Martin CR (1996) Chem Mater 8:1739
4. Whitesides GMB (2002) Grzybowski. Science 295:2418
5. Ma PXR, Zhang J (1999) Biomed Mat Res 46:60
6. Ondarcuhu TC (1998) Joachim. Europhys Lett 42:215
7. Fabbricante AG, Ward T (2000) Fabbricante. US Patent 114:017
8. Mazoochi T, Jabbari V (2011) Int J Polymer Anal Charact 16:277
9. Gopal R, Kaur S, Feng C, Chan C, Ramakrishna S, Tabe S, Matsuura T (2007) J Membr Sci 289:210
10. Xu Z, Gu Q, Hu H, Li F (2008) Environ Technol 29:13
11. Choi J, Lee KM, Wycisk R, Pintauro PN, Mather PT (2010) J Electrochem Soc 157:B919
12. Yoon KH, Kim KS, Wang XF, Fang DF, Hsiao BS, Chu B (2006) Polymer 47:2434
13. Wang H-S, Guo-Dong Fu, Li X-S (2009) Recent Pat Nanotechnol 3:21
14. Ma ZW, Masaya K, Ramakrishna S (2006) J Memb Sci 282:237
15. Lannutti J, Reneker DH, Ma T, Tomasko D, Farson D (2007) Materials Sci Eng C 27:504
16. Schiffman JD, Elimelech M (2011) Appl Mater Interfaces 3:462
17. Min BMG, Lee SH, Kim YS, Nam TS, Lee WH (2004) Park. Biomaterials 25:1289
18. Lee SW, Choi SW, Jo SM, Chin BD, Kim DY, Lee KY (2006) J Power Sources 163:41
19. Li G, Li P, Yu Y, Jia X, Zhang S, Yang X, Ryu S (2008) Mater Lett 62:511
20. Wang ZG, Wan LS, Liu Z, Huang XJ, Xu ZK (2009) J Mole Catalysis B: Enzymatic 56:189
21. Kim C (2005) J Power Sources 142:382
22. Ma ZW, Ramakrishna S (2006) J Appl Polym Sci 101:3835
23. Kim GT, Hwang YJ, Ahn YC, Shin HS, Lee JK, Soung CM (2005) Korean J Chem Eng 22:147
24. Zhao SL, Wu XH, Wang LG, Huang Y (2004) J Appl Polym Sci 91:242
25. Ramakrishna S, Fujihara K, Teo WE, Lim TC, Zuwei M (2005) An introduction to electrospinning and nanofibers. World Scientific Publishing Company, Singapore, p 101
26. Wang Y, Su Y, Sun Q, Ma X, Jiang Z (2006) J Membr Sci 282:44
27. Megelski SJ, Stephens S, Chase DB, Rabolt JF (2002) Macromolecules 35:8456
28. Renuga G, Satinderpal K, Chao YF, Casey C, Seeram R (2007) J Memb Sci 289:210
29. Deitzel JM, Kleinmeyer JD, Hirvonen JK, Tan NCB (2001) Army Research Laboratory ARL-TR-2415. Aberdeen Proving Grounds, March
30. Jarusuwannapoom T, Hongrojjanawiwat W, Jitjaicham S, Wannatong L, Nithitanakul M, Pattamaprom C, Koombhongse P, Rangkupan R, Supaphol P (2005) Eur Polymer J 41:409
31. Ayutsede J, Gandhi M, Sukigara S, Micklus M, Chen H-E, Ko F (1995) Polymer 46:1625
32. Reneker DH, Yarin AL, Fong H, Koombhongse S (2000) J Appl Phys 87:4531
33. Lee JS, Choi KH, Ghim HD, Kim SS, Chun DH, Kim HY, Lyoo WS (2004) J Appl Polym Sci 93:1638

2

Cadmium chloride ($CdCl_2$): a mild and efficient catalyst for the synthesis of benzimidazoles

B Sammaiah, D Sumalatha, GS Satyanarayana Reddy, M Rajeswari and LN Sharada*

Abstract

One-pot synthesis of benzimidazoles has been carried out using ortho-phenylenediamine and aldehydes. The condensation reaction smoothly took place in the presence of a mild Lewis acid cadmium chloride. All the reactions were carried out in acetonitrile at 80°C to 85°C.

Keywords: Benzimidazoles, Aldehydes, Ortho-phenylenediamine, Cadmium chloride

Background

Molecules having benzimidazole as a basic structural unit are known to exhibit a wide range of biological properties [1-3]. The potency and wide applicability of imidazole pharmacophore can be attributed to its hydrogen bond donor/acceptor capability as well as its high affinity for metals, which are present in many protein active sites [4-6]. The benzimidazole molecular skeleton is found in a number of clinical therapeutic agents such as anticancer, antiulcerative, antihypersensitive, antiviral, antifungal, antitumor and antihistaminic [7-9] (Figure 1). The synthesis of benzimodazoles and its derivatives occupies pivotal position in the field of organic synthesis. In this contest, several efforts have been developed for the synthesis of benzimidazole derivatives. One of the most common methods for the preparation of benzimidazole derivatives involves the condensation of arylenediamines and carbonyl compounds such as aldehydes and acid derivatives [10-12]. The condensation of phenylene diamine and carboxylic acid (derivatives) often requires strong acidic conditions and/or high temperature [13-18]. The other way involves the oxidative cyclodehydrogenation of Schiff bases, which is generated from phenylene diamine and aldehyde in the presence of various oxidative and catalytic reagents. This is the most popular approach in general for the synthesis of benzimidazole derivatives. The reagents are CAN, K_3PO_4, oxone, sulfamic acid, DDQ, $PhI(OAc)_2$, iodine and $KHSO_4$ [19-24]. In addition, several catalysts such as metal halides and metaloxychlorides [25-30], metal oxides, PTSA, metal triflate, ionic liquid, heteropoly acid, BDSB [31-36], proline, solid-supported catalysts, polymer-supported catalysts and microwave-promoted [37-41] reactions have been reported in the literature.

Unfortunately, many of these methods suffer from drawbacks such as drastic reaction conditions [14], low yields [16], tedious workup procedures [22] and co-occurrence of several side reactions [20]. As a consequence, the introduction of an efficient and mild method is still needed to overcome these limitations.

Methods

Results and discussion

Herein, we report a simple and efficient protocol for the synthesis of benzimidazoles using the catalyst cadmium chloride. In a preliminary study, we have examined the reaction of 1,2-phenylenediamine or ortho-phenylenediamine (OPD) (1 in Scheme 1) and benzaldehyde (2 in Scheme 1) in the presence of cadmium chloride to optimize the reaction conditions. The optimal conditions were found to involve the use of OPD and aldehydes in 1:1 molar ratios, with the catalyst loading of 10 mol% in acetonitrile at 80°C to 85°C. As per the optimized conditions, the OPD and benzaldehyde reaction was completed within 3 h to give the corresponding product of 2-phenylbenzimidazole (3a) in excellent yield, as shown in Scheme 1.

Encouraged by the result obtained with benzaldehyde and OPD, the method was applied to various aldehydes to establish the generality of the protocol. As shown in the Table 1, aromatic, heteroaromatic, α- and β-unsaturated aldehyde and aliphatic aldehydes reacted very well

* Correspondence: sumalathadandu@yahoo.com
Department of Chemistry, Osmania University, Koti, Hyderabad 500195, India

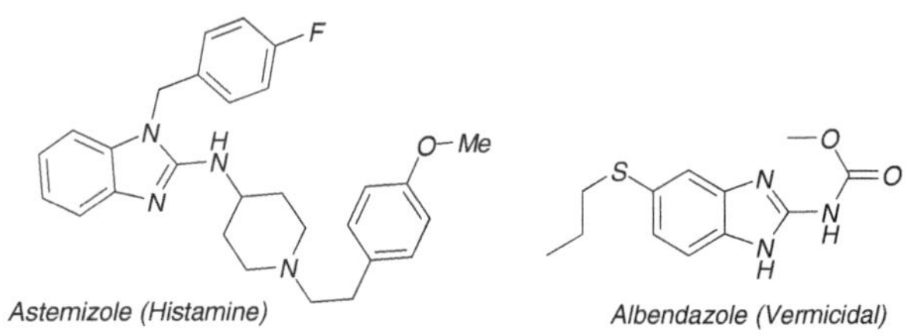

Figure 1 Benzimidazole molecular skeleton found in some clinical therapeutic agents.

to afford the corresponding products of benzimidazole derivatives in very good to excellent yields. In general, the aromatic aldehydes having electron-donating groups and heteroaromatic compounds reacted faster when compared with other aldehydes. Also, the aliphatic aldehydes and aromatic aldehydes containing electron-withdrawing groups reacted comparatively slower in terms of reaction rate and yields. Among several organic solvents tested for this condensation reaction, such as dichloromethane, DMF, methanol, dioxane, THF and acetonitrile, the better conversion and easy isolation of products were found to be with acetonitrile. To test the solvent's role in the reaction, one experiment was carried out with benzaldehyde and OPD in the presence of the catalyst cadmium chloride, but in the absence of solvent. This reaction did not go to completion (80%) even after stirring for a period of 10 h. In a similar manner, one blank experiment was carried out without the use of catalyst, and no product was formed after 10 h of stirring. Finally, it was decided that the suitable conditions for this condensation is in a solvent and in the presence of an activator or promoter. In general, under the optimized condition, all the reactions went to completion within 3 to 5 h, with the desired product in the range of 80% to 90%. All of the products were characterized by their proton nuclear magnetic resonance (^{1}H NMR), infrared (IR) and mass spectroscopy data.

Experimental

General methods

Melting points were recorded on a Buchi R-535 apparatus (BUCHI India Private Ltd., Mumbai, India) and were uncorrected. IR spectra were recorded on a Perkin-Elmer FT-IR 240-c spectrophotometer (PerkinElmer, Inc., Waltham, MA, USA) using a KBr disk. ^{1}H NMR spectra were recorded on a Gemini-200 spectrometer (HORIBA India Private Ltd., New Delhi, India) in $CDCl_3$ using TMS as internal standard. Mass spectra were recorded on a Finnigan MAT 1020 mass spectrometer (THERMO Scientific, Waltham, MA, USA) operating at 70 eV.

General procedure

A mixture of ortho-phenylenediamine (216 mg, 2 mmol) and benzaldehyde (212 mg, 2 mmol) in the presence of cadmium chloride (18.3 mg, 10 mol%) was stirred in acetonitrile (5 mL) at 80°C to 85°C. The progress of the reaction was monitored by thin layer chromatography (TLC). After completion of the reaction as indicated by TLC, the solvent was removed under reduced pressure. The residue was dissolved in ethyl acetate and washed with water and brine. The organic layer was dried over Na_2SO_4 and concentrated under reduced pressure. The crude products were purified by column chromatography using ethyl acetate-hexane mixture in 3:7 ratio. All the products were identified by their ^{1}H NMR, IR and mass spectroscopy data and compared with literature reports.

Spectral data for selected compounds

2-Phenylbenzimidazole (3a)

Melting range, 291°C to 293°C. IR (KBr): υ 3,296, 3,084, 2,957, 2,839, 1,651, 1,579, 1,523, 1,438, 1,396, 1,267, 1,159, 1,083, 967, 842 and 739 cm^{-1}. ^{1}H NMR (DMSO-d_6): δ 6.91 to 6.98 (m, 2 H), 7.18 to 7.28 (m, 4 H), 7.42 to 7.52 (m, 1 H), 7.90 to 7.96 (m, 1 H) and 11.85 (brs, 1 H, NH). EIMS m/z (%): 194 (m^{+}100), 192 (12), 179 (10), 167 (20), 117 (15), 103 (35), 77 (25), 76 (30) and 51 (40).

2-(4-Chlorophenyl) benzimidazole (3e)

Melting range, 292°C to 293°C. IR (KBr): υ 3,256, 3,061, 2,979, 2,847, 1,626, 1,548, 1,425, 1,372, 1,253, 1,139, 1,081, 1,015, 978, 835 and 743 cm^{-1}. ^{1}H NMR (DMSO-d_6): δ 7.02 to 7.15 (m, 2 H), 7.25 to 7.35 (m, 3 H), 7.50 to 7.60 (m, 1 H), 8.01 to 8.16 (m, 1 H) and 12.50 (brs, 1 H, NH). EIMS m/z (%): 230 (m^{+}10), 228 (15), 215 (15), 193 (30), 139 (52), 117 (100), 113 (12), 91 (15), 76 (70) and 51 (18).

2-(Naphthalen-2-yl) benzimidazole (3f)

IR (KBr): υ 3,312, 3,054, 2,969, 2,831, 1,623, 1,560, 1,431, 1,356, 1,249, 1,122, 1,132, 1,073, 1,006, 984, 827 and 749 cm^{-1}. ^{1}H NMR (DMSO-d_6): δ 7.20 to 7.30 (m, 2 H), 7.55 to 7.70 (m, 4 H), 7.90 to 8.10 (m, 3 H), 8.35 to 8.45 (m, 1 H), 8.75 (brs, 1 H) and 11.85 (brs, 1 H, NH). EIMS m/z (%): 244 (m^{+}100), 229 (10), 153 (40), 127 (65), 102 (18), 97 (21), 77 (20), 76 (22) and 51 (30).

2-Propylbenzimidazole (3 k)

Melting range, 154°C to 155°C. IR (KBr): υ 3,291, 3,079, 2,963, 2,845, 1,561, 1,433, 1,342, 1,263, 1,156, 1,108, 1,093, 1,021, 971, 834 and 751 cm^{-1}. ^{1}HNMR

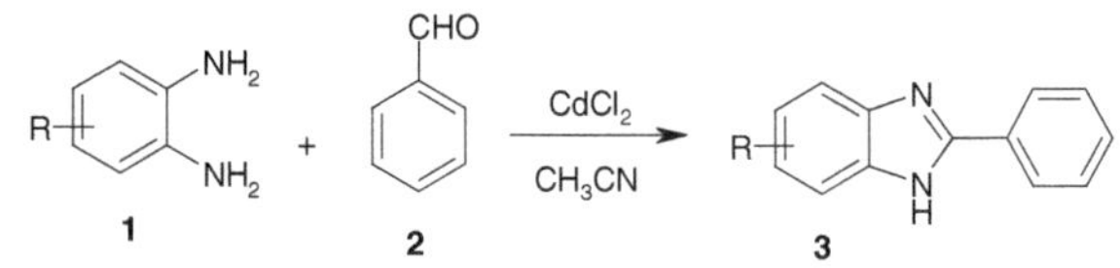

Scheme 1 The OPD and benzaldehyde reaction producing 2-phenylbenzimidazole.

Table 1 Cadmium chloride-catalyzed synthesis of benzimidazoles

Serial number	Aldehyde	Product (3a to 3o)	Reaction time (h)	Yield (%)
a	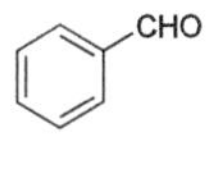	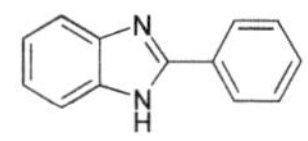	3.0	88
b	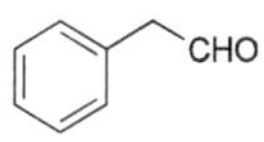	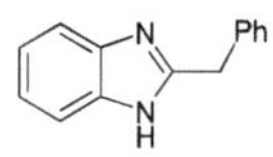	4.5	85
c	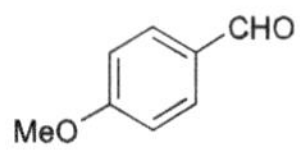	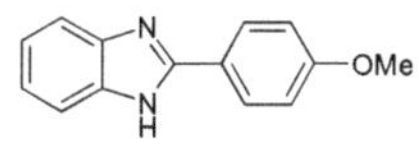	3.0	90
d	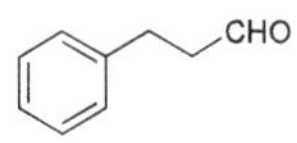	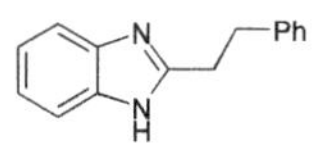	5.0	81
e	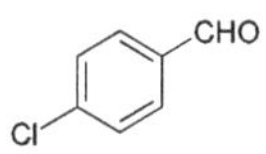	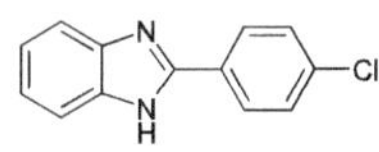	4.0	85
f	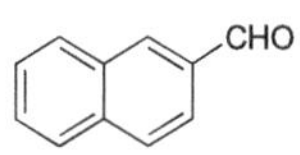	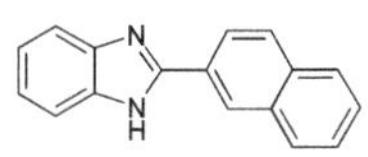	4.5	83
g	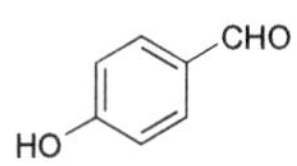	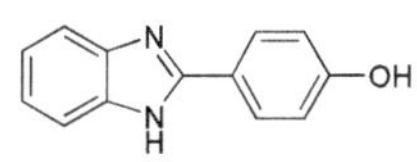	3.5	81
h	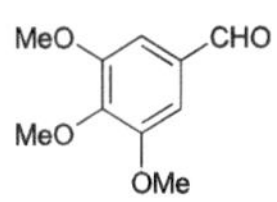	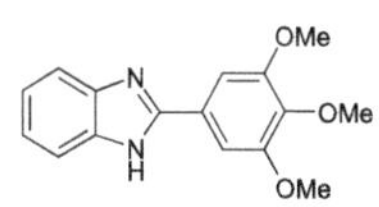	3.0	90
i	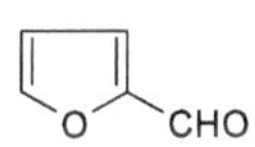	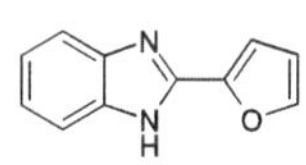	3.0	90
j	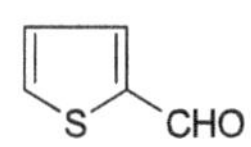	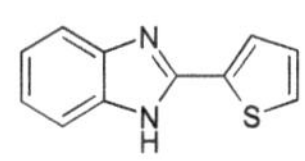	4.0	85
k	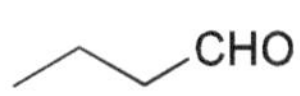	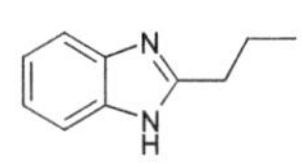	4.5	80
l	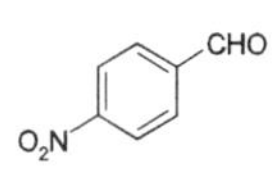	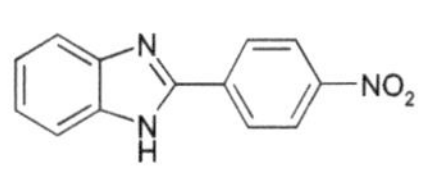	5.0	85
m		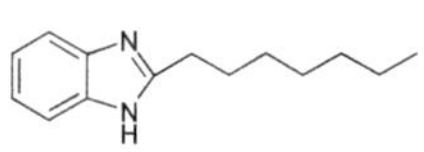	5.0	82
n	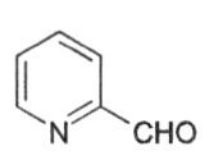	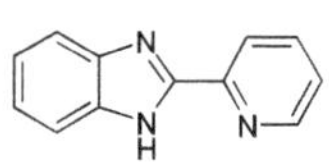	3.5	80
o	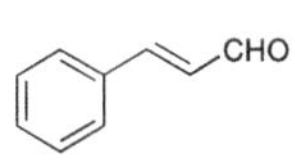	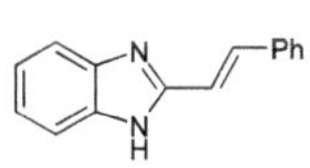	4.0	81

(DMSO-d_6): δ 0.98 (t, 3 H, J = 7.5 Hz), 1.80 to 1.90 (m, 2 H), 3.05 (t, 2 H, J = 7.5 Hz), 7.20 to 7.30 (m, 2 H), 7.50 to 7.60 (m, 2 H) and 12.5 (brs, 1 H, NH). EIMS *m/z* (%): 160 (m^+ 30), 131 (12), 116 (10), 90 (15), 76 (100) and 51 (25).

Conclusions

In conclusion, the catalyst cadmium chloride has been demonstrated as a novel and efficient promoter for the synthesis of benzimidazoles in high yields, from ortho-phenylenediamine and a wide variety of aldehydes. All the reactions were carried out at 80°C to 85°C, while using the catalyst cadmium chloride in 10 mol%. The reaction conditions were very mild, and the isolation of products was also very easy.

Acknowledgement
The authors BS and DS are thankful for the fellowship provided by UGC-New Delhi.

References

1. Sondhi SM, Singhal N, Johar M, Reddy BSN, Lown JW (2002) Curr Med Chem 1045
2. Battershill AJ, Scott LJ (2006) Drugs 51
3. Horton DA, Bourne GT, Smythe ML (2003) Chem Rev 893
4. Haul NH, Nar H, Priepke H, Ries U, Stassen J, Wienen W (2002) J Med Chem 45:1757
5. Nakano H, Inoue T, Kawasaki N, Miyataka H, Matsumoto H, Taguchi T, Inagaki N, Nagai H, Satoh T (2000) Bioorg Med Chem 373
6. Falco JL, Pique M, Gonza M, Buira I, Mendez E, Terencio J, Perez C, Princep M, Palomer A, Guglietta A (2006) Eur J Med Chem 985
7. Patil A, Ganguly S, Surana S, Patil A, Ganguly S, Surana S (2008) Rasayan J Chem 447
8. Valdez J, Cedillo R, Campos AH, Yepez L, Luis FH, Vazquez GN, Tapia A, Cortes R, Hernandez M, Castillo R (2002) Bioorg Med Chem 222
9. Villanueva JP, Santos R, Campos AH, Giulianotti MA, Castilloa R, Franco JLM (2011) Med Chem Commun 44
10. He Y, Wu B, Yang J, Robinson D, Risen L, Ranken R, Blyn L, Sheng S, Swayze EE (2003) Bioorg Med Chem Lett 3253
11. Baudy RB, Fletcher H, Yardley JP, Zaleska MM, Bramlett DR, Tasse RP, Kowal DM, Katz AH, Moyer JA, Gharbia MA (2001) J Med Chem 1516
12. Goker H, Kus C, Boykin DW, Yildiz S, Altanlar N (2002) Bioorg Med Chem 2589
13. Tidwell RR, Geratz JD, Dann O, Volz G, Zeh D, Loewe H (1978) J Med Chem 26:613
14. Perumal S, Mariappan S, Selvaraj S (2004) Arkivoc 8:46
15. Hisano T, Ichikawa M, Tsumoto K, Tasaki M (1982) Chem Pharm Bull 2996
16. Reddy GV, Rao NSR, Narsaiah B, Rao PS (2002) Synth Commun 32:2467
17. Kumar BVS, Vaidya SD, Kumar RV, Bhirud SB, Mane RB (2006) Eur J Med Chem 599
18. Salehi P, Dabiri M, Zolfigol MA, Otokesh S, Baghbanzadeh M (2006) Tetrahedron Lett 47:2557
19. Du LH, Wang YG (2007) Synthesis 675
20. Bahrami K, Khodaei M, Naali F (2008) J Org Chem 73:6835
21. Zheng N, Anderson KW, Huang X, Nguyen HN, Buchwald SL (2007) Angew Chem 46:7509
22. Chakrabarty M, Karmakar S, Mukherji A, Arima S, Harigaya Y (2006) Heterocycles 68:967
23. Gogoi P, Konwar D (2006) Tetrahedron Lett 47:79
24. Eynde JJV, Delfosse F, Lor P, Van YH (1995) Tetrahedron 15:5813
25. Vliet DSV, Gillespie P, Scicinski JJ (2005) Tetrahedron Lett 46:6741
26. Singh MP, Sasmal S, Lu W, Chatterjee MN (2000) Synthesis 1380
27. Baltork IM, Khosropour AR, Hojati SF (2007) Catal Commun 1865
28. Wang XJ, Zhang L, Xu Y, Murthy DK, Senanayake CS (2004) Tetrahedron Lett 45:7167
29. Yang D, Fu H, Hu L, Jiang Y, Zhao Y (2008) J Org Chem 73:7841
30. Alloum AB, Bougrin K, Soufiaoui M (2003) Tetrahedron Lett 44:5935
31. Trivedi R, De SK, Gibbs RA (2006) J Mol Catal A 245:8
32. Selvam K, Swaminathan M (2007) Chem Lett 1060
33. Lin S, Yang L (2005) Tetrahedron Lett 46:4315
34. Tandon VK, Kumar M (2004) Tetrahedron Lett 45:4185
35. Nadaf RN, Siddiqui SA, Daniel T, Lahoti RJ, Srinivasan KV (2004) J Mol Catal A 214:155
36. Heravi MM, Sadjadi S, Oskooie HA, Shoar RH, Bamoharram FF (2008) Catal Commun 504
37. Zou B, Yuan Q, Ma D (2007) Angew Chem 46:2598
38. Varala R, Nasreen A, Enugala R, Adapa SR (2007) Tetrahedron Lett 48:69
39. Srinivas U, Srinivas C, Narender P, Rao VJ, Palaniappan S (2007) Catal Commun 8:107
40. Wang Y, Sarris K, Sauer DR, Djuric SW (2006) Tetrahedron Lett 47:4823
41. Wu CH, Sun CM (2006) Tetrahedron Lett 47:2601

Characterization of modified silica aerogel using sodium silicate precursor and its application as adsorbent of Cu^{2+}, Cd^{2+}, and Pb^{2+} ions

HR Pouretedal[1*] and M Kazemi[2]

Abstract

Sodium silicate ($Na_2O/SiO_2 = 1:3.3$) as a precursor is used to prepare silica aerogel. The prepared silica aerogel is modified by mercaptopropyl trimethoxysilane. The modified silica aerogel is characterized using Fourier transform infrared spectroscopy, Brunauer, Emmett, Teller method, and transmission electron microscopy. The adsorption and removal of Cu^{2+}, Cd^{2+}, and Pb^{2+} ions from aqueous samples using modified silica aerogel are studied. The experimental conditions such as pH of samples, initial concentration of cations, contact time, and adsorbent dose are optimized to attain the maximum adsorption capacity. The optimum values are obtained as follows: contact time = 30 min, adsorbent dose = 0.05 g for $Cd^{2+}$and Pb^{2+} ions and 0.1 g for Cu^{2+} ions, pH = 6 for Cd^{2+} and Pb^{2+} ions, and pH = 4 for Cu^{2+} ions. The equilibrium data are fitted well by the Langmuir and Freundlich isotherm models. The adsorption capacities for Cu^{2+}, Cd^{2+}, and Pb^{2+} ions are found 90.1, 181.8, and 250.0 mg/g, respectively.

Keywords: Silica aerogel, Adsorbent, Sodium silicate, Metal ion, Modification

Background

Silica aerogels are materials that show unusual properties. The high specific surface area (500 to 1,200 m^2/g), high porosity (80 to 99.8 %), low density (approximately 0.003 g/cm^3), high thermal insulation value (0.005 W/mK), ultra low dielectric constant ($k = 1.0$ to 2.0), and low index of refraction (approximately 1.05) are the advantages of these materials. The nanoporous network of interconnected primary particles is due to these properties [1]. Because of their unique texture, silica aerogels are promising materials as super-thermal insulators, catalytic supports, adsorbents, and host materials for drug delivery systems. The synthesis of silica aerogels has received significant attention especially during the last two decades. Some investigators have studied the use of different precursors, and many have focused on the modification of synthesis parameters. Recently, many research works have been devoted to ambient pressure drying which makes the production commercial and industrial [2-4].

* Correspondence: HR_POURETEDAL@mut-es.ac.ir
[1]Faculty of Applied Chemistry, Malek-ashtar University of Technology, Shahin-Shahr, 1387763681 Iran
Full list of author information is available at the end of the article

During this decade, many researchers have attempted to reduce the cost of aerogel synthesis. In brief, these attempts took one of two approaches. The first involved the use of cheap precursors, and the second is concerned with the development of novel drying techniques that made it possible to synthesize aerogels at ambient pressure (known as an ambient-drying or evaporative drying) instead of supercritical drying [5-7]. Most of the recent reports of various forms of silica aerogels such as powder, granules, films, and bulks utilize at least one of the above approaches and sometimes more than one approach [8-10].

The applications of the silica aerogel have expanded into many fields [11,12]. They are used as (1) fillers for paints, varnishes, etc.; (2) thermal and acoustic insulation materials; (3) adsorbents and catalyst supports; and (4) electronic materials such as Cerenkov detectors and sensor materials.

The existence of heavy metals in wastewaters contributes to water toxicity and represents an increasing danger for the environment, human beings, and other living organisms. In addition, to rock leaching due to some external effects, these effluents discharge from various anthropogenic sources such as power plants, chemical manufacturing, painting, mining, metallurgy, electroplating, and many other industries [13-15].

In the present study, we report the synthesis of silica aerogel by using sodium silicate as precursor. The prepared silica aerogel is modified and applied as adsorbent for the removal of Cu^{2+}, Cd^{2+}, and Pb^{2+} ions from aqueous samples. The characterization of prepared silica aerogel is also reported in this study.

Results and discussion

Characterization of silica aerogel

The Fourier transform infrared spectroscopy (FT-IR) spectra of unmodified and modified silica aerogel are shown in Figure 1. The wide peak with strong intensity was 1,000 to 1,100 cm^{-1}, and the weak peak near 800 cm^{-1} was assigned to the asymmetric and symmetric bending of Si-O-Si bonds, respectively [16-18]. The strong peak near 470 cm^{-1} is attributable to the bending of the O-Si-O bonds. In the spectra of unmodified aerogel, the wide bond at 3,426 cm^{-1} and the peaks at 1,640 and 944 cm^{-1} correspond to the hydroxyl groups adsorbed on the surface as well as the bending of the H-O-H bonds and the stretching of the Si-OH bonds, respectively. The stretching vibrations of S-H and C-H appeared at 2,564 and 2,932 cm^{-1}, respectively, in the spectra of modified silica aerogel which confirmed the modification of silica aerogel surface with mercaptopropyl trimethoxysilane (MPTMS) [17,18].

The N_2 adsorption-desorption isotherms and pore size distribution of unmodified silica aerogel are shown in Figures 2 and 3, and the obtained data are collected in Table 1. The type IV isotherm curves' well-defined step clearly indicates that these materials possess mesoporous structure. The isotherms of nitrogen adsorption exhibited an abrupt increase at P/P_0 is approximately 0.70, which is characteristic for capillary condensation within the uniform mesopores of the materials (Figure 2a). The values of specific surface area, pore volume, and pore diameter considerably decrease after modification of silica aerogel with MPTMS (Table 1). The average pore diameter of the unmodified and modified silica aerogel is 7.6 and 3.9 nm, respectively. The mean pore diameter of the unmodified silica aerogel is obtained from Brunauer, Emmett, Teller (BET) plot (Figure 2b). The decrease of pore volume has been observed for modified silica aerogel by the filling of micropores in the substrate with the modifier [19,20].

Figure 4 shows the TEM images of unmodified and modified silica aerogel. The synthesized aerogel exhibit a porous network structure which contains 50 to 60 nm spherical solid clusters and pores below 120 nm between them. The particle distribution of aerogels is uniform. Silica aerogels exhibit a porous structure with fine particulate morphology, and the modified aerogel also showed porous structure but with less pores size in comparison with unmodified silica aerogels.

The TGA curve of modified silica aerogel is shown in Figure 5. A little weight loss at temperature 320 °C is seen, and the decomposition of aerogel with weight loss of 20.2 % occurred at higher temperature of up to 800 °C. An XRD pattern (Figure 6) shows an amorphous structure for synthesized silica aerogel.

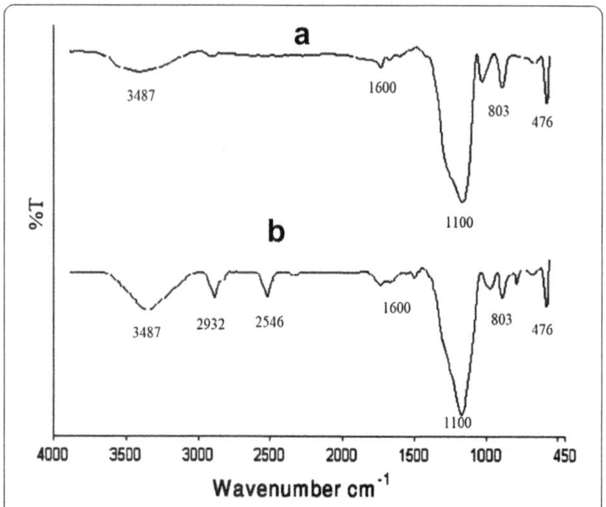

Figure 1 FT-IR spectra of unmodified (a) and modified (b) silica aerogel.

Adsorption by modified silica aerogel

Effect of pH

The effect of the pH of aqueous samples on the adsorption of Cu^{2+}, Cd^{2+}, and Pb^{2+} ions by modified silica aerogel is shown in Figure 7. As seen, the adsorption capacity of adsorbent increased with increasing of the samples' pH. The characteristic of chelation mechanism and the less insignificant competitive adsorption of hydrogen ions at higher pH are due to increase of the removal yield [18-20]. The following reactions may take place at the solid-solution interface of modified silica aerogel with MPTMS at different pHs [21].

$$SH + H^+ \rightleftarrows SH_2^+ \quad (1)$$

$$SH + M^{2+} \rightleftarrows SHM^{2+} \quad (2)$$

$$SH + OH^- \rightleftarrows SHOH^- \quad (3)$$

Equation 1 indicates the protonation and deprotonation reactions of the SH groups of MPTMS in silica aerogel. Equation 2 shows the formation of surface complexes of M^{2+} ion with the SH groups. The adsorption of OH^- ions from the solution through a hydrogen bond on the SH groups occurred at alkaline pHs (Equation 3). The protonation of SH groups (Equation 1) was favored versus the

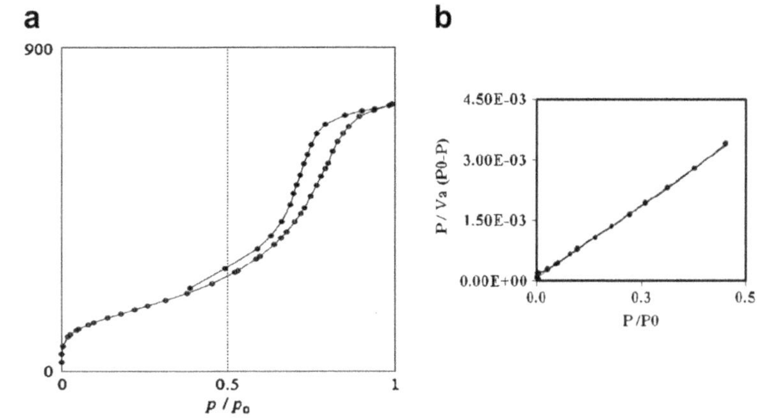

Figure 2 Nitrogen adsorption-desorption isotherms (a) and BET plot of unmodified silica aerogel.

complex formation of M^{2+} (Equation 2) at lower pH. The available sites of adsorbent for complex formation were reduced with protonation of SH groups. On the other hand, the electrostatic repulsion between the cations and the surface of modified silica aerogel increased with protonation of SH groups. Thus, the adsorption of cations decreased with decreasing aqueous sample pH. By increasing the pH of solutions, the protonation of SH groups was reduced, and the number of SH sites on the surface of modified silica aerogel for cations adsorption (Equation 2) increased as well as its adsorption capacity.

However, the adsorption capacity of adsorbents was reduced at the upper pH of 6 for Cd^{2+} and Pb^{2+} and pH of 4 for Cu^{2+} ion. The increasing adsorption of Cd^{2+}, Pb^{2+}, and Cu^{2+} with increasing pH shows the ion exchange mechanism for these ions. A decrease in the removal yield at upper optimum pH for each cation is due to the formation of soluble hydroxy complexes [22]. The difference in the adsorption behavior of different heavy metal ions may be because of the difference in their ion exchange capacity on the surface depending on their charge density, extent of hydrolysis, and solubility of hydrolyzed metal ions in solution under the present experimental condition [22-24].

Effect of initial concentration

Figure 8 indicates the effect of initial concentration of cations on the adsorption capacity. The reduction of the removal yield was observed with the increase of initial concentration of M^{2+} ions in range of 50 to 400 mg/L. The sufficient adsorption sites are available for adsorption of cations at a lower initial concentration. Also, the fraction of adsorption for each ion is independent of its initial concentration. However, the available sites of

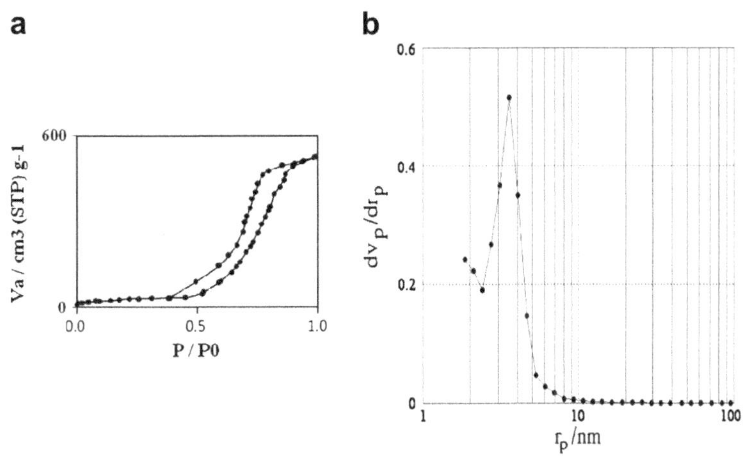

Figure 3 Nitrogen adsorption-desorption isotherms (a) and pore size distribution (b) of modified silica aerogel.

Table 1 The N_2 adsorption-desorption data of unmodified and modified silica aerogel

Silica aerogel	BET data		
	S (m^2/g)	P (cm^3/g)	D (nm)
Unmodified	602.0	1.1455	7.59
modified	387.9	0.4561	3.89

S, Specific surface area; P, pore volume; D, average pore diameter.

adsorbent were reduced at higher concentrations of the adsorbate [21,22].

Effect of contact time

The influence contact time of adsorbent and adsorbates in the range of 10 to 60 min is indicated in Figure 9. The conditions are optimum pH and 50 mg/L of M^{2+} ions. The removal yield of cations increased with the increasing contact time until equilibrium is attained at 30 min. Hence, the time of 30 min is experimentally required to attain equilibrium of the modified silica aerogel with MPTMS.

Effect of adsorbent dose

Figure 10 shows the removal yield of cations versus the adsorbent dose in range of 0.01 to 0.12 g. A sharp increase in the capacity adsorption was observed for Cd^{2+} and Pb^{2+} with the increase of adsorbent dose from 0.01 to 0.05 g. Meanwhile, increase in the adsorption is seen up to 0.10 g of adsorbent for Cu^{2+} ion. Hence, the optimal dose is 0.05 g for Cd^{2+} and Pb^{2+} and 0.10 g for Cu^{2+} ion. However, the greater availability of functional groups on the surface area with the increase of adsorbent dose is due to the higher removal percent.

The difference in the removal yield of different cations at the same conditions such as initial concentration, adsorbent dose, and contact time may be attributed to the difference in their chemical affinity with respect to SH groups on the surface of the adsorbent [23].

The adsorption of Cu^{2+}, Cd^{2+}, and Pb^{2+} ions is also studied in a sample with initial concentration of 50 mg/L of each ion, 0.1 g/L adsorbent, and pH of 5. The removal efficiencies of 48.2, 60.7, and 76.1 % are obtained for Cu^{2+}, Cd^{2+}, and Pb^{2+} ions, respectively, at 30 min. However, the decreasing removal efficiencies of cations are expected because of the limitation of the adsorption capacity of the adsorbent.

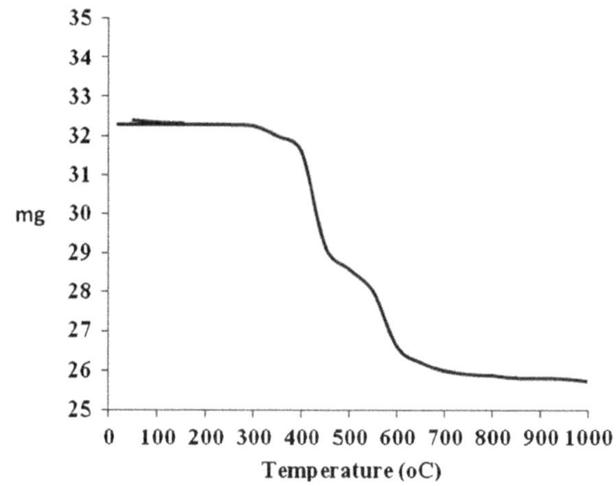

Figure 5 Thermogravimetric curve of modified silica aerogel.

Study of adsorption model

The equilibrium adsorption of Cu^{2+}, Cd^{2+}, and Pb^{2+} is studied as a function of initial concentration in the range of 50 to 400 mg/L. The Langmuir and Freundlich adsorption isotherms are used to this study. The linear equations of Langmuir and Freundlich isotherms are represented in Equations 4 and 5, respectively,

$$C_{eq}/q_{eq} = l/K_L q_m + C_{eq}/q_m \tag{4}$$

$$\log q_{eq} = \log K_F + (l/n)\log C_{eq} \tag{5}$$

Study of adsorption modelwhere q_{eq} (milligram/gram) is the amount of cation adsorbed per unit mass of adsorbent particles at equilibrium, C_{eq} (milligram/liter)

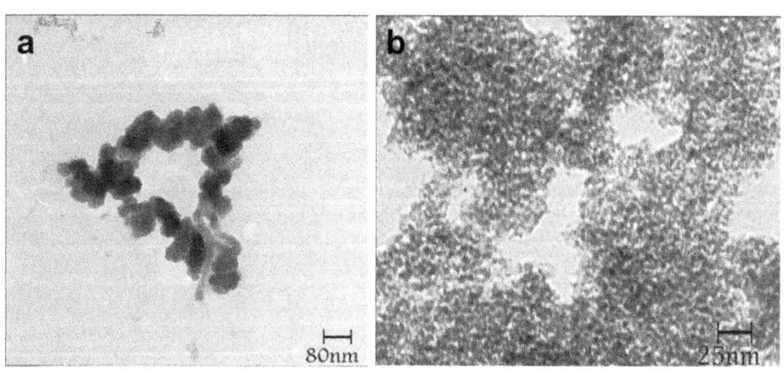

Figure 4 TEM images of (a) unmodified and (b) modified silica aerogel.

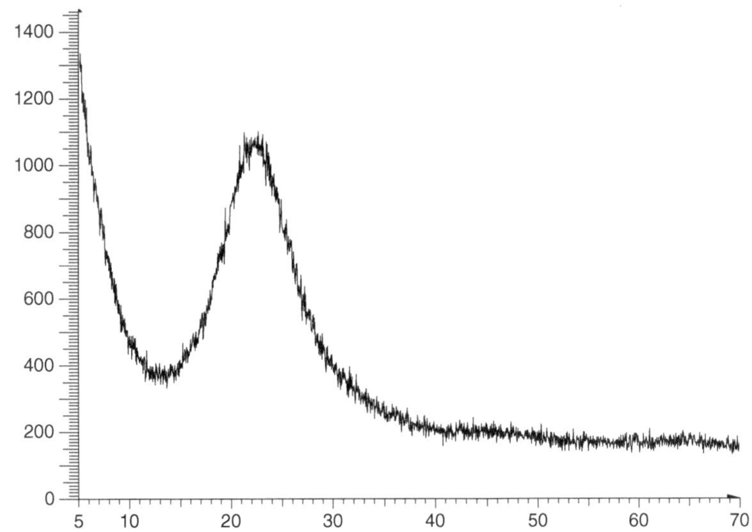

Figure 6 XRD pattern of silica aerogel.

is the equilibrium concentration of the cation in the aqueous phase, K_L is the equilibrium constant (liter/milligram) related to the energy of the adsorption and q_m is the maximum adsorption capacity (milligram/gram), K_F and n are Freundlich isotherm constants related to the adsorption capacity and adsorption intensity, respectively [24,25].

Therefore, in Langmuir isotherm, a plot C_{eq}/q_{eq} versus C_{eq} should be a straight line with a slope $1/q_m$ and intercept as $l/K_L q_m$. Also, a plot of log q_{eq} versus log C_{eq} is linear with the slope $1/n$ and intercepts log K_F accordance with Freundlich isotherm. The obtained data from linear isotherms are collected in Table 2.

The data of adsorption isotherms show that the adsorptive behavior of cations on modified silica aerogel satisfies Langmuir and Freundlich assumptions in accordance with the R^2 values. As seen, the maximum adsorption capacities for Cu^{2+}, Cd^{2+}, and Pb^{2+} ions are found to be 90.1, 181.8, and 250.0 mg/g, respectively. The n values are greater than in unity and indicate chemisorptions of adsorbates on the adsorbent. Isotherms with $n > 1$ are classified as L-type isotherms that show a high affinity between adsorbate and adsorbent [26,27].

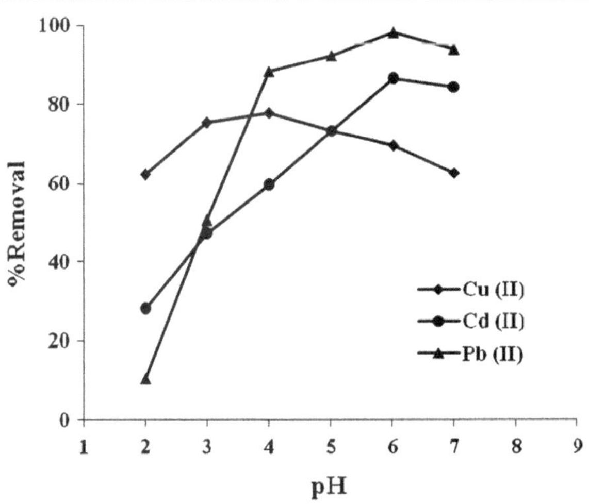

Figure 7 Effect of pH on the removal yield of cations; contact time = 30 min; adsorbent dose = 0.1 g for Cu(II) and 0.05 g for Cd(II) and Pb(II); initial concentration = 50 mg/L.

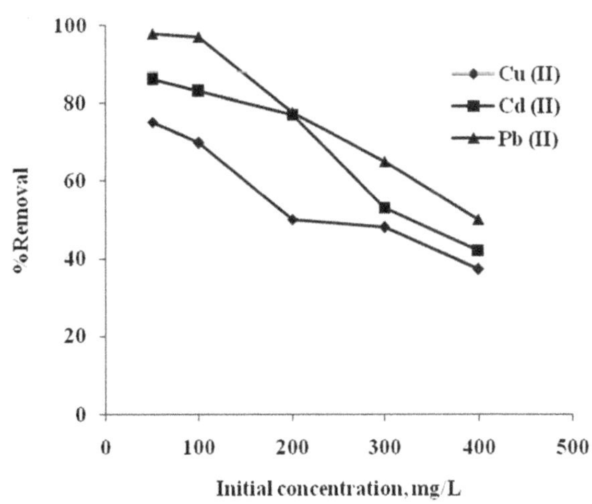

Figure 8 Effect of initial concentration on the removal yield of cations; contact time = 30 min; adsorbent dose = 0.1 g for Cu(II) and 0.05 g for Cd(II) and Pb(II); pH = 4 for Cu(II) and 6 for Cd(II) and Pb(II).

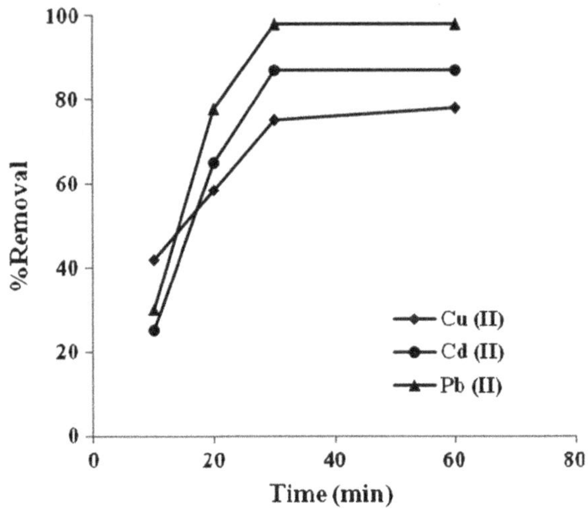

Figure 9 Effect of contact time on the removal yield of cations; initial concentration = 50 mg/L; adsorbent dose = 0.1 g for Cu(II) and 0.05 g for Cd(II) and Pb(II); pH = 4 for Cu(II) and 6 for Cd(II) and Pb(II).

Study of kinetic adsorption

The study of adsorption kinetics describes the adsorbate uptake rate and evidently, this rate controls the residence time of the adsorbate at a solid liquid interface [27-29]. The kinetics of the adsorption of Cu^{2+}, Cd^{2+}, and Pb^{2+} on modified silica aerogel is studied using pseudo-first order and pseudo-second order models. The pseudo-first order equation [27] is generally expressed as follows:

$$\log(q_{eq}-q_t) = \log(q_{eq})-k_1t/2.303 \tag{6}$$

where q_{eq} and q_t are the sorption capacities at equilibrium and at time t, respectively (mg.g^{-1}), and k_1 is the rate

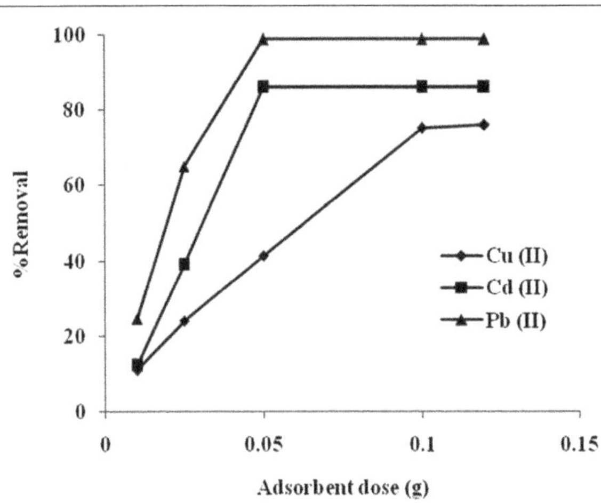

Figure 10 Effect of adsorbent dose on the removal yield of cations; initial concentration = 50 mg/L; contact time = 30 min; pH = 4 for Cu(II) and 6 for Cd(II) and Pb(II).

Table 2 Langmuir and Freundlich isotherm constants for adsorption of Cu^{2+}, Cd^{2+}, and Pb^{2+} ions

Adsorbents	Langmuir constants			Freundlich constants		
	q_m(mg/g)	K_L (L/mg)	R^2	K_F	n	R^2
Cu(II)	83.3	0.039	0.971	16.1	3.62	0.897
Cd(II)	200.0	0.052	0.996	25.1	2.62	0.845
Pb(II)	250.0	0.143	0.997	48.3	3.41	0.883

constant of pseudo-first order sorption (L.min^{-1}). The plot of log $(q_{eq} - q_t)$ versus t will give a linear relationship from which k_1 and q_{eq} can be determined from the slope and intercept in the graph, respectively. The pseudo-first order equation (Lagergren's equation) describes adsorption in solid–liquid systems based on the sorption capacity of solids [28].

The pseudo-second order chemisorption kinetic equation [28,29] is expressed as Equation 7,

$$t/q_t = 1/k_2{q_{eq}}^2 + t/q_{eq} \tag{7}$$

where q_{eq} and q_t are the sorption capacity at equilibrium and at time t, (mg.g^{-1}), respectively, and k_2 is the rate constant of the pseudo-second order sorption (g.mg^{-1}. min^{-1}). The pseudo-second-order rate expression has been applied for analyzing chemisorption kinetics from liquid solutions [29]. If the pseudo-second order kinetics is applicable to the experimental data, the plot of t/q_t versus t should give a linear relationship from which q_{eq} and $k_2q_{eq}^2$ can be determined from the slope and intercept of the plot, respectively.

The obtained kinetic rate constants of pseudo-first and pseudo-second order of sorption of Cu^{2+}, Cd^{2+},and Pb^{2+} on the modified silica aerogel are collected in Table 3. Fitted equilibrium adsorption capacities derived from Equation 7 for each cation are similar those observed experimentally. However, correlation coefficients values for the pseudo-second-order kinetic and pseudo-first order model are close to 1.0. Given the good agreement between model fit and experimentally observed equilibrium adsorption capacity, in addition to the large correlation coefficients, this suggests that Cu^{2+}, Cd^{2+}, and Pb^{2+} ions adsorption followed pseudo-second order kinetics and were adsorbed onto modified silica aerogel via chemical interaction.

Comparison of adsorption capacity (qm) of Cu, Cd and Pb ions on several adsorbents is given in Table 4.

Synthesis of silica aerogel

Industrial grade of sodium silicate (Bahman Gostar Investment Co., Isfahan, Isfahan, Iran) is used as a source of cost-effective silica sols. An 83.37 g of sodium silicate with molar ratio of SiO_2:Na_2O = 3.3:1 is dissolved in 316.62 ml deionized water. The resulting solution is stirred

Table 3 Kinetic rate constants of pseudo-first and pseudo-second order of sorption of Cu^{2+}, Cd^{2+} and Pb^{2+}

Adsorbent	q_{exp}(mg.g^{-1})	Pseudo-first order			Pseudo-second order		
		$k_1 \times 10^{-2}$ (min^{-1})	q_{eq} (mg.g^{-1})	R^2	$k_2 \times 10^{-5}$ (g.mg^{-1}.min^{-1})	q_{eq} (mg.g^{-1})	R^2
Cu(II)	72.6	2.1	70.5	0.999	15.7	105.3	0.999
Cd(II)	159.3	3.2	165.6	0.980	7.6	263.2	0.965
Pb(II)	165.4	4.1	174.6	0.997	8.2	277.8	0.995

Kinetic rate constants of pseudo-first and pseudo-second order of sorption of Cu^{2+}, Cd^{2+} and Pb^{2+} on the modified silica aerogel with initial concentration 300 mg/L.

in the presence of sulfonated copolymer strong cationic resin as H^+-form for 10 min to remove sodium ions. The pH of silica sol is in the range of 2 to 3 after this ion exchange, then 1 mol/L NaOH solution is added to the silica sol to raise its pH to the range of 4 to 5 for gelation. The obtained silica sols are stirred for 2 min and then transferred into a plastic beaker where the sols aged into hydrogels within about 30 min. Then, the hydrogels are immersed in deionized water to age them for 24 h at 60 °C to strengthen the networks of the gels. The obtained gel is also aged for 48 h in acetone solvent for two times and for 72 h in *n*-hexane solvent at room temperature. Finally, the aged gels are separated and dried at 50 ℃ in a saturated atmosphere of *n*-hexane. The chemicals *n*-hexane, acetone, and NaOH (with highest purity) are prepared from petrochemical company of Iran (Petrochemical Commercial Co., Tehran, Iran).

Modification of silica aerogel

The surface of synthesized of silica aerogel is modified by MPTMS (purchased from Merck & Co., Inc., Whitehouse Station, NJ, USA). An 11.3 g MPTMS is dissolved in 56.85 g *n*-hexane. The synthesized silica aerogel is added to MPTMS solution and the obtained mixture is stirred for 3 h. Then, the modified silica aerogel is separated and washed with *n*-hexane. Finally, the modified silica aerogel is derided in a rich atmosphere of *n*-hexane for 48 h at room temperature and for 5 h at 50 °C.

Methods of characterization

The N_2 adsorption-desorption isotherms were measured by the BET method using nitrogen as an adsorption gas at 77 K using Belsorp Mini II instrument (BEL Japan, Inc., Sumida-ku, Tokyo, Japan). Samples were out gassed at 300 °C for 4 h prior to surface area measurements. The FT-IR spectrum recorded from KBr pellets (99 wt.% of KBr) on Nicolet Impact 400D FT-IR spectrophotometer (Spectralab Scientific Corporation, Markham, Toronto, Canada). The morphology of synthesized silica aerogel before and after surface modification is studied by Philips CM10 transmission electron microscope (TEM) operating at 120 kV. The supporting grids were formvar-covered, carbon-coated, and 200-mesh copper grids. A diffractometer Bruker D8 Advance (Bruker Daltonik GmbH, Bremen, Germany) with anode of Cu (λ = 1.5406 Å of Cu K_α) and filter of Ni was applied to record the X-ray diffraction (XRD) pattern of silica aerogel. Thermogravimetric analysis (TGA) was carried out with Shimadzu DT-40 thermal analyzer (Tokyo, Japan).

Adsorption of metal ions

The adsorption of Cu^{2+}, Cd^{2+}, and Pb^{2+} ions by modified silica aerogel as an adsorbent is studied in a batch procedure. The synthetic stock solutions of metals ions are prepared by dissolving required quantity of analytical grade of nitrate salts of Cu^{2+}, Cd^{2+}, and Pb^{2+} ions in deionized water. The stock solutions were further diluted with deionized water to the desired concentration for obtaining the test solutions. All chemical nitrate salts and nitric acid are used from Merck (Merck & Co., Inc., Whitehouse Station, NJ, USA).

A known weight of adsorbent (e.g., 0.01 to 0.12 g) was equilibrated with 50 mL of the cation solution of known concentration (e.g., 50 to 400 mg/L) in a polyethylene vessel at room temperature in a thermostatic mechanical shaker for a known period (10 to 60 min) of time. Mechanical shaker is used for all the adsorption experiments for agitating the sample for a desired contact time. The pH of samples is controlled in the range 2 to 7 with addition of

Table 4 Comparison of adsorption capacity (q_m) of Cu, Cd and Pb ions on several adsorbents

Adsorbent	Ions	q_m (mg/g)	References
Synthesized silica aerogel from tetramethoxysilane modified with mercapto	Cu	51.0	[24]
Activated carbon prepared from cashew nut shells	Cd, Pb	14.3, 28.9	[30]
Nano zerovalent iron particles	Cd	769.2	[26]
Polyazomethineamides	Cu, Cd, Pb	470.7, 462.3, 452.1	[31]
Chitosan crosslinked with both epichlorohydrin and triphosphate,	Cu, Cd, Pb	130.7, 83.7, 199.9	[32]
In this work	Cu, Cd, Pb	83.3, 200.0, 250.0	

1.0 mol/L of HNO_3 and NaOH solutions by pH-analyzer of Metroholm 661 (Metroholm Company, Ionenstrasse, Herisau, Switzerland). After equilibration the suspension was centrifuged for 15 min at 7,000 rpm, and the sample solution then was analyzed using Atomic Adsorption Spectrometer (AAS, Perkin Elmer 300, PerkinElmer Inc., Waltham, MA, USA) operating with an air acetylene flame. The effect of several parameters, such as pH, adsorbate concentrations, contact time, and adsorbent dose is optimized.

The percent of removal is calculated using Equation 8,

$$\%\text{Removal} = [(C_O - C_{eq})/C_O] \times 100 \tag{8}$$

where C_0 and C_{eq} are the initial and equilibrium concentration of metal ions of sample solution (mg/L).

Conclusions

Sodium silicate as a safe and inexpensive precursor can be used to prepare silica aerogel with a high surface of 600 m^2/g. The modification of prepared silica aerogel is due to the application of it as a suitable adsorbent to remove some heavy metals such as Cu(II), Cd(II) and Pb(II) ions from aqueous samples in the range of 50 to 400 mg/L. The high removal yields (76 % to 99 %) in short contact time of 30 min at the presence of low dose of adsorbent (0.05 to 0.1 g) are the precious advantages of modified silica aerogel.

Competing interests

The authors declare that they have no competing interests.

Author details

[1]Faculty of Applied Chemistry, Malek-ashtar University of Technology, Shahin-Shahr, 1387763681 Iran. [2]Department of Chemistry, Islamic Azad University, Shahreza Branch, Shahreza and Fadak Group, Isfahan 37541-13115, Iran.

References

1. Dorcheh AS, Abbasi MH (2008) Silica aerogel; synthesis, properties and characterization. J Mater Processing Technol 199:10–26
2. Schwertfeger F, Frank D, Schmidt M (1998) Hydrophobic waterglass based aerogels without solvent exchange or supercritical drying. J Non-Cryst Solids 225:24–29
3. Lee CJ, Kim GS, Hyun SH (2002) Synthesis of silica aerogels from waterglass via new modification ambient drying. J Mater Sci 37:2237–2241
4. Rao AP, Pajonk GM, Rao AV (2005) Effect of preparation conditions on the physical and hydrophobic properties of two step processed ambient pressure dried silica aerogel. J Mater Sci 40:3481–3489
5. Bhagat SD, Park K-T, Kim Y-H, Kim J-S, Han J-H (2008) A continuous production process for silica aerogel powders based on sodium silicate by fluidized bed drying of wet-gel slurry. Solid State Sci 10:1113–1116
6. Gao G-M, Liu D-R, Zou H-F, Zou L-C, Gan SC (2010) Preparation of silica aerogel from oil shale ash by fluidized bed drying. Powder Technol 197:283–287
7. Shi F, Liu J-X, Song K, Wang Z-Y (2010) Cost-effective synthesis of silica aerogels from fly ash via ambient pressure drying. J Non-Crystalline Solids 356:2241–2246
8. Hwang SW, Kim T-Y, Hyun S-H (2010) Effect of surface modification conditions on the synthesis of mesoporous crack-free silica aerogel monoliths from waterglass via ambient-drying. Micropor Mesopor Mat 130:295–302
9. Hwang SW, Jung HH, Hyun SH, Ahn YS (2007) Effective preparation of crack-free silica aerogels via ambient drying. J. Sol–gel. Sci Technol 41:139
10. Rao AP, Rao AV, Pajonk GM (2007) Hydrophobic and physical properties of the ambient pressure dried silica aerogels sodium modification agents. Appl Surf Sci 253:6032–6040
11. Rao AV, Rao AP, Kulkarni MM (2004) Influence of gel aging and Na_2SiO_3/H_2O molar ratio. J Non-Cryst Solids 350:224–229
12. Tang Q, Wang T (2005) Preparation of silica aerogel from rice hull ash by supercritical carbon dioxide drying. J Supercritical Fluids 35:91–94
13. Ramadan H, Ghanem A, El-Rassy H (2010) Mercury removal from aqueous solutions using silica, polyacrylamide and hybrid silica–polyacrylamide aerogels. Chem Eng J 159:107–115
14. Meena AK, Mishra GK, Rai PK, Rajagopal C, Nagar PN (2005) Removal of heavy metal ions from aqueous solutions using carbon aerogel as an adsorbent. J Hazard Mater 122:161–170
15. Xie XLJ, Liang J (2007) Absorbency and adsorption of poly(acrylic acid co-acrylamide) hydrogel. J Appl Polymer Sci 106:1606–1613
16. Rao AV, Kulkarni MM, Amalnerkar DP, Seth T (2003) Superhydrophobic silica aerogels based on methyltrimethoxysilane precursor. J Non-Cryst Solids 330:187–195
17. Rao AV, Nilsen E, Einarsrud MA (2001) Effect of precursors, methylation agents and solvents on the physicochemical properties of silica aerogels prepared by atmospheric pressure drying method. J Non-Cryst Solids 296:165–171
18. Reynolds JG, Coronado PR, Hrubesh LW (2001) Hydrophobic aerogels for oil-spill clean up synthesis and characterization. J Non-Cryst Solids 292:127–137
19. Li T, Wang T (2008) Preparation of silica aerogel from rice hull ash by drying at atmospheric pressure. Mater Chem Phys 112:398–401
20. Zhu J, Xie J, Lü X, Jiang D (2009) Synthesis and characterization of superhydrophobic silica and silica/titania aerogels by sol–gel method at ambient pressure. Colloid Surface A 342:97–101
21. Chang YC, Chen DH (2005) Preparation and adsorption properties of monodisperse chitosan-bound Fe_3O_4 magnetic nanoparticles for removal of Cu(II) ions. J Colloid Interf Sci 283:446–451
22. Ngah WSW, Endud CS, Mayanar R (2002) Removal of copper(II) ions from aqueous solution onto chitosan and cross-linked chitosan beads. Functional Polymers 50:181–190
23. Qu R, Wang C, Ji C, Sun C (2005) Preparation, characterization, and metal binding behavior of novel chelating resins containing sulfur and polyamine. J Appl Polymer Sci 95:1558–1656
24. Standeker S, Veronovski A, Novak Z, Knez Z (2011) Silica aerogels modified with mercapto functional groups used for Cu(II) and Hg(II) removal from aqueous solutions. Desalination 269:223–230
25. Meena AK, Mishra GK, Rai PK, Rajagopal C, Nagar PN (2005) Removal of heavy metal ions from aqueous solutions using carbon aerogel as an adsorbent. J Hazard Mater B 122:161–170
26. Boparai HK, Joseph M, O'Carroll DM (2011) Kinetics and thermodynamics of cadmium ion removal by adsorption onto nanozerovalent iron particles. J Hazard Mater 186:458–465
27. Banerjee SA, Chen DH (2007) Fast removal of copper ions by gum arabic modified magnetic nano-adsorbent. J Hazard Mater 147:792–799
28. Ho YS (2004) Citation review of Lagergren kinetic rate equation on adsorption reactions. Scientometrics 59:171–177
29. Ho YS (2006) Review of second-order models for adsorption systems. J Hazard. Mater. 136:681–689
30. Tangjuank S, Insuk N, Tontrakoon J, Udeye V (2009) Adsorption of lead(II) and cadmium(II) ions from aqueous solutions by adsorption on activated carbon prepared from cashew nut shells. World Academy of Sci, Eng Technol 52:1–7
31. Murugesan A, Ravikumar L, Sathya Selva Bala V, Senthil Kumar P, Vidhyadevi T, Dinesh Kirupha S, Kalaivani SS, Krithiga S, Sivanesan S (2011) Removal of Pb(II), Cu(II) and Cd(II) ions from aqueous solution using polyazomethineamides: equilibrium and kinetic approach. Desalination 271:199–208
32. Laus R, Costa TG, Szpoganicz B, Fávere VT (2010) Adsorption and desorption of Cu(II), Cd(II) and Pb(II) ions using chitosan crosslinked with epichlorohydrin-triphosphate as the adsorbent. J Hazard Mater 183:233–241

4

Application of soft and hard modeling methods to resolve the three competitive complex formation of 13 lanthanide-Arsenazo III complexes

Javad Fadaee Kakhki* and Mohammad Reza Abedi

Abstract

The applicability of soft and hard modeling methods was used to determine the formation constants of complexes between 3,6-bis[(2-arsonophenyl)azo]-4,5-dihydroxy-2,7-naphtalenedisulphonic acid disodium salt (Arsenazo III) and 13 lanthanides. The results showed that all the 13 lanthanides (M) studied form a similar type of complexes and spectral profiles with Arsenazo III (L) with three formation constants corresponding to M/L ratios of 1:1, 1:2, and 2:2.

Keywords: Arsenazo III, Chemometrics, Hard modeling, Lanthanides, Soft modeling

Background

Spectrophotometric titrations are important methods for the investigation of solution equilibria. The titration consists of a collection of spectra of a solution measured as a function of the reagent added which influences the equilibrium under the investigation [1,2].

The proper chemometric algorithms can be used in evaluating the equilibrium information such as the stability constant through analysis of the spectroscopic data. Several soft and hard modeling algorithms have been developed to analyze bilinear data obtained from chemical systems.

Soft modeling methods range from very general approaches, such as evolving factor analysis (EFA) [3], which is particularly used to estimate the number of species involved in equilibrium studies and the kinetic process by repeated factor analysis of rationally selected subsets of spectra. Factor analysis investigates the rank of the subset of spectra by determining its number of significant eigenvalues [4].

Multivariate curve resolution-alternating least squares (MCR-ALS) [5] is an iterative soft modeling resolution method that has been successfully applied to solve many mixed dynamic processes monitored spectrometrically, such as chromatographic runs [6,7], biomacromolecular reactions [8,9], voltammetric data, and environmental data [10-12]. In fact, MCR is a self-modeling chemometric discipline that comprises several techniques in establishing an initial model on the data and extracting maximum amount of information from the data. These results are useful to validate hard modeling results and investigate the complex chemical systems [13].

Hard modeling approaches of fitting multivariate response data are based on mathematical relationships, which describe the measurements quantitatively [14-16]. In chemical equilibria, the analysis is based on the equilibrium model which quantitatively describes the reaction and all concentrations in the solution under investigation. Within the hard modeling methods, analysis of spectral data has shown an optimal performance on many occasions [16-18].

Hard modeling or model-based method was used to analyze a given measurement based on a predetermined model. The model can be a simple mathematical function or a complex chemical system that is fitted to a data set. So, it needs to determine the right model to be fitted, such as the exact chemical reaction mechanism in kinetics or the correct equilibrium model in a titration experiment. Unfortunately, it is a very difficult one. There are no general applicable tools available to guide towards finding the model that correctly describes the chemical process under investigation because the model fitting is much easier than model finding. There is a good collection of methods available in performing model-free or soft modeling analyses. Typically, these methods (soft-model) deliver the shapes of the concentration profiles of all reacting components as

* Correspondence: ja_fadaee@yahoo.com
Department of Applied Chemistry, Quchan Branch, Islamic Azad University, Quchan, Iran

well as the shapes of their absorption spectra. Such information can be very useful in supplying preliminary information about the system under investigation and ultimately could guide towards the correct model. In many instances, there is no model or mathematical function at all that could be used to quantitatively describe the process under investigation. Then, the concentration profiles, in conjunction with the pure component spectra, are all there which are to be extracted from the data.

In this work, the soft and hard modeling methods are suggested for the spectrophotometric study of three competitive complex formations. The applicability of the methodology serves as an example in the estimation of the complexation constants of lanthanide cations (M) by Arsenazo III (L).

Arsenazo III, with the common descriptive name of 3,6-bis[(2-arsonophenyl)azo]-4,5-dihydroxy-2,7-naphtalenedisulphonic acid disodium salt, is a disodium salt with the general formula of $C_{22}H_{16}As_2Na_2N_4O_{14}S_2$. This bisazo derivative of chromotropic acid forms stable complexes with large variety of different cations [19]. The chromophore groups in the ligand promote the spectrophotometric study in the visible field. Specifically, this multipurpose complexing ligand is extensively used for spectrophotometric analysis. Interactions of Arsenazo III with lanthanides were investigated in some detail [20-23]. In particular, Arsenazo III was shown to make the similar type of complex with lanthanides and is a very useful remark to minimize the number of models for this series.

Our interest in the behavior of this ligand is stimulated by at least two reasons. Firstly, formation constant values of the above complexes were reported for 11 lanthanides by Lu [24] which makes use of the data from the first stages of the complexometric titration and is based on the Taylor expansion of the concentration of the ligand as a function of the titrant concentration. This method is rather complicated and is not user-friendly.

The second reason was the fact that Arsenazo III was shown to build up a similar type of complex with lanthanides. In this context, two different methodologies were proposed to evaluate formation constants. These methodologies are based on soft and hard modeling that leads to the evaluation of stoichiometry of the main species and formation constants. Meanwhile, we have presented the spectral and concentration profiles for each species in the titration process.

Methods

Apparatus and software

An Agilent photodiode array spectrophotometer (model 8453, Santa Clara, CA, USA) equipped with a 1-cm path length glass cell was used in recording absorbance spectra. A Pentium 4 personal computer was used in controlling the spectrophotometer and collecting the data from its interface. All spectral measurements were performed using a blank solution as a reference. Measurements of pH were made with a Metrohm 632 pH meter (Labequip Ltd. Markham, Canada) using a combined glass electrode. The computations were made with a Pentium 4 computer. All the programs in the computing process were written in MATLAB 6.5 for Windows (MathWorks, Inc., Natick, MA, USA).

Reagents and solutions

All reagents were of analytical reagent grade, and distilled water was used throughout. All the 13 lanthanide stock solutions (0.01 M) were prepared by dissolving the following weight and kind of salts in 10 mL of distilled water.

The reagents were as follows: La, 43.30 mg lanthanum nitrate (Merck, Darmstadt, Germany); Ce, 56.84 mg cerium(III) sulfate (Reacton, Moffat, UK); Pr, 17.02 mg praseodymium oxide (Reacton); Nd, 16.82 mg neodymum oxide (Reacton); Sm, 25.67 mg samarium chloride (Reacton); Eu, 42.81 mg europium(III) nitrate (Reacton); Gd, 18.13 mg gadolinium oxide (Reacton); Tb, 18.69 mg terbium oxide (Reacton); Dy, 18.65 mg dysprosium oxide (Reacton); Ho, 18.89 mg holium oxide (Reacton); Er, 19.10 mg erbium oxide (Reacton); Yb, 19.70 mg ytterbium(III) oxide (Reacton); and Lu, 19.89 mg lutetium oxide (Reacton)

A stock solution of monochloroacetic acid (BDH, Dorset, UK) buffer solution (0.1 M) was prepared daily by dissolving 9.44 g of $CH_2ClCOOH$ in water and diluting to 1,000 mL after adjusting the pH to 3.2. A stock solution of Arsenazo III (0.001 M) was prepared by dissolving 0.0820 g of arsenazo chloride as a disodium salt (Merck) in water and diluting to 100 mL.

Procedure

A 5.0-mL Arsenazo III (0.001 M) solution and 50.0-mL monochloroacetic acid buffer (pH = 3.2; 0.1 M) were added in a 100-mL volumetric flask and diluted to the mark with distilled water after adjusting the pH to 3.2. A 3-mL Arsenazo III solution (5×10^{-5} M at pH = 3.2) was transferred to a spectrophotometric cell, and the titration was started by successive 0.2-μL additions of each standard stock lanthanide solution (0.01 M) to the cell. The absorbance data of this solution were recorded (at 380 to 780 nm) after each addition, against a reagent blank. The titration was followed until the absorbance variations became small and nearly constant.

Results and discussion

The three-dimensional absorbance plot (absorbance vs. wavelength vs. mole ratio) of Arsenazo III titrated with europium cation in a solution of pH = 3.2 is shown in Figure 1. This lanthanide is selected as a representative example because basically similar observations are made

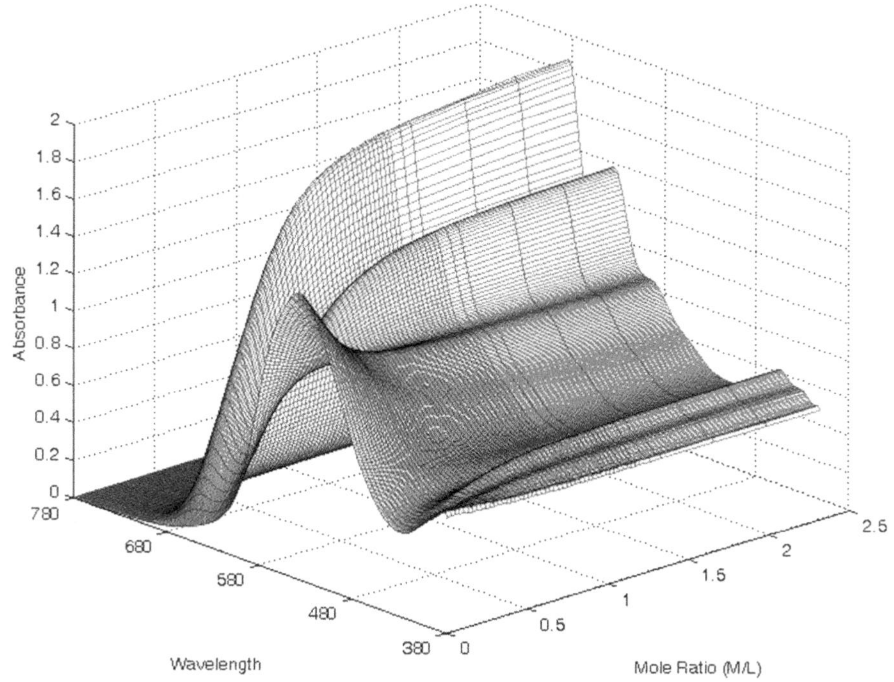

Figure 1 Three-dimensional absorbance plot of Arsenazo III titrated with europium cation.

with all other 12 investigated lanthanides in this research. The λ_{max} of Arsenazo III is 530 nm, and a shift with two narrow and distinct isosbestic points is observed in its spectrum to longer wavelengths upon addition of metal cations.

After building up the titration data matrices for Arsenazo III with each of the 13 lanthanide-Arsenazo III complex systems, the next step was to determine the number of species and models for the complexation to estimate the conditional constants for the complexes formed based on mole ratio data.

Determination of the number of efficient factors and species by factor analysis

In order to estimate the number of coexisting absorbing species in the reaction systems (i.e., the number of produced complexes), principal component analysis (PCA) was performed on the digitized absorbance mole ratio data matrices, which were obtained from the spectrophotometric titration of Arsenazo III with each lanthanide cation, separately.

According to Figure 2, the structure of the residuals shows that only noise has remained after the subtraction of the contributions of the first four eigenvectors. Therefore, according to this figure and some other methods based on PCA, there are four absorbing species created in the titration process.

Models for the complexation

Studies of multiple equilibria in solutions are essential for the solution of many analytical and speciation problems. There are different analytical procedures in this field [25,26], using least-squares methods, to estimate the number of species present in the reaction system, their stoichiometries, and their stability constants. The experimental data needed usually arise from the successive steps of a potentiometric or spectrometric titration. Least-squares methods are hard models and require knowledge about the chemistry of the reaction system, i.e., an initial set of stoichiometric coefficients and stability constants for the postulated species. Moreover, these methods, using a systematic search of the possible models and with the aid of chemical reasoning, can determine the composition of the species and the stability constants, provided that mass-law and mass-balance equations are fulfilled. According to Figure 2, each solution at a particular mole ratio can be considered as a mixture of four components.

Therefore, based on the results obtained by factor analysis methods ('Determination of the number of efficient factors and species by factor analysis' section) and MCR-ALS (Figure 3), we confined our studies on the systems containing four absorbing species, and by using the mass balance equations and equilibrium expressions, a hard model was generated for lanthanide-Arsenazo III systems using the following simple equilibria

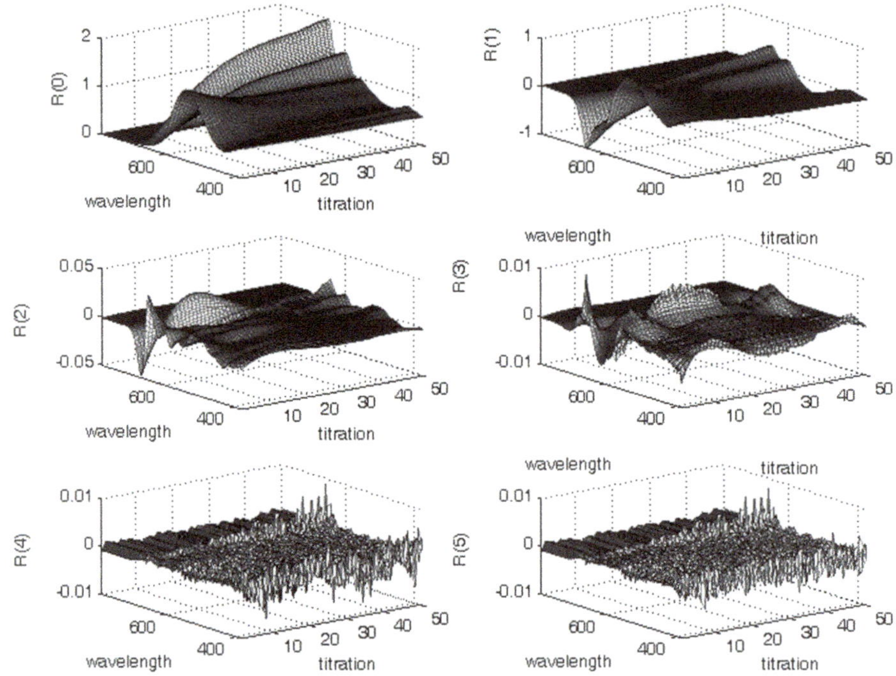

Figure 2 Residual's structure after subtraction of the first 0, 1, 2, 3, 4 and 5 eigenvectors from the main data matrix of the europrium -Arsenazo (III) titration process.

(Equations 1 to 3) and ML, ML_2, and M_2L_2 complexes were tested and defined.

So, a system which contains M, L, ML, ML_2, and M_2L_2 species with three competitive conditional complex formations is defined by three different chemical equilibria (Equations 1 to 3).

$$M + L \rightleftarrows ML \qquad K_{11} = \frac{[ML]}{[M][L]} \tag{1}$$

$$M + 2L \rightleftarrows ML_2 \qquad K_{12} = \frac{[ML_2]}{[M][L]^2} \tag{2}$$

$$2M + 2L \rightleftarrows M_2L_2 \qquad K_{22} = \frac{[M_2L_2]}{[M]^2[L]^2} \tag{3}$$

Simulated data

In checking the potential of the proposed method, the algorithm was tested using the simulated data set that was created according to mole ratio method and was very similar to the real data set. In simulating the three different competitive complexation systems, four simulated spectra of species were formed according to the spectral profiles

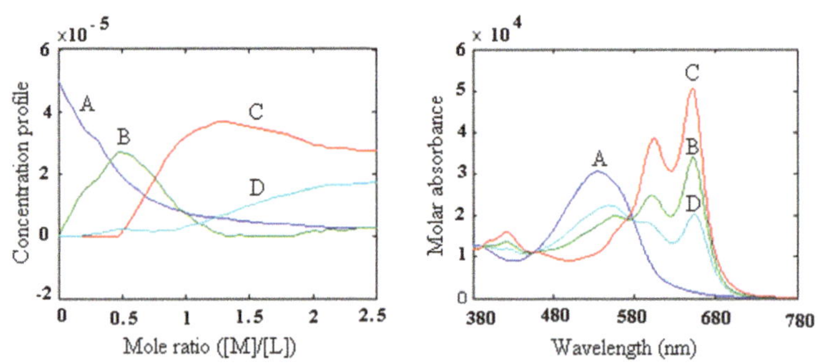

Figure 3 Results of the data analysis in the absorption region of Arsenazo III titration with europium data matrix by MCR-ALS. Left panel: concentration profiles. Right panel: pure absorption spectra. A, L; B, ML_2; C, ML; D, M_2L_2.

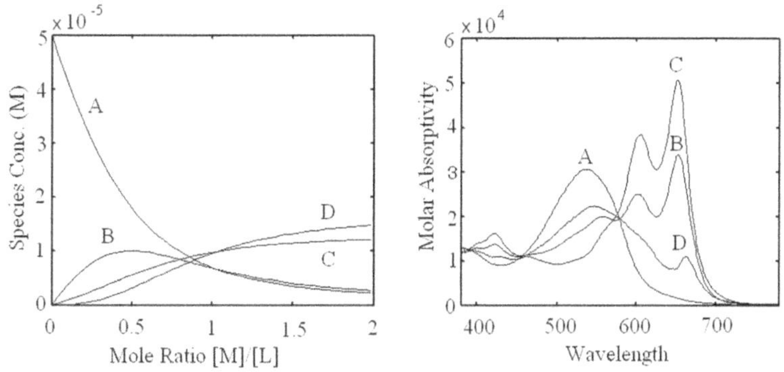

Figure 4 Results of the data analysis in the absorption region of data matrix fitting. Left panel: concentration profiles. Right panel: pure absorption spectra. A, L; B, ML_2; C, ML; D, M_2L_2.

obtained by MCR-ALS (Figure 3) and summed together in a known proportion in such a way that the model of mole ratio method is mimicked using the equilibrium constants $K_{11} = 10^5$, $K_{12} = 10^{10}$, and $K_{22} = 10^{15}$ and the concentration of Arsenazo III, CL = 5.0×10^{-5} M.

Random error was added to the set of generated artificial data. The error is a set of noise drawn from a normal distribution with mean zero and standard deviation equal to 0.25% of absorbance value.

Experimental

The spectra measured from a chemical equilibrium process can form a second-order data matrix, which contains both the equilibrium information and the pure spectrum of each component involved in the chemical process. Using the initial concentration profiles estimated by EFA and MCR-ALS for the mole ratio method with a reasonable lack-of-fit error of 0.13802% has resolved the pure absorption spectra and concentration profiles and allowed the identification of some species equilibria (Equations 1 to 3).

Model-based fitting is readily explained as the direct consequence of the much smaller number of parameters to be fitted. In essence, the number of nonlinear parameters equals the number of unknown equilibrium constants. They are the only parameters that need to be fitted iteratively. The unknown molar absorptivities are linear parameters and are, as mentioned before, effectively eliminated from the iterative fitting process, and the parameters that should be changed are the equilibrium constants (K_{11}, K_{12}, and K_{22}).

The simple equilibrium (Equations 1 to 3) models which were confirmed by previous methods (MCR-ALS) were first tested. They turn out to give simple provisional inequalities that can be used to predict the main complexes that will form.

Using the equilibrium models and calculated concentration profile (Figure 4) obtained by the fitting procedure, the formation constants of lanthanide-Arsenazo III complexes were measured and presented in Table 1.

In addition, the estimated concentration profile which was obtained by the MCR-ALS procedure was used to calculate the formation constants of lanthanides-Arsenazo III complexes (Table 2). Tables 1 and 2 show a good convergence of the results obtained by the two soft and hard modeling approaches, which confirms the previous equilibrium models (Equations 1 to 3).

Figure 5a,b,c illustrates the results obtained by two different methods: MCR-ALS and fitting, which were used separately to find the complex formation constants and to compare with the Taylor expansion results reported earlier [24].

Conclusions

The proposed work uses chemometric methods to extract complex formation constants from a system with

Table 1 Formation constants of Lanthanide–Arsenazo (III) complexes obtained by model based fitting procedure

No.	Lanthanide	Log K_{11}	Log K_{12}	Log K_{22}
1	La	5.4	10.56	15.24
2	Ce	5.12	9.83	14.85
3	Pr	5.37	10.64	15.18
4	Nd	5.81	10.97	15.99
5	Sm	6.03	11.53	16.17
6	Eu	5.35	10.73	15.32
7	Gd	5.18	10.38	14.95
8	Tb	4.88	10.44	15.14
9	Dy	5.21	10.55	15.03
10	Ho	5	10.27	14.69
11	Er	4.85	10.18	14.63
12	Yb	4.84	10.08	14.39
13	Lu	4.94	10.21	14.59

Table 2 Formation constants of lanthanide–Arsenazo (III) complexes obtained by MCR-ALS procedure

No.	Lanthanide	Log K_{11}	Log K_{12}	Log K_{22}	Lack of fit
1	La	4.45	10.31	14.72	0.19452
2	Ce	4.93	9.54	16.54	0.26166
3	Pr	5.26	9.64	15.75	0.22931
4	Nd	5.96	10.93	16.64	0.13079
5	Sm	5.31	9.66	16.12	0.19001
6	Eu	5.23	10.17	15.02	0.16428
7	Gd	5.24	9.82	14.97	0.20298
8	Tb	4.79	9.99	14.69	0.22175
9	Dy	4.51	9.27	15.36	0.16630
10	Ho	4.86	9.85	14.81	0.24462
11	Er	4.13	9.38	14.65	0.22119
12	Yb	5.02	9.73	14.62	0.24372
13	Lu	4.74	10.38	15.3	0.16491

several chemical equilibria from the analysis of the data obtained by spectrophotometric mole ratio titration method, and there are comparable results obtained by MCR-ALS and curve fitting method.

The knowledge about the chemical behavior of the analytes which was obtained by factor analysis and soft modeling method was used to implement a hard modeling equilibrium constraint in hard modeling that ensures the correct complex formation constants. The proposed method makes it possible to obtain the stability equilibrium constants, pure absorption spectra, and species concentration profiles in several ligand-metal ion complex formation systems by severe spectral overlapping. In addition, the proposed method was tested with simulated data sets which were created according to mole ratio method and were very similar to the real data set. The proposed method can be applied in resolving the real three competitive complex formation processes, with full spectral overlapping of three complex species. The results open up a new way to study this kind of

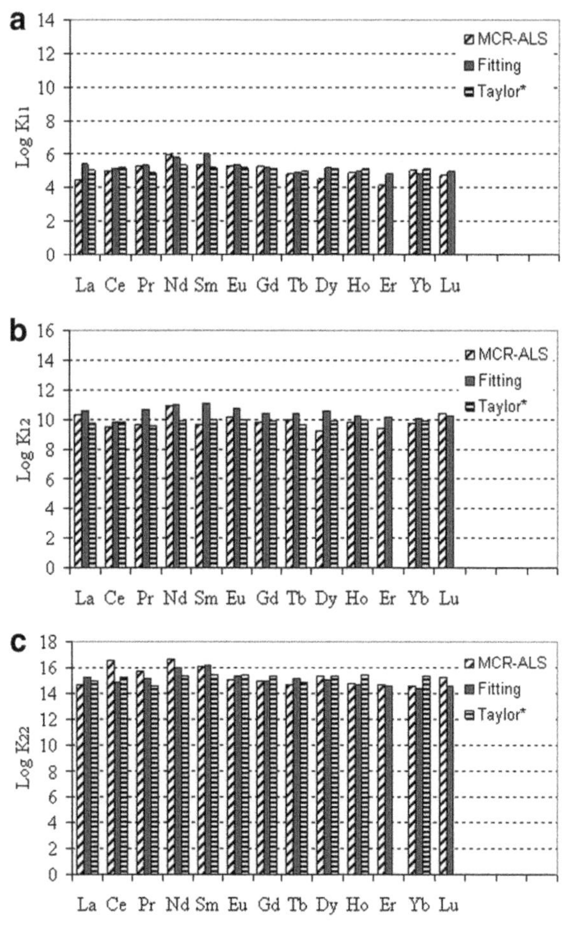

Figure 5 Results obtained by the MCR-ALS, fitting, and taylor expansion [24] methods for (a) logK_{11}, (b) logK_{22}, and (c) logK_{22}.

competitive complexation processes based on the following achievements:

1. The applicability of chemometric methods was shown to resolve complex systems with three competitive complex formation processes in extracting complex formation constants of species with severe spectral overlapping.
2. MCR-ALS was used in the initial estimation of parameters related to the equilibrium concentration and spectra, and then they were confirmed using the hard model.
3. Complexation constants were measured by two different methods (MCR-ALS and fitting) with satisfactory results and good agreement with the earlier published work [24].

Competing interests

The authors declare that they have no competing interests.

Acknowledgement

The authors gratefully acknowledge Dr. Z. Zeinali for her theoretical support.

References

1. Martell AE, Motekaitis RJ (1988) Determination and use of stability constants. VCH, Weinheim
2. Polster J, Lachmann H (1989) Spectrometric titrations. Analysis of chemical equilibria. VCH, Weinheim
3. Maeder M, Zuberbuhler AD (1986) Anal Chim Acta 181:287
4. Malinowski ER (2002) Factor analysis in chemistry, 3rd ed. Wiley-VCH, New York
5. de Juan A, Maeder M, Martinez M, Tauler R (2000) Chemom Intell Lab Syst 54:123
6. Tauler R, Barceló D (1993) Trends Anal Chem 12:319
7. Johnson K, de Juan A, Rutan SC (1999) J Chemom 13:331
8. de Juan A, Izquierdo-Ridorsa A, Tauler R, Fonrodona G, Casassas E (1997) Biophys J 73:2937
9. Vives M, Gargallo R, Tauler R (1999) Anal Chem 71:4328
10. Mendieta J, Diaz-Cruz MS, Tauler R, Esteban M (1996) Anal Biochem 240:34
11. Salau JS, Tauler R, Bayona JM, Tolosa I (1997) Environ Sci Technol 37:3482
12. Nigam S, de Juan A, Cui V, Rutan SC (1999) Anal Chem 71:5225
13. Hirschfeld T (1980) Anal Chem 52:297A
14. Gui M, Rutan SC (1994) Anal Chem 66:1513
15. Molloy KJ, Maeder M, Schumacher MM (1999) Chemom Intell Lab Syst 46:221
16. Janssens LD, Boens N, Ameloot M, De Schryver FC (1990) J Phys Chem 94:3564
17. Maeder M, Zuberbühler AD (1990) Anal Chem 62:2220
18. Bugnon P, Chottard JC, Jestin JL, Jung B, Laurenczy G, Maeder M, Merbach AE, Zuberbühler AD (1994) Anal Chim Acta 298:193
19. Basargin NN, Ivanov VM, Kuznetsov VV, Mikhailova AV (2000) J Anal Chem 55:204
20. Rohwer H, Collier N, Hosten E (1995) Anal Chim Acta 314:219
21. Hosten E, Rohwer H (1997) Anal Chim Acta 345:227
22. Rohwer H, Hosten E (1997) Anal Chim Acta 339:271
23. Budesinsky B (1968) Talanta 15:1063
24. Lu YW, Laurent G, Pereira H (2004) Talanta 62:959
25. Leggett J (1985) Computational methods for the determination of formation constants. Plenum, New York
26. Gans P (1976) Chem Rev 19:99

Rheological study of partially hydrolyzed polyacrylamide-hexamine-pyrocatechol gel system

Upendra Singh Yadav* and Vikas Mahto

Abstract

Background: The cross-linked polymer gels exhibit non-Newtonian rheological behavior which can be described well by the different types of rheological models. This study investigates the onset of gelation behavior of partially hydrolyzed polyacrylamide-hexamine-pyrocatechol polymer gel system which may be used to control excessive water production in the oil fields. Rheological measurements of this system have been performed at different time intervals and pH at 90°C. Attempts have been made to validate the onset of gelation behavior with Bingham plastic, Herschel-Bulkley, Mizrahi-Berk, and Robertson-Stiff model.

Results: It was observed that the developed polymer gel system under the present study has better agreement with the Robertson-Stiff model.

Conclusions: The viscosity of cross-linked polymer gel solution increases with temperature with the passage of time. This increased viscosity leads to gel formation which in turn behaves as flow diverting agent or blocking agent for controlling excessive water production in the oil fields.

Keywords: Polymer gel system, Rheology, Excessive water production, Viscoplastic flow models

Background

The excessive water production in association with crude oil is one of the major production difficulties for the oil industries worldwide. The costs of lifting, handling, separation, and disposal of large amounts of produced water increase the operating cost of the crude oil production and decrease the economic life of a well. Therefore, there is a need to reduce excessive water production [1-4].

The polymer gel treatment is one of the most useful chemical methods to reduce water production [5-9]. Polymer gels are typically composed of a water soluble polymer and cross-linking agents which are dissolved in brine. After allowing sufficient time, the gelant solution sets into a semisolid mass and behaves as flow-diverting or blocking agent [10-12]. The selection of a polymer gel system for a given well treatment strongly depends on reservoir conditions such as temperature, salinity, hardness, and the pH of the water used for the preparation of the gelant [13-19]. Other parameters to be considered for the proper selection of a given polymer gel system include salinity of the formation water, permeability of the target zone, and the lithology of the formation [20,21].

The different polymers used for the development of polymer gel in the oil fields are polyacrylamide with different degrees of hydrolysis (partially hydrolyzed polyacrylamide or PHPA) and polysaccharide such as xanthan biopolymer. These polymers can be cross-linked with metallic and organic cross-linkers to produce a three-dimensional polymer structure of the gel [22].

Gelation behavior is a fundamental parameter in oil-field application. The gelation behavior will determine the injection period and how deep into the formation the gel solution can be placed. For the study of gelation behavior, several methods like bottle testing method, sealed tube method, dynamic shear method (rheometer), and steady shear method (viscometer) are reported in the literatures [23]. Different equations have been used to describe the flow behavior of polymer gel system. The rheological behavior of polymer gel system is non-Newtonian, in which

* Correspondence: yadavupend@gmail.com
Department of Petroleum Engineering, Indian School of Mines, Dhanbad Jharkhand 826004, India

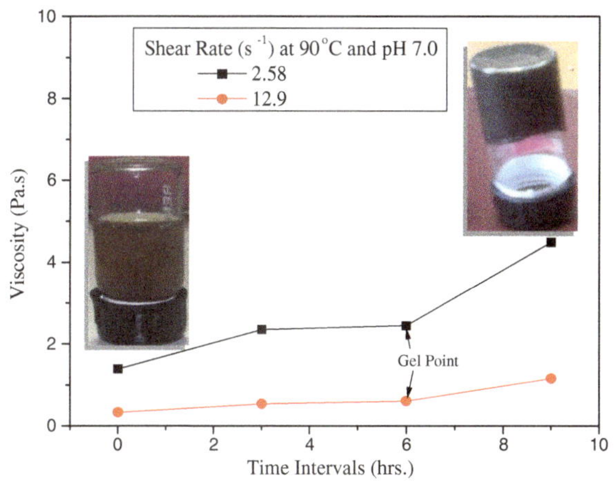

Figure 1 Change in viscosity with time intervals at constant shear rate, 7.0 pH, and 90°C. PHPA 0.9 wt.%, HMTA 0.4 wt.%, and pyrocatechol 0.4 wt.%.

there is a nonlinear relationship between shear stress and shear rate. These polymer gel systems do not display simple behavior and require more complex models for their characterization [24].

In this work, the rheological behavior of PHPA, hexamine, and pyrocatechol gel solution was studied and was simulated using different rheological models. The rheological and gelation behaviors of polymer solution due to the effect of temperature and pH were evaluated using the equipment of Physica Rheolab MC1 (Anton-Paar GmbH, Stuttgart, Germany).

Methods

Materials

The materials used in this work are partially hydrolyzed polyacrylamide, hexamine, pyrocatechol, sodium chloride, hydrochloric acid, and sodium hydroxide. Partially hydrolyzed polyacrylamide in the form of white crystalline powder (procured from Oil and Natural Gas Corporation Limited, Mumbai, India) was used as the water soluble polymer in carrying out the experimental work. The organic cross-linkers hexamine and pyrocatechol were obtained in solid form from renowned manufacturers: Otto Kemi Pvt. Ltd., Mumbai, India and Central Drug House (P) Ltd., New Delhi, India, respectively. Sodium chloride is used to maintain the salinity of polymer gelant, purchased from Nice Chemical Pvt. Ltd., Kerala, India. Hydrochloric acid (Central Drug House Ltd.) and sodium hydroxide (S. D. Fine–Chem Ltd., Mumbai, India) are used for the adjustment of pH.

Equipment

All rheological measurements were performed in a stress-controlled Physica Rheolab MC1. It is a rotational rheometer based on the Searle principle. It features a DC motor drive and optical encoder which provides excellent speed regulation, dynamic range, and transient response. These features provide superior performance in comparison with rheometers using gear-driven (indirect) or stepper motor drives. It enables precise measurements of fluid viscosity over the widest range of conditions available today. This machine is not equipped with a high pressure cell, and measurement at high temperature is not allowed. Hence, the gel had to be cured in hot air oven. The cured samples were transferred to the rheometer, and tests were conducted at 30°C. If these experiments were done at higher temperature, the viscosity of the polymer gel may have increased. Viscosity measurement tests were performed at different shear rates ranging from 1 to 1,200 s^{-1} using cone-and-plate geometry with a logarithmically increasing scale. Gelled samples were placed on the lower plate, and the upper spindle was brought slowly to the 0.2-mm gap. The reasons why a cone-and -plate sensor configuration was chosen as a test geometry are (1) cleaning is very easy after each measurement, (2) the cone and plate can be easily covered with tissue paper, and (3) there is a relatively smaller gap error due to a larger gap size between the cone and plate (0.2 mm in this experiment) compared to a cone-and-plate sensor (50-mm diameter, 1° angle).

Rheological measurements

The steady shear flow properties of prepared partially hydrolyzed polyacrylamide-hexamine-pyrocatechol gel solutions were measured using a stress-controlled Physica Rheolab MC1. All measurements were performed at a fixed temperature of 30°C over a wide range of shear rates from 1 to 1,200 s^{-1} with a logarithmically increasing scale. The 2 ml of the sample was used for each measurement of the polymer gel solution. The cone and plate were covered with tissue paper in order to remove a wall

Table 1 Viscoplastic flow models used in this study and their characteristics

Flow model	Equation	n_1	n_2	γ_∞	Shear-thinning condition
Bingham plastic	$\tau = \tau_o + k\gamma$	1	1	k	-
Herschel-Bulkley	$\tau = \tau_o + k\gamma^n$	1	n	0 k, if $n = 1$	$0 < n \leq 1$
Mizrahi-Berk	$\tau^{1/2} = \tau_o^{1/2} + k\gamma^n$	0.5	n	0 k^2, if n=0.5	$0 < n \leq 0.5$
Robertson-Stiff	$\tau = k(\gamma + C)^n$	1	n	k, if $n = 1$	$0 < n \leq 1$

Table 2 Calculated flow model parameters for polymer gel solution (PHPA 0.9 wt.%, HMTA 0.4 wt.%, and pyrocatechol 0.4 wt.%) at 0 h, pH 7.5, and 30°C

Flow models	Parameters			
	τ_o or C	k (Pa·s^{n-1})	n	R^2
Bingham plastic	4.6845	0.0068	-	0.8671
Herschel-Bulkley	2.130	0.2704	0.5018	0.7942
Mizrahi-Berk	2.4372	0.0994	0.4136	0.8510
Robertson-Stiff	205.32	0.136	0.631	0.8788

Table 4 Calculated flow model parameters for polymer gel solution (PHPA 0.9 wt.%, HMTA 0.4 wt.%, and pyrocatechol 0.4 wt.%) after 6 h at pH 7.5 and 90°C

Flow models	Parameters			
	τ_o or C	k (Pa·s^{n-1})	n	R^2
Bingham plastic	7.2198	0.0181	-	0.9544
Herschel-Bulkley	2.5593	0.3632	0.5882	0.8785
Mizrahi-Berk	3.4865	0.1097	0.4828	0.9313
Robertson-Stiff	155.268	0.131	0.751	0.9554

slippage between the test material and the cone and plate. Through a preliminary test using a direct visualization technique in which a straight line marker was drawn from the cone to the lower plate passing through the free surface of the sample solution, it was confirmed that a wall slip effect could almost be eliminated over a shear rate range tested by covering the plate surfaces with tissue paper. Special care was taken to minimize the effect of work softening when the sample solution was initially loaded on the plate each time. In all measurements, a fresh sample solution was used and rested for 20 min after loading to allow material relaxation and temperature equilibration. It was found from a preliminary test that 10 min of resting time is enough for all sample solutions to be completely relaxed and to be thermally equilibrated. All measurements were made at least three times for each test, and highly reproducible data were obtained within the coefficients of variation of ±5% in all cases.

Results and discussions

In the partially hydrolyzed polyacrylamide structure, the amide group ($CONH_2$) is converted to hydrophilic carboxylate group ($COONa^+$) by hydrolysis reaction, and further cross-linking agent builds up a complex network with carboxylate groups of polymer and forms a three-dimensional gel network structure.

The gel point is defined as the time needed to reach the inflection point on the viscosity vs. time curve (Figure 1). This method has been widely used by several authors [7,16]. Before the gel formation, the viscosity of the gelling solution is relatively low; therefore, it can be measured accurately. After the gel formation, it will be hard to get an accurate viscosity value. Generally speaking, the strength of the cross-linking polymer solution is hard to measure and also hard to define. When the gel has reached a certain high strength, the gel will deform and is easily transferred in various mechanical viscometers, which will bring some damage to the gel performance. Unless this gelling solution allows gelation in the viscometer, the value of the measurement cannot represent its real strength.

The change in shear stress with shear rate at different reaction time intervals and pH was studied using Physica Rheolab MC1. It was observed that the polymer gel solution is non-Newtonian, and the values of shear stress increase with increasing shear rate which indicates that the viscosity of solution is shear rate-dependent. Also, it was observed that the values of shear stress increases with temperature at constant shear rate, which reflects that the viscosity of cross-linked polymer gel solution increases with temperature with the passage of time. This increased viscosity leads to gel formation which in turn behaves as flow-diverting agent or blocking agent for controlling excessive water production in the oil fields. There is no strong effect of pH on the rheological behavior of HMTA and pyrocatechol cross-linked partially hydrolyzed polyacrylamide gel system. Different polymer gel systems have different ranges of pH over which they can maintain their stability. All the experiments were carried out at pH 7.5.

Gelation mechanism

The general mechanism involved in the polymer gel formation at any gelation temperature and pH (in this

Table 3 Calculated flow model parameters for polymer gel solution (PHPA 0.9 wt.%, HMTA 0.4 wt.%, and pyrocatechol 0.4 wt.%) after 3 h at 7.5 pH and 90°C

Flow models	Parameters			
	τ_o or C	k (Pa·s^{n-1})	n	R^2
Bingham plastic	7.1737	0.0146	-	0.9563
Herschel-Bulkley	4.5052	0.1399	0.6937	0.9285
Mizrahi-Berk	4.8104	0.0561	0.5399	0.9512
Robertson-Stiff	218.571	0.111	0.747	0.9631

Table 5 Calculated flow model parameters for polymer gel solution (PHPA 0.9 wt.%, HMTA 0.4 wt.%, and pyrocatechol 0.4 wt.%) after 9 h at pH 7.5 and 90°C

Flow models	Parameters			
	τ_o or C	k (Pa·s^{n-1})	n	R^2
Bingham plastic	14.118	0.0224	-	0.9398
Herschel-Bulkley	7.0577	0.7382	0.5172	0.8368
Mizrahi-Berk	8.1910	0.1232	0.4640	0.8939
Robertson-Stiff	232.539	0.33	0.664	0.9489

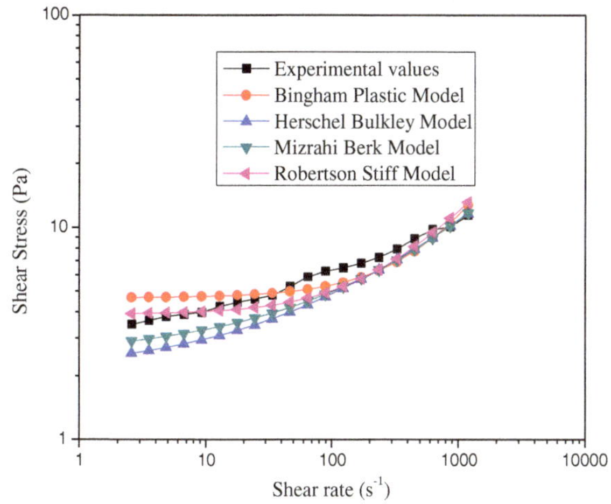

Figure 2 Rheogram of polymer gel solution (PHPA 0.9 wt.%, HMTA 0.4 wt.%, and pyrocatechol 0.4 wt.%). After 0 h at 7.5 pH and temperature 30°C.

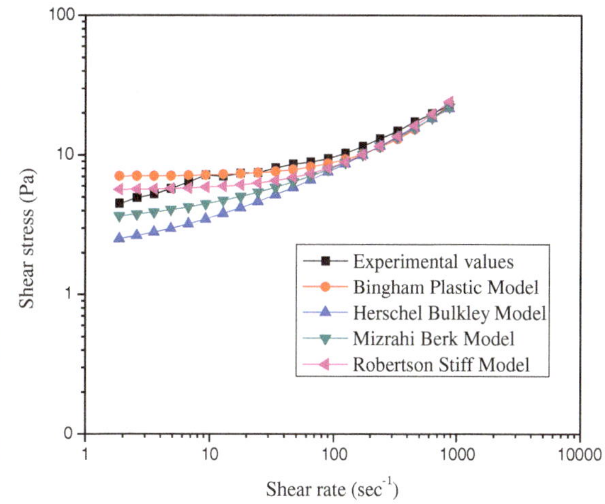

Figure 4 Rheogram of polymer gel solution (PHPA 9,000 ppm, HMTA 4,000 ppm, and pyrocatechol 4,000 ppm). After 6 h at 7.5 pH and temperature 90°C.

study, the temperature is 90°C and pH 7.5) is the cross-linking of polymer solution with organic or inorganic cross-linkers. In this case, the hexamethylenetetramine cross-linker is produced by the condensation reaction of formaldehyde with ammonia which then hydrolyses to form an unstable compound methanediol which dissociates to yield formaldehyde at room temperature. The hexamine cross-linker on hydrolysis yields formaldehyde which then combines with pyrocatechol cross-linker and forms 3, 4, 5, 6-tetramethylol pyrocatechol. Further, partially hydrolyzed polyacrylamide reacts with 3, 4, 5, 6-tetramethylol pyrocatechol and forms three-dimensional networks of polymer gel [25].

Applicability of viscoplastic flow models

The developed polymer gel behaves like a non-Newtonian fluid; hence, viscoplastic flow models may be applied in this system. Here, four inelastic-viscoplastic flow models including a yield stress parameter were employed to make a quantitative evaluation of the steady shear flow behavior of polymer gel solution, and then the applicability of these models was also examined in detail.

A general viscoplastic flow model having a yield stress term may be expressed by the following form:

$$\tau^{n_1} = \tau_o{}^{n_1} + k\gamma^{n_2} \qquad (1)$$

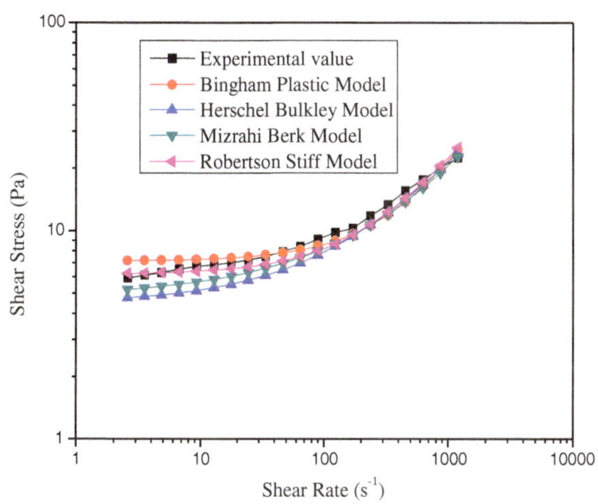

Figure 3 Rheogram of polymer gel solution (PHPA 0.9 wt.%, HMTA 0.4 wt.%, and pyrocatechol 0.4 wt.%). After 3 h at 7.5 pH and temperature 90°C.

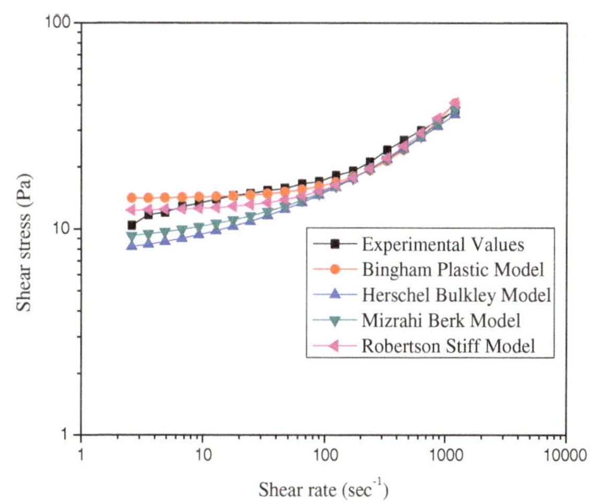

Figure 5 Rheogram of polymer gel solution (PHPA 9,000 ppm, HMTA 4,000 ppm, and pyrocatechol 4,000 ppm). After 9 h at 7.5 pH and temperature 90°C.

where τ is the shear stress, τ_o is the yield stress, γ is the shear rate, k is the consistency index, and n_1 and n_2 are material parameters related to the material's flow behavior, and consequently, each flow model is determined according to the conditions of n_1 and n_2.

Table 1 summarizes the viscoplastic flow models adopted in this study and their flow characteristics. Here, τ_o is the yield stress of each model, and k is the consistency index related to a high-shear limiting viscosity, and γ_∞ as the shear rate is increased towards infinity. n is the flow behavior index, a material parameter that determines the shear-thinning nature of a material. In order that the models summarized in Table 1 could predict a shear-thinning behavior, both n_1 and n_2 must be positive values, and n_1 should be larger or equivalent than n_2.

In order to determine the polymer gel solution parameters of each model, a linear regression analysis was used for the Bingham model, while a nonlinear regression analysis method was used for the Herschel-Bulkley, Mizrahi-Berk, and Robertson-Stiff models. The calculated values of the polymer gel solution parameters from the viscoplastic flow models along with those of the determination coefficients and its comparison with the experimental values are reported in Tables 2, 3, 4, and 5 and Figures 2, 3, 4, and 5.

All the polymer gel solutions exhibit a marked shear-thinning behavior regardless of their pH. As expected, both the yield stress and consistency index values are not the same for the flow models. The value of the flow behavior indices provides a reference for the assessment of a shear-thinning nature, and it was observed that developed gelant is shear-thinning in nature.

The Bingham plastic, Herschel-Bulkley, Mizrahi-Berk, and Robertson-Stiff models give an excellent ability to describe the flow behavior of polymer gel solution. All of these four models have the values to determine coefficients near to 1 and show a clear tendency that the yield stress values remain unchanged with change of pH of the polymer gel solution.

The similar values of the determination coefficients are obtained for different flow models. The Bingham plastic, Herschel-Bulkley, Mizrahi-Berk, and Robertson Stiff models are all in good agreement with the experimentally measured data over a whole range of shear rates tested. But Robertson stiff model has better agreement with the experimental values of shear stress and shear rate data of the developed polymer gel system than the other models tested.

Experimental

Initially, the stock solution of partially hydrolyzed polyacrylamide in brine solution (1.0 wt.%) was prepared and kept for aging at ambient temperature for 24 h. Further, fresh solutions of hexamine and pyrocatechol were prepared in brine ahead of mixing the polymer and cross-linkers to form gelant. The appropriate solution of partially hydrolyzed polyacrylamide, hexamine, pyrocatechol, and brine were thoroughly mixed at room temperature by magnetic stirrer. The pH of the gelant solution was measured by Century CP-901 digital pH meter (Century Instruments Private Limited, Chandigarh, India). The pH of the gelant was maintained using 1N sodium hydroxide and 1N hydrochloric acid. Finally, the solution was transferred into small glass bottles and kept at the desired temperature in the hot air oven (90°C). At regular intervals of time, rheological behavior at different shear rates was measured using Physica Rheolab MC1. The concentration of polymer was 0.9 wt.%, whereas those of the cross-linkers, i.e., pyrocatechol and HMTA, were 0.4 and 0.4 wt.%, respectively.

Conclusions

The following conclusions are drawn from the present study:

- The developed polymer gel solution behaves like a non-Newtonian shear-thinning fluid.
- The rheological behavior of polymer gel varies with time as the gelation time increases; the gel solution will be more viscous, and the increase in viscosity is due to the cross-linking reaction.
- The nonlinear rheological models like Herschel-Bulkley, Mizrahi-Berk, and Robertson-Stiff models are found to be more suitable compared to the linear rheological model like the Bingham plastic flow model.
- Out of the three nonlinear rheological models (Herschel-Bulkley, Mizrahi-Berk, and Robertson-Stiff), the Robertson-Stiff model has better agreement with the experimental values of shear stress and shear rate data of the developed polymer gel system.

Competing interest

The authors declare that they have no competing interests.

Authors' contributions

USY carried out the experiments using the Physica Rheolab MC1 and analyzed the data with non-Newtonian viscoplastic flow models using nonlinear regression method. VM has supervised the research work. USY and VM participated in the interpretation of results and drafted the manuscript. All authors read and approved the final manuscript.

Author' information

USY is a Ph.D. student of the Department of Petroleum Engineering, Indian School of Mines. VM is an associate professor of the same department.

Acknowledgment

The authors are very thankful to the Council of Scientific and Industrial Research (CSIR), New Delhi, India for the financial support to carry out the research work.

References

1. Urbissinova TS, Trivedi J, Kuru E (2010) Effect of elasticity during viscoelastic polymer flooding: a possible mechanism of increasing the sweep efficiency. J Can Pet Technol 49:49–56
2. Stokke BT, Elgsaeter A, Smidsrod O, Christensen BE (1995) Carboxylation of scleroglucan for controlled crosslinking by heavy metal ions. Carbohydr Polym 27:5–11
3. Moradi-Araghi A (2000) A review of thermally stable gels for fluid diversion in petroleum production. J Pet Sci Eng 26:1–10
4. Yadav US, Mahto V (2013) Modeling of partially hydrolyzed polyacrylamide-hexamine-hydroquinone gel system used for profile modification jobs in the oil field. Journal of Petroleum Engineering. doi:10.1155/2013/709248
5. Mortensen K, Brown W, Jorgensen E (1994) Phase behavior of poly (propylene oxide)-poly (ethylene oxide)-poly (propylene oxide) triblock copolymer melt and aqueous solutions. Macromolecules 27:5654–5666
6. Perez D, Fragachan FE, Barrera RR, Feraud JP (2001) Applications of polymer gel for establishing zonal isolations and water shutoff in carbonate formations. SPE Drill Completion 16:182–189
7. Chatterji J, Borchardt JK (1981) Applications of water-soluble polymers in the oil field. J Petrol Technol 33:2042–2056
8. Al-Muntasheri GA, Nasr-El-Din HA, Hussein IA (2007) A rheological investigation of a high temperature organic gel used for water shut-off treatments. J Pet Sci Eng 59:78–83
9. Woodcock JW, Jiang X, Wright RAE, Zhao B (2011) Enzyme-induced formation of thermo reversible micellar gels from aqueous solutions of multiresponsive hydrophilic ABA triblock copolymers. Macromolecules 44:5764–5775
10. Kavanagh GM, Ross-Murphy SB (1998) Rheological characterization of polymer gel. Prog Polym Sci 23:533–562
11. Taylor KC, Nasr-El-Din HA (1998) Water-soluble hydrophobically associating polymers for improved oil recovery: a literature review. J Pet Sci Eng 19:265–280
12. Needham RB, Doe PH (1987) Polymer flooding review. J Petrol Technol 39:1503–1507
13. Storm ET, Paul JM, Phelps CH, Sampath K (1991) A new biopolymer for high-temperature profile control: part 1 - laboratory testing. SPE Reservoir Eng 6:360–364
14. Wever DAZ, Picchioni F, Broekhuis AA (2011) Polymers for enhanced oil recovery: a paradigm for structure–property relationship in aqueous solution. Prog Polym Sci 36:1558–1628
15. Cai W, Huang R (2001) Study on gelation of partially hydrolyzed polyacrylamide with titanium (IV). Eur Polym J 37:1553–1559
16. Grattoni CA, Al-Sharji HH, Yang C, Muggeridge AH, Zimmerman RWJ (2001) Rheology and permeability of crosslinked polyacrylamide gel. J Colloid Interface Sci 240:601–607
17. Al-Muntasheri GA, Hussein IA, Nasr-El-Din HA, Amin MB (2007) Viscoelastic properties of a high temperature cross-linked water shut-off polymeric gel. J Pet Sci Eng 55:56–66
18. Seright RS, Martin FD (1993) Impact of gelation pH, rock permeability, and lithology on the performance of a monomer-based gel. SPE Reservoir Eng 8:43–50
19. Mortimer S, Ryan AJ, Stanford JL (2001) Rheological behavior and gel-point determination for a model Lewis acid-initiated chain growth epoxy resin. Macromolecules 34:2973–2980
20. Park EK, Song KW (2010) Rheological evaluation of petroleum jelly as a base material in ointment and cream formulations: steady shear flow behavior. Arch Pharmacal Res 33:141–150
21. Scott T, Roberts LJ, Sharpe SR, Clifford PJ, Sorbie KS (1987) In-situ gel calculations in complex reservoir systems using a new chemical flood simulator. SPE Reservoir Eng 2:534–646
22. Hao J, Weiss RA (2011) Viscoelastic and mechanical behavior of hydrophobically modified hydrogels. Macromolecules 44:9390–9398
23. Ohen HA, Bick EF (2011) Golden section search method for determining parameters in Robertson-Stiff non-Newtonian fluid model. J Pet Sci Eng 4:309–316
24. Kelessidis VC, Maglione R (2006) Modeling rheological behavior of bentonite suspensions as Casson and Robertson–Stiff fluids using Newtonian and true shear rates in Couette viscometry. Powder Technol 168:134–147
25. Yadav US, Mahto V (2012) Experimental studies, modeling and numerical simulation of gelation behavior of a partially hydrolyzed polyacrylamide-hexamine-pyrocatechol polymer gel system for profile modification jobs. International Journal of Advanced Petroleum Engineering and Technology 1:1–16

Optical properties of pure and modified poly (vinyl chloride)

Emad Yousif[1*], Mustafa Abdallh[1], Hassan Hashim[2], Nadia Salih[3], Jumat Salimon[3], Bashar Mudhaffar Abdullah[3] and Yip-Foo Win[4]

Abstract

Background: In this study, poly(vinyl chloride) (PVC) is considered one of the most versatile plastics. It is the second largest manufactured resin by volume worldwide.

Results: To study the optical properties of modified PVC containing 1,3,4-thiadiazole and phthalyl groups in the repeating units, thin films of the prepared polymers (P I, P II, P III, P IV, P V, and PVC) were casted from tetrahydrofuran solvent with a thickness of 30 μm. The optical properties were studied in the range of 200 to 600 nm, and the optical data were analyzed and interpreted in terms of the theory of phonon-assisted direct electronic transitions; the measured E_g of pure and modified PVC increases in the following order:

$$\overrightarrow{\text{P II, PIV, PV, PI, PIII, and PVC}}.$$

Also, the refractive index was measured.

Conclusions: The results showed that for all compositions, the refractive index decreases with increasing wavelength.

Keywords: PVC, Optical properties, Energy band gap

Background

Modification of polymers has received greater attention as it brings specific properties to the modified material, such as enhanced thermal stability, multiphase physical responses, compatibility, impact response, flexibility, and rigidity [1]. Poly(vinyl chloride) (PVC) is one of the most versatile plastics. It is the second largest manufactured resin by volume worldwide [2].

The modification of PVC lies basically in the dechlorination process, including both substitution and elimination. Because of the peculiarities of the material relevant to its applications, the chemical functionalization of PVC seems permanent to improve and modify the polymer properties [3].

Synthesis of polymer-bound chelating ligands has been reported. A number of ligands including polydentate amines, crown ethers, phosphines, and bipyridines have been bound with mainly polystyrene divinylbenzene copolymers [4,5].

These studies are mostly concerned with ion-binding and catalytic aspects. Very little information appears to exist as to whether, and to what extent, such organic compounds can influence the properties of the macromolecule. Recently, scientists were able to modify PVC by introducing aromatic and heterocyclic moieties through halogen displacement reaction. PVC, thus modified, showed improved overall photochemical stability and optical properties [6,7]. In this paper, we present results of such a study on the optical properties of pure and modified PVC films.

Methods

Optical measurements

The best definition of the energy gap is the minimum energy difference between the lowest minimum of the conduction band and the highest maximum of the

* Correspondence: emad_yousif@hotmail.com
[1]Department of Chemistry, College of Science, Al-Nahrain University, Baghdad, 10072, Iraq
Full list of author information is available at the end of the article

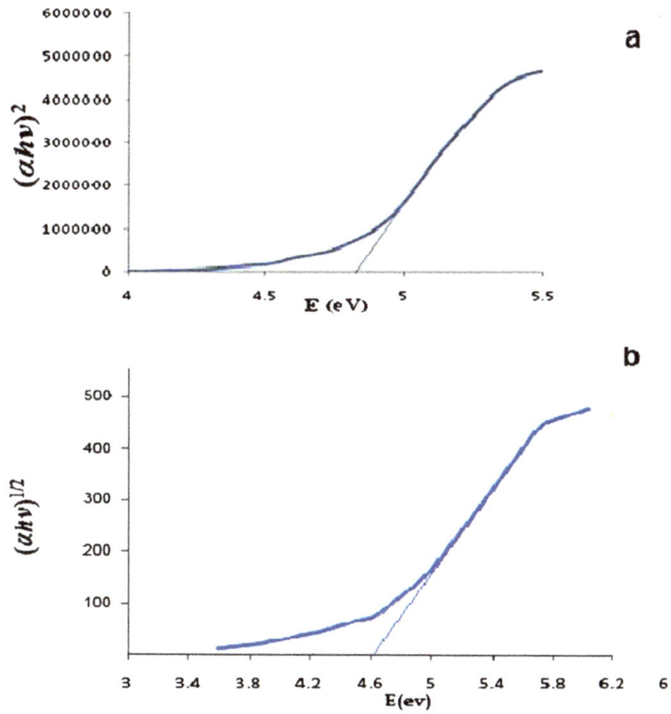

Figure 1 Direct and indirect allowed energy gap for PVC. (a) Direct allowed energy gap for PVC $(ahv)^2$. **(b)** Indirect allowed energy gap for PVC $(ahv)^{1/2}$.

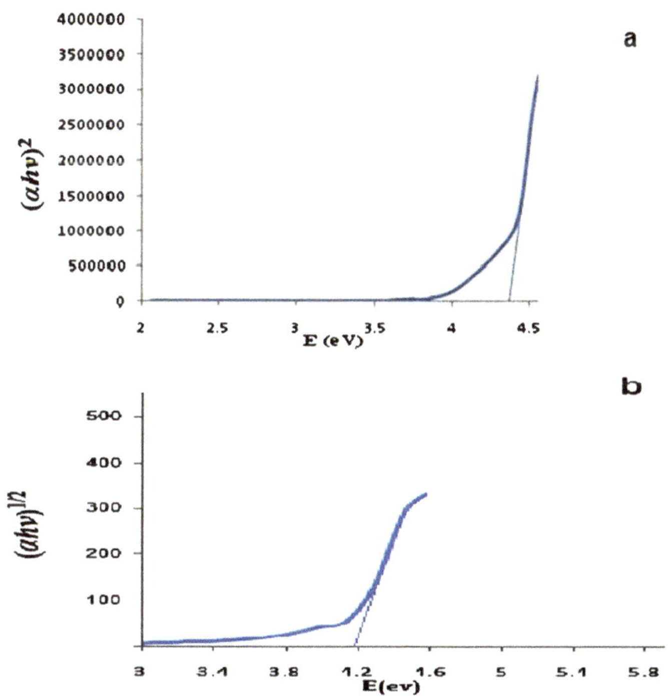

Figure 2 Direct and indirect allowed energy gap for P I. (a) Direct allowed energy gap for P I $(ahv)^2$. **(b)** Indirect allowed energy gap for P I $(ahv)^{1/2}$.

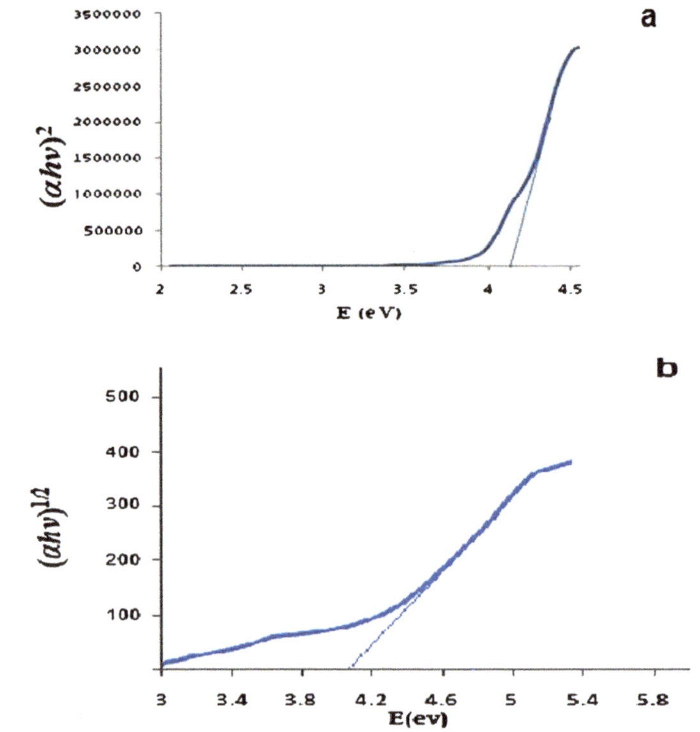

Figure 3 Direct and indirect allowed energy gap for P II. (a) Direct allowed energy gap for P II $(ahv)^2$. **(b)** Indirect allowed energy gap for P II $(ahv)^{1/2}$.

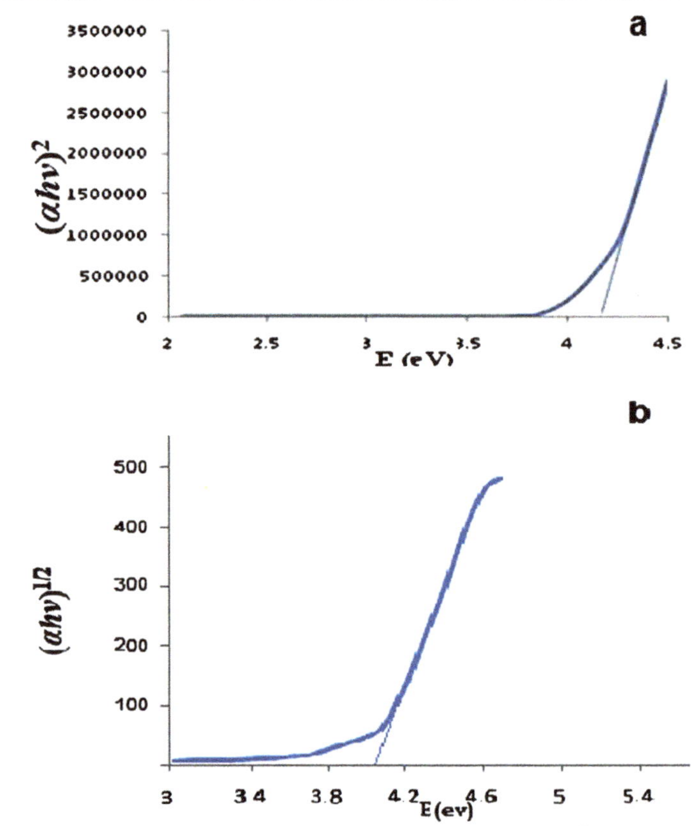

Figure 4 Direct and indirect allowed energy gap for P III. (a) Direct allowed energy gap for P III $(ahv)^2$. **(b)** Indirect allowed energy gap for P III $(ahv)^{1/2}$.

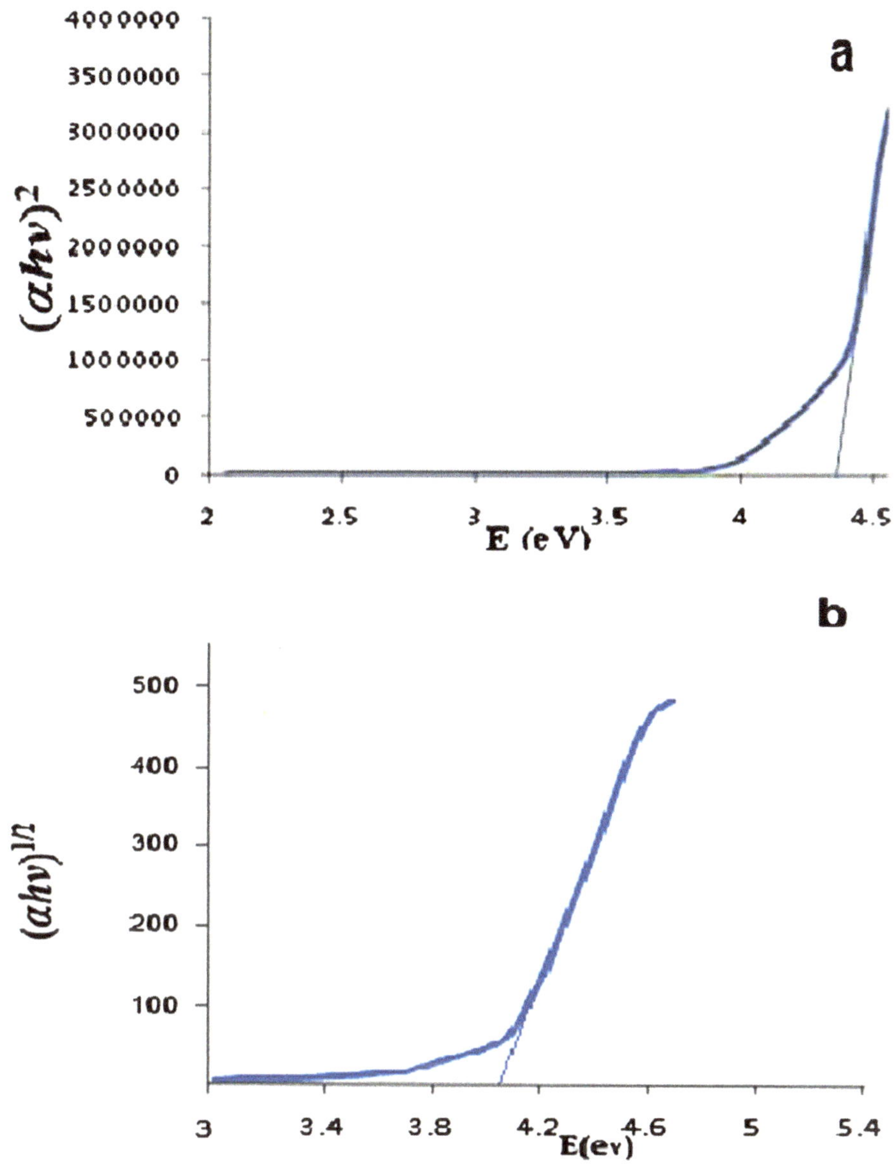

Figure 5 Direct and indirect allowed energy gap for P IV. (a) Direct allowed energy gap for P IV $(\alpha h\nu)^2$. **(b)** Indirect allowed energy gap for P IV $(\alpha h\nu)^{1/2}$.

valance band [8]. The value and shape of the mobility gap in PVC depend on the proportion conditions such as substrate temperature, degree of impurity, and defect of the material. Any variation in such parameters leads to a shift in the absorption edge towards higher or lower energy. The absorption coefficient can be determined from the normal incidence transmission through plane-parallel plate of thickness (*d*), and interference and multiple reflections are neglected. The reflectance (*R*) and transmittance (*T*) are related by Equation 1:

$$T = B(1-R)^2 \exp(-\alpha d) \tag{1}$$

where *B* is a constant. This equation can be rewritten as follows:

$$\alpha d = 2.303\, A + \mathrm{Ln}\left[B(1-R)^2\right], \tag{2}$$

where *A* is the optical absorbance ($A = -\log T$).

So, the absorption coefficient $\alpha(\omega)$ in terms of absorbance becomes

$$\alpha(\omega) = 2.303(A/d). \tag{3}$$

For band-to-band transitions, the energy dependence of the absorption coefficient is of the form

$$\alpha = \alpha_0 \left(h\nu - E_g\right)^r / h\nu, \tag{4}$$

where $h\nu$ is the photon energy, E_g is the energy gap, *r* is

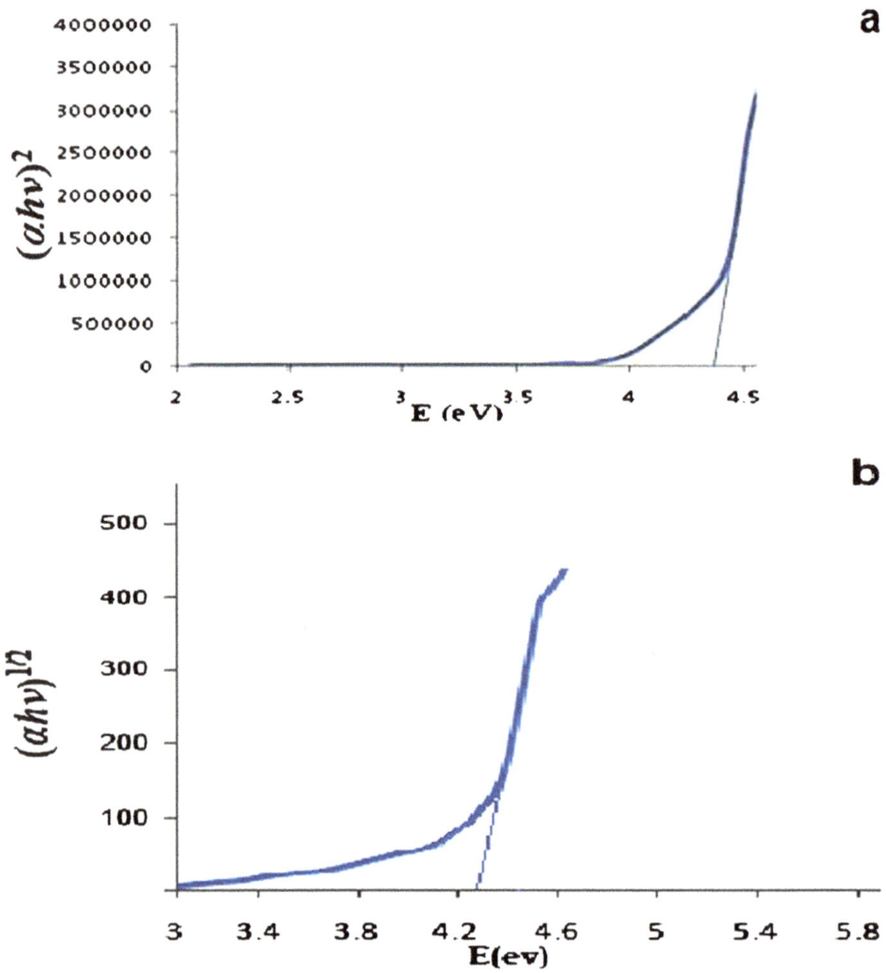

Figure 6 Direct and indirect allowed energy gap for P V. (a) Direct allowed energy gap for P V $(\alpha h\nu)^2$. **(b)** Indirect allowed energy gap for P V $(\alpha h\nu)^{1/2}$.

a constant depending on the type of electronic transitions which takes the value 1/2 for allowed direct transition and 3/2 for forbidden direct transition, and the absorption coefficient $\alpha \geq 10^4$ cm^{-1}. The energy dependence absorption coefficient is in the form

$$\alpha = \alpha_0\left(h\nu - E_g \pm E_p\right)^r / h\nu, \quad (5)$$

where E_g is the minimum energy gap, E_p is the phonon energy, (+) represents absorbance, (−) represents emission, r takes the values 2 for allowed indirect transition and 3 for forbidden indirect transition, and the absorption coefficient $\alpha < 10^4$ cm^{-1}. By plotting $(\alpha h\nu)^{1/r}$ vs. $h\nu$ for a fixed r value, the extrapolation of the linear part could be used to define E_g [7].

Table 1 The energy band gap according to the direct allowed and indirect allowed transitions

Polymer	Direct allowed	Indirect allowed
PVC	4.84	4.65
P I	4.35	4.18
P II	4.13	4.05
P III	4.38	4.23
P IV	4.17	4.07
P V	4.31	4.28

Results and discussion

The relation of $(\alpha h\nu)^2$ and $(\alpha h\nu)^{1/2}$ vs. photon energy for the PVC sample is shown in Figure 1 for allowed transition. For modified samples shown in Figures 2, 3, 4, 5, and 6, the shift in the energy gap could be attributed to the formation of polarons in the modified films. The evidence of polaron formation is that the reaction in the band-to-band transition is due to the shifting of the band density of state toward the energy gap. The effect of the organic addition on the values of phonon energies is also investigated, and the results are shown in Table 1.

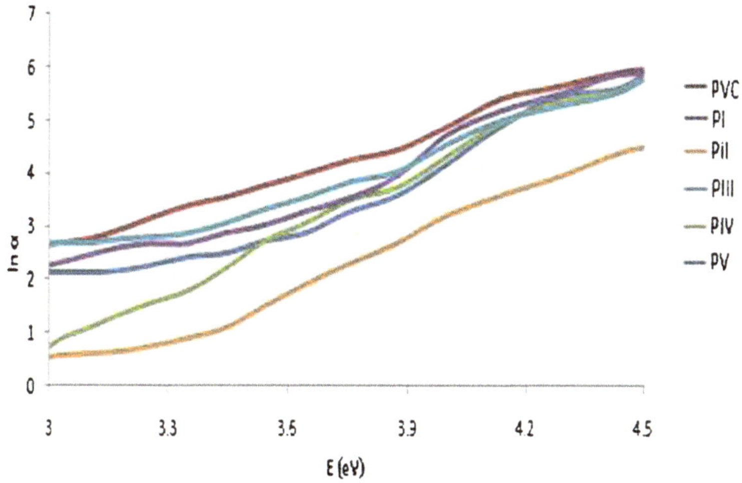

Figure 7 Plots of *α* vs. *E* of the pure and modified PVC.

The energy gaps were measured by adopting the data of energy gap (see Table 1 and Figures 1, 2, 3, 4, 5, and 6). The E_g for PVC in the presence of additives increases in the following order:

$$\overrightarrow{P\,\text{II}, P\,\text{IV}, P\,V, P\,I, P\,\text{III}, \text{and PVC}.}$$

The presence of an ion pair of electrons at different atoms, such as sulfur and nitrogen, for the stereochemistry of the polymer leads to the change of the value of energy gap rather than PVC.

The tail of the absorption edge is exponential, indicating the presence of localized states in the energy band gap. The amount of tailing can be predicted to a first approximation by plotting the absorption edge data in terms of an equation originally given by Urbach. The absorption edge of nonmetallic materials gives a measure of the energy band gap, and the exponential dependence of the absorption coefficient, α, on photon energy, $h\nu$, is found to hold over several decades for a polymeric material and takes the following form:

$$\alpha = \alpha_0 e^{\left(\frac{h\nu}{E_t}\right)}, \qquad (6)$$

where α_0 is a constant and E_t is interpreted as the width of the tails of localized states in the gap region. To evaluate the values of α_0 and E_t, α should be in the logarithmic scale as a function of photon energy $h\nu$ as shown in Figure 2. The reciprocal of the slope of each line yields the magnitude of E_t and its values for different polymeric films. It is clear that modification increases the width of the tail of localized states and decreases the energy gap of PVC thin films. Using Equation 6 at a constant temperature, a graph representing α on

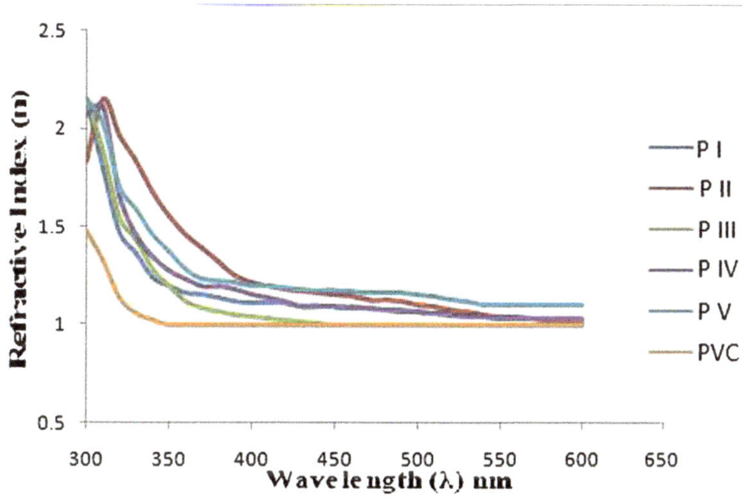

Figure 8 The refractive index dispersion curves of pure and modified PVC.

Table 2 The refractive index of pure and modified PVC at λ = 600 nm

Pure and modified PVC	Refractive index
P I	1.00814
P II	1.0036434
P III	1.0000000
P IV	1.0315588
P V	1.1048604
PVC	1.0000000

the y-axis and energy E (eV) on the x-axis (Figure 7) in the range of the Urbach tail would yield a straight line.

The refractive index is a suitable state parameter directly correlated to the material density. The refractive index, $n(\lambda)$, was determined from the absolute values of the transmittance and reflectance of the investigated films using the following formula:

$$n = \left[\frac{1+R}{1-R}\right] + \left[\frac{4R}{(1-R)^2} - k^2\right]^{1/2}, \quad (7)$$

where k is the extinction coefficient and R is the optical reflectance. The extinction coefficient can be obtained from the relation where $k = \alpha\lambda/4\pi$. Plots in Figure 8 represent the dispersion in the refractive index for pure and modified PVC thin films in the investigated range of wavelengths. Inspection of Figure 8 indicates that for all compositions, the refractive index decreases with increasing wavelength.

Table 2 illustrates the refractive index values for all polymers at a wavelength of 600 nm (for the values of other polymer wavelength, see Additional file 1).

Experimental

All the polymers were prepared as described by [9] (Scheme 1). A solution of PVC or modified PVC (5%) in tetrahydrofuran (THF) was used to prepare the films. The polymeric films were prepared by casting the polymers (P I, P II, P III, P IV, P V, and PVC) onto leveled glass plates. The films were prepared by evaporation technique at room temperature for 24 h. To remove the possible residual THF solvent, the film samples were further dried at room temperature for 3 h under reduced pressure. The samples have the same thickness of 30 μm (measured using a micrometer type 2610 A, Mainz, Germany). The optical

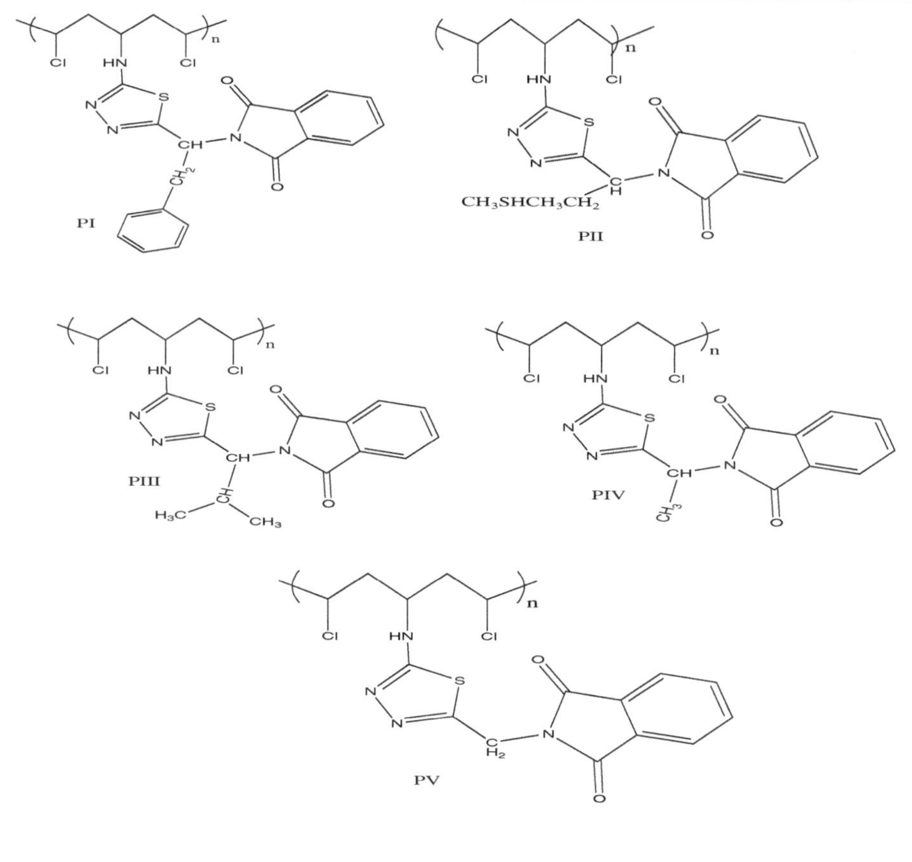

Scheme 1 Structures of polymers.

absorbance (A) of the samples was measured as a function of the wavelength (λ) which ranged from 200 to 600 nm using a computerized Shimadzu UV-160A UV–vis spectrophotometer (Kyoto, Japan) with a full-scale absorbance up to 2.5. The light sources are halogen lamp and socket deuterium lamp, and the detector is a Si photodiode. All measurements were performed at room temperature.

Conclusions

The study on the effect of PVC modification with 1,3,4-thiadiazole and phthalyl groups in the repeating units on its optical properties has shown that the energy gap of modified polymers decreases in comparison with that of pure PVC, and for all compositions, the refractive index decreases with increasing wavelength.

Additional file

Additional file 1: Table. The refractive index of pure and modified PVC.

Competing interests

The authors declare that they have no competing interests.

Authors' contributions

EY developed the concept, analyzed the data, and drafted the manuscript. MA, JS, and NS performed the PVC modification, optimization study, and statistical analysis. BMA, HH, and YW advised on the methods of tests. All authors read and approved the final manuscript.

Acknowledgments

The authors would like to thank Al-Nahrain University, Universiti Tunku Abdul Rahman, and Universiti Kebangsaan Malaysia for the research grants UKM-GUP-NBT-08-27-113 and UKM-OUP-2012-139 and for the financial support and technical assistance on this work.

Author details

[1]Department of Chemistry, College of Science, Al-Nahrain University, Baghdad, 10072, Iraq. [2]Department of Physics, College of Science, Al-Nahrain University, Baghdad 10072, Iraq. [3]School of Chemical Sciences and Food Technology, Faculty of Science and Technology, Universiti Kebangsaan Malaysia, Bangi, Selangor 43600, Malaysia. [4]Department of Chemical Science, Faculty of Science, Universiti Tunku Abdul Rahman, Perak Campus, Jalan Universiti, Bandar Barat, Kampar, Perak 31900, Malaysia.

References

1. Amit B, Rawlins JW, Ray P (eds) (2009) Polymer crafting and crosslinking. Wiley, Hoboken
2. Claudiu L, Edina R, Bogdan M, Teodora Z, Gheorghe H (2010) UPB Sci Bull. Series B 72:127–140
3. Moulay S (2010) Prog Polym Sci 35:303–331
4. Gustafsson B, Hakansson M, Jagner S (2003) Inorg Chim Acta 350:209–214
5. Genhua Z, Pan C (2005) Polymer 46:2802–2810
6. Yousif E, Hameed A, Bakir E (2007) Journal of Al-Nahrain University (Science) 10:7–12
7. Mohammed A, Emad Y, Atheel A, Yang F (2009) Iraqi J Polymers 13(2):27–36
8. Tauc J (1974) Amorphous and liquid semiconductor. Plenum, New York
9. Yousif E, Ayad H, Adel K, Yang F, Noora A, Abdualbasit G (2009) Aust J Basic Appl Sci 3:1786–1794

7

The manufacturing process parameters affecting color and brightness of TiO_2 pigment

Riyas Sharafudeen

Abstract

Background: Originally derived from colored minerals, inorganic pigment titanium dioxide is now highly engineered particles that imparts color or functionality to the objects in which they are used. Optical properties such as color, opacity, brightness, and gloss are important to users of paper and board grades. The color and brightness of TiO_2 sometimes get affected during its manufacture. These phenomena were observed in some commercial production plants. In order to investigate the effect of oil contamination on the pigment during its manufacturing process, such as from sand milling or larox filtration, the measurement on the brightness and color of titanium dioxide is made in this study. A correlation between the color and brightness values measured using different instruments is also included in this paper.

Results: It is observed that whenever there is an oil contamination in the pigment, the vehicle b values measured in acrylic lacquer paint film of the pigment is affected. BYK L* is poorly correlated with the HunterLab vehicle L, and the dry Hunter b is in correlation with the DBC BYK b*, whereas L has no correlation with DBC BYK L.

Conclusion: The contamination of pigment with lubricating oil from the machinery used for the manufacturing process influences both dry and vehicle b values. It is not possible to use HunterLab in place of BYK for the dry L and b measurements.

Keywords: Color, Brightness, TiO_2, Oil contamination, Opacity, Scattering of light

Background

Titanium dioxide is the most widely used white pigment because of its brightness and very high refractive index, in which it is surpassed only by a few other materials. TiO_2 is unique because it efficiently scatters visible light, thereby imparting whiteness, brightness, and opacity when incorporated into a coating. Dry compacted TiO_2 samples are characterized by their brightness and whiteness and exhibit reflectance properties approaching that of the perfect reflecting diffuser. 'Color' is carefully controlled during the TiO_2 manufacturing process through the removal of the trace amounts of metal oxide contaminants. These light-absorbing contaminants will detract from the brightness and whiteness of the pigment and can affect the appearance of white and near-white paints containing that pigment. The opacity of white pigments mainly derives from the scattering, while the opacity of black pigments is a result of absorption ability [1-5]. It was observed that if the TiO_2 is produced in a commercial production plant on particular days when there is oil leak in the larox filter, the color and brightness of the produced pigment failed when measured the same in acrylic lacquer paint (vehicle b). Also, this failure is not seen when the color and brightness are measured in dry mode (dry b). These observations were not a single case but were seen occasionally in the quality analysis. The failure in the vehicle color, once commenced, was seen continued consecutively with three to four final products. The sources of oil contamination during the manufacturing process can be as follows: In the chloride process of manufacturing TiO_2, rutile or high-grade ilmenite is converted to titanium tetrachloride ($TiCl_4$) gas. The conversion takes place in a chlorinator (i.e., fluidized bed reactor) in the presence of chlorine gas at 850°C to 950°C, with petroleum coke added as a reductant. The chief reaction products are volatile metal chlorides, including $TiCl_4$, which are collected. The nonvolatile chlorides and the unreacted solids that remain in the chlorinator are wasted,

Correspondence: riyaschem@rediffmail.com
Department of Chemical and Process Engineering Technology, Jubail Industrial College, P. O. Box 10099, Jubail Industrial City 31961, Kingdom of Saudi Arabia

forming the special waste stream 'chloride process waste solids'. The gaseous raw product stream is purified to separate the titanium tetrachloride from other chlorides. Separation is by fractional condensation, double distillation, and chemical treatment. Ferric chloride ($FeCl_3$) is removed as a major acidic liquid waste stream through fractional condensation. Additional trace metal chlorides are removed through double distillation. Finally, vanadium oxychloride ($VOCl_3$), which has a boiling point close to that of $TiCl_4$ (136°C), is removed as a low-volume non-special waste by complexing with mineral oil and reduced with hydrogen sulfide to $VOCl_2$ or by complexing with copper. The purified $TiCl_4$ is then oxidized to TiO_2 at 985°C, driving off chlorine gas, which is recycled to the chlorinator. Aluminum chloride is added in the oxidation step to promote formation of the rutile crystal which is the TiO_2 product. The next stage of TiO_2 manufacturing is surface treatment, where the TiO_2 from oxidizer is converted to slurry in water, which is then subjected to sand milling to the required particle size. The clarified slurry from the sand mill is treated with alumina or zirconia for coating the surface of bare titanium dioxide with amorphous aluminum or zirconium oxide. The surface-treated product is then filtered through larox by pressing and followed by organic treatment and spin flash drying. Finally, the dried product is subjected to micronization process to get the surface-treated product. The brightness and color of this product is important for its marketing.

The possible sources of oil contamination are during sand milling and larox filtration process. The oil leak is due to lack of monitoring system and improper maintenance. The oil used in the sand mill rotator for avoiding friction can leak due to its continuous use. Likewise, the larox oil can leak during the pressing process, which can contaminate the filtered cake. In order to investigate this issue, it was decided to study the impact of oil contamination on the pigment, color, and brightness values and correlate the color and brightness values measured using different methods.

This paper explains how the pigment L and b values measured in dry and vehicle are influenced by oil contamination during the manufacturing process and determine a correlation between the HunterLab (Reston, VA, USA) L and b with the BYK DBC L and b in order to identify the pigments that failed in vehicle b using HunterLab with the dry b* measured using BYK instrument (BYK-Gardner, Columbia, USA). The correlation between the dry HunterLab values with the dry BYK is also included in this study; the TiO_2 samples which failed in acrylic lacquer paint b values were collected and measured in terms of brightness and color by both BYK and HunterLab instrument as per the above procedure for this correlation study.

Methods

Materials

Chemicals used in the study includes the raw slurry of TiO_2 from the commercial plant, Larox oil RENOLIN B (Fusch Petrolube, Shah Alam, Selangor Darul Ehsan, Malaysia), grade 46, acrylic resin A21 (Rohm and Hass, The Dow Chemical Company, Bristol, PA, USA), Santicizer 160 (Monsanto Chemical Company, St. Louis, MO, USA), commercial grade toluene, commercial grade methyl ethyl ketone, methyl *n*-amyl ketone, standard TiO_2, Ektasolve EB acetate (Eastman Chemical Company, Kingsport, TN, USA).

Apparatus and equipments

The experiments were carried out using Wig-L-Bug grinder with timer (Sigma-Aldrich Corporation, St. Louis, MO, USA), 50 ml polyethylene Nalgene vials (Nalge Nunc International Corporation, Rochester, NY, USA), glass beads, white cards, bird blade film applicator, 6 mm and 6 in. (15.2 cm) path, vacuum plate, Mylar film, cut from roll of approximately 2–3 in.^{2} area, metal cylinder, metal plunger, metal base with a smooth top, hydraulic press, HunterLab and BYK colorimeter and laboratory larox (Labox).

HunterLab color measurement in acrylic lacquer paint

Paint mixtures and the standard TiO_2 were prepared for each sample by weighing the following ingredients into a 50-ml polyethylene vial: 20.0-g size (203) glass beads, 5.0 g of TiO_2 sample, and 26.9 g of master batch prepared in acrylic lacquer-13.2% PVC and grinded for 15 min on the Wig-L-Bug. The beads were allowed to settle for 5 min, and one white card was placed on the vacuum plate. A side-by-side drawdown of the sample and standard was made. The second card was placed on the vacuum plate and another drawdown was made, reversing the order of the sample and the standard. The standard and the sample on each card were labeled. There are no void streaks and each film is wide enough to fill the color meter light port, and a sufficient amount of paint has been used to make the drawdown to produce the desired film thickness. The samples were air-dried for 1 h on a flat, draft-free, dust-free surface and were measured for brightness and color using standardized HunterLab color difference meter for L and b scales. The average the L and b values for each standard/sample pair was calculated as

Average value = average sample reading − average standard reading + standard value.

Color measurement of dry compressed TiO_2

An organization called Commission Internationale de l'Eclairage (CIE) determined the standard values that are

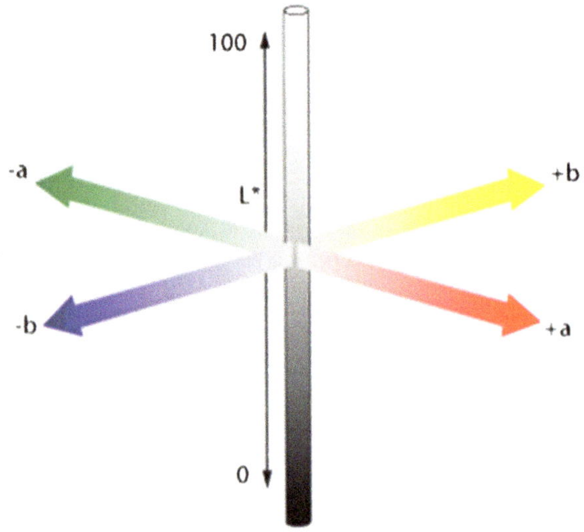

The CIELAB color coordinate system

Figure 1 The CIELAB color coordinate system.

used worldwide to measure color. The values used by CIE are called L*, a*, and b*, and the color measurement method is called CIELAB.

Symbol L* represents the difference between light (where L* = 100) and dark (where L* = 0). A* represents the difference between green (–a*) and red (+a*), and b* represents the difference between yellow (+b*) and blue (–b*). Using this system any color that corresponds to a place on the CIELAB coordinate system was shown in Figure 1. The variables of L*, a*, b*, or E* are represented as delta L*, delta a*, delta b*, or delta E*, where delta E* = delta (delta L^{*2} + delta a^{*2} + delta b^{*2}). It represents the magnitude of the difference in color, but does not indicate the direction of the color difference.

A 2–3 $in.^2$ of Mylar film was placed on top of a smooth metal base, and a metal cylinder was filled about full with the TiO_2 sample to be tested, covered with a metal seal, and applied 5,000 psi for 30 s. The pressure was released, the cylinder was removed, and the face of the pellet was inspected for smoothness and measured for dry color and brightness using HunterLab. The standard TiO_2 sample was also prepared by the above procedure. For comparison, the dry L and b values were measured using BYK color-view meter as well.

Results and discussion

An important observation during the filtration was that the oil was coming out from the system along with the filtrate. The filtrate collected appeared to be a suspension of oil in water. Even though the majority of the oil was visually seen passing with the filtrate, yet a trace amount of oil was found adsorbed on the pigment surface, giving a distinct odor to the pigment. The dry and vehicle L and b values of the samples were tested as mentioned above and the results are tabulated in Table 1.

HunterLab results

It is clear from the above results that there is no considerable change in the dry L value of the pigment upon oil contamination, while the dry b values increased by 0.12 and 0.16 units, respectively, in oil-contaminated samples. Still, the values are within the specification limits for the final marketing product and therefore, the oil-contaminated pigment passed the dry b test using HunterLab. However, when the same samples were subjected to the vehicle b test, an increase of 0.08 and 0.11 units was seen. This increase, though less compared to the increase in dry b value, leads to the failure of the pigment in the vehicle b test. Since the specification for vehicle b value is very narrow, the weight of smaller difference in vehicle b will be really higher than that of dry b values. Also, the gloss of the paint film measured was found to be more than expected. This may be due to the increase in specular reflection due to the presence of oil, and the diffused reflections may be reduced, which appear as a low-brightness value. The specular and diffuse reflection is shown in Figure 2. The brightness of a pigment is due to diffuse reflection while the gloss is due to specular reflection.

Effect of temperature on oil-contaminated samples

When the air-micronized samples were subjected to heating in air oven at 100°C and 180°C for 1 h, the oil-contaminated pigment was found to change in color towards yellow. This effect was more when the temperature was raised from 100°C to 180°C. Both dry and vehicle L

Table 1 Dry and vehicle L and b values of the samples

Experiment No	Sample	Dry				Vehicle	
		BYK		HunterLab		HunterLab	
		L	b	L	b	L	b
1	Reference sample	98.33	1.63	97.24	1.29	95.40	0.68
	Oil-contaminated sample	98.37	1.73	97.27	1.41	95.45	0.76
2	Reference sample	98.18	1.60	97.34	1.27	95.20	0.66
	Oil-contaminated sample	98.24	1.71	97.49	1.43	95.22	0.77

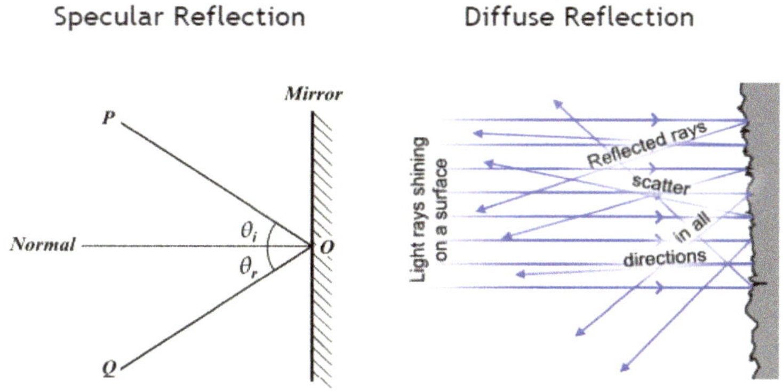

Figure 2 The specular and diffuse reflection.

and b values were measured by HunterLab and BYK. The results of dry and vehicle L and b values of the laboratory-prepared samples that are subjected to heating in air oven at different temperatures are given in Table 2.

In all the samples, the L values remained more or less the same upon heating; while both dry and vehicle b values were increased noticeably upon heating. It is also observed that the increase in dry b value is more pronounced in HunterLab than BYK. However, the increase in dry as well as vehicle b values upon heating is more or less the same in the reference sample and the sample contaminated by oil. So it is not possible to distinguish between reference and oil-contaminated samples from this result. In other words, both reference and oil-contaminated samples heated at 100°C passed the dry b test, while when the temperature was increased to 180°C, both samples failed the dry b test. Unlike this observation, all the four samples failed in the vehicle b test irrespective of the nature of the sample. So, it is obvious from this result that the heating did not make any special change in the sample contaminated by oil; rather, it changed the b value of both reference and oil-contaminated sample to the same extent. The brightness and color of oil-contaminated pigment were compared with the pure, uncontaminated pigment prepared from the same base pigment under similar conditions.

One of the reasons for this was expected to be the oil contamination during the filtration stage due to the possible leakage of lubricant oil from the larox filter pressing stage; hence, the same larox lubricating oil, RENOLIN B, grade 46 was doped with the pigment during filtration. Another source for oil contamination is from the sand mills during the grinding process. The gear oil used in such mills can leak and contaminate the slurry which proceeds to the surface treatment process. In order to avoid this oil contamination, an effective monitoring of all the sand mills and larox filter for oil leak is required.

For comparison, all these samples were subjected to dry L and b tests using the new BYK color-view meter. Similar to HunterLab results, there was no noticeable change in dry L value; while the dry b value was increased in oil-contaminated samples by 0.1 and 0.11, respectively. Unlike HunterLab dry b values, the b values measured by BYK were above the present specification limit and, therefore, the oil contaminated samples failed the dry b test by BYK. So correlations between the two instruments readings are necessary.

Correlation between color and brightness measured with Hunter and BYK colorimeter

Correlation of dry color and brightness measured with BYK color-view with HunterLab

Correlation between vehicle b with BYK b* It is found from the study that almost all the TiO_2 samples failed in vehicle b showed dry b value of around 2 and above.

Table 2 Values of the laboratory-prepared samples that are subjected to heating in air oven at different temperatures

Sample	Dry				Vehicle	
	BYK		HunterLab		HunterLab	
	L	b	L	b	L	b
Reference sample heated at100°C/1 h	98.11	1.63	97.16	1.30	95.08	0.77
Reference sample heated at 180°C/1 h	98.14	1.67	97.11	1.71	94.96	0.99
Oil-contaminated sample heated at 100°C/1 h	98.15	1.74	97.46	1.41	95.11	0.83
Oil-contaminated sample heated at 180°C/1 h	98.14	2.02	97.35	1.84	95.06	1.08

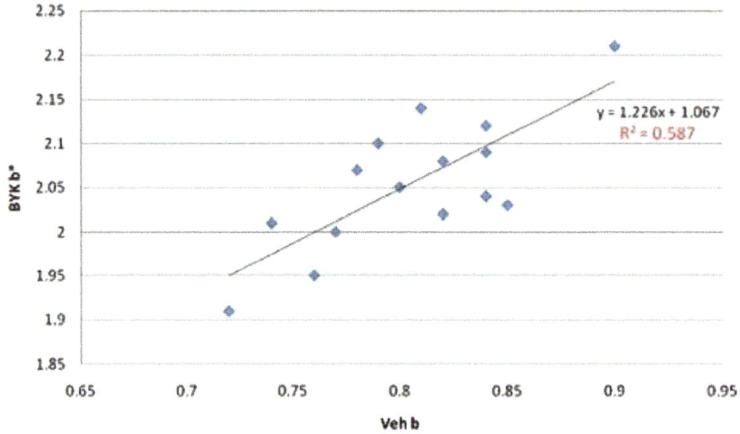

Figure 3 Variation of vehicle b values with BYK b* for samples (indicated by blue diamonds).

The same trend was observed for lot samples also. Figure 3 represents the variation of vehicle b values with BYK b* for samples. It is clear from the graphs that for the bagging bin samples R^2 was found to be 0.583, hence, it can be inferred that there exist a good correlation between vehicle b and BYK b*for samples.

*Correlation between vehicle L with BYK L**

As a variation of vehicle L with BYK L* for the TiO_2 samples in Figure 4, the vehicle L of Figure 3 samples were plotted against the BYK L*. It is evident from the figure that the R^2 value of the bagging bin samples is 0.162 which indicates that the correlation between vehicle L and BYK L* is very poor. Hence, the correlation is poor in the case of L value.

Correlation of dry block color BYK with dry HunterLab L and b

Correlation between dry HunterLab b with BYK b* It is observed that the dry HunterLab b is in good correlation with BYK b*. As it is clear from the Figure 5, the variation of dry HunterLab b with BYK b* for the TiO_2 samples of R^2 was found to be 0.837 which can be inferred that there exist a good correlation between dry HunterLab b and BYK b*.

*Correlation between dry HunterLab L with BYK L**

Figure 6 represents the variation in dry hunter lab L with the BYK L*. It is evident that the R^2 is 0.382 which indicates that the correlation between them is poor.

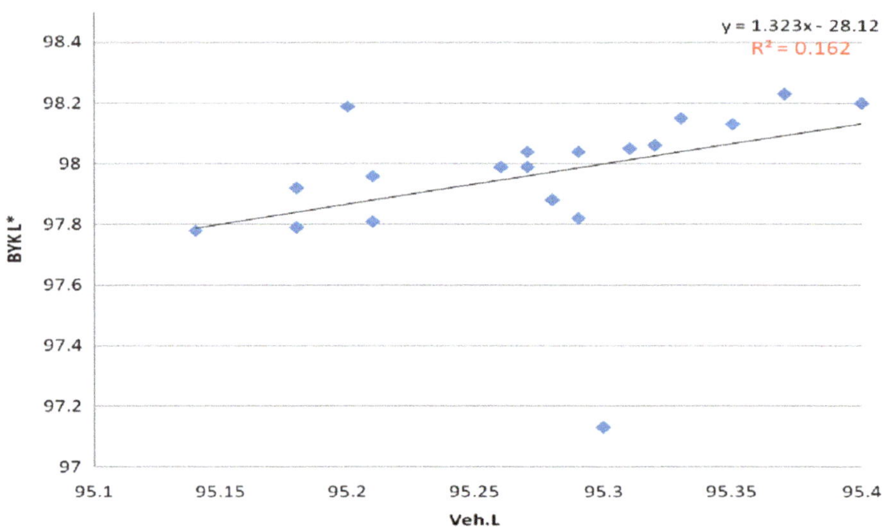

Figure 4 Variation of vehicle L with BYK L* for samples (indicated by blue diamonds).

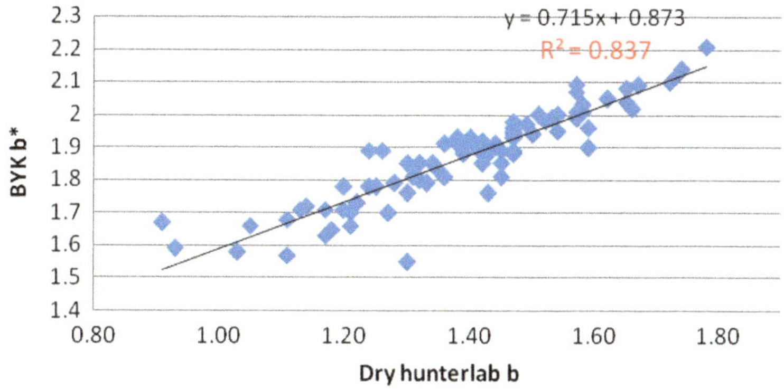

Figure 5 Variation of dry HunterLab b with BYK b* for samples (indicated by blue diamonds).

Experimental

Experimental

Sample used in this study was prepared by the following procedure. A 0.5 ml of oil was contaminated with RENOLIN B, grade 46 which is the same oil used in the hydraulic press in the plant larox filter. A 300 ml of surface-treated slurry of titanium dioxide was filtered using laboratory larox (Labox), was dried after filtration, and was micronized in the laboratory air micronizer then measured in its brightness and color. Another reference sample was prepared without oil contamination. Since the laboratory micronizer utilizes air while the plant micronizer utilizes super-heated steam, it was decided to heat the air-micronized samples at 100°C and 180°C for 1 h to see the effect of temperature on oil-contaminated samples. The experiment was repeated for confirmation. The brightness and color were measured using instruments HunterLab and BYK. The dry acrylic lacquers L and b were measured using HunterLab and BYK colorimeters [6,7].

Conclusion

The contamination of pigment with lubricating oil from the machinery used for the manufacturing process influences both dry and vehicle b values. The oil contamination can be identified by dry b test, if the HunterLab dry b specification limits are made narrow. Since part of this oil can be carried with the filtrate during filtration process, the effect can impact the succeeding batches of treatment, since the filtrate is recycled to the treatment tank. This effect would be continued until the oil is completely removed from the system. The dry b test by BYK is promising. It is clear that while setting the upper specification limit of BYK b*, it is good to consider in-process samples; if the upper specification limit for BYK b* is set at 2.0, then it is possible to identify samples which fail in the present vehicle b test due to any process contaminations.

From the correlation study, it can concluded that BYK b* is in good correlation with the HunterLab vehicle b values and when it exceeded 0.70, the corresponding BYK b* reaches 1.9 or higher, and from this it can be inferred that the sample which failed in vehicle b test can be identified with the BYK b* value. the BYK L* is poorly correlated with the HunterLab vehicle L, and the dry Hunter b is in correlation with the DBC BYK b*, whereas L has no correlation with DBC BYK L*; hence, it is not possible to use HunterLab in place of BYK for the dry L and b measurements.

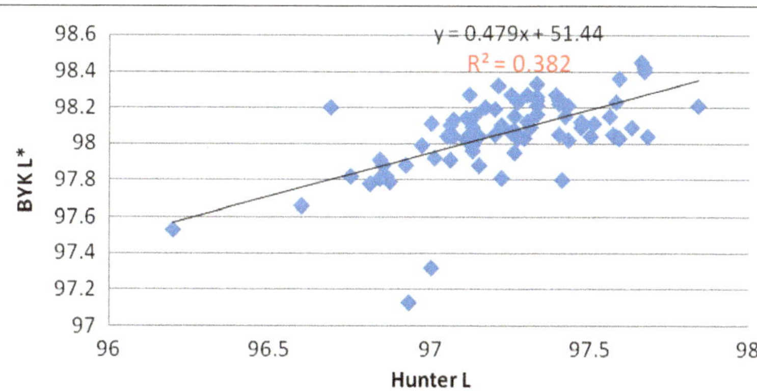

Figure 6 Variation in dry HunterLab L with the BYK L* (indicated by blue diamonds).

Competing interest

The author declares that there is no competing interest.

Acknowledgment

I express my sincere thanks to Mr. Hassan Mones El-Dekki, the general manager of the National Titanium Dioxide Company, Saudi Arabia for providing the necessary facilities and resources for this study, and to all members of the research and development department for their valuable help.

References

1. Kirk O (1997) Kirk-Othmer encyclopedia of chemical technology, 24th edn. Wiley, New York, p 4
2. Lambourne R (1987) Paint and surface coatings. Camelot Press, Southampton
3. Billmeyer FW Jr, Salzman M (1981) Principles of color technology. Wiley, New York
4. Wyszecki G, Stiles WS (1982) Color science. Wiley, New York, p 2
5. Patton TC (1973) Pigment handbook. Wiley, New York
6. HunterLab (2008) Application note. Test methods for color measurement. HunterLab 8:4
7. Buxbaum G, Pfaff G (2005) Industrial inorganic pigments. Wiley-VCH, Weinheim

8

A recyclable PEG-crowned potassium tribromide $[\{K^{+}PEG\}\cdot Br_3^{-}]$ for the synthesis of tetrabromobisphenol A

Sanny Verma, Suman L Jain* and Bir Sain*

Abstract

Polyethylene glycol-bound potassium tribromide $[\{K^{+}PEG\}\cdot Br_3^{-}]$ was found to be an efficient brominating agent for the bromination of bisphenol A to tetrabromobisphenol A of a very high purity in excellent yield. The brominating agent was easily prepared by reacting the PEG_{400} with KBr_3 using the host-guest chemistry concept. The effect of various reaction parameters such as solvent, temperature, and concentration of brominating agent on the yield and purity of tetrabromobisphenol A was studied. The key advantages of the developed method are the use of inexpensive, non-volatile poly(ethylene)glycols, facile regeneration, and inbuilt recycling of generated HBr waste.

Keywords: Bromination, Bisphenol A, Tetrabromobisphenol A, PEG, Recyclable reagent

Background

The formation of host-guest complexes between crown ethers and alkali metal cations is well known and is being used worldwide in various fields such as supramolecular chemistry, biomimetic chemistry, and materials science [1]. However, the expensive nature and limited accessibility of the crown ethers make their utility limited. Compared with the crown ethers, poly(ethylene) glycols (PEGs) have many advantages such as having a much lower cost; being nontoxic, biodegradable, and easily accessible; and displaying complexation properties which are similar to crown ethers [2,3]. Therefore, PEGs are known as a 'poor chemist's crown ethers' and can successfully replace crown ethers for various applications [4,5]. These advantageous features inspired us to use PEGs in the formation of host-guest type complexes for industrially important chemical reactions.

Tetrabromobisphenol A (TBBPA) with the chemical name 4,4′-isopropylidine-bis-(2,6-dibromophenol) is a versatile chemical which has been used worldwide in the preparation of brominated flame retardants in electronic appliances, automotives, textiles, and furniture industries [6,7]. It is also used to provide flame retardancy for styrenic thermoplastics and for some thermoset resins. Conventionally, the synthesis of TBBPA involves the direct bromination of bisphenol (BPA) with molecular bromine in methanol [8]. This method suffers from many drawbacks such as the cumbersome handling of liquid bromine and the generation of the substantial amounts of methyl bromide as by-product.

Consequently, some improved methods for the bromination of bisphenol at a low temperature, e.g., 0°C to 20°C, using methanol/water mixture have been developed [9-11] since the presence of water and low temperature attenuates the production of methyl bromide and therefore overcomes the limitations of the conventional bromination method. However, the longer reaction times and requirement of additional steps for the final precipitation of tetrabromobisphenol A are the major drawbacks associated with these methods.

Recently, the use of organic ammonium tribromides, molecular bromine in conjunction with oxidants, and $HBr\text{-}H_2O_2$ has been reported to be efficient for the bromination of bisphenol A [12,13]. Despite their potential applicability, most of the methods have some limitations such as high cost, difficult handling, longer reaction times, and, moreover, less purity of the desired product. Very recently Kumar et al [14] reported that potassium tribromide (KBr_3) as a highly efficient brominating agent provided instantaneous bromination of bisphenol to TBBPA. This method is advantageous in many ways

* Correspondence: suman@iip.res.in; birsain@iip.res.in
Chemical Sciences Division, Indian Institute of Petroleum (Council of Industrial and Scientific Research) Mohkampur, Dehradun, 248005, India

as it describes the inbuilt recycling of generated HBr, cost-effective nature of the brominating agent, and high purity of the desired product. However, difficult solvent recovery and neutralization of the alkali metal halide prior to the recycling of brominating agent make the utility of this method limited. Therefore, the development of an efficient, recyclable, and cost-effective brominating agent for the bromination of TBA is still required. With continuous interest towards developing novel synthetic methodologies for organic reactions, herein, we report for the first time a highly efficient, readily synthesized, and recyclable brominating agent, i.e., poly(ethylene)glycol-crowned potassium tribromide [{K^+PEG}·Br_3^-] **1** (Figure 1) for the bromination of BPA to TBBPA.

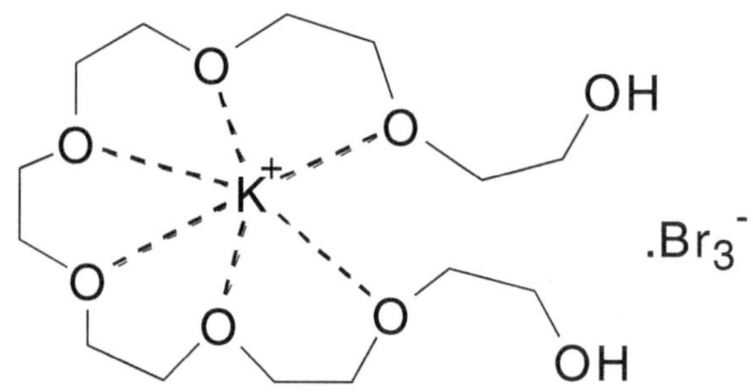

1

Figure 1 Poly(ethylene)glycol-crowned potassium tribromide 1.

Methods

The melting points were determined in open capillaries on a Buchi apparatus and are uncorrected. The ^{1}H NMR Spectra were recorded on Bruker 300 MHz spectrometer and the chemical shifts are expressed in δ parts per million relative to tetramethylsilane (TMS) as internal standard. The IR spectra were recorded on a Perkin Elmer FTIR X 1760 instrument. Elemental analysis was done by using ASTM D-3828 (Kjeldhal method). HPLC analyses were conducted using a Waters 2695 instrument, Meadow Instrumentation with PDA detector using solvent system (70% CH3CN+ 30% H2O), and a flow rate of 1 ml/min.

Result and discussion

The required PEG-bound tribromide reagent **1** could readily be synthesized by following the published procedure [15]. The loading of the reagent was found to be 1.2 mmol/g as determined by elemental analysis. The corresponding brominating agents of PEG_{200} and PEG_{600} were formed successfully, albeit the solid PEG, i.e, PEG_{2000}, gave a sticky material which was difficult in handling; thus, we used compound **1** for further studies. The prepared reagent was dark red in color which disappeared during the bromination of BPA. After completion of the reaction, no bromine could be detected in the reaction mixture as analyzed by UV-Vis spectroscopy; this established the *in situ* generation of bromine from the reagent during the reaction. The bromination of BPA with tribromide **1** was performed at 25°C using water as a reaction media (Scheme 1). The progress of reaction was monitored by thin layer chromatography (TLC). After isolating the brominated product from the reaction mixture, the aqueous layer was treated with bromine in a controlled manner to regenerate the brominating reagent **1**.

The regenerated reagent could easily be recovered from the aqueous layer by extraction with dichloromethane. The combined organic layer was under reduced pressure; the viscous oil was dried under vacuum and reused for next run. The remaining aqueous layer containing HBr waste could be used for the bromination using H_2O_2, resulting in the formation of activated brominating species *in situ*, i.e., hypobromous acid (HOBr) according to the following equation [16]:

$$HBr + H_2O_2 \rightarrow HOBr + H_2O \quad (1)$$

In order to establish the utility and versatility of the present method, we also studied the effect of various reaction parameters to the yield and purity of the desired product. To evaluate the effect of the solvent, bromination of bisphenol was carried out with **1** in different solvents under identical reaction conditions. The results of these experiments are plotted in Figure 2. The purity of the product could easily be analyzed by the high-performance liquid chromatography (HPLC) analysis. As depicted in Figure 2, water was found to be the best reaction medium in the bromination of bisphenol

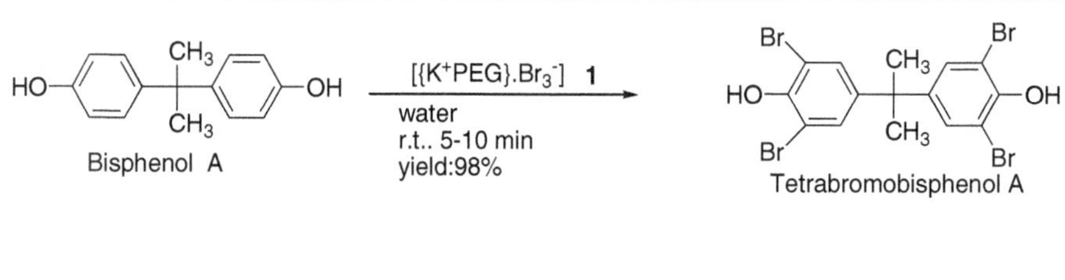

Scheme 1 Bromination of bisphenol A.

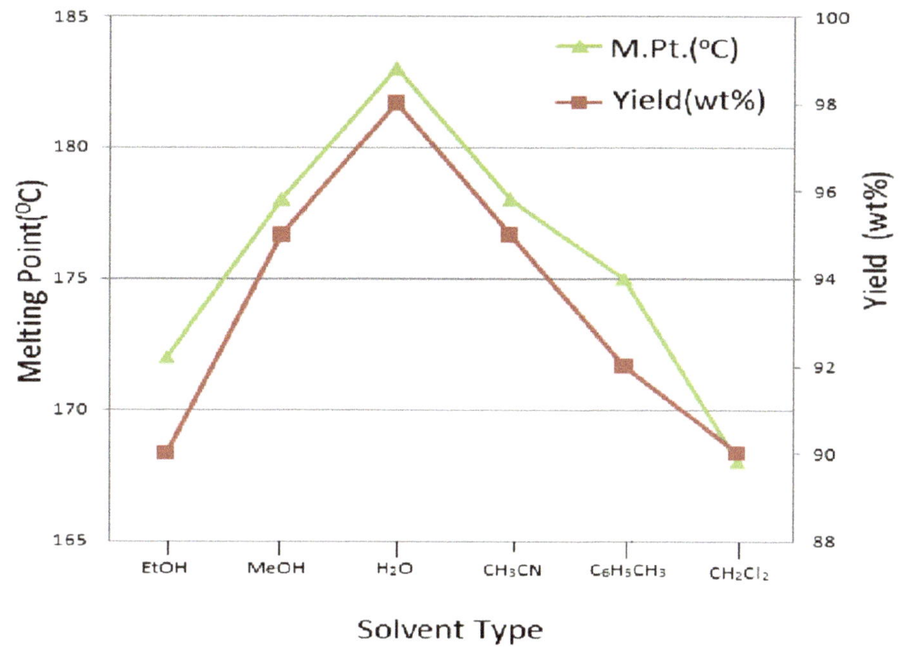

Figure 2 Effect of solvent on the yield and melting point of TBBPA.

A to TBBPA in terms of yield (98%), melting point (M.Pt.) (182°C), and color (white crystalline solid), making the method advantageous from environmental viewpoints. Other solvents such as methanol, ethanol, acetonitrile, toluene, and dichloromethane were also found to be less effective and afforded poor product yield, suggesting the presence of an under-brominated product. The presence of water enhanced the rate of reaction, probably due to the conversion of hydrogen bromide to hydrobromic acid, which allows thorough bromination of bisphenol A, which otherwise was inhibited by the presence of HBr. Another reason for the significant acceleration in water is due to the formation of hypobromous acid in the system ($Br_2 + H_2O = HOBr + HBr$), which serves to be the active brominating species.

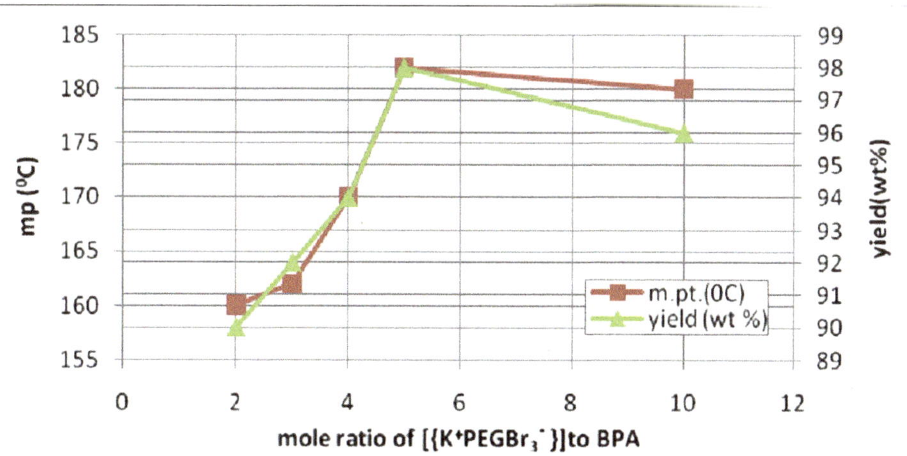

Figure 3 Effect of mole ratio of 1/BPA on the yield and melting point of TBBPA.

Similarly, we studied the bromination of BPA by changing the mole ratio of **1** to BPA from 2:1 to 10:1 under similar reaction conditions. As shown in the Figure 3, the quality of the product is strongly dependent on the mole ratio of **1**/BPA. It has been observed that the optimum yield of TBBPA (98%) with desired purity (99.2% as analyzed by HPLC) was obtained when the mole ratio of **1**/BPA was 5.0/1, whereas, the lower concentration of **1** afforded the brominated product of less purity, probably due to the formation of under-brominated products. The purity of the product was found to be increased as the mole ratio was increased from 2 to 5 mol%; however, a further increase in the mole ratio of **1** affected the quality as well as yield of the product adversely, and we could get less yield of the desired product with decreased purity.

The effect of reaction temperature was studied by performing the bromination of BPA at different temperatures varying from 15°C to 75°C. The results of these experiments are summarized in Table 1. At the lower reaction temperature (15°C), the yield and purity of the product remained comparable albeit the reaction required a longer reaction time (30 to 60 min) in completion. However, 25°C to 30°C was found to be the optimum temperature for this reaction and afforded excellent product yield with high HPLC purity. The lower reaction temperature is another advantage, which further makes this method superior over the existing ones, which employed temperatures >55°C.

Next, we checked the regeneration and recycling of the brominating agent. At the end of the reaction, the brominated product TBBPA was isolated by simple filtration and dried under reduced pressure. The M.Pt. of the white crystalline powder of TBBPA was found to be sharp (182°C) with the 99.2% purity (as analyzed by HPLC). The filtrate so obtained containing {PEG·KBr} and HBr was treated with bromine in a controlled manner to regenerate the reagent **1**. The dark colored reagent **1** was recovered from the aqueous solution by extraction with dichloromethane, dried under vacuum, and reused for the subsequent run. Also, the remaining aqueous layer containing HBr was treated with aqueous H_2O_2 to generate *in situ* HOBr and used for the bromination of bisphenol A. The results of recycling experiments are summarized in Table 2. As shown, the yield, melting point, and purity of the obtained TBBPA during the recycling experiments were found to be almost the same, establishing the successful regeneration, recycling of **1**, and inbuilt utilization of HBr for the bromination of BPA to TBBPA.

Table 2 Results of recycling experiments

Parameters of TBBPA	Run 1	Run 2	Run 3	Run 4	Run 5
M.Pt. (°C)	182	183	181	182	182
Yield (%)[a]	98	97.5	97.5	98	97
Purity (%)[b]	99.2	99.3	99.2	99.2	99.2

Experiment conditions: BPA (1 mmol), reagent **1** (5 mmol), and water (5 ml) at 25°C; [a]isolated yield of TBBPA; [b]purity of the product was analyzed by HPLC.

The exact mechanism of this reaction is not clear at this stage; however, we could assume that the reaction probably involves the *in situ* generation of bromine from $[\{K^+PEG\}\cdot Br_3^-]$, which reacts with water to give HOBr **2**. The electrophilic addition of Br^+ to BPA yields TBBPA along with the formation of 4 mol of HBr as by-product. The generated HBr could easily be recycled by treating with H_2O_2 to give *in situ* HOBr, which subsequently reacts with BPA to afford TBBPA (Scheme 2).

Experimental

General

All the substrates and reagents used were commercially available and were used without further purification. HPLC analyses were conducted using a Waters 2695 instrument (Meadow Instrumentation Inc, 10780, Bristol Rd, Bristol)with PDA detector, solvent system (70% CH_3CN + 30% H_2O), and a flow rate of 1 ml/min.

Preparation of $[\{K^+PEG\}\cdot Br_3^-]$1

In a round-bottomed flask (100 ml), bromine (68.6 mmol, 10.8 g, 3.5 ml) was added to a 34% aqueous solution of KBr (25 g; prepared by adding the 9 gm KBr in 135 g water) to prepare the solution of potassium tribromide. Polyethylene glycol (PEG_{400}; 27.4 g, 68.5 mmol) was added to the above solution, and the mixture was stirred for 5 h at room

Table 1 Effect of reaction temperature

Entry	Temperature (°C)	Yield (%)	Purity of TBBPA (%)[a]
1	15	97.5	98.6
2	25	98	99.2
3	30	97	99.0
4	40	98	94.0
5	50	98	85.0
6	75	97	80.0

Reaction temperature conditions: BPA (1mmol), **1** (5 mmol), and water (2 ml); [a]purity was analyzed by HPLC.

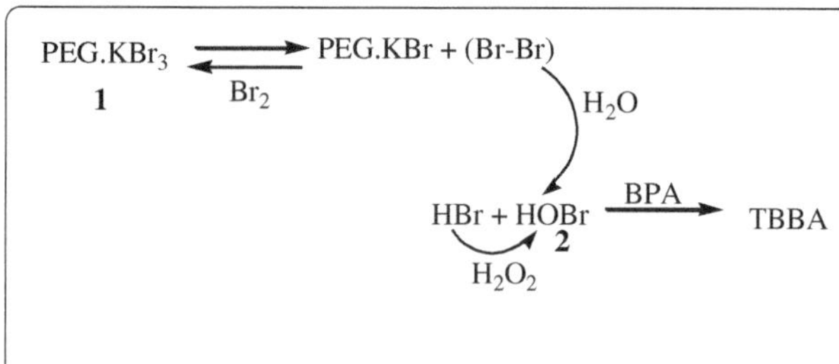

Scheme 2 Probable mechanistic pathway.

temperature. The dark orange-red solution was extracted with dichloromethane, and the combined organic layer was dried over anhydrous sodium sulfate, concentrated under reduced pressure. The dark orange-red viscous liquid **1** was obtained in quantitative yield (44.5 g, 97%). Infrared (IR) (cm^{-1}): 3,410 (OH), 1,112 (CH_2-O-CH_2), and 548 (Br-Br).; analytical calculations of [K^+{PEG·Br_3^-}] are as follows: calculated C, 31.85%; H, 5.60%; K, 5.75%; found C, 31.82%; H, 5.66%; K, 5.72%. The loading of the KBr_3 in the prepared reagent **1** was found to be 1.2 mmol/g.

General procedure for the synthesis of TBBPA

In to a 100-ml round-bottomed double-necked flask containing bisphenol A (0.23 g, 1 mmol) and water (5 ml), [{K^+PEG}·Br_3^-]**1** (3.4 g, 5 mmol) was added with the mole ratio of 1:5, and the resulting mixture was stirred at room temperature until the color of the bromine disappeared. The progress of the reaction was monitored by TLC; after completion, the product was separated by filtration, and the residue so obtained was used for the regeneration of reagent **1** by the addition of molecular bromine in a controlled manner. The crude product was purified by crystallization with hot water. The yield of the TBBPA (98%) and purity of the product were found to be 99.2 (as analyzed by HPLC). The identity of the product was confirmed by its spectroscopic analysis. The characterization data of TBBPA are given as follows: M.Pt., 182°C; ^{1}H-NMR ($CDCl_3$, 300 MHz) δ, 7.25 (4H, s), 5.79 (2H, s), and 1.58 (6H, s). IR: (KBr) 3,514; 3,479; 2,986; 1,552; 1,470; 1,392; 1,321; 1,270; 1,160; 1,127; 865; 776; 731; 707; and 615 cm^{-1}. The analytical calculation is as follows: calculated C, 33.08%; H, 2.20%.; found C, 32.78%; H, 2.24.

Conclusion

In summary, we have developed for the first time an efficient, simple, and easily recyclable brominating agent for the selective bromination of BPA to TBBPA in excellent yield and of high purity using [{K^+PEG}·Br_3^-] as brominating agent. The developed reagent has many advantages such as the use of cheap and environmentally benign PEGs, facile regeneration, nonproduction of HBr as waste, efficient recycling, and higher yields of the desired product in mild reaction conditions.

Competing interests

Authors have filed a patent in India on the process described in the manuscript.

Authors' contributions

SV has done all the experiments mentioned in the manuscript, SLJ and BS provided the technical supported and drafted the mansucript. All authors have read and approved the manuscript.

Authors' information

SV have received an Indo-French Research fellowship under sanwich research programme. SLJ is currently holding a position of Senior Scientist at CSIR-Indian Institute of Petroleum and has received many prestigious National Awards for her work in Synthetic Organic Chemistry.

Acknowledgement

We kindly acknowledge Director IIP for his kind permission to publish these results. SV would like to thank CSIR, New Delhi, for his research fellowship. DST is acknowledged for funding.

References

1. Turro NJ (2005) Proc Nat Acad Sci U S A 102:10766–10770
2. Han CY, He CL, Zhou QZ, Zhang JQ, Wang C, Li N, Huang FH (2008) J Org Chem 73:5872–5880
3. Harris JM, Zalipsky ZS (1997) Poly(ethylene glycol) Chemistry and biological applications;ACS Symp. Ser. No. American Chemical Society, Washington, DC
4. Xu DQ, Luo SP, Wang YF, Xia AB, Yue HD, Wang LP, Xu ZY (2007) Chem Commun :4393–4395
5. Luo S, Zhang S, Wang Y, Xia A, Zhang G, Du X, Xu D (2010) J Org Chem 75:1888–1891
6. Lindemann RF (1969) Ind Eng Chem 61:70–75
7. Gribble GW (1999) Chem Soc Rev 28:335–346
8. Mc Kinnie BG, Sharp GL, Williams RE (1994) Process for high purity tetrabromobisphenol A. U.S. Patent. 5:283–375
9. Sanders DC, Mack AG, Timberlake LD (1995) Reduced methyl bromide process for making tetrabromobisphenol A. U.S. Patent 5,446,212
10. Mc Kinnie BG, Holub RA, Elnagar HY (2000) Process for the preparation of tetrabromobisphenol A. U.S. Patent 6,084,137
11. Mc Kinnie BG, Holub RA, Elnagar HY (2000) Process for the preparation of tetrabromobisphenol A. U.S. Patent 6,162,953
12. Kavala V, Naik S, Patel BK (2005) J Org Chem 70:4267–4271
13. McKinnie BG (1999) Process for the preparation of tetrabromobisphenol A. U.S. Patent 6,002,050
14. Kumar L, Sharma V, Mahajan T, Agarwal DD (2010) Org Proc Res Dev 14:174–179
15. Verma S, Jain SL, Sain B (2011) Ind Eng Chem Res. doi:10.1021/ie101748w
16. Daniel RB, de Visser SP, Shaik S, Newmann R (2003) J Am Chem Soc 125:12116–12117

9

Synthesis and chemical modification of maleic anhydride copolymers with phthalimide groups

Mohammad Hossein Nasirtabrizi*, Zeinab Mohammadpoor Ziaei, Aiyoub Parchehbaf Jadid and Leila Zare Fatin

Abstract

Background: Maleic anhydride (MAN) copolymers with methyl methacrylate, ethyl methacrylate, methyl acrylate, ethyl acrylate, and butyl acrylate monomers at 1:1 mole ratios were synthesized by free radical polymerization in the presence of α,α′-azobis(isobutyronitrile) as initiator and dry ethyl acetate as solvent.

Results: Copolymer compositions were obtained using related proton nuclear magnetic resonance (^{1}HNMR) spectra, and the polydispersity of the copolymers was determined by gel permeation chromatography. Then, a solution of phthalimide salt as a nucleophilic reagent reacted through the ring opening of anhydride groups in copolymers resulting to modified copolymers I_{Pht} to V_{Pht}. All the prepared polymers were characterized by Fourier transform infrared and ^{1}H NMR spectroscopic techniques. The glass transition temperature (T_g) of all copolymers before and after modification was determined by dynamic mechanical thermal analysis (DMTA).

Conclusions: It was shown that chemical modification of MAN copolymers with phthalimide substituents as side chains decreased the free volume of the polymers, and therefore, the rigidity and T_g are increased. Solubility of the copolymer products was tested in some organic solvents.

Keywords: Maleic anhydride copolymers, Chemical modification, DMTA, T_g, MAN, Phthalimide

Background

Maleic anhydride (MAN) is an excellent monomer which can provide reactive anhydride or carboxylic groups with nucleophilic molecules, as discussed by Zhou et al. [1]. MAN has an extremely low tendency to homopolymerize in a radical polymerization condition; on the contrary, it copolymerizes with a variety of donor monomers, as discussed by Nieuwhof et al. [2]. The anhydride groups, in the polymer chain, make the MAN polymer very reactive, and therefore, it is commonly used in various fields, as discussed elsewhere [3,4]. The anhydride moiety constitutes a nice handle to modify the polymer [5]. One can modify the polymer via addition of low molecular weight compounds such as water, alcohols, or amines because of the high reactivity of the anhydride group, as discussed by Bruch et al. [6]. The chemical modification of synthetic polymers allows the control of their mechanical and thermal properties [7] and expands their applicabilities, as discussed elsewhere [8-10].

Polymers containing phthalimide groups are found to possess excellent heat resistance and transparency. Copolymers containing phthalimide derivatives have been used as optical brightening agents, as discussed by Jayakumar et al. [11]. The incorporation of the phthalimide group remarkably enhanced the thermal stability of polymers. Polymers bearing the phthalimide group as a pendant group will exhibit relatively high thermal stability and good solubility [12,13]. Recently, we reported the synthesis and properties of MAN polymers modified with carbazole groups [14]. In this research work, we report the synthesis and properties of MAN copolymers modified with highly sterically hindered phthalimide groups. Firstly, copolymers of MAN with different methacrylate and acrylate monomers were synthesized by the free radical polymerization method. Then, phthalimide groups were linked to the resulting copolymers by the ring-opening reaction of the anhydride groups with sodium phthalimide salt. Dynamic mechanical thermal analysis (DMTA) showed that the incorporation of phthalimide groups as side chains decreased the free volume of the polymers, and therefore, the rigidity and the glass transition temperature (T_g) are increased.

* Correspondence: nasirtabrizi@gmail.com
Department of Applied Chemistry, Islamic Azad University, Ardabil Branch, Ardabil 5614633167, Iran

Methods

Materials

Phthalimide, methyl methacrylate (MMA), ethyl methacrylate (EMA), methyl acrylate (MA), ethyl acrylate (EA), and butyl acrylate (BA) (Merck, Whitehouse Station, NJ, USA) were distilled under reduced pressure to remove inhibitors before use. Maleic anhydride (Aldrich, St. Louis, MO, USA) was recrystallized from chloroform. Ethyl acetate was stirred over calcium hydride for 24 h and distilled in argon atmosphere. The radical initiator of α,α′-azobis (isobutyronitrile) (AIBN; Merck) was purified by crystallization from methanol. Sodium hydride (60%) was obtained from Aldrich. *N,N*-Dimethylformamide (DMF) was dried over anhydrous $MgSO_4$ for 2 days and later with phosphine anhydride overnight. After drying, DMF was distilled under reduced pressure.

General procedure for solubility

The polymer (0.1 g) and solvent (10 ml) were placed in an airtight vial and agitated about 25°C for 1 h. The samples were considered to be soluble if a single-phase, clear, gel-free solution was observed. This procedure was carried out again with the same condition as mentioned but for a much longer time (overnight) using a shaker.

Instruments

Infrared spectra were recorded using a Bruker-IR Scientific spectrophotometer as KBr pellets (Madison, WI, USA). ^1H NMR spectra were run on a Bruker 250 MHz spectrometer at room temperature using deuterated dimethyl sulfoxide (DMSO-d_6) as solvent. The molecular weights (M_w and M_n) were determined using a Waters 501 gel permeation chromatograph fitted with 10^3-nm Waters Styragel columns (Milford, MA, USA). T_g was determined using Tritec 2000 DMA (Triton Technology, Ltd., Keyworth, UK) at a heating rate of 5°C/min in air.

For I: ^1H NMR (DMSO-d_6, 250 MHz) ppm: 0.76 to 1.29 (3H, in CH_3-C), 1.89 to 2.28 (2H, in -CH_2-C-), 3.70 (2H, in -*CH*-CH-), 3.50 (3H, in -OCH_3); FT-IR (KBr, cm^{-1}): 2,999, 2,959 (aliphatic C-H); 1,854, 1,781 (C=O stretching of anhydride unit); 1,726 (C=O ester); 1,240, 1,148 (C-O ester); 1,061, 1,022 (C-O-C).

For II: ^1H NMR (DMSO-d_6, 250 MHz) ppm: 0.80 to 1.20 (3H, in CH_3-C), 1.70 to 2.80 (2H, in -CH_2-C-), 4.05 (2H, in -CH-CH-), 3.66 (2H, in -OCH_2); FT-IR (KBr, cm^{-1}): 2,988, 2,956 (aliphatic C-H); 1,857, 1,784 (C=O of anhydride unit); 1,726 (C=O ester); 1,240, 1,148 (C-O ester); 1,020 (C-O-C).

For III: ^1H NMR (DMSO-d_6, 250 MHz) ppm: 1.58 to 2.85 (3H, in -CH_2-CH-), 3.84 (2H, in -CH-CH-), 3.63 (3H, in -OCH_3); FT-IR (KBr, cm^{-1}): 3,005, 2,957 (aliphatic C-H); 1,849, 1,781 (C=O of anhydride unit); 1,733 (C=O ester); 1,169, 1,223 (C-O ester); 1,101, 1,059 (C-O-C).

For IV: ^1H NMR (DMSO-d_6, 250 MHz) ppm: 1.70 to 2.84 (3H, in -CH_2-C-), 4.20 (2H, in -CH-CH-), 4.00 (2H, in -OCH_2); FT-IR (KBr, cm^{-1}): 2,965, 2,876 (aliphatic C-H); 1,843, 1,770 (C=O of anhydride unit); 1,732 (C=O ester); 1,148, 1,223 (C-O ester); 1,097, 1,022 (C-O-C).

For V: ^1H NMR (DMSO-d_6, 250 MHz) ppm: 1.29 to 1.50, 1.5 to 1.80 (4H, in -CH_2-CH_2-), 2.21 to 2.86 (3H, in -CH_2-C-), 3.94 (2H, in -CH-CH-), 3.64 (2H, in -OCH_2); FT-IR (KBr, cm^{-1}): 2,957, 2,989 (aliphatic C-H); 1,854, 1,781 (C=O of anhydride unit), 1,728 (C=O ester); 1,148, 1,240 (C-O ester); 1,087, 1,020 (C-O-C).

For I_{Pht}: ^1H NMR (DMSO-d_6, 250 MHz) ppm: 0.77 to 1.163 (3H, in CH_3-C-), 2.49 to 2.86 (2H, in CH_2-CH), 7.44 to 8.11 (Ar-H of phthalimide); 11.31 (1H, in O-H); FT-IR (KBr, cm^{-1}): 3,059 (Ar-H of phthalimide); 2,400 to 3,400 (OH acid); 1,774, 1,750 (carbonyls of phthalimide).

For II_{Pht}: ^1H NMR (DMSO-d_6, 250 MHz) ppm: 0.76 to 1.15 (3H, in CH_3-C-), 2.03 to 2.85 (CH_2-C-), 7.47 to 7.91 (4H, in Ar-H of phthalimide), 11.31 (1H, in O-H); FT-IR (KBr, cm^{-1}): 3,058 (Ar-H of phthalimide); 2,400 to 3,400 (OH acid), 1,771, 1,748 (carbonyls of phthalimide).

For III_{Pht}: ^1H NMR (DMSO-d_6, 250 MHz) ppm: 0.74 to 2.83 (3H, in CH_2-CH), 7.5 to 7.83 (4H, in Ar-H of phthalimide), 11.28 (1H, in O-H); FT-IR (KBr, cm^{-1}): 3,061 (Ar-H of phthalimide), 2,400 to 3,400 (OH acid); 1,770, 1,733 (carbonyls of phthalimide).

For IV_{Pht}: ^1H NMR (DMSO-d_6, 250 MHz) ppm: 0.84 to 1.49 (3H, in CH_2-CH), 7.44 to 8.10 (4H, in Ar-H of phthalimide), 11.30 (1H, in O-H); FT-IR (KBr, cm^{-1}): 3,060 (Ar-H of phthalimide); 2,400 to 3,400 (OH acid), 1,770, 1,750 (carbonyls of phthalimide).

For V_{Pht}: ^1H NMR (DMSO-d_6, 250 MHz) ppm: 7.45 to 8.11 (4H, in Ar-H of phthalimide), 11.31 (1H, in O-H); FT-IR (KBr, cm^{-1}): 3,060 (Ar-H of phthalimide); 2,400 to 3,400 (OH acid); 1,772, 1,720 (carbonyls of phthalimide).

Results and discussion

Preparation of MAN copolymers

The MAN monomer was copolymerized with methyl methacrylate, ethyl methacrylate, methyl acrylate, ethyl acrylate, and butyl acrylate monomers in ethyl acetate at 70°C ± 1°C using AIBN as the radical polymerization initiator.

Solubility

The resulting copolymers I to V were soluble in ethyl acetate, *N,N*-dimethylformamide, and dimethylsulfoxide and insoluble in *n*-hexane, chloroform, tetrahydrofuran, and hydroxyl group-containing solvents such as methanol, ethanol, or water.

Molecular weights

The number and weight average molecular weights (M_n and M_w, respectively) and polydispersity index of copolymer samples were determined by gel permeation chromatography and are presented in Table 1.

Table 1 Molecular weights and T_g of MAN copolymers

Polymer	M_w (10^3)	M_n (10^3)	M_w/M_n	T_g (°C)
I	37.4	19.35	1.94	77.3
II	39.2	19.8	1.97	94
III	28	15.5	1.8	36
IV	32.1	16.05	2.0	39
V	47	24.73	1.9	52

Copolymer composition

In the past few decades, ^{1}H NMR spectroscopic analysis has been established as a powerful tool for the determination of copolymer compositions because of its simplicity, rapidity, and sensitivity [15-17]. The copolymer compositions were calculated by the ^{1}H NMR spectral data (Table 2).

The compositions of copolymers synthesized using various monomer feed ratios were determined by the ^{1}H NMR method and were achieved by comparing the integrals of the maleic anhydride, methyne, and methoxy group regions in the spectra of the comonomer units (m_1 and m_2) in copolymers using ^{1}HNMR analysis data that were calculated according to the following equations:

$$\frac{Am_1(\mathrm{CH})}{A_{\mathrm{total}}} = \frac{n_1 m_1}{a_1 m_1 + b_2 m_2} \quad (1)$$

$$\frac{Am_2(\mathrm{OCH_3 \ or \ OCH_2})}{A_{\mathrm{total}}} = \frac{n_2 m_2}{a_1 m_1 + b_2 m_2} \quad (2)$$

$$m_1 + m_2 = 1 \quad (3)$$

where Am_1 and Am_2 are the normalized areas per H from the corresponding functional groups of the monomer unit regions in ^{1}H NMR spectra, A_{total} is the total area of protons in the copolymer, n_1 and n_2 are the integers of proton (s) in the functional group of the monomers, and a and b are integers of protons in the monomer units (m_1 and m_2). Monomer unit ratios can be calculated from Equations 1 and 2 using the following simplified formula:

$$\frac{m_1}{m_2} = f = \frac{n_2 Am_1(\mathrm{CH})}{n_1 Am_2(\mathrm{OCH_3 \ or \ OCH_2})} \quad (4)$$

Table 2 The molar composition of copolymers (I to V)

Polymer	MAN (mol%)	MMA (mol%)	EMA (mol%)	MA (mol%)	EA (mol%)	BA (mol%)
I	45	45	-	-	-	-
II	45	-	55	-	-	-
III	38	-	-	62	-	-
IV	45	-	-	-	55	-
V	39	-	-	-	-	61

Glass transition temperature analysis

The thermal behavior of all the copolymers was investigated by DMTA. The glass transition temperature of MAN copolymers with methacrylate monomers is higher than that of MAN copolymers with acrylate monomers since the α-methyl group in the main chain of methacrylate copolymers has reduced the flexibility of the polymer chains and increased T_g.

The modification of copolymers with phthalimide

The maleic anhydride functional group which grafts on the copolymers' backbone acts as the chemical link between the anhydride group and the sodium phthalimide in a nucleophilic substitution reaction. Copolymers of MAN are receiving great attention due to the presence of a reactive anhydride ring, which offers it an opportunity to enter into a wide range of chemical reactions. The anhydride group possesses a higher reactivity for phthalimide even though this coupling reaction is very rapid at −5°C. A solution of phthalimide salt as a nucleophilic reagent reacted with anhydride groups in copolymers I to V with a ring opening of all anhydride groups to give new modified copolymers. These new modified copolymers I_{Pht} to V_{Pht} were isolated in high yields at distilled HCl (0.2 N) as a non-solvent. Compared to the previous coupling methods for the preparation of modified copolymers of maleic anhydride in the literature [16], the present method is simple - the coupling reaction is fast.

Although the methods of phthalimide connection to initial copolymers and their modification has been carried out using similar methods that we have worked previously [14], phthalimide connection has been done in the present paper for the first time. The ring-opening reaction of maleic anhydride copolymer at low temperature (−5°C) is attempted for the first time in our research group.

Solubility

Copolymers I_{Pht} to V_{Pht} were soluble in solvents the same as those in which copolymers I to V are dissolved. The solubility of copolymers before and after modification with phthalimide salt was tested with polar and non-polar solvents, and it was observed that the resulting polymers are soluble in polar aprotic solvents and insoluble in polar protic and non-polar solvents. The results showed that the solubility of the modified copolymers with phthalimide has not changed compared to the initial polymers. The results of the solubility test for the longer time (overnight) were similar to those for the short time (1 h).

Infrared spectra

In the IR spectra of the copolymers (I to V), asymmetrical and symmetrical stretching due to the methyl and methylene groups are observed at 3,005 and 2,957 cm^{-1}, respectively. The two bands at 1,849 and 1,781 cm^{-1} are attributed to symmetric and asymmetric stretching of anhydride carbonyl of copolymers, respectively. The band at 1,733 cm^{-1} is attributed to the ester carbonyl stretching of methacrylate and acrylate monomer units. The two bands at 1,059 and 1,101 cm^{-1} are attributed to the asymmetrical stretching of the C-O-C anhydride ring. The bands at 1,223 and 1,169 cm^{-1} are due to C-O stretching. The FT-IR spectra of copolymer II is shown in Figure 1a.

Modification of poly(MAN-co-EMA) is confirmed by the infrared spectrum in Figure 1b. The absorbance intensity is substantially diminished at 1,857 and 1,784 cm^{-1}, whereas a broad band arises between 2,400 and 3,400 cm^{-1} of O-H acid, indicating that the ring-opening reaction of the maleic anhydride unit with phthalimide was completed. The band at 3,058 cm^{-1} is due to C-H stretching of the aromatic ring. The ring stretching vibration of the aromatic nuclei was observed at 1,600 and 1,491 cm^{-1}.

^{1}H NMR spectra

^{1}H NMR spectra of the compounds (I to V) show that the proton resonances of C-H in anhydride groups in copolymers I, II, III, IV, and V are located at 3.70, 4.05, 3.84, 4.20, and 3.94 ppm, respectively. Aliphatic protons of CH_3 and CH_2 in copolymers I and II are located at 0.80 to 1.20 and 1.70 to 2.85 ppm, respectively, and aliphatic protons of CH and CH_2 in copolymers III to V are located at 1.53 to 2.85 ppm. The resonance signal at 3.66 ppm was attributed to two methylene protons of $COOCH_2$ in copolymers II, IV, and V, and three methoxy protons of $COOCH_3$ in copolymers I and III appeared at 3.60 ppm. In the ^{1}H NMR spectrum of the copolymers (I to V) modified by phthalimide groups, the aromatic resonances between 7.50 and 7.90 ppm are due to the phthalimide protons. The peaks at 11.28 ppm are due to the hydrogen of the acidic group. A typical ^{1}H NMR spectrum of copolymer III is given in Figure 2.

Glass transition temperature

The thermal behavior of all the copolymers was investigated by DMTA. DMTA analysis showed that the presence of phthalimide groups led to an increase in T_g from 92°C for copolymer I to 120°C for copolymer I_{Pht}, from 94°C for copolymer II to 133°C for copolymer II_{Pht}, from 36°C for copolymer III to 85°C for copolymer III_{Pht}, from 39°C

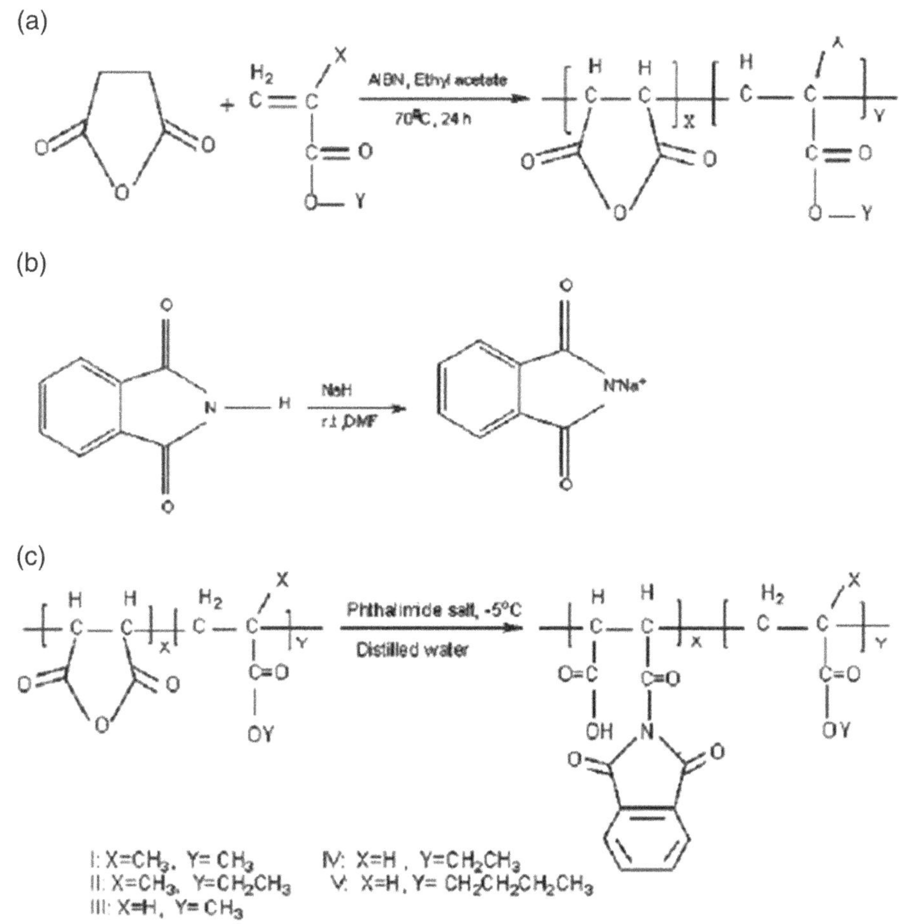

Figure 1 Infrared spectra of copolymers (a) II and (b) II_{Pht}.

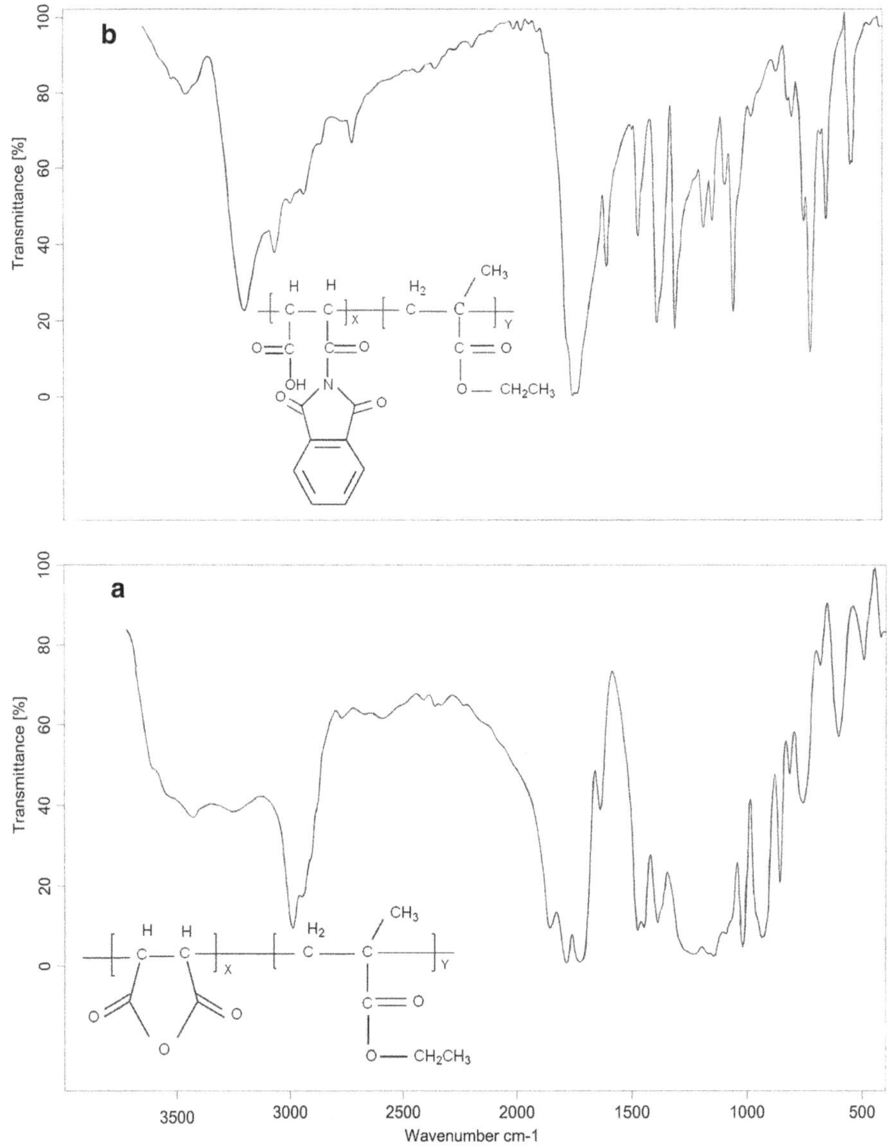

Figure 2 ^{1}**H NMR spectra of copolymers (a) III and (b) III$_{Pht}$ in DMSO-d$_6$.**

for copolymer IV to 96°C for copolymer IV$_{Pht}$, and from 52°C for copolymer V to 102°C for copolymer V$_{Pht}$ (Table 3, Figure 3a,b). Incorporation of phthalimide substituents into the polymer increased T_g as expected because it reduces the free volume of macromolecules; in addition, the π-stacking and ring structure of the aromatic groups increased the rigidity of the polymer. This reduced the flexibility of polymer chains, and hence, more energy has to be induced in order to overcome the interactions. The presence of a number of carbonyl groups also adds to the rigidity of the polymer. Because the bulky group is pendant and not a part of the polymer backbone, the rise in T_g was slight; however, it withstands semiconductor processing [7]. Therefore, novel polymer systems with new physical and chemical properties and new applications can be obtained.

Experimental

Copolymerization of maleic anhydride with methacrylate and acrylate monomers

Poly(MAN-co-MMA) (I), poly(MAN-co-EMA) (II), poly (MAN-co-MA) (III), poly(MAN-co-EA) (IV), and poly

Table 3 The result of reaction between 2 mmol of each polymers (I to V) with Pht

Modified copolymer	Weight (g)	Yield (%)	T_g (°C)
I$_{Pht}$	0.68	89	120
II$_{Pht}$	0.72	91	133
III$_{Pht}$	0.65	85	85
IV$_{Pht}$	0.66	87	96
V$_{Pht}$	0.60	76	102

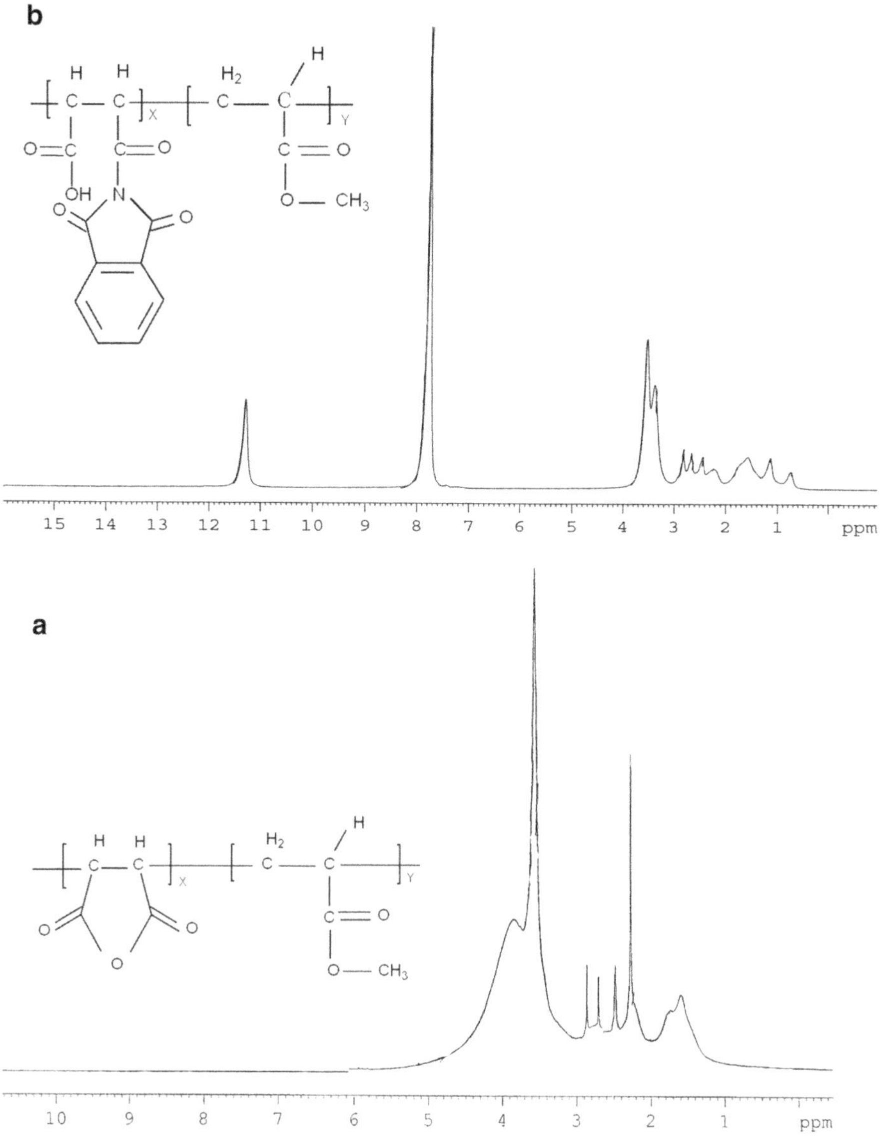

Figure 3 DMTA curves of copolymers (a) III and (b) III_{Pht} at 5°C/min in air.

(MAN-co-BA) (V) were synthesized based on the following general procedure: In five different Pyrex glass ampoules, a mixture of 1.96 g (20 mmol) of MAN, 0.065 g (0.4 mmol) of AIBN, and 2.00 g (20 mmol) of methyl methacrylate, 2.28 g (20 mmol) of ethyl methacrylate, 1.73 g (20 mmol) of methyl acrylate, 2.00 g (20 mmol) of ethyl acrylate, or 2.56 g (20 mmol) of butyl acrylate was dissolved in 15 ml of dry ethyl acetate, respectively. Then, the ampoules were degassed, sealed under vacuum condition, and maintained at 70°C ± 1°C in a water bath and shaken using a shaker for about 24 h. Then, the viscous solutions were poured from the ampoules into 150 ml of cooled methanol separately. The precipitates were collected and washed with methanol for several times and dried under vacuum at room temperature to give 3.1 g of copolymer I, 3.39 g of copolymer II, 3.07 g of copolymer III, 3.1 g of copolymer IV, and 3.38 g of copolymer V (Table 4, Figure 4a).

Preparation of sodium phthalimide salt

For preparing sodium phthalimide, sodium hydride (4 mmol, 0.096 g) was slowly added to phthalimide (4 mmol, 1 g) dissolved in 15 ml of DMF at room temperature. Then, the mixture was stirred under nitrogen atmosphere for 30 min (Figure 4b).

Attaching phthalimide groups to copolymers (I to V)

Poly(MAN-co-MMA), poly(MAN-co-EMA), poly(MAN-co-MA), poly(MAN-co-EA), and poly(MAN-co-BA) containing phthalimide groups were prepared as follows: In a

Table 4 The conditions of preparation of polymers (I to V) at 70°C

Sample	Monomer 1	Monomer 2	Amount of 1 (mmol)	Amount of 2 (mmol)	Time (h)	Weight (g)	Yield (%)
I	MAN	MMA	20	20	24	3.1	78
II	MAN	EMA	20	20	24	3.39	80
III	MAN	MA	20	20	24	3.07	83
IV	MAN	EA	20	20	24	3.1	78
V	MAN	BA	20	20	24	3.38	75

100-ml two-necked round-bottom flask equipped with a dropping funnel, a solution of sodium phthalimide in DMF (15 ml) was prepared. After this time, the mixture was cooled to −5°C in an ice-salt mixture under vigorous stirring. A solution of copolymer I (0.198 g, 2 mmol), copolymer II (0.230 g, 2 mmol), copolymer III (0.200 g, 2 mmol), copolymer IV (0.230 g, 2 mmol), or copolymer V (0.226 g, 2mmol) in DMF (15 ml) was prepared and transferred into the dropping funnel under argon atmosphere. Then, a solution of copolymers (I to V) was added dropwise with stirring to the solution of sodium phthalimide into the flask at −5°C within 1 h. The reaction mixture was kept at −5°C temperature for another 3 h under stirring. Then, the solution was poured into an excess cold HCl (0.2 N), and the white precipitate was filtered and washed with HCl (0.2 N). The resulting polymer was washed sequentially with NaOH (0.2 N) and water (50 ml). Finally, the obtained white polymers containing phthalimide substituents in the

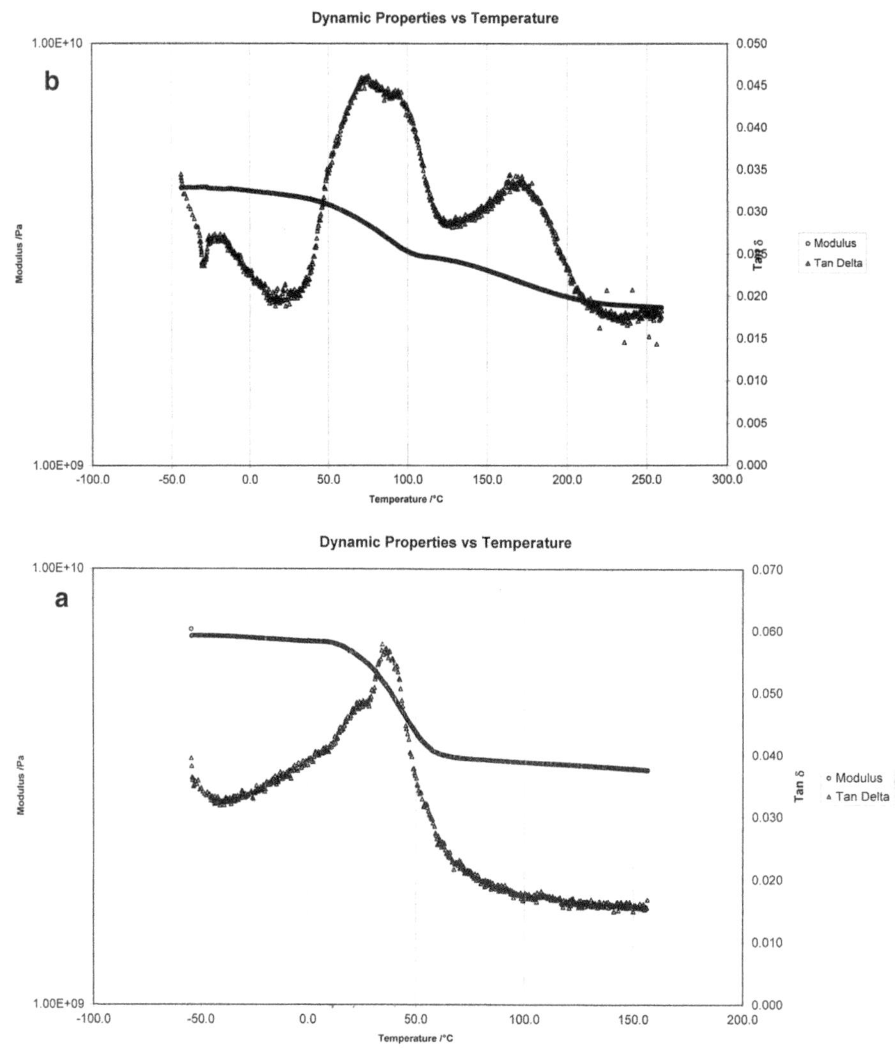

Figure 4 MAN copolymerization (a), sodium phthalimide salt synthesis (b), reaction of MAN copolymer with phthalimide salt (c).

side chains were collected and dried under vacuum at room temperature (Figure 4c). The yields of modified copolymers are presented in Table 3.

Conclusions

We have detailed the preparation and characterization of novel MAN copolymers bearing phthalimide groups of side chain substituents. The work described in this paper deals with the nucleophilic ring opening of anhydrides by $Pht^- Na^+$. The copolymers of MAN with different methacrylate and acrylate monomers were synthesized by free radical solution polymerization. The molar compositions of the obtained copolymers were calculated by the corresponding ^1H NMR analysis. The anhydride group possessed a higher reactivity for the phthalimide in the modification process. Comparison of the ^1H NMR and IR spectra of copolymers before and after modification showed disappearance of the peaks assigned to the anhydride ring and appearance of a new peak characteristic of phthalimide moieties, thus proving the modification of MAN copolymers. The MAN copolymers prepared here should be useful for the construction of well-defined functional polymers. DMTA indicated that the T_g of copolymers increases with incorporation of bulky phthalimide groups in side chains of MAN units. The presence of the phthalimide groups in the polymer structure reduces the free volume of macromolecules with modification of their properties and applications.

Competing interests

The authors declare that they have no competing interests.

Authors' contributions

MHN developed the concept. ZMZ drafted the manuscript. APJ advised on the methods of tests. LZF analyzed the data. All authors read and approved the final manuscript.

Acknowledgements

The authors are thankful to Islamic Azad University, Ardabil Branch, for the technical support during this research.

References

1. Zhou J, Wang LI, Wang CH, Chen T, Yu H, Yang Q (2005) Synthesis and self-assembly of amphiphilic maleic anhydride–stearyl methacrylate copolymer. J Polymer 46:1157–1164
2. Nieuwhof R, Marcelis A, Sudholter E (1999) Side-chain liquid-crystalline polymers from the alternating copolymerization of maleic anhydride and 1-olefins carrying biphenyl mesogens. J Macromol 32:1398–1406
3. Al-Sabagh A, Noor MR, Din EL, Morsi RE, Elsabee MZ (2009) Styrene-maleic anhydride copolymers esters as flow improvers of waxy crude oil. J Pet Sci Eng 62:139–146
4. Zhu LP, Yi Z, Liu F, Wei XZ, Zhu BK, Xu YY (2008) Amphiphilic graft copolymers based on ultrahigh molecular weight poly(styrene-alt-maleic anhydride) with poly(ethylene glycol) side chain for surface modification of polyethersulfone membranes. Eur Polym J 44:1907–1914
5. Nieuwhof R, Koudijs A, Marcelis A, Sudholter E (1999) Modification of side-chain liquid-crystalline poly(maleic anhydride-co-alt-1-alkene)s with mesogen-containing alcohols. Macromol 32:6499–6506
6. Bruch M, Mader D, Bauers F, Loontjens T, Mulhaupt R (2000) Melt modification of poly(styrene-co-maleic anhydride) with alcohols in the presence of 1,3-oxazolines. J Polym Sci Part A: Polym Chem 38:1222–1231
7. Safa KD, Babazadeh M (2004) Glass transition temperature modification of acrylic and dienic type copolymers of 4-chloromethyl styrene with incorporation of $(Me_3Si)_3C-$ groups. Eur Polym J 40:1659–1669
8. Boztug A, Basan S (2007) The modification and characterization of maleic anhydride-styrene-methyl methacrylate terpolymer by poly(ethylene adipate). J Mol Struct 830:126–130
9. Wang K, Huang W, Xia P, Gao C, Yan D (2002) Fluorescent polymer made from chemical modification of poly(styrene-co-maleic anhydride). React Funct Polym 52:143–148
10. Safa KD, Eram HA, Nasirtabrizi MH (2006) Chemical modification of some 4-chloromethylstyrene polymers with highly sterically hindered tris (trimethylsilyl)methyl groups. Iran Polym J 15:1249–1257
11. Jayakumar R, Balaji R, Nanjundan S (2009) Studies on coplymers of 2-(N-phthalimido) ethyl methacrylated with methyl methacrylate. Eur Polym J 36:1659–1666
12. Liaw DJ, Hung CC, Wu PL (2001) Synthesis and characterization of novel diblock copolymers of 5-(N-carbazoyl methyl)bicycle [2.2.1] hept-2-ene and 5-(phthalimide methyl)bicyclo [2.2.1] hept-2-ene via living ring-opening metathesis polymerization. Polymer 42:9371–9377
13. Liaw DJ, Tsai CH (1999) Synthesis and characterization of poly(norbornene) substituted with phthalimide and ammonium groups via living ring-opening metathesis polymerization. J Mol Cat A: Chem 147:23–31
14. Nasirtabrizi MH, Ziaie ZM, Parchehbaf Jadid A (2011) The modification and characterization of maleic anhydride copolymers by carbazole groups. J Iran Chem Res 4:59–68
15. Dincer S, Koseli V, Kesim H, Rzeav Z, Piskin E (2002) Radical copolymerization of N-isopropylacrylamide with anhydrides of maleic and citraconic acids. Eur Polym J 38:2143–2152
16. Kesim H, Rzeav Z, Pincer S, Piskin E (2003) Functional bioengineering copolymers. II. Synthesis and characterization of amphiphilic poly(N-isopropylacrylamide-co-maleic anhydride) and its macrobranched derivatives. Polymer 44:2897–2909
17. Nicolescu FA, Jerca VV, Albu AM, Vuluga DM, Vasilescu DS (2009) Poly (maleic-amic) structures for potential nonlinear optical applications. U P B Sci Bull 71:1454–2331

Kinetic modeling: a predictive tool for the adsorption of zinc ions onto calcium alginate beads

Morteza Hasanzadeh Kafshgari[1], Mohsen Mansouri[1], Mohammad Khorram[2*] and Shahab Rahimi Kashani[3]

Abstract

Background: The main purpose of this study is to develop a nonlinear model of a batch adsorption and to evaluate the model's capability in the prediction of experimental adsorption data. Additionally, results of the nonlinear model were compared with data of pseudo first- and second-order models. Experimental data were extracted from the adsorption of zinc ions using calcium alginate beads prepared via electrospray method. In order to study the effects of biosorbent porosity and the initial concentration of the zinc solution on adsorption kinetics, calcium alginate beads were prepared with two different porosities and zinc concentrations.

Results: The results revealed that the nonlinear model offers a much more accurate prediction compared with other models as the average root mean square deviation of the nonlinear model was calculated to be only 2.66%, which was smaller at least four times than that of others. Furthermore, the nonlinear model showed that diffusive transport through beads was limited by pore diffusion.

Conclusions: The nonlinear model provided a good fit to the experimental data as the calculated equilibrium adsorptions were shown to be in good agreement with their experimental counterparts.

Keywords: Nonlinear kinetic model, Pseudo-kinetic models, Zinc adsorption, Calcium alginate beads

Background

In certain industries, the adsorption of heavy metals from various waste streams which are cleaned using a sorbent is a crucial separation process [1]. Although there are numerous commercial absorbents that are available for this purpose, biodegradable polymers such as chitosan and sodium alginate can be considered to be more beneficial, cost-effective, and more abundant in nature than others [2,3]. Sodium alginate extracted from brown algae is a linear copolymer of a-L-guluronate and a-D-mannuronate [4]. In addition, sodium alginate could be ionically gelated by the divalent cations such as Ca^{2+} and be converted to a stable and cross-linked structure [5], which has been used for the removal of some heavy metals such as cadmium [6,7], chromium [8], lead [6], cobalt [9], iron, copper [7], mercury [6], and zinc [10] from aqueous solutions. Furthermore, prediction of the adsorptive behavior of sorbents is important as a consequence of the investigation of adsorption mechanism [11-17]. In order to investigate the mechanisms of the adsorption of heavy metals into adsorbents, pseudo first- and second-order models have been usually used to assess experimental data [2,12]. Since the model should be fitted by existing experimental data due to the model's nature, they can only predict the rate-controlling steps [2]. On the other hand, in almost all previous research in the adsorption processes, no mathematical model such as nonlinear ones has been applied [18,19] as the adsorption data have been extracted and undergone a simple investigation in their mechanisms via some pseudo models. The nonlinear model is instrumental and rational for the modeling and predicting of adsorption systems. The physiochemical conditions such as the type of ionic solutions (liquid diffusion coefficient), bead size, agitation rate, and porosity of beads are effective factors in an adsorption process that can be considered in the nonlinear model while their effects could be investigated in detail [14,20]. Implementing the mathematical model, once it has been endorsed, has the advantages of minimizing the number of experiments related to new operating

* Correspondence: mkhorram@shirazu.ac.ir
[2]School of Chemical, Petroleum and Gas Engineering, Shiraz University, Shiraz, Iran
Full list of author information is available at the end of the article

conditions. In the nonlinear model, the interactive behavior between solutes and adsorbent is certainly described using essential isotherm equations such as that of Freundlich and Langmuir for modeling an adsorption system [2,3,9,21].

In the present study, biosorbents were produced with sodium alginate using the electrospray method, and their capability in removing zinc ions from an aqueous solution under batch adsorption was investigated. The main objectives were to investigate the zinc adsorption mechanism onto the calcium alginate beads, to develop a mathematical model for description of the adsorption, and to assess the flexibility of the mathematical model in comparison with some pseudo-kinetic models in the prediction of absorptive behavior. In addition, some effective parameters such as porosity and concentration of aqueous solution could not be clearly investigated on adsorption processes with pseudo models; on the contrary, the nonlinear model is able to provide a framework of analysis of these parameters. Therefore, the effects of some parameters such as porosity of the beads and initial zinc concentration in the adsorption process were considered on the adsorptive behavior of the models.

Methods

General assumptions

For quantitative depiction of the sorption process, the following assumptions have been made:

1 Each prepared calcium alginate bead was idealized as a sphere.
2 The mass transfer into calcium alginate beads was only considered in the radius direction, and other directions were deemed negligible.
3 Transport of the ions was from the bulk solution to the bead surface and across the boundary layer.
4 Adsorption was considered at the inner site of the beads.
5 The physical properties of the calcium alginate beads and the solutions' properties were constant in the isotherm of the adsorption processes.
6 Pore diffusion was considered onto calcium alginate sorbent.
7 Radius of the beads was constant with the adsorption of the water and the heavy metal (i.e., no swelling).
8 Bulk convection was neglected onto the beads.

Model description

Mass balance development

Based on the assumptions, the nonlinear model for the prepared sphere beads is as follows [7,20]:

$$\frac{\partial C}{\partial t} + \frac{\rho_b}{\varepsilon}\frac{\partial P}{\partial t} = \frac{D_l}{\tau}\left(\frac{\partial^2 C}{\partial r^2} + \frac{2}{r}\frac{\partial C}{\partial r}\right), \tag{1}$$

where C (mg-solute/L) is the concentration of the dissolved solute inside the porous beads, t (s) is the time, ρ_b (g-alginate/L) is the medium bulk density, ε (dimensionless) is the porosity, P (mg-solute/g-alginate) is the adsorbed concentration, D_l is the liquid diffusion coefficient (m^2/s), τ (dimensionless) is the tortuosity factor, and r (m) is the radial distance. The characteristics of calcium alginate beads and initial zinc solution are presented in Table 1. The adsorbed concentration can be defined using the Freundlich and Langmuir isotherm [2,22].

Langmuir isotherm

The Langmuir isotherm assumes that the monolayer coverage of sorption of each molecule onto the surface has equal sorption activation energy [13]. It has been usually applied to determine the quantity and compare the performance of different biosorbents and is expressed as follows:

$$P = \frac{P_{\max} K_f C}{1 + K_f C}, \tag{2}$$

where $P_{\max}$ (mg/g) is the maximum amount of adsorption, and K_f (L/mg) is the sorption equilibrium constant. Substitution of the Langmuir isotherm into Equation 1, the nonlinear model for a solute inside a porous bead becomes Equation 3:

$$\frac{\partial C}{\partial t} = D_a\left(\frac{\partial^2 C}{\partial r^2} + \frac{2}{r}\frac{\partial C}{\partial r}\right), \tag{3}$$

where D_a is the apparent diffusion coefficient, which is determined as follows:

$$D_a = \frac{D_l}{\tau\left(1 + \frac{\rho_b P_{\max} K_f}{\varepsilon\left(1+K_f C\right)^2}\right)} \tag{4}$$

Table 1 Characteristics of calcium alginate beads and initial zinc solutions

Characteristic	Sample number			
	I	II	III	IV
Solid density (g/cm^3) [23]	0.8755			
Sodium alginate concentration (*w/v*,%)	1	2	1	2
Average wet weight (mg)	2.5	2.75	2.5	2.75
Average dry weight (mg)	0.335	0.5	0.335	0.5
Porosity	0.85	0.794	0.85	0.794
Average radius (cm)	0.055			
Initial zinc concentration (ppm)	100	100	300	300
Agitation rate (rpm)	250			
Diffusion coefficient (m^2/s) [24]	3.6×10^{-10}			
Solution bulk density (g-alginate/cm^3)	≈1			
Water density (g/cm^3)	≈1			
Solution pH	4.5			
Temperature (°C)	27			

Solution technique
Boundary and initial conditions, which were determined based on general assumptions and experimental conditions, were needed to solve Equation 1. Accordingly, the following boundary and initial conditions were used:

$$C(r,0) = 0 \rightarrow at\,t = 0, \quad (5)$$

$$D_a \frac{\partial C}{\partial r} = 0 \quad at\,r = 0, at\,the\,center\,of\,the\,sphere, \quad (6)$$

$$D_a \frac{\partial C}{\partial r}|_{r=R} = k_c(C_b - C_R) \quad at\,the\,outside\,boundary. \quad (7)$$

In the derived model, the bead radius was divided into N stages using implicit finite difference method. According to the initial condition, it was considered that the prepared calcium alginate beads were devoid of zinc at the initial time of adsorption. Concentration of the center was parallelized to the next stage that was shown in Equation 6. As shown in Equation 7, the flow of solute onto the bead on the surface and the flow through the boundary layer into the bead as k_c which is the mass transfer coefficient (m/s), and C_b and C_R are the bulk solute concentration and the solute concentration at the bead surface, respectively [14,20]. Because the nonlinear model was applied to calculate solution concentration at different times, the implicit method was used to find the concentration profiles in the bead. Solving the numerical equation needs an assumption of the value of fitting parameters in Equation 3. The fitting parameters are τ and k_c which depend on the bead size, initial concentration of heavy metal in the solution, and agitation rate [14,20]. The mass transfer coefficient (k_c) can be regulated using the amount of equilibrium adsorption. In order to adjust the fitting parameters, k_c was first determined by calculating the minimum deviation between experimental and calculated adsorption equilibrium data at a constant τ, and then τ was estimated as the last and only remaining fitting parameter, which is independent from the calculated data of k_c as the estimated amount of equilibrium adsorption was constant during the determination of τ. The satisfied amount of tortuosity factor was confirmed when the deviation between experimental and calculated data was minimum.

Results and discussion
Adsorption isotherm
The adsorption capacities of the calcium alginate beads were compared at different concentrations in Figure 1. The beads were prepared with two different polymer concentrations, 1% and 2% w/v. It indicates that the saturation adsorption capacity increases with the increasing amount of initial zinc concentration at both polymer concentrations. On the other hand, regarding to the same diameter of all beads, the change in the polymer concentration was effective on the porousness. Therefore, the saturation adsorption capacities of the prepared beads with the different concentrations were slightly dissimilar, owing to their near porosities. The Langmuir model was adaptively applied to describe the equilibrium data of zinc adsorption onto the beads. Results show that Langmuir isotherm has been much perfectly fitted to the equilibrium data compared with other isotherms. The Langmuir equilibrium constants and corresponding R^2 adjustment coefficients, which were calculated through the application of the least squares method, are shown in Table 2.

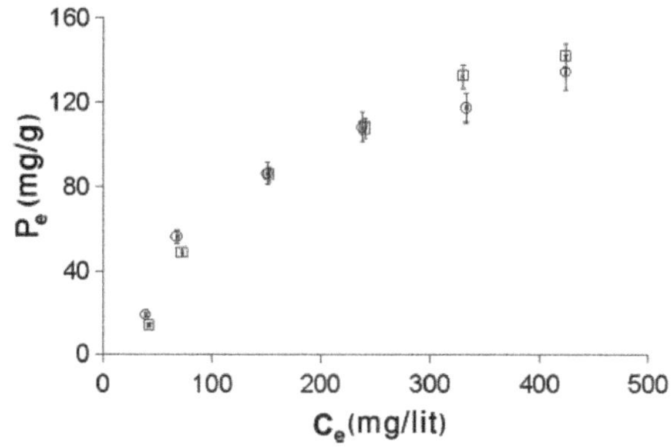

Figure 1 Equilibrium adsorption of zinc ions onto prepared calcium alginate beads. With polymer concentrations at (empty circle) 1% and (empty square) 2% *w/v*.

Pseudo first- and second-order models
The pseudo first- and second-order models were applied to test the experimental data for investigating the mechanism of the adsorption. The first-order rate expression of Lagergren [22] is given as follows:

$$\log(P_e - P) = \log P_e - \frac{k_1 t}{2.303}, \quad (8)$$

where P_e and P (mg/g) are the amounts of zinc ions, which were adsorbed into the adsorbents at equilibrium and at the time t, respectively, and k_1 (1/min) is the rate

Table 2 Langmuir isotherm constants for the prepared beads with the different polymer concentrations

Concentration	K_f (L/mg)	P_{max} (mg/g)	R^2
Alginate 1% *w/v*	0.0058	200	0.991
Alginate 2% *w/v*	0.0035	250	0.994

constant of the pseudo first-order model. The experimental data could be fitted with a straight line of log ($P_e - P$) versus t, which can determine the flexibility of the model to predict the data. On the other hand, the second-order kinetic model is expressed as follows [22]:

$$\frac{t}{P} = \frac{1}{k_2 P_e^2} - \frac{t}{P_e}, \tag{9}$$

where k_2 (g/mg min) is the rate constant of the pseudo second-order model. There should be a linear relationship in the plot of t/P versus t if second-order kinetics is applicable. The slopes and intercepts of the plots give the second-order rate constant k_2 and P_e. The model is capable of predicting the behavior of all range of adsorption data without any parameters in advance and to show a rapport with chemical adsorptions, which are the rate-controlling steps [2,22,25]. Table 3 represents the calculated results obtained from the first- and second-order kinetic models and a comparison of the results with correlation coefficients (R^2). The correlation coefficients for the first-order model were low. Figure 2 shows both pseudo-model behaviors in prediction of the experimental data.

The pseudo-kinetic models are not in good agreement with the pre-equilibrium times of adsorptions, but the prediction of the second-order model shows a better convergence at equilibrium time rather than the first-order model.

The second-order kinetic model generally presents a more or less rational prediction as it showed high correlation coefficients in certain samples. The calculated P_e values resulted from second-order kinetic model were in a good fit with the experimental P_e values.

Furthermore, the first-order model of Lagergren could not support perfectly all ranges of contact time and was not generally adjustable in the initial stage of the adsorption processes [2,22,25]. Figure 2 illustrates that second-order model tends to follow the experimental data better than first-order model as the correlation coefficients for the second-order kinetic model (that are displayed in Table 3) are higher than those of first-order adsorption model. In order to determine the deviations between all calculated and experimental data, root mean square deviation (σ_{RMSD}) was calculated as follows:

$$\sigma_{\mathrm{RMSD}} = \left[\frac{1}{n}\sum_{i=1}^{n}\left(\frac{P_{\mathrm{Exp}} - P_{\mathrm{Cal}}}{P_{\mathrm{Exp}}}\right)_i^2\right]^{0.5} \times 100. \tag{10}$$

Table 4 shows that the first-order model could not follow the data as it has an average σ_{RMSD} of 15.61%, but the second-order model has a comparatively more rational agreement. Although a better agreement was shown for samples I and II, the average σ_{RMSD} was calculated to be 10.59%. Overall, Table 4 shows a high value of σ_{RMSD} of the models, indicating a poor agreement with the experimental data. It is obviously alleged that the adsorption of zinc ions onto the prepared calcium alginate beads could be the rate-limiting step, owning to the better agreement of the second-order model [2].

Nonlinear model

The nonlinear diffusion model was applied to model the experimental data. The adsorption behavior has been described with the Langmuir adsorption equation. The Langmuir equation constants are shown in Table 2. Therefore, the results indicate a single molecule layer adsorption [2,22]. The nonlinear model was numerically solved with the implicit, finite difference approximation using the Newton-Raphson's method [26]. The data were fitted using τ and k_c as a lumped parameter applied to describe the boundary layer effect as the fitting parameters. Figure 2a shows that the calculated amount of adsorbed zinc ions increased as the time increases and reached an equilibrium value after 130 min, which is approximately equal to the experimental equilibrium time. In addition, a reasonable agreement is shown in the following experimental data and in fitting the balanced adsorption amount. The fitting parameters (which depend on some effective conditions such as the agitation rate and initial concentration of the solutions [14,20]) were determined with the calculation of σ_{RMSD} between the predicted and experimental data. Furthermore, Figure 2b,c,d shows the same adaptive behaviors in estimating the experimental data with high accuracies as the predicted data are fitted

Table 3 Comparison of first- and second-order adsorption rate constant and calculated/experimental P_e values at different samples

Sample number	$P_{e,\ \mathrm{Exp}}$ (mg/g)	First-order kinetic model			Second-order kinetic model		
		k_1 (1/min)	$P_{e,\ \mathrm{Cal}}$ (mg/g)	R^2	k_2 (g/mg min)	$P_{e,\ \mathrm{Cal}}$ (mg/g)	R^2
I	53.4	0.032	54.70	0.959	0.0025	53.66	0.9984
II	38.87	0.019	37.58	0.964	0.0011	38.22	0.9911
III	96.14	0.024	89.53	0.964	0.0011	97.12	0.9953
IV	94.74	0.028	96.2	0.995	0.0008	96.98	0.995

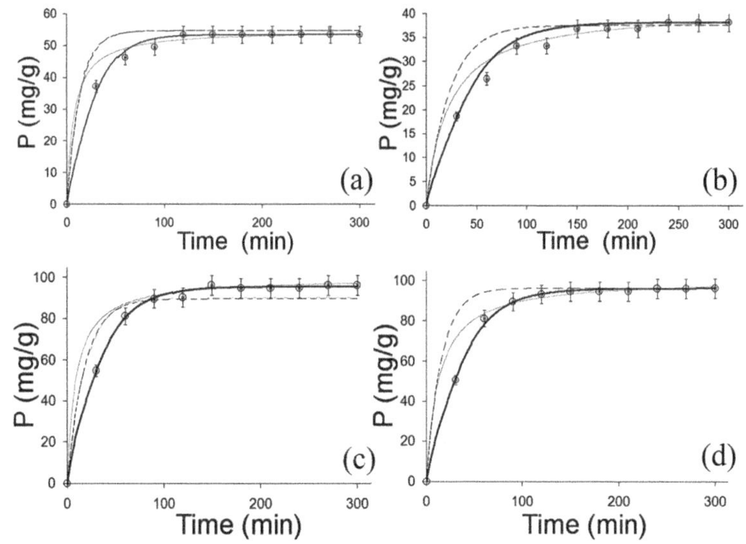

Figure 2 Comparison of first-order (dotted line), second-order (thin line), and the non-linear (thick line) adsorption kinetics. These are zinc ions onto the prepared calcium alginate beads with experimental data (empty circle): **(a)** sample I, **(b)** sample II, **(c)** sample III, and **(d)** sample IV.

at the balanced adsorption data. Table 5 shows the estimated fitting parameters and the equilibrium adsorption data in different conditions. The factor of tortuosity was calculated from 9.52 to 16.8 with the changing of porosity and initial concentration of the zinc solution. The value of τ depends on the characteristics of the beads, and the high value of τ indicates a slow diffusion into the adsorbent. It is worth mentioning that the calculated τ is independent on the conditions of initial solution and sorbent loading characteristics [14,20]. It is important to notice that increasing the porosity of the beads from 0.794 to 0.85 decreases the factor of tortuosity, but the changes of porosity do not have any significant effect on the τ values at high zinc concentration in the initial solution (300 mg/L). The fitted values of k_c were estimated to be 1.39×10^{-7} to 2.21×10^{-7} m/s, showing the independence of the uptake kinetics on agitation rate. Therefore, the diffusive transport in the prepared calcium alginate beads at high-speed agitation rate (250 rpm) was limited by pore diffusion that indicated the rate-limiting step [14,20]. Although it seems that the kinetic pseudo models are flexible for the data, the nonlinear model comes with lower σ_{RMSD} values compared with the pseudo models. The calculated σ_{RMSD} values of the pseudo models showed a disparity in the prediction of experimental data (Table 4). Interestingly, the nonlinear model has an average σ_{RMSD} value of 2.66% which is at least four times smaller than that of the pseudo models, indicating a more reliable prediction capability of the nonlinear model in the adsorption process. In all samples, the mathematical model provided an adjustable behavior and perfectly fitted the data.

Experimental

Materials

Sodium alginate (medium viscosity, 3,500 cps, 2% *w*/*v* aqueous solution at 25°C) was purchased from Sigma-Aldrich Corporation (St. Louis, MO, USA). Zinc sulfate pentahydrate ($ZnSo_4{\cdot}5H_2O$), sodium hydroxide (NaOH), hydrochloride acid (HCl), and calcium chloride ($CaCl_2$)

Table 4 Comparison of the calculated σ_{RMSD} of the kinetic models for the different samples

Kinetic models	σ_{RMSD} samples				
	I	II	III	IV	Average
First-order kinetic model	11.68	18.71	11.39	20.67	15.61
Second-order kinetic model	6.46	9.82	13.11	12.96	10.59
Nonlinear Model	2.85	5.11	1.35	1.31	2.66

Table 5 Calculated fitting parameters of the nonlinear model in adsorption of zinc ions

Fitting parameters in the nonlinear model	Sample number			
	I	II	III	IV
τ	9.52	14.54	16.53	16.8
k_c (m/s)	2.21×10^{-7}	1.39×10^{-7}	1.44×10^{-7}	1.92×10^{-7}
$P_{e,\ Cal}$ (mg/g)	53.454	38.217	95.496	96.099

Figure 3 Image of size and shape of the prepared calcium alginate beads.

were purchased from Merck AG (Darmstadt, Germany) and were used as received without further purification.

Preparation of the porous calcium alginate beads

Calcium alginate beads were prepared according to the method of Moghadam et al. [27]. Briefly, sodium alginate solution was prepared by dissolving 1 and 2 g of the polymer in 99 and 98 mL of distillated water, respectively. The solution of the sodium alginate flowed through a stainless steel nozzle via a syringe pump at a 100-mL/h flow rate. Electric field strength was set using a high voltage, DC power supply. The liquid meniscus at the tip of the nozzle was affected by the adjusted electric field where droplets were formed and inundated. The droplet was dripped into a container where ionotropic gelation occurred. The solution made in the container was calcium chloride (3% *w*/*v*). The solution was gently stirred for 5 min while the droplets were cured. The prepared calcium alginate beads were washed with distillated water and acetone and dried until the water completely evaporated. The size of the dried calcium alginate beads was measured randomly using the images processor (SAIRAN Co., Iran). Figure 3 shows the size and shape of the prepared calcium alginate beads.

Calculation of the porosity of calcium alginate beads

Porosity of the calcium alginate beads was calculated from the density of the alginate and the weight change before and after drying [1]; the following formula was applied:

$$\varepsilon = \frac{(W_W - W_D)/\rho_W}{W_D/\rho_{AL} + (W_W - W_D)/\rho_W}, \qquad (11)$$

where ε is the porosity of the calcium alginate beads, W_W (g) is the weight of the wet calcium alginate bead before drying, W_D (g) is the weight of the dried calcium alginate bead, ρ_W is the density of water, and ρ_{AL} is the density of the alginate (Table 1).

Adsorption process

Standard zinc solutions with certain concentrations were prepared by dissolving $ZnSO_4{\cdot}H_2O$ into 500 mL of distillated water. In experiments of the batch kinetic adsorption, 50 mL of standard solution with an adjusted pH (4.5) and 25 mg of the calcium alginate beads were shaken with the use of a shaker (250 rpm) in a thermostat water bath. The concentration of zinc was determined by an atomic absorption spectrophotometer (SPCA-626D, Shimadzu Corporation, Kyoto, Japan). The amount of zinc absorbed onto the calcium alginate beads, P (mg/g dried beads), was calculated using the following equation:

$$P = \frac{V(C_0 - C)}{W_D}, \qquad (12)$$

where V (L) is the volume of the zinc solution, C_0 (mg/L) is the initial zinc concentration, C (mg/L) is the zinc concentration at a certain time, and W_D (g) is the weight of the dried calcium alginate beads [2,3].

Conclusions

In this study, calcium alginate beads were prepared using the electrospray method as a biosorbent, and adsorption of zinc ions was investigated using the batch equilibrium method. In addition, the effective parameters on adsorption such as porosity were analyzed with the suggested nonlinear model, and the predicted behavior of the nonlinear model in the prediction of experimental data was compared with that of the pseudo-kinetic models. The pseudo models could not follow the experimental data at pre-equilibrium time spans of the adsorption processes as high deviations from the experimental data were observed by the calculation of σ_{RMSD}.

The fitting parameters (τ and k_c) of the nonlinear model were estimated using minimum σ_{RMSD} between the experimental and calculated data. According to porosity and solution concentration, τ and k_c (m/s) were calculated from 9.52 to 16.8 and from 1.39×10^{-7} to 2.21×10^{-7} m/s, respectively. The nonlinear model demonstrated that diffusive transport through the beads was limited by pore diffusion. The nonlinear model was shown to possess a superb adaptive behavior in the prediction of zinc adsorption in the different samples as its average σ_{RMSD} was calculated to be 2.66%, which was the lowest σ_{RMSD} compared to all other models. The nonlinear model provided a good fit to the experimental data as the calculated equilibrium adsorptions were shown to be in good agreement with their experimental counterparts.

Abbreviations

C: Zinc concentration (mg/L); C_b: Bulk solute concentration (mg/L); C_e: Equilibrium zinc concentration (mg/L); C_R: Solute concentration at the surface of the sphere (mg/L); C_0: Initial zinc concentration (mg/L); D_a: Apparent diffusion coefficient; D_l: Diffusion coefficient (m^2/s); k_1: Rate constant of pseudo first-order model (1/min); k_2: Rate constant of pseudo second-order model (g/mg·min); k_c: Mass transfer coefficient (m/s); K_f: Langmuir constant (mg/g); *n*: The number of experimental data; P: Adsorbed concentration (mg/g); P_{Cal}: Calculated adsorbed concentration (mg/g); P_e: Equilibrium adsorbed concentration (mg/g); P_{Exp}: Experimental adsorbed concentration (mg/g); P_{max}: Maximum monolayer amount of adsorption (mg/g); R: Radius (m); R^2: Correlation coefficient; T: Time (s); W_D: Weight of dried calcium alginate bead (g); W_W: Weight of wet calcium alginate bead before drying (g); V: Volume of zinc solution (L); τ: Tortuosity factor; ρ_b: Bulk density (g/cm); ε: Porosity; ρ_W: Density of water; ρ_{AL}: Density of sodium alginate; σ_{RMSD}: Root mean square deviation.

Competing interests

The authors declare that they have no competing interests.

Authors' contributions

MHK carried out the experiments, studied the mechanisms of the adsorption, wrote the adsorption nonlinear model program, analyzed the data, and drafted the manuscript. MM helped in some parts of the experiments. MK participated in the interpretation of results. SRK read and edited the manuscript and commented on it. All the authors read and approved the final manuscript.

Authors' information

MHK is a PhD student in Material Science. MM is a PhD student in Chemical Engineering. MK is an associate professor of Chemical Engineering. SRK is a PhD student in Material Science.

Acknowledgments

The authors gratefully thank the helpful comments of Dr. Abdolreza Samimi and Hamid Moghadam and the facilities provided by the University of Sistan and Baluchestan.

Author details

[1]Department of Chemical Engineering, University of Sistan and Baluchestan, PO Box 98164-161, Zahedan, Iran. [2]School of Chemical, Petroleum and Gas Engineering, Shiraz University, Shiraz, Iran. [3]Materials Science Centre, School of Materials, University of Manchester, Grosvenor Street, Manchester M1 7HS, UK.

References

1. Zhao F, Yu B, Yue Z, Wang T, Wen X, Liu Z, Zhao C (2007) Preparation of porous chitosan gel beads for copper(II) ion adsorption. J Hazard Mater 14:67–73
2. Liang ZP, Meng YQFSX, Liang ZY (2005) Equilibrium and kinetic modeling of adsorption of urea nitrogen onto chitosan coated dialdehyde cellulose. Process Biochem 40:3218–3224
3. Vieira RS, Beppu MM (2006) Dynamic and static adsorption and desorption of Hg(II) ions on chitosan membranes and spheres. Water Res 40:1726–1734
4. Kafshgari MH, Khorram M, Mansouri M, Samimi A, Osfouri S (2012) Preparation of alginate and chitosan nanoparticles using a new reverse micellar system. Iran Polym J 21:99–107
5. George M, Abraham TE (2006) Polyionic hydrocolloids for the intestinal delivery of protein drugs: alginate and chitosan. J Controlled Release 114:1–14
6. Bayramoğlu G, Tuzun I, Celik G, Yilmaz M, Arica MY (2006) Biosorption of mercury(II), cadmium(II) and lead(II) ions from aqueous system by microalgae *Chlamydomonas reinhardtii* immobilized in alginate beads. Int J Miner Process 81:35–43
7. Papageorgiou SK, Kouvelos EP, Katsaros FK (2008) Calcium alginate beads from *Laminaria digitata* for the removal of Cu^{+2} and Cd^{+2} from dilute aqueous metal solutions. Desalination 224:293–306
8. Bajpai J, Shrivastava R, Bajpai AK (2004) Dynamic and equilibrium studies on adsorption of Cr(VI) ions onto binary bio-polymeric beads of cross linked alginate and gelatin. Colloids Surf A 236:81–90
9. Ozdemir G, Ceyhan N, Manav E (2005) Utilization of an exopolysaccharide produced by *Chryseomonas luteola* TEM05 in alginate beads for adsorption of cadmium and cobalt ions. Bioresour Technol 96:1677–1682
10. Choi J-W, Yang K-S, Kim D-J, Lee CE (2009) Adsorption of zinc and toluene by alginate complex impregnated with zeolite and activated carbon. Curr Appl Phys 9:694–697
11. Al-Dujaili AH, Awwad AM, Salem NM (2012) Biosorption of cadmium (II) onto loquat leaves (*Eriobotrya japonica*) and their ash from aqueous solution, equilibrium, kinetics, and thermodynamic studies. Int J Ind Chem. doi:10.1186/2228-5547-3-22
12. Chen Z, Ma W, Han M (2008) Biosorption of nickel and copper onto treated alga (*Undaria pinnatifida*): application of isotherm and kinetic models. J Hazard Mater 155:327–333
13. Clercq JD (2012) Removal of mercury from aqueous solutions by adsorption on a new ultra stable mesoporous adsorbent and on a commercial ion exchange resin. Int J Ind Chem 3:1–6
14. Evans JR, Davids WG, MacRae JD, Amirbahman A (2002) Kinetics of cadmium uptake by chitosan-based crab shells. Water Res 36:3219–3226
15. Moon DJ, Chung MJ, Kim H, Lee BG, Lee SD, Park KY (1998) Adsorption equilibria of chloropentafluoroethane on activated carbon powder. Korean J Chem Eng 15:619–624
16. Pouretedal H, Kazemi M (2012) Characterization of modified silica aerogel using sodium silicate precursor and its application as adsorbent of Cu^{2+}, Cd^{2+}, and Pb^{2+} ions. Int J Ind Chem 3:1–8
17. Sheibani A, Shishehbor MR, Alaei H (2012) Removal of Fe(III) ions from aqueous solution by hazelnut hull as an adsorbent. Int J Ind Chem 3:1–3
18. Crank J (1970) Mathematics of Diffusion. Clarendon, Oxford
19. Ruthven D (1984) Principles of Adsorption and Adsorption Processes. Wiley, New York
20. Merrifield JD, Davids WG, MacRae JD, Amirbahman A (2004) Uptake of mercury by thiol-grafted chitosan gel beads. Water Res 38:3132–3138
21. Ozer A, Gürbüz G, Calimli A, Körbahti BK (2008) Investigation of nickel(II) biosorption on *Enteromorpha prolifera*: optimization using response surface analysis. J Hazard Mater 152:778–788
22. Malkoc E (2006) Ni(II) removal from aqueous solutions using cone biomass of *Thuja orientalis*. J Hazard Mater 137:899–908
23. Gao C, Liu M, Chen J, Zhang X (2009) Preparation and controlled degradation of oxidized sodium alginate hydrogel. Polym Degrad Stab 94:1405–1410
24. Guerra E, Bestetti M (2006) Physicochemical properties of $ZnSO_4$-SO_4-H_2O electrolytes of relevance to zinc electrowinning. J Chem Eng Data 51:1491–1497
25. Saygideger S, Gulnaz O, Istifli ES, Yucel N (2005) Adsorption of Cd(II), Cu(II) and Ni(II) ions by *Lemna minor* L.: effect of physicochemical environment. J Hazard Mater 126:96–104
26. Lindfield G, Penny J (1995) Numerical Methods Using MATLAB. Ellis Horwood, New York
27. Moghadam H, Samimi M, Samimi A, Khorram M (2008) Electro-spray of high viscous liquids for producing mono-sized spherical alginate beads. Particuology 6:271–275

11

Pre-concentration of ultra-trace furan in beverage samples and its determination by high performance liquid cromatography

Hossein Sid kalal[1*], Ali Reza Khanchi[1], Elahe Konoz[2], Sara Khaki[2], Mohammad Taghiof[1], Majid Baghdadi[3] and Homayon Ahmad Panahi[2]

Abstract

Background: Furan has been considered to be toxic and carcinogenic. In this research, a new method was applied for the pre-concentration of furan in beverage samples followed by high-performance liquid chromatography-ultraviolet detection (HPLC-UV) determination.

Results: Determination was carried out by a UV detection at 205 nm. Some of the most important parameters, including extraction solvent, extraction time, sample temperature and volume, helium flow rate and salt effect were investigated and optimized. Linear calibration curve was obtained over a concentration range of 0.005 to 0.3 μg/mL with coefficient of determination ($R^2 = 0.998$). Limit of detection and relative standard deviation were obtained as 0.001 μg/mL and 2.21% ($n = 5$), respectively.

Conclusions: The proposed method was applied successfully for the determination of furan in beverage samples. Among them, cola had a detectable amount of furan.

Keywords: Furan, Beverage samples, Pre-concentration, HPLC-UV

Background

Chemical food safety is a topic with an increasing concern within the respective community, although due to the present risk management procedures, the acute threat posed by harmful chemicals is very uncommon. Instead, there has been much more interest focused on the minute concentrations and consequent potential chronic symptoms. It has been postulated that chemical contaminants found in foods (and feeds) might be etiological factors behind several diseases (e.g., cancer), the causes of which are still greatly unknown. In addition to the environmental and agrochemical residues as well as natural toxins, contaminants produced during the food processing or storage, such as acrylamide, polycyclic aromatic hydrocarbons (compounds), benzene and furan, are becoming an issue.

* Correspondence: hsidkalal@aeoi.org.ir
[1]NFCRS, Nuclear Science and Technology Research Institute, Atomic Energy Organization of Iran, P.O. Box 11365-3486 Kargar Shomali street, Tehran, Iran
Full list of author information is available at the end of the article

Furan (C_4H_4O), a lipophilic contaminant, is formed during heating process for the manufacture of foods. Furan is an aromatic heterocyclic compound, which is highly volatile with a boiling point of 31°C. Furan has been identified in hot-air dried, baked, fried and roasted food items, such as cereal products and coffee as well as can or jar-prepared foods, at low concentration levels (<200 ng/g) [1]. Its presence in food is known since the 1960s as one of the Maillard reaction products formed in the thermal processes of foods such as birch syrup, caramel, coffee, bread and cooked meat [2-10]. Nowadays, the formation of furan remains unclear, although it is known that its generation is related to the thermal degradation of carbohydrates, thermal oxidation of polyunsaturated fatty acids and/or thermal decomposition of ascorbic acid or its derivatives [11-14]. In recent years, the study of furan in foods has received special attention by several international food organizations such as the US Food and Drugs Administration and the European Food Safety Authority [15-17]. Due to its toxicity and carcinogenicity, furan has been included by the

International Agency for Research on Cancer in the Group 2B as possibly carcinogenic to humans [18-20] and has been classified by the US Department of Health and Human Services as a human pathogen [21]. Therefore, there is a need to propose reliable methods for its determination in foods in order to obtain information about its occurrence as well as the source of exposure, mechanisms of formation and toxicology. The analysis of furan in food is not easy due to its high volatility, low molecular weight and low concentration levels in food products.

The most common analytical technique used to determine the concentration of furan is headspace extraction combined with gas chromatography-mass spectrometric (GC-MS) detection [15,22,23]. Later on, solid-phase microextraction followed by GC-MS has been developed [22,24]. Both approaches are convenient sample preparation techniques as they are very simple and demand no expensive equipment for sample extractions/concentration.

In the present study, purge-and-trap method combined with HPLC was applied for the determination of the amount of furan in food samples. The detection wavelength for the absorption of furan was set at 205 nm [25]. A systematic study of the effective parameters on the extraction and desorption processes was used to obtain maximum sensitivity and selectivity. In addition, we investigated the applicability of the proposed method for the quantification of furan in solid and liquid food samples. The proposed method provides relatively low limits of detection (LOD; 0.001 μg/mL).

Methods

Experimental

Materials and instruments

Furan, water, methanol and acetonitrile of HPLC grade were purchased from Merck and Co. Inc. (Darmstadt, Germany). All Other reagents were of analytical grade and obtained from Merck.

Chromatographic conditions for sample analysis

Chromatographic Separation Were Carried out on a HPLC Instrument (Waters, Model 600, USA) equipped with a Waters 2487 UV-Vis Detector. A Zorbax Extend C18 column (250 mm × 5 mm, with 5 μm particle size) was purchased from Agilent Technologies Inc. (Wilmington, DE, USA). A mixture of water/methanol/acetonitrile (70:20:10 vol.%) at the flow rate of 1 mL/min was used as the mobile phase in the isocratic elution mode. The injection volume was 20 μL, and the detection wavelength was set at 205 nm.

Extraction procedure

A balloon containing standard solutions was connected to the trapping unit which was placed in the ice bath (Figure 1). Trapping unit was filled with glass beads to increase its trappig efficiency. From the other side of the balloon, helium gas was purged through the heated sample solution under stiring condition. The interface between the balloon and trapping unit was heated by industrial drier to prevent furan condensation and it's return to the balloon.

Results and discussion

Selection of an appropriate extraction solvent

The selection of an appropriate solvent is important to achieve the highest sensitivity. Different solvents such as dimethy formamide, acetonitrile, methanol were studied as the extraction solvents. Among them, methanol showed the largest peak area and was used as the extraction solvent (Figure 2A).

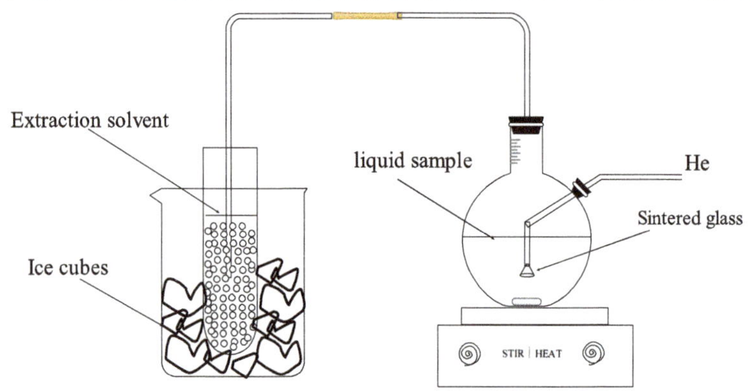

Figure 1 Schematic view of experimental setup.

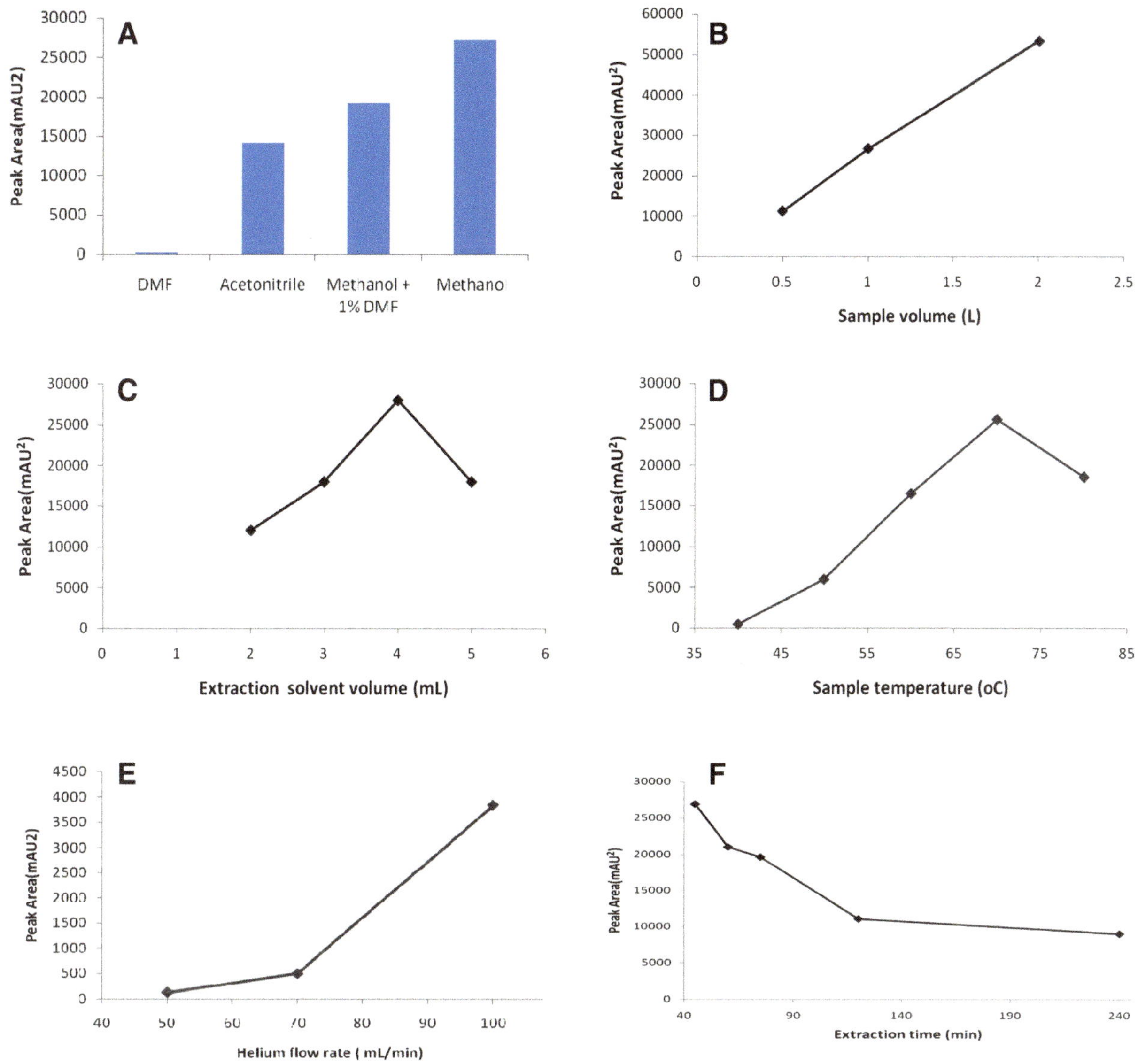

Figure 2 Utilized conditions: furan 0.01 μg mL^{-1}, temperature 40°C, time 45 min, 1,000 rpm, helium flow rate 100 mL min.$^{-1}$. Effect of extraction solvent on the peak area (**A**); effect of the sample volume (**B**); effect of extraction solvent volume (**C**); effect of sample solution temperature, extraction solvent 4 mL methanol (**D**); effect of helium flow rate, extraction solvent 4 mL methanol (**E**); effect of extraction time, extraction solvent 4 mL methanol (**F**).

Optimization of the studied parameters

Sample volume

Sample volume was studied in the range of 0.5 to 2 L. Peak area increased with the increase of the sample volume. The volume of 1 L was chosen as the optimum sample volume. More than 1 L leads to poor purging capacity and precision (Figure 2B).

Volume of the extraction solvent

The volume of the extraction solvent was studied in the range of 2 to 5 mL. Four milliliter was chosen as an optimum volume for the extraction solvent. More than 4 mL leads to a decrease in analytical sensitivity basically due to dilution (Figure 2C).

Extraction temperature

Temperature affects the purging efficicency and transfer rate of analyte from sample solution to gas phase. Purging efficicency increases with increase in the temperature of the sample solution mainly because of increase in the vapor pressure. Extraction temperature was studied in the range of 40°C to 80°C. At temperatures above 80°C,

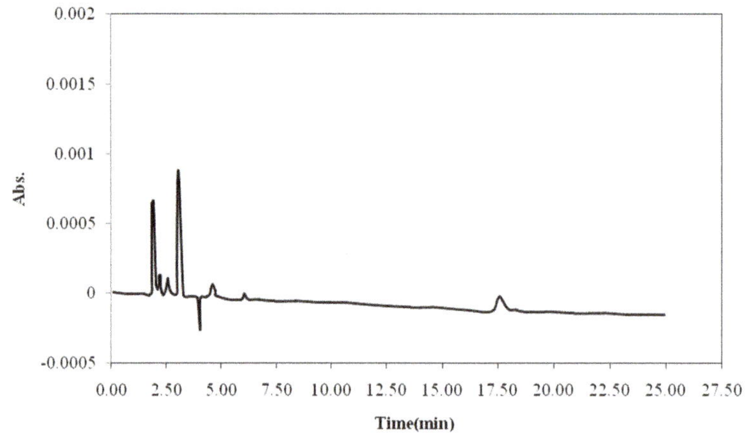

Figure 3 Chromatogram of 0.001 μg/mL furan standard solution in optimum condition.

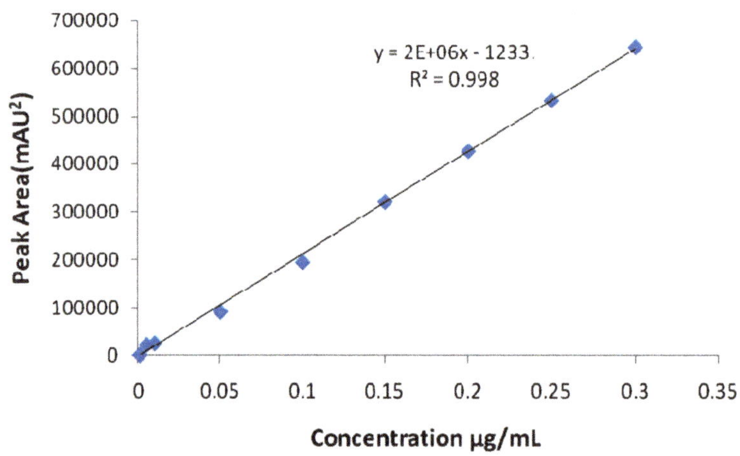

Figure 4 Calibration curve for furan determination using HPLC.

helium temperature increases which subsequently decreases the trapping efficiency and precision. The optimum temperature was obtained as 70°C (Figure 2D).

Flow rate

Helium flow rate was studied in the range of 50 to 100 mL/min. By increasing the helium flow rate, volatile species were transferred more rapidly into the trapping solvent, and as a result, the concentration was increased. At higher flow rate of helium, trapping solvent splashed out. The optimum value was obtained as 100 ml/min (Figure 2E).

Extraction time

Furan cannot remain in the trapping solvent for a long time due to its volatility. Extraction time was studied in the range of 45 to 240 min. Forty-five minutes was chosen as the optimum extraction time due to the increase peak area. The peak area was decreased after 45 min (Figure 2F).

Salt effect

The effect of salt was investigated by adding the sodium chloride to the aqueous samples. Peak areas were slightly increased because of salting-out effect. A concentration

Table 1 Analytical characteristics of the method

Characteristic	Value
Slope	2×10^6
Intercept	−1,233
Coefficient of determination (R^2)	0.998
LOD[a] (μg mL^{-1})	0.001
LOQ (μg mL^{-1})	0.005
LOL (μg mL^{-1})	0.300
RSD[b] ($n = 5$)	2.21 (0.005), 2.07 (0.100)

[a]Calculated based on the signal-to-noise ratio of three. [b]Values in parentheses are the furan concentrations (μg/mL) for which the RSD was obtained. LOD, limit of detection; LOQ, limit of quantification; LOL, limit of linearity; RSD, relative standard deviation.

Table 2 Determination of furan in beverage samples

Sample	Concentration (mean ± SD[a], μg/mL)	Added (μg/mL)	Found (mean ± SD[a], μg/mL)	Recovery (%)
Cola (Fanta[b], 1.5 L[g])	0.020 ± 0.002	0.030	0.040 ± 0.003	67
Coffee (Klassno Company Malaysia[c], 18 g[g])	nd[h]	0.030	0.026 ± 0.003	87
Beer (Patrik Company (Iran)[d], 1 L[g])	nd	0.030	0.026 ± 0.002	87
Grape (Takdaneh[e], 1 L[g])	nd	0.030	0.023 ± 0.002	77
Sour cherry (Pakban[f], 1.5 L[g])	nd	0.030	0.025 ± 0.003	83

[a]Standard deviation ($n = 3$); [b]The Coca-Cola Company (Iran); [c] ; [d]; [e]Takdaneh (Iran); [f]Pakban Dairy Company (Alborz, Iran); [g]packaging; [h]not detected. SD, standard deviation.

ration of 0.3% NaCl was selected throughout the experiments.

Analytical performance of the method

According to the above optimum conditions obtained for each factor, the furan chromatogram can be achieved as shown in Figure 3 and in order to evaluate the linearity of the method, a calibration curve over a concentration range of 0.001 to 0.300 μg/ml of furan was obtained (Figure 4). In order to evaluate the linearity of the method, a calibration curve over a concentration range of 0.001 to 0.300 μg/mL of furan was obtained. Linear dynamic range, coefficient of determination, LOD, limit of quantification (LOQ) and relative standard deviation (% RSD) are shown in Table 1. For calculation of LOD, it is considered to be the concentration of analyte that gives a signal that is equal to three times the standard deviation of the blank. This is equivalent to defining the LOD as that concentration at which the signal-to-noise ratio equals 3 at 99% confidence limit:

$$\mathrm{LOD} = \bar{x}_{\mathrm{blank}} \pm 3\sigma_{\mathrm{blank}}$$

Similarly, LOQ in the above equation is defined as $10\sigma_{\mathrm{blank}}$. As shown in Table 1, only small deviations between sequential determinations were found.

Validation of the method and analysis of beverage samples

The proposed methodology was applied for the determination of furan in several beverage samples (cola, coffee, beer, grape, sour cherry) in which, among them, cola was found to have a significant amount of furan (Table 2). The beverage samples were spiked to assess

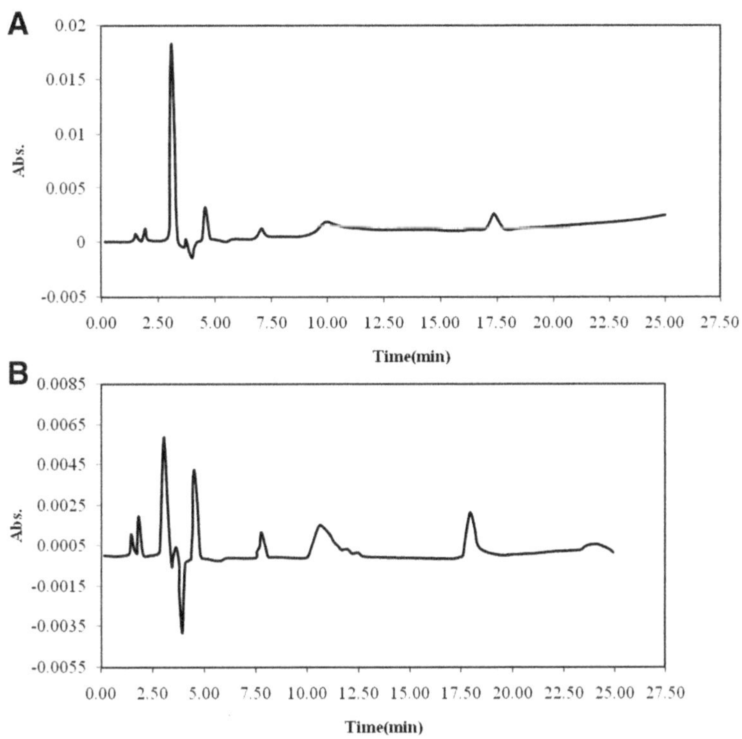

Figure 5 HPLC chromatograms of cola without spike (a) and with spike (b).

the matrix effect. The high recoveries of furan at a level of 0.03 μg/mL showed no matrix interferences. The chromatograms of cola without spiking (a) and with spiking (b) are illustrated in Figure 5.

Conclusions

In this study, a new pre-concentration method combined with HPLC technique was applied for extraction and determination of furan in beverage samples. Some effective parameters such as the extraction solvent, extraction time, sample temperature and volume, helium flow rate and salt effect were investigated and optimized. LOD and RSD were obtained as 0.001 μg/mL and 2.21% ($n = 5$), respectively. The proposed methodology was applied successfully for the determination of furan in cola, coffee, beer, grape and sour cherry which, among them, cola was found to have a significant amount of furan.

Competing interests

The authors declare that they have no competing interests.

Authors' contributions

HK conceived of the study and participated in coordination. AK participated in experimental design and optimization. EK participated in preconcentration experiments. MT participated in graphical and tables preparation. SK participated in HPLC experiments. MB and HP participated in the statistical analysis of data. All authors read and approved the final manuscript.

Acknowledgment

The authors would like to thank Dr. M.H. Mousazadeh for his helpful comments.

Author details

[1]NFCRS, Nuclear Science and Technology Research Institute, Atomic Energy Organization of Iran, P.O. Box 11365-3486 Kargar Shomali street, Tehran, Iran. [2]Department of Chemistry, Islamic Azad University, Central Tehran Branch, Tehran, Iran. [3]Graduate Faculty of Environment, University of Tehran, Tehran, Iran.

References

1. Zoller O, Sager F, Reinhard H (2007) Furan in food: headspace method and product survey. Food Addit Contam 24(S1):91–107
2. Merritt C, Bazinet ML, Sullivan JH, Robertson DH (1963) Mass spectrometric determination of the volatile components from ground coffee. J Agric Food Chem 11:152–155
3. Stoffelsma J, Simpa G, Kettenes DK, Pypker J (1968) New volatile components of roasted coffee. J Agric Food Chem 16:1000–1004
4. Walter RH, Fagerson IS (1968) Volatile compounds from heated glucose. J Food Sci 33:294–297
5. Tatum JH, Shaw PE, Berry RE (1969) Degradation products from ascorbic acid. J Agric Food Chem 17:38–40
6. Persson T, von Sydow E (1973) Aroma of canned beef: gas chromatographic and mass spectrometric analysis of the volatiles. J Food Sci 38:377–385
7. Maga JA (1979) Furans in foods. Pre-Concentration of Ultra-Trace Furan in Beverage Samples and its determination by high performance liquid cromatography. 11:355–400
8. Kallio H, Leino M, Salorinne L (1989) Analysis of the headspace of foodstuffs near room temperature. J High Resolution Chromatogr 12:174–177
9. Shimoda M, Shibamoto T (1990) Isolation and identification of headspace volatiles from brewed coffee with an on-column GC/MS method. J Agric Food Chem 38:802–804
10. Holscher W, Steinhart H (1992) Investigation of roasted coffee freshness with an improved headspace technique. Z Lebensm Unters Forsch 195:33–38
11. Locas CP, Yaylayan VA (2004) Origin and mechanistic pathways of formation of the parent furan: a food toxicant. J Agric Food Chem 52:6830–6836
12. Becalski A, Seaman S (2005) Furan precursors in food: a model study and development of simple headspace method for determination of furan. J Assoc Anal Communities Int 88(1):102–106
13. Fan XJ (2005) Formation of furan from carbohydrates and ascorbic acid following exposure to ionizing radiation and thermal processing. J Agric Food Chem 53:7826–7831
14. Mark J, Pollien P, Lindinger C, Blank I, Mark T (2006) Quantitation of furan and methylfuran formed in different precursor systems by proton transfer reaction mass spectrometry. J Agric Food Chem 54:2786–2793
15. US FDA (2004b) (2008) Determination of furan in foods. 7 May 2004 (updated 27 October 2006) [cited: 2008 Jul 21], Available from http://www.cfsan.fda.gov/dms/furan.html
16. US Food and Drug Administration (FDA) (2004) Furan in foods - thermal treatment., www.cfsan.fda.gov/~lrd/fr040510.html Accessed in 10/8/2011
17. European Food Safety Authority (EFSA) (2004) Report of the scientific panel on contaminants in the food chain on provisional findings on furan in food. EFSA J 137:1–20
18. Glatt H, Schneider H, Liu Y (2005) A cell line for the sensitive detection of genotoxic effects induced by carbohydrate pyrolysis products and other food-borne chemicals. Mutat Res 580:41–52
19. Byrns MC, Vu CC, Neidigh JW, Abad JL, Jones RA, Peterson LA (2006) Detection of DNA adducts derived from the reactive metabolite of furan, cis-2-butene-1,4-dial. Chem Res Toxicol 19:414–420
20. International Agency for Research on Cancer (IARC) (1995) Dry cleaning some chlorinated solvents and other industrial chemicals. Monogr Eval Carcinog Risks Hum 63:393
21. US Department of Health and Human Service (2005) 11th Report on Carcinogens, Public Health Service, National Toxicology Program, Research Triangle Park., NC, www.ntp.niehs.nih.gov/ntp/roc/eleventh/profiles/s090fura.pdf Accessed in 10/8/2011
22. Bianchi F, Careri M, Mangia A, Musci M (2006) Development and validation of a solid phase micro-extraction-gas chromatography-mass spectrometry method for the determination of furan in baby-food. J Chromatogr A 1102:268–272
23. Hasnip S, Crews C, Castle L (2006) Some factors affecting the formation of furan in heated foods. Food Addit Contam 23(3):219–227
24. Goldmann T, Périsset A, Scanlan F, Stadler RH (2005) Rapid determination of furan in heated foodstuffs by isotope dilution solid phase micro-extraction-gas chromatography-mass spectrometry (SPME-GC-MS). Analyst 130:878–883
25. Hyman D, Sluiter A, Crocker D, Johnson D, Sluiter J, Black S, Scarlata C (2008) Determination of acid soluble lignin concentration curve by UV-Vis spectroscopy. Natl Renewable Energy Laboratory (NREL) Analytical Proced: 1–13, Technical Report NREL/TP-510-42617

12

The effect of feldspar and kaolin on mechanical performance of SBR/LDPE composites

Ahmad Mousa

Abstract

Background: Feldspar and kaolin was ball milled at 320 rpm for hours with the aid of a dry grinding agent. Both minerals were combined with inorganic salt namely sodium chloride as a grinding agent. The grinding agent and mineral were mixed at the ratio of 1:1.

Results: The application of such grinding agent led to effectively reduce the particle size in a short time. The mineral materials were viewed under a scanning electron microscope (SEM) before and after grinding. SEM graphs revealed that a significant reduction in average particle size was recorded for both minerals. The reduction percentage for kaolin was about 82%, and in the case of feldspar, it was about 93%. The BET surface area of both minerals increased after milling. The potential of the ground filler on the tensile properties of low-density polyethylene/styrene butadiene rubber (LDPE/SBR) composites has been reported.

Conclusions: The ball milling of the raw fillers in the presence of inorganic salt was found to be able to reduce the particle size of the fillers significantly in a short time. The application of dry grinding agent did not display any agglomeration or cold welding between the mineral particles as evidenced by the SEM images. The incorporation of the ground filler into LDPE/SBR composites showed a positive impact on the tensile strength at the cost of elongation at break except for the feldspar.

Keywords: Ball milling, Feldspar, Kaolin, Dry grinding

Background

Comminution is the process whereby particular materials are reduced from coarse feed to fine product size for end use [1]. Comminution can be performed either by wet or dry means [2]. Such operation is utilized in various industrial processes such as ceramics, agricultural products, food, pigments, and composite materials. To recall, ultrafine grinding is an energy-intensive process as far as fine-sized particle achievement is concerned. Nevertheless, grinding has been applied over the years due to the fact that various mineral fillers can be made into useful products or ingredients when crushed to a smaller size range [3-6]. It is important to note that such powders are cheap materials that can be incorporated into polymers to improve the technical utility of the mineral powder. The potential of such minerals will be great as long as their particle size is small enough, since fine powders will have higher surface area to utilize in producing polymer composites. Such composites are expected to display enhanced mechanical properties and thermal performance. Furthermore, the barrier properties are expected to increase with the use of the proper mineral filler powder. The effect of inorganic electrolytes such as sodium hydroxide and soap on grinding of minerals has been investigated by several workers [7,8]. The current investigation has dual purpose; firstly, it aims to produce fine particle-sized minerals with increased specific surface area in the presence of sodium chloride as grinding agent via ball milling. Secondly, it aims to examine the potential of the powders as a reinforcing agent for low-density polyethylene/styrene butadiene rubber (LDPE/SBR) composites compared to raw powders.

Methods

Materials

LDPE, with a grade of L705 with MFI 0.20 g/min and a density of 0.918 g/ml, was obtained from Polyolefin Company (Singapore, Singapore). Styrene butadiene

Correspondence: mousa@rocketmail.com
Department of Materials and Metallurgical Engineering, Al-Balqa Applied University, Salt 19117, Jordan

rubber (Buna Hüls 1502) produced by Bayer (M) Ltd, (Petaling Jaya, Malaysia) is obtained from a local rubber processing factory. Commercial grade coarse sodium chloride was purchased from a local market (Jordan). Kaolin and feldspar minerals in raw form were obtained from a mine at the Kingdom of Saudi Arabia, and the initial particle size was 173 and 495 μm, respectively.

Filler treatments

The mineral filler was tumbled ahead with a dry grinding agent to load into a ball mill. The tumbled powder was loaded into the ball mill. The powder was ground at a speed of 320 rpm for 4 h. The powder was removed from the mill and soaked in water to dissolve the salt used as grinding agent. The suspension that contains powder was filtered and washed with water several times. The washed powder was dried in a hot air oven for 24 h at 100°C. The specific surface area of the powder was measured using the Brunauer, Emmet, and Teller (BET) method [9]. The average particle size of the raw and milled powder was determined using the software of the XL-30 scanning electron microscope (SEM).

Results and discussion

The effect of ball milling on filler particle size and surface area

The effect of ball milling for hours at 320 rpm in the presence of the inorganic grinding agent on the particle size of kaolin and feldspar is presented in Figure 1; it is compared with the untreated counterpart. It is clear that the average particle size for both minerals was reduced after milling from 495 to 35 μm for feldspar, whereas the size was down from 195 to 38 μm for kaolin. In terms of percentage reduction, it was about 93% for feldspar mineral and about 82% in the case of kaolin mineral. The observed trend should be due to the influence of the milling process in the presence of inorganic salt. The grinding agent is expected to increase the friction between the balls and the grinding media since the grinding agent is harder than the mineral to be milled. As a result, the stress conferred to the mineral is expected to increase, causing the bulky size to rupture to a smaller size as shown in Figure 1. The reduction in the particle size obtained in Figure 1 is confirmed by the SEM graphs depicted in Figures 2 and 3. It is obvious that the particle size for kaolin and feldspar was minimized after milling, which is in conformity with the data shown in Figure 1. Note that the micrographs do not display any agglomeration of the minerals after the milling process. Hence, this is a clue that the addition of the dry grinding agent did not cause any cold welding between the particles of the minerals. The implication is that the grinding agent has dual functions: the former is as grinding aid, and the latter is as matrix separation agent that prohibits the formation of agglomerates. To further support the enhancement of

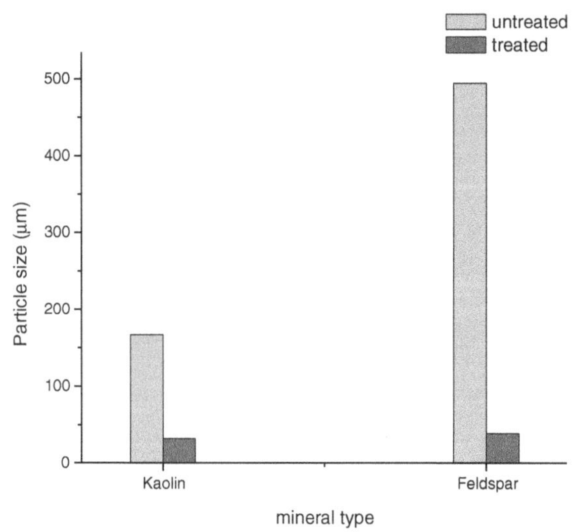

Figure 1 Effect of ball milling on the average particle size of kaolin and feldspar. Kaolin and feldspar are milled for 4 h at 320 rpm.

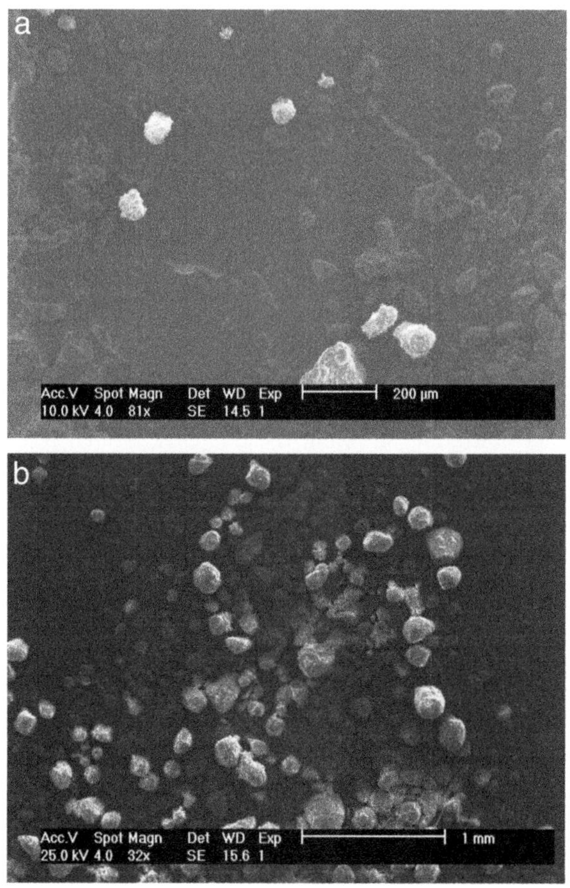

Figure 2 SEM images of kaolin. (**a**) Before and (**b**) after ball milling for 4 h and 320 rpm.

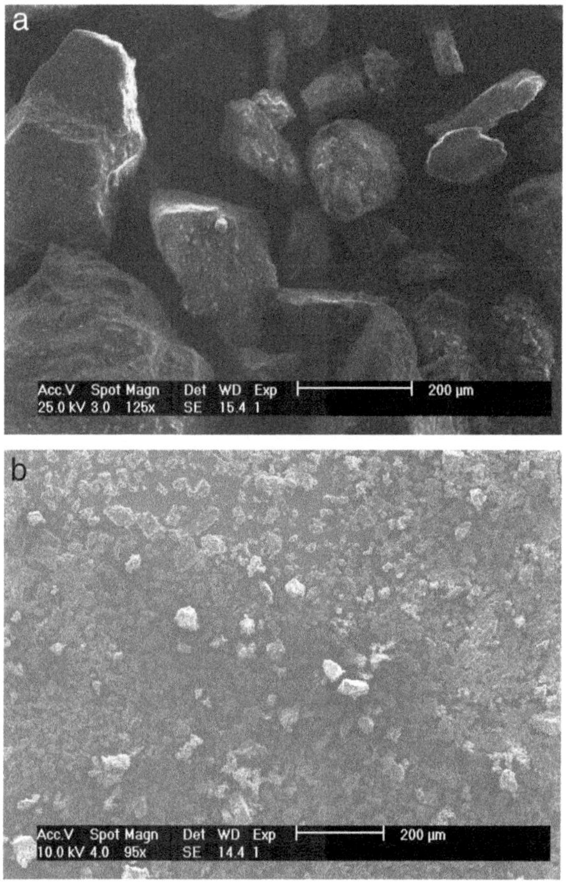

Figure 3 SEM images of feldspar. (**a**) Before and (**b**) after ball milling for 4 h and 320 rpm.

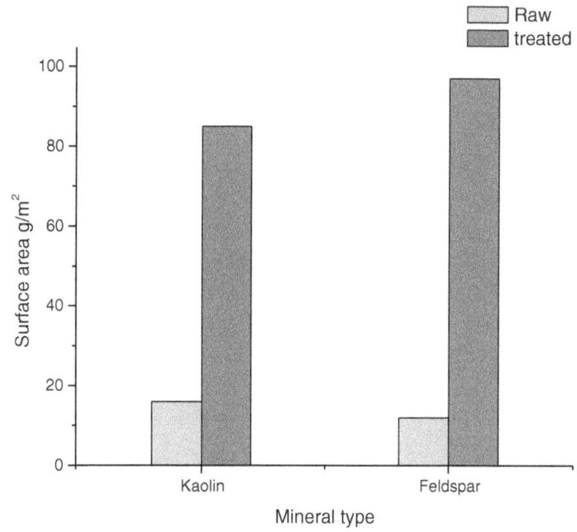

Figure 4 BET surface area of kaolin and feldspar. Before and after ball milling 4 h and 320 rpm.

surface area, the BET surface of both feldspar and kaolin was checked as reflected in Figure 4. Obviously, due to the milling process, the surface area is increased as shown in Figure 4. This should be related to the creation of new surfaces after milling. The new surface will have larger surface areas.

The effect of fine fillers on the mixing torque of LDPE/SBR composite

Figure 5 shows the effect of the untreated and treated mineral fillers on the equilibrium torque (torque at the end of the mixing process) of the LDPE/SBR composites obtained from the Brabender plasticorder. It is clear that the mixing torque of the composite filled with treated fillers is higher than the composite with untreated fillers. This is likewise to be attributed to the enhanced degree of interaction between the treated fillers and the matrix. Since it is well known that the smaller the particle size, the larger the surface area, hence, a better degree of interaction with improved dispersion state was achieved. As a result, the mixing torque increased as demonstrated in Figure 5. This is in line with previous works on filled similar polymer composites [10,11]. It is worth mentioning that the equilibrium torque displayed by the composite-based feldspar is less than that of the kaolin-based feldspar, although the surface area of feldspar is larger. This is most likely due to the platy shape of the feldspar which acts as internal lubricant for the composite. The last conclusion from Figure 5 is that the equilibrium torque of the plain composite is the least among the whole composites. Similar findings were reported earlier in the case of the platy shape

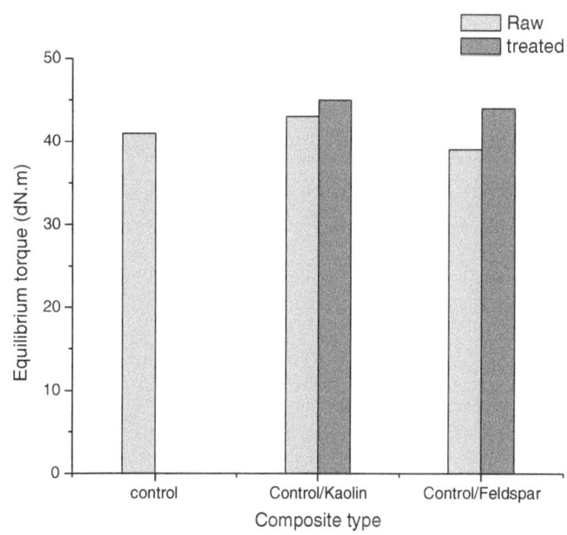

Figure 5 Influence of untreated and treated mineral filler on the equilibrium mixing torque of SBR/LDPE composites.

including nanoclay mineral fillers on the torque and viscosity of polymer composites [12,13].

Tensile properties

Figure 6 reflects the influence of the untreated and treated filler on the tensile strength at the breaking of the prepared samples. Note that the incorporation of the treated fillers increased the tensile strength at the breaking as compared with the pristine filler. This is due to the improved interaction between the treated fillers and the matrix, which in turn, increased the work of adhesion of the composites against the external stress. Furthermore, the treated filler is expected to contribute to the stress transfer mechanism, which means that the filler will sustain some of the external load that leads to the observed trend. Considering the influence of the treated filler on the modulus of elasticity of the prepared composites shown in Figure 7, it is clear that the modulus increased after the incorporation of the treated filler. This indicates quite well that the stiffness and rigidity of the composites increased after the addition of mineral fillers. This could be due to the fact that the mineral filler is a highly rigid material. Thus, the rigidity will be conferred to the matrix. In addition to this, the improved degree of interaction mentioned earlier is also responsible for the increased value of the modulus. Again, the rigidity of the feldspar-based composites is lower than that of the kaolin-based composite; this is due to the lubrication action practiced by the platelike feldspar discussed earlier, which is in line with the equilibrium torque results. Figure 8 shows the course of the percentage elongation at the breaking (%EB) of the LDPE/SBR composites after the incorporation of the treated fillers. It is clear that the %EB decreased after the incorporation of kaolin. This should be due to the occupation of the free volume of the matrix by the kaolin mineral, which increased the rigidity and decreased the compliance of the composites. Therefore, the extensibility of the matrix decreased prior to failure. On the other side, the addition of feldspar increased the %EB as compared with the kaolin-based composite. Such discrepancy might be due to the platy structure of the feldspar, which, in turn, might act as the internal lubricant that allows the matrix to be elongated, to a better extent, to rupture [11,14]. It is worth to recall that both fillers displayed lower strain as compared with the control sample. The observed trend is in line with the earlier report on the

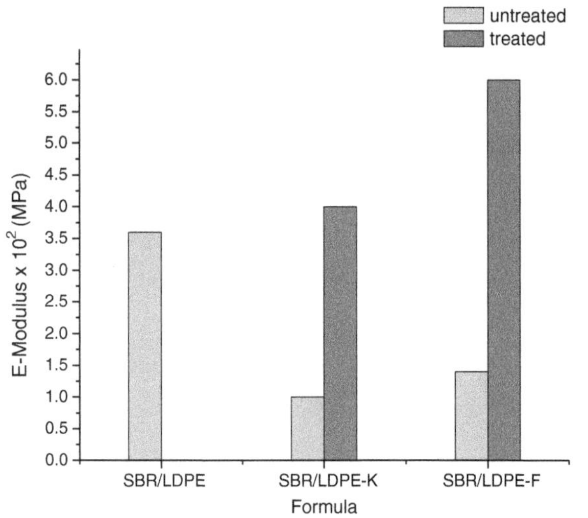

Figure 7 **Influence of untreated and treated mineral filler on tensile modulus of SBR/LDPE composites.**

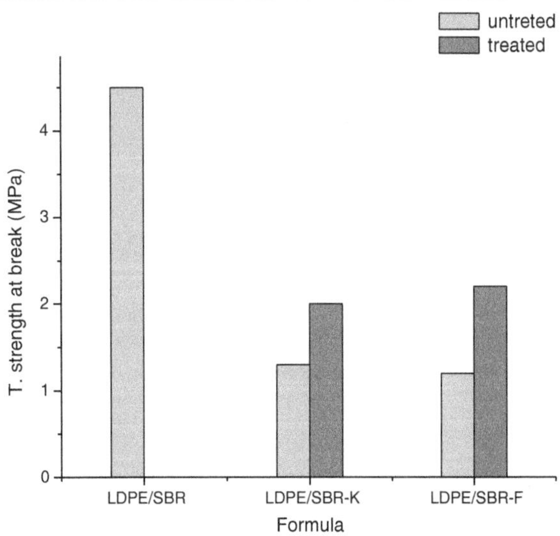

Figure 6 **Influence of untreated and treated mineral filler on tensile strength at the break of SBR/LDPE composites.**

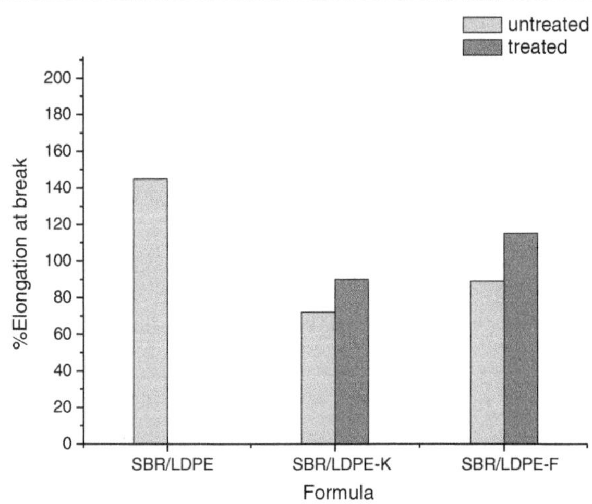

Figure 8 **Influence of untreated and treated mineral filler on percentage elongation at the break of SBR/LDPE composites.**

effect of feldspar on the performance of recycled PET/SBR composites [11,12].

Conclusions

Based on the aforementioned results, the following conclusions can be drawn:

1. The ball milling of the raw fillers in the presence of inorganic salt was found to be able to reduce the particle size of the fillers significantly in a short time. This was due to the increased stress on the minerals in the presence of the dry grinding agent. Simultaneously, the BET surface area of the minerals increased as such.
2. The application of the dry grinding agent did not display any agglomeration or cold welding between the mineral particles as evidenced by the SEM images.
3. The incorporation of the ground filler into LDPE/SBR composites showed a positive impact on the tensile strength at the cost of elongation at the break except for the feldspar. This was due to the enhanced interaction between the matrix and the fillers.

Competing interests

The author declares that he has no competing interests.

Acknowledgments

The author acknowledges, for his final year project, Feras al Tayeb for his help in running some experiments.

References

1. Russell A (1989) Fine grinding - a review. Industrial Minerals 259:59–70
2. Gupty CK (2003) Extraction flowsheets. In: Chemical metallurgy: principles and plastics. Wiley-VCH Verlag GmbH & Co. KGaA, Weinheim
3. Lynch A, Rowland C (2005) Hand stones. In: The history of grinding. Society for Mining, Metallurgy and Exploration, Littleton, CO
4. Florez-Zamora MI (2008) Comparative study of Al-Ni-Mo alloys obtained by mechanical alloying in different ball mills. Rev Adv Mater Sci 18:301
5. Kwade A, Bernotat B (2002) Present and future treads in dry fine grinding. Industrial Minerals 412:69–71
6. Jankovic A, Dundar H, Mehta R (2010) Relationships between comminution energy and product size for a magnetite ore. J South Afr Inst Min Metall 110:141–146
7. Brown JH (1955) M.I.T. progress report. NYO 7172:97
8. Stanzyk MH, Feld IL (1981) Comminution by the attrition grinding process 1981. US Bureau of Mines, Washington DC
9. Digabel L, Boquillon F, Dole N, Monties P, Averous BL (2004) Properties of thermoplastic composites based on wheat-straw lignocellulosic fillers. J Appl Polym Sci 93:428–436
10. Mousa A, Heinrich G, Wagenknecht U (2008) The potential of O-MMT as a reinforcing filler for uncured and dynamically cured PVC/XNBR composites. J Macromolecular Science, Part A Pure and Applied Chemistry 45:733–741
11. Mousa A (2008) On the potential of feldspar sand for recycled PET/SMRL composites. J of Elastomers and Plastics 40:179–190
12. Mousa A, Karger-Kocsis J (2001) Rheological and thermodynamic behavior of SBR/organoclay nanocomposites. Macromolecules Materials Engineering 286:260–266
13. Mousa A, Banerjee S (2009) The effect of nano layered silicate on mechanical properties of polypropylene composites. University of Sharjah Journal of Pure & Applied Sciences 6:21–34
14. Teh PL, Mohd Ishak ZA, Hashim AS, Karger-Kocsis J, Ishiaku US (2004) On the potential of organoclay with respect to conventional fillers (carbon black, silica) for epoxidized natural rubber compatibilized natural rubber vulcanizates. J Appl Polym Sci 94:2438–2445

13

Effect of the ratio of catalyst to carbon source on the growth of vertically aligned carbon nanotubes on nanostructured porous silicon templates

Noor Asnida Asli[1,2*], Muhammad Salleh Shamsudin[1,2,3], Suriani Abu Bakar[4], Mohamad Rusop Mahmood[1,5] and Saifollah Abdullah[1,2]

Abstract

Background: The effect of the ratio of catalyst to carbon source on the growth of vertically aligned carbon nanotubes (VACNTs) has been studied.

Results: Dense VACNTs were successfully synthesised on optimised nanostructured porous silicon templates using modified floated carbon source-catalyst in a two-stage hot filament thermal chemical vapour deposition system with different amounts of ferrocene as the catalyst at 800°C. The surface morphologies of the VACNTs were analysed using field emission scanning electron microscopy, and the crystallinity of the nanotubes was observed using micro-Raman spectroscopy.

Conclusions: These data revealed that the amount of catalyst used significantly affected the diameter, crystallinity and growth rate of the synthesised nanotubes. The average diameter of the nanotubes ranged from ≈ 9 to 30 nm with lengths of ≈ 110 μm when 0.5 g ferrocene was used.

Keywords: Porous silicon, Carbon nanotubes, Catalyst ratio, Botanical hydrocarbon, Diameter.

Background

Carbon nanotubes (CNTs) are tube-shaped structures made of carbon atoms and can have diameters on the nanometre and micrometre scales [1]. CNTs have attracted much attention since their discovery 20 years ago [2] due to their properties. They can be used as components in light-emitting diodes [3], electron field emitters for optical displays and electrode materials for fuel cells and lithium ion batteries due to their extraordinary properties [4]. CNTs have superior mechanical, chemical, electronic and optoelectronic properties [3].

Camphor ($C_{10}H_{16}O$) is a valued material in biotechnology and nanotechnology research, and it was first reported that CNTs could be synthesised using camphor in 2001. This environmentally friendly carbon source for the synthesis of CNTs can be used to produce multiwall carbon nanotubes, single-wall carbon nanotubes (SWCNT) and vertically aligned nanotubes. In this context, the striking feature of camphor as a precursor is that a small amount of catalyst is needed compared to what is needed in other similar kinds of CNT syntheses that employ hydrocarbons [5].

Carbon nanotubes are commonly synthesised using one of three techniques: arc discharge, laser ablation and chemical vapour deposition (CVD) [6]. CVD is the most popular and extensively used method because it is easy to scale-up, can be performed at a lower temperature [6], has a lower set-up cost and higher production yield than other approaches [1]. Most of the CNT literature claims that the additional template synthesis used in the CVD technique is a promising way to synthesise well-aligned CNT structures with control over diameter, size

* Correspondence: asnidasli@yahoo.com
[1]Centre of Nanoscience and Nanotechnology (NANO-SciTech Centre), Institute of Science, Universiti Teknologi MARA, 40450 Shah Alam, Selangor D.E., Malaysia
[2]School of Physics and Material Studies, Faculty of Applied Sciences, Universiti Teknologi MARA, 40450 Shah Alam, Selangor D.E., Malaysia
Full list of author information is available at the end of the article

uniformity and density [6-9]. The pore diameter of the template can be controlled which consequently enables tube diameter and orientation to the dictated [10]. For example, anodic aluminium oxide (AAO) has been used as a template to grow aligned carbon nanotubes (ACNTs). ACNTs are defined as quasione-dimensional carbon cylinders oriented in a parallel fashion perpendicular to a substrate [11]. Thess et al. [12] first reported the synthesis of ACNTs in bundle form, where 70% of the volume produced consisted of nanotubes. Ali et al. [13] reported that the pore sizes of the substrates used played a role in controlling the growth directions and diameters of the carbon nanotubes. The literature also reports the use of porous silicon templates with pore sizes ranging from 10 to 100 nm [14].

By using an AAO template, ACNTs were successfully grown, as reported by Sui [15]; however, fabrication of AAO is a time-intensive and expensive process that takes approximately 10 h to complete [10]. In a different approach to synthesising well-aligned CNTs, Cao et al. [16] reported the use of a gold-patterned quartz substrate. According to Li et al. [6], porous silicon templates have been loaded with catalytic particles by sputtering iron (Fe) as a route to the synthesis of ACNTs. In addition, Kordás et al. [17] reported that carbon nanotubes have been synthesis on oxidised porous silicon. Catalysts play an important role in the synthesis of CNTs; common catalysts that are widely used for the preparation of CNTs include nickel (Ni), iron (Fe), cobalt (Co) and combinations thereof [18].

This paper will report the use of a porous silicon template and an Fe catalyst to produce vertically aligned carbon nanotubes (VACNTs) using a novel method that involves modifying the catalyst-nanotemplate interaction in the CVD technique. In this study, we focused on understanding how the ratio of the catalyst to carbon source affected the nanotube growth rate and size, and the carbon and ferrocene distributions.

Methods

NPSiT were prepared as outlined by Asli et al. [19]. Briefly, boron-doped, p-type (100) Si wafers that were 525 ± 25-μm thick were subjected to a photo-electrochemical anodisation process with 30-min etching time and 20 mA/cm^2 of current density. These Si wafers were placed in Teflon cells with an electrolyte consisting of a 1:1 ratio of ethanol and HF (48%) and illuminated with a halogen lamp. Then, the resulting templates were annealed in ambient air at 300°C overnight. A quartz tube with a length of 80 cm and an inner diameter of 3 cm served as a thermal CVD reactor and was kept horizontal in a two-stage zone furnace. The samples were prepared by varying the weight of ferrocene (obtained from Sigma-Aldrich, MO, USA) from 0.3 to 0.8 g, while holding the weight of camphor oil used at 5 g. The catalyst and carbon source were then placed in separate alumina boats and positioned side-by-side in furnace zone 1. NPSiT was placed in the centre of furnace zone 2. Zone 2 was heated to 800°C and was left to stabilise for 10 min after it was purged with nitrogen gas (0.05 to 1.00 L min^{-1}). Then, zone 1 was heated to 180°C for the precursor vaporisation process. Once it reached the required temperature, the synthesis process continued for 60 min. The nitrogen gas was continuously flowed before, during and after synthesis to prepare the ambient nitrogen environment. Zone 1 was turned off after the synthesis process was complete, and zone 2 was left for 30 min after the annealing process. The samples were taken out for characterisation after the furnaces cooled down to room temperature.

Results and discussion

Figure 1 is a typical field emission scanning electron microscope (FESEM) top-view image of a nanostructured porous silicon template (NPSiT) fabricated by the photo-electrochemical anodisation technique that was described in a previous study [19]. In the top-view image, the dark areas are pores that have an average diameter of approximately 20 nm. Figure 2a,b,c,d,e,f shows the morphology of CNTs grown on the optimised NPSiT using different amounts of catalyst. FESEM images at low (×100) and high magnifications (×100,000, insets) are shown for CNTs prepared with varying amounts of ferrocene (from 0.3 to 0.8 g) and a constant amount of camphor oil (5 g). The low magnification image in Figure 2a shows that a low yield of randomly entangled CNTs was obtained when 0.3-g ferrocene was used. The high magnification FESEM image in the inset of Figure 2a shows randomly entangled nanotubes with an abundance of amorphous carbon (a-C) content.

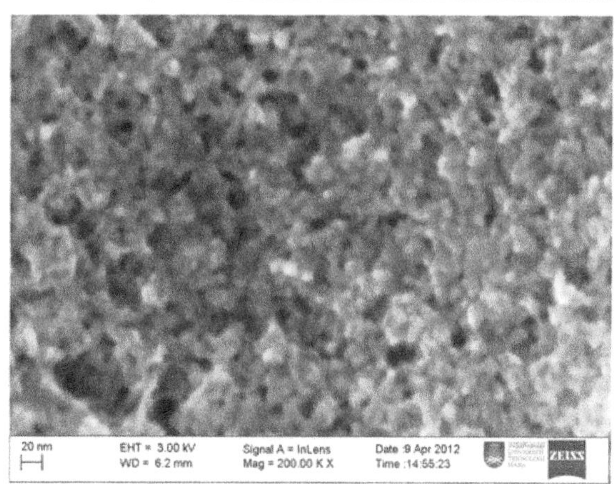

Figure 1 Nanostructured porous silicon template anodized at 30 min and 20 mA/cm^2 of current density.

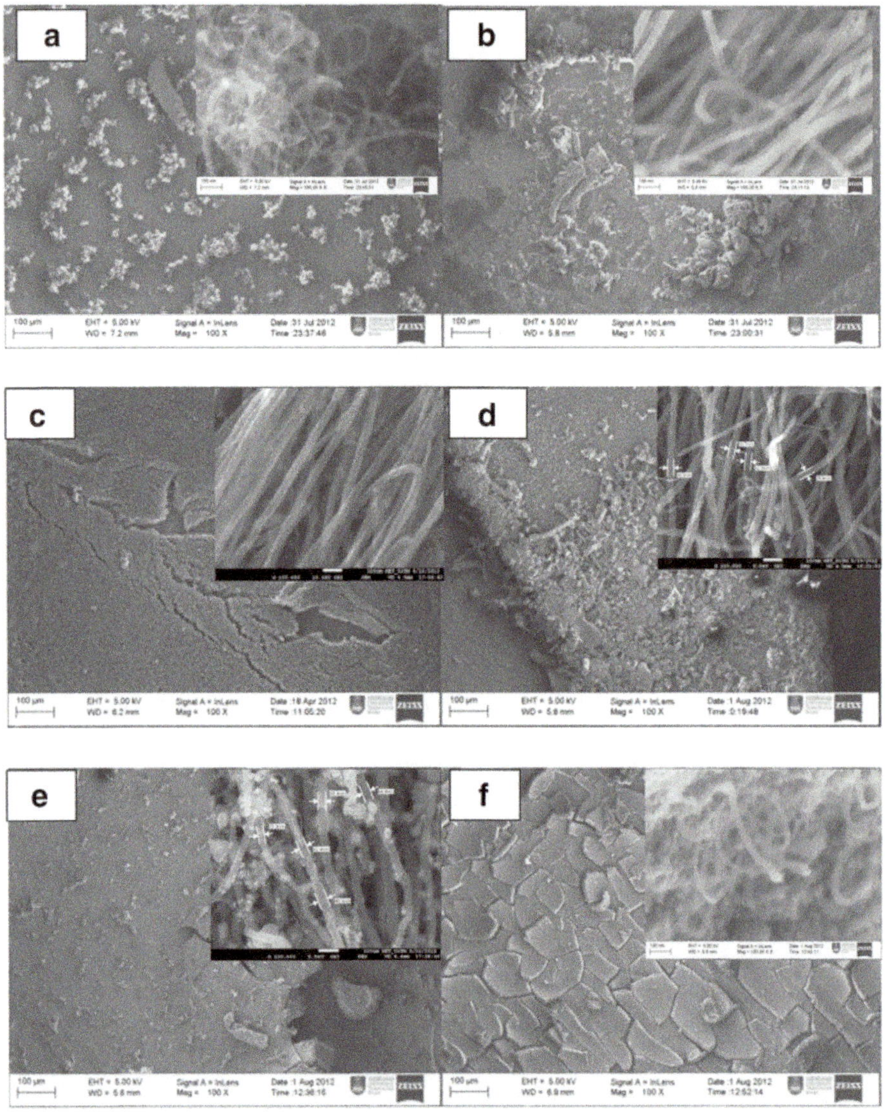

Figure 2 FESEM images of CNT at different Fe ratio for 5 g camphor oil. (**a**) 0.3 g, (**b**) 0.4 g, (**c**) 0.5 g, (**d**) 0.6 g, (**e**) 0.7 g, (**f**) 0.8 g.

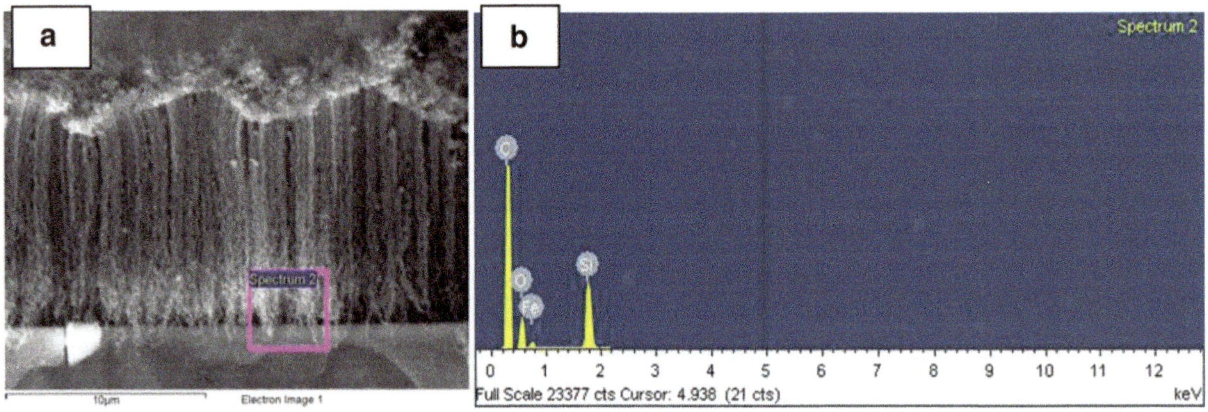

Figure 3 Carbon signals both inside and outside the pores. (**a**) FESEM cross-section image and (**b**) EDX analysis.

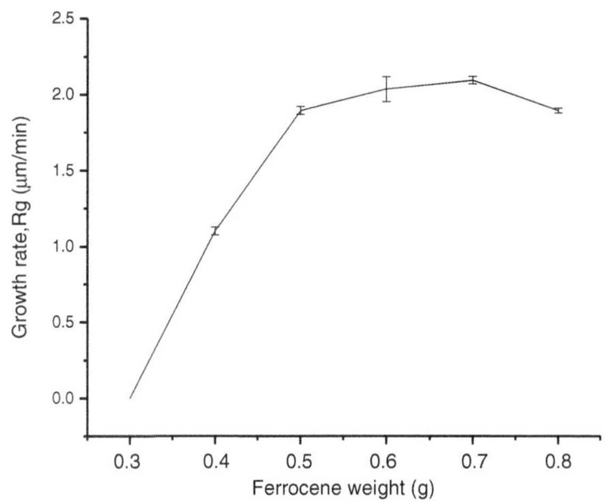

Figure 4 The VACNT growth rate at different ferrocene ratios.

When a low amount of ferrocene was used, the nanotubes form in low yield, bend and grow in random directions. The nanotubes grow without support from neighbouring nanotubes, resulting in the growth of randomly oriented carbon nanotubes. When the amount of ferrocene used is less than 0.4 g, aligned CNTs are not produced. By increasing the ferrocene amount, the CNT yield increases substantially, as seen in Figure 2b,c,d,e,f. Nevertheless, carbonaceous crusts were present and covered the top part of the vertically aligned nanotubes, as seen in Figure 2d,e,f. The formation of this crust might be due to the higher amount of ferrocene present relative to the amount of carbon source and related to the non-uniformity in the evaporation processes of ferrocene and camphor oil during synthesis [8]. The CNTs produced using 0.5 g ferrocene and 5 g camphor oil are shown in Figure 2c. In this image, the nanotubes are densely packed uniform vertical columns; CNTs, as long as 110 to 113 µm are evident. In this case, the growing carbon nanotubes are densely packed on the NPSiT in vertical columns and support each other to grow in the same direction (perpendicular to the template surface). Therefore, a low amount of catalyst resulted in less dense CNTs with a-C content and poor alignment, as seen in Figure 2a,b. Disproportionate catalyst addition results in the agglomeration of the catalyst, leaving unwanted catalyst impurities in the final product (Figure 2d,e,f).

The diameters of the CNTs were also significantly affected by the amount of ferrocene used. In addition, the size of catalyst particles determined the diameter of the tubes. The carbon atoms diffused into the iron nanoparticles during the CVD process and formed FeC. Fe catalyst particles with various dimensions were found encapsulated within the tubes. Therefore, when catalyst particles become supersaturated with carbon atoms, the diameters of the tubes become bigger, and precipitate carbon atoms were found attached to the nanotube bundles [20].

Nanotube sizes in the range of 14.0 to 25.0 nm were observed when a low amount of ferrocene (0.3 g) was used. By increasing the ferrocene amount to 0.4 g, the tube size increased greatly to between 30.1 to 39.9 nm. Interestingly, the smallest diameter (9.9 nm) was observed when 0.5 g ferrocene was used. However, when this amount of ferrocene was used, the tube diameters ranged from 9.9 to 30.5 nm. The inconsistency in the size of the tubes is due to their growth from inside the pores and on the top surface of the NPSiTs. It was clearly observed that CNTs grew from both sites, as verified by examining the cross-sectional FESEM images and energy dispersive X-ray spectrometer (EDX) analysis showing carbon signals both inside and outside the pores (Figure 3). In contrast, the nanotube diameters continue to increase when the amount of ferrocene used was greater than 0.5 g. In summary, ferrocene amounts greater or less than 0.5 g are not capable of producing nanotubes with small diameters when 5 g camphor oil is used.

Using the data collected from the FESEM images, the growth rate of the nanotubes can be calculated. The relationship between the growth rate of the VACNTs and the amount of ferrocene used (ranging from 0.3 to 0.8 g for 5 g camphor oil) is shown in Figure 4. The growth rate can be calculated as $\alpha = \beta$ (µm) / γ (min) [15], where β is the distribution length of the tubes, and γ is the deposition time. The lengths, diameters and growth rates of CNTs synthesised using at 0.3 to 0.8 g of ferrocene and 5 g of camphor oil are shown in Table 1. As

Table 1 Lengths, diameters and growth rates of CNTs

Camphor oil (g):ferrocene (g)	Length (µm)	Diameter (nm)	Growth rate (µm/min)
5.0:0.3	No side view	14.0 to 25.0	-
5.0:0.4	60 to 72	30.1 to 38.0	1.1
5.0:0.5	110 to 113	9.9 to 30.5	1.9
5.0:0.6	115 to 120	20.7 to 33.0	2.0
5.0:0.7	125 to 129	22.3 to 37.1	2.1
5.0:0.8	100 to 125	35.0 to 39.9	1.9

The CNTs were synthesised using 0.3 to 0.8 g ferrocene catalyst and 5.0 g camphor oil.

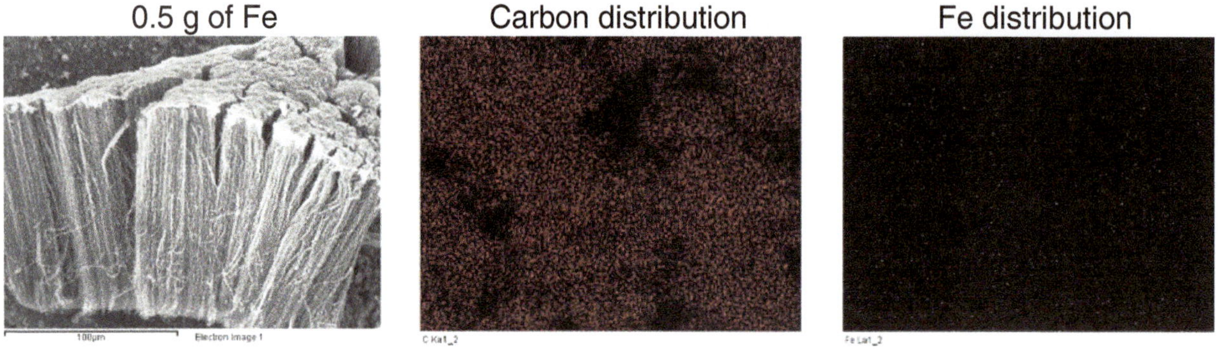

Figure 5 FESEM image and its corresponding EDX dot mapping of carbon and Fe catalyst.

the amount of ferrocene used increased from 0.3 to 0.5 g, the growth rate of the VACNTs increased rapidly from 0 to 1.9 μm/min. The growth rates continued to increase gradually, and a maximum rate of 2.1 μm/min was found for a sample catalysed with 0.7 g ferrocene. Then, the growth rate decreased to 1.9 μm/min when 0.8 g of ferrocene was used. Figure 2a shows entangled tubes with short length and poor alignment which resulted from an insufficient number of Fe particles to catalyse the growth of VACNTs. However, the reduced growth rates for samples synthesised with 0.8 g ferrocene were due to low catalytic activity that resulted from an increment of Fe particle size. Figure 2e,f shows that the CNTs produced contain impurities that stick to the nanotubes and result in nanotube with larger diameters. This phenomenon happens at high ferrocene concentrations when ferrocene decomposes to an abundance of Fe particles, which tend to agglomerate into a larger cluster. The catalytic effect is reduced when the catalyst size increases, resulting in larger nanotube diameters as discussed before. According to Bai et al. [21], the catalyst acts as a nucleation site and dictates the tube diameter.

Figure 5 shows the FESEM images and corresponding EDX maps for CNTs synthesised using 0.5 g ferrocene and 5 g camphor oil. EDX mapping was utilised to pinpoint catalyst location and distribution, and relate it to carbon production. The optimised parameters were found to be 0.5 g ferrocene and 5 g carbon precursor because this synthesis showed tolerable ferrocene catalyst in the final product. This conclusion is supported by the top-view image shown in Figure 2c and the Raman spectra that will be discussed later.

Figure 6 shows the Raman spectra of VACNTs synthesised using different amounts of ferrocene. The

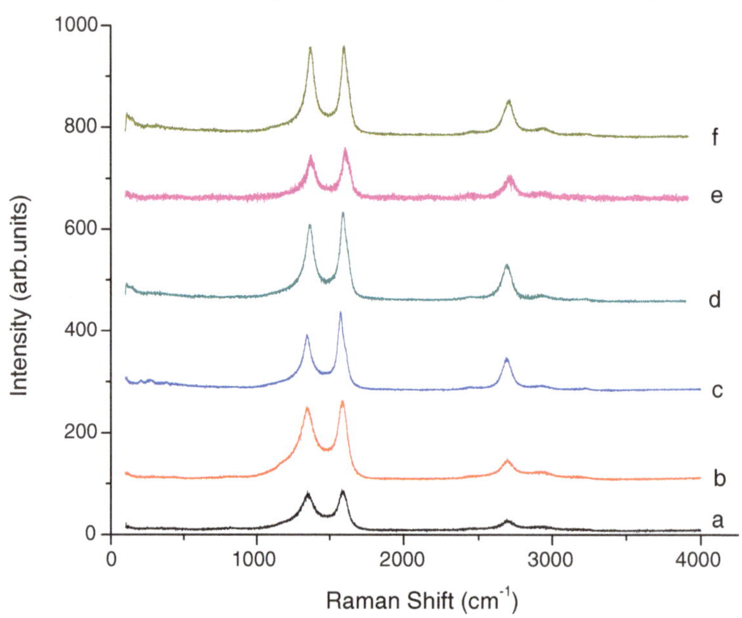

Figure 6 Raman patterns at different Fe ratios for 5 g camphor oil. The peaks are (**a**) 0.3 g, (**b**) 0.4 g, (**c**) 0.5 g, (**d**) 0.6 g, (**e**) 0.7 g, (**f**) 0.8 g.

Table 2 Raman peak positions and D-to-G peak intensity ratios

Sample (g) (amount of ferrocene used for 5 g camphor oil)	G-peak (cm^{-1})	D-peak (cm^{-1})	I_D/I_G ratio
0.3	1,583.21	1,346.17	0.97
0.4	1,585.17	1,346.92	0.93
0.5	1,571.57	1,346.95	0.72
0.6	1,588.15	1,361.49	0.86
0.7	1,607.02	1,363.49	0.88
0.8	1,596.69	1,366.41	0.98

Raman peak positions and I_D/I_G ratios are summarised in Table 2. Significant radial breathing mode (RBM) peaks were detected in the samples synthesised using 0.5 g ferrocene and 5 g camphor oil, as shown in Figure 7. Peaks were observed at 205.8, 272.9 and 382.9 cm^{-1}, which correspond to SWCNT with diameters of 1.21, 0.91 and 0.64 nm, respectively. RBM peaks were not observed for the samples synthesised at catalyst concentrations lesser or greater than 0.5 g ferrocene. A low I_D/I_G ratio of 0.72 was detected for samples synthesised using 0.5 g ferrocene; the ratio increases to a maximum value of 0.98 for samples synthesised using 0.8 g catalyst. The poor graphite crystallinity observed for the sample prepared using 0.8 g ferrocene resulted from the low catalytic properties owing to the bigger catalyst size. A low ferrocene concentration results in an imbalance of carbon atoms compared to Fe catalyst; the huge amount of carbon is not completely decomposed by the low amount of Fe catalyst. In this situation, the catalyst is rendered inactive and low quality nanotubes are produced, as detected for samples prepared with 0.3 g ferrocene.

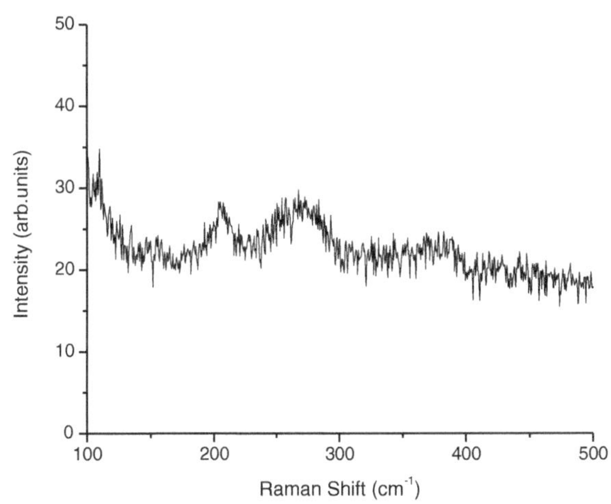

Figure 7 Low frequency peaks associated with RBM for sample catalysed at 0.5 g ferrocene.

Experimental

As-obtained VACNTs deposited on NPSiT were visualised using a FESEM (SUPRA 40 VP, Carl Zeiss SMT, Oberkochen, Germany), and an EDX was used to identify the elements constituting the specimen (qualitative analysis) and the weight concentration of the contained elements (quantitative analysis). Structural properties were observed by micro-Raman spectroscopy using argon ion laser excitation at 514.532 nm (DU420A-OE-325, Horiba Jobin Yvon, Kyoto, Japan).

Conclusion

In summary, CNT growth was detected when as little as 0.3 g ferrocene was used. However, at least 0.4 g ferrocene must be used to produce VACNTs. As the catalyst concentration increases, the density and growth rate of VACNTs were found to increase (up to 2.1 μm min^{-1}) until 0.7 g ferrocene was reached. The VACNTs produced when 0.5 g ferrocene was used have small diameters and were found to have low defect levels, as indicated by the calculated I_D/I_G ratio of 0.72. Only at this concentration, RBM peaks were clearly observed for the as-synthesised SWCNTs. The growth of vertically aligned carbon nanotubes has been successfully reported using a porous template and the appropriate amount of catalyst such that the pores can physically hold the nanotubes in the vertical direction and the bundles of nanotubes can support each other during growth. Further studies should be carried out to assess the prospective applications of these structures; for example, VACNT exhibits excellent properties for use as field electron emission cathodes.

Competing interests

The authors declare that they have no competing interests.

Authors' contributions

NAA and MSS carried out the experiments and drafted the manuscript. SAB built the experimental setup. MRM and SA reviewed the manuscript. All authors read and approved the final manuscript.

Authors' information

NAA is currently a Ph.D. student in the Faculty of Applied Sciences and doing research at NANO-SciTech Centre, Institute of Science, University Teknologi MARA. Her studies focus on the growth of carbon nanotubes on nanoporous materials .She received a BS degree in Materials Technology from Universiti Teknologi MARA, Malaysia in 2009. She has authored and co-authored over 15 peer-reviewed papers in journals and conference proceedings in the field of nanostructured carbon materials and serves as a reviewer for several national and international journals. MSS is currently a graduate research assistant in the Faculty of Engineering and the Environment at University of Southampton Malaysia Campus, where his studies focus on the growth of graphene. He is also one of the members in Centre of Nanoscience and Nanotechnology, UiTM, where his research interest involves the synthesis and characterization of carbon nanotubes. He received a BS degree in Physics from Universiti Teknologi MARA, Malaysia in 2010. He has authored and co-authored over 15 peer-reviewed papers in journals and conference proceedings in the field of nanostructured carbon materials and serves as a reviewer for several national and international journals. SAB is currently a senior lecturer in the Department of Physics, Faculty of Science and Mathematics, Universiti Pendidikan Sultan Idris. She is also head of the Centre of Nanoscience and Nanotechnology, UiTM, where her research interest involves the synthesis and characterization of carbon nanotubes based on palm oil. She received a Ph.D. degree in nanomaterials from the Universiti Teknologi MARA, Malaysia in 2010. She has authored and co-authored over 25 peer-reviewed papers in journals and conference proceedings in the field of nanostructured carbon materials and serves as a reviewer for several national and international journals. MRM is currently a professor and founder in NANO-Electronic Centre, Faculty of Electrical Engineering, Universiti Teknologi MARA. He is also head of the Centre of Nanoscience and Nanotechnology, UiTM, where his research interest involves the synthesis and characterization of nanomaterials and their application in electronic devices. He received a Ph.D. degree in Optoelectronics Engineering from Nagoya Institute of Technology, Japan in 2004. He has authored and co-authored over 100 peer-reviewed papers in journals and conference proceedings in the field of nanostructured carbon materials and serves as a reviewer for several national and international journals. SA is currently a professor and lecturer in the Faculty of Applied Sciences, Universiti Teknologi MARA. He is also one of the members of the Centre of Nanoscience and Nanotechnology, UiTM, where his research interest involves the synthesis and characterization of carbon nanotubes and nanoporous materials. He received his Ph.D. degree in Physics from the University of Malaya, Malaysia in 2002. He has authored and co-authored over 100 peer-reviewed papers in journals and conference proceedings in the field of nanostructured carbon materials and serves as a reviewer for several national and international journals.

Acknowledgements

The authors would like to thank the staff of the NANO-SciTech Centre, Universiti Teknologi MARA for their technical support. This work was financially supported by the Excellent Fund (600-RMI/ST/DANA 5/3/Dst (20/2011) and the Fundamental Research Grant Scheme (600-RMI/ST/FRGS 5/3/Fst (204/2010), Malaysian Ministry of Higher Education.

Author details

[1]Centre of Nanoscience and Nanotechnology (NANO-SciTech Centre), Institute of Science, Universiti Teknologi MARA, 40450 Shah Alam, Selangor D.E., Malaysia. [2]School of Physics and Material Studies, Faculty of Applied Sciences, Universiti Teknologi MARA, 40450 Shah Alam, Selangor D.E., Malaysia. [3]University of Southampton Malaysia Campus (USMC), No. 3, Persiaran Canselor 1, Kota Ilmu Educity@Iskandar, 79200 Nusajaya, Johor D.T., Malaysia. [4]Department of Physics, Faculty of Science and Mathematics, Universiti Pendidikan Sultan Idris, 35900 Tanjung Malim, Perak D.R., Malaysia. [5]NANO-Electronic Centre, Faculty of Electrical Engineering, Universiti Teknologi MARA, 40450 Shah Alam, Selangor D.E., Malaysia.

References

1. Kumar M, Ando Y (2010) Chemical vapor deposition of carbon nanotubes: a review on growth mechanism and mass production. J Nanosci Nanotechnol 10:3739–3758
2. Iijima S (1991) Helical microtubules of graphitic carbon. Nature 354:56–58
3. Lee WY, Weng CH, Juang ZY, Tsai CH (2005) Lateral growth of single-walled carbon nanotubes across electrodes and the electrical property characterization. Diamond Relat Mater 14:1852–1856
4. Somani SP, Somani PR, Tanemura M, Lau SP, Umeno M (2009) Carbon nanofibers and multiwalled carbon nanotubes from camphor and their field electron emission. Curr Appl Phys 9:144–150
5. Mukul K, Tsugio O, Mineo H, Ando Y (2007) The use of camphor-grown carbon nanotube array as an efficient field emitter. Carbon 45:1899–1904
6. Li J, Lei W, Zhang X, Wang B, Ba L (2004) Field emission of vertically-aligned carbon nanotube arrays grown on porous silicon substrate. Solid State Electron 48:2147–2151
7. Lee WY, Liao TX, Juang ZY, Tsai CH (2004) Patterned aligned growth of carbon nanotubes on porous structure templates using chemical vapor deposition methods. Diamond Relat Mater 13:1232–1236
8. Musso S, Porro S, Rovere M, Giorcelli M, Tagliaferro A (2008) Fluid dynamic analysis of gas flow in a thermal-CVD system designed for growth of carbon nanotubes. J Cryst Growth 310:477–483
9. Lee TJ, Seo J, Lee H, Lee JW, Yi W (2010) Fabrication of single-walled carbon nanotube three-dimensional networks inside the pores of a porous silicon structure. Carbon 48:1473–1479
10. Merchan-Merchan W, Saveliev AV, Kennedy L, Jimenez WC (2010) Combustion synthesis of carbon nanotubes and related nanostructures. Prog Energy Combust Sci 36:696–727
11. Afre RA, Soga T, Jimbo T, Kumar M, Ando Y, Sharon M (2005) Growth of vertically aligned carbon nanotubes on silicon and quartz substrate by spray pyrolysis of a natural precursor: turpentine oil. Chem Phys Lett 414:6–10
12. Thess A, Lee R, Nikolaev P, Dai H, Petit P, Robert J (1996) Crystalline ropes of metallic carbon nanotubes. Science 273:483–487
13. Ali NK, Hashim MR, Aziz AA, Hamammu I (2008) Method of controlling spontaneous emission from porous silicon fabricated using pulsed current etching. Solid State Electron 52:249–254
14. Guo T, Nikolaev P, Thess A, Colbert DT, Smalley RE (1995) Catalytic growth of single-walled nanotubes by laser vaporization. Chem Phys Lett 243:49–54
15. Sui YC, González-León JA, Bermúdez A, Saniger JM (2001) Synthesis of multi branched carbon nanotubes in porous anodic aluminum oxide template. Carbon 39:1709–1715
16. Cao A, Zhang X, Xu C, Liang J, Wu D, Wei B (2001) Synthesis of well-aligned carbon nanotube network on a gold-patterned quartz substrate. Appl Surf Sci 181:234–238
17. Kordás K, Pap AE, Vähäkangas J, Uusimäki A, Leppävuori S (2005) Carbon nanotube synthesis on oxidized porous silicon. Appl Surf Sci 252:1471–1475
18. Azira AA, Zainal NFA, Nik SF (2009) Properties of carbon nanotubes synthesised using fluidized floating catalyst method from camphor oil. Malayas J Sci 28:123–128
19. Asli NA, Yusop SFM, Rusop M, Abdullah S (2011) Surface and bulk structural properties of nanostructured porous silicon prepared by electrochemical etching at different etching time. Ionics 17:653–657
20. Sharma SN, Sharma RK, Bhagavannarayana G, Samanta SB, Sood KN, Lakshmikumar ST (2006) Demonstration of the formation of porous silicon films with superior mechanical properties, morphology and stability. Mater Lett 60:1166–1169
21. Bai X, Li D, Wang Y, Liang J (2005) Effects of temperature and catalyst concentration on the growth of aligned carbon nanotubes. Tsinghua Science & Technology 10:729–735

14

Synthesis and application of CNT-supported MoO_3-K_2O nanocatalyst using microemulsion technique: role of nanoparticle size on catalyst activity and selectivity in higher alcohol synthesis

Ahmad Tavasoli*, Saba Karimi and Morteza Shoja

Abstract

Background: Alkalized MoO_3 nanocatalyst supported on carbon nanotubes (CNTs) is prepared using microemulsion technique. The nanocatalyst was extensively characterized by different methods, and its activity and selectivity in higher alcohol synthesis (HAS) have been assessed in a fixed bed micro-reactor. The physico-chemical properties and performance of the nanocatalyst were compared with the catalyst prepared by impregnation method.

Results: The transmission electron microscopy images showed that small Mo nanoparticles are confined inside the CNTs (1 to 5 nm) and on the outer surface of the CNTs (1 to 7 nm). Using microemulsion technique, the average MoO_3 particle sizes decreased from 7 to 4.8 nm, the percentage of dispersion increased from 51 to 59 and the percentage of reduction increased by 37%. Also, the chemical interactions between K-Mo-O species increased, enhancing the conditions for the formation of alcohols.

Conclusion: The percentage of CO conversion increased from 27.7% to 43.8%. The hydrocarbon selectivity decreased from 32.4% to 24.5%, and alcohol selectivity increased from 45.3% to 61.4%.

Keywords: Higher alcohols, Molybdenum, Carbon nanotubes, Microemulsion, Particle size

Background

Higher alcohol synthesis from syngas, an economically attractive method for making fuels and chemicals, is of interest due to the enhancement of petroleum price, environmental concerns, and additive for gasoline to increase octane number [1]. There are three different methods for making higher alcohols from syngas: (1) using modified catalysts for synthesis of methanol including Cu-, Zn-, and Cr-based catalysts, (2) using modified catalysts for Fisher-Tropsch synthesis including Fe-, Ni-, and Co-based catalysts, and (3) using promoted molybdenum sulfide catalysts [2-9]. Among them, molybdenum-based catalysts are more attractive due to their high activity for the water-gas shift reaction [10]. Molybdenum disulfide (MoS_2) catalyst mainly produces hydrocarbons, but when it is promoted with alkali metals, it can produce alcohols from syngas [11,12]. The function of alkali metals is to reduce the hydrogenation ability of alkyl species to form alkanes and increase the sites active for the formation of alcohols [13].

Some investigations have been carried out to study the influence of catalyst physico-chemical properties on higher alcohol synthesis and to have a better understanding of the structure-sensitive effects in higher alcohol synthesis (HAS) catalysis [14-16]. A sub-category of the structure-sensitive reactions regards the dependence of both catalytic activity and selectivity on catalytic metal particle size. Microemulsion, a novel technique for catalyst preparation, enables the control of metal particle size with a narrow particle size distribution, regardless of metal content [17-20]. Briefly, a microemulsion consists of nanosized water droplets surrounded by an oil phase, which is stabilized by a surfactant [20]. The size of the metal

* Correspondence: tavassolia@khayam.ut.ac.ir
School of Chemistry, College of Science, University of Tehran, Tehran 1417614411, Iran

particles formed in water-in-oil (w/o) microemulsions is controlled by changing the micelle size (the water-to-surfactant ratio) [20]. Hayashi et al. found that the activity of a Fe/SiO_2 catalyst prepared by microemulsion was higher than that of the catalyst prepared by incipient wetness impregnation with the same metal loading [21]. However, it has been also shown that metal nanoparticles obtained by microemulsion techniques interact strongly with oxygen-carrying ceramic supports (such as Al_2O_3, SiO_2, and TiO_2). Such interactions lead to a decrease of the catalyst reduction efficiency [17,22,23]. Therefore, oxidic carriers impose serious limitations to the investigation of structure-sensitive effects in catalysis because of the co-existence of incompletely reduced metal phase caused by strong interaction with the oxidic carrier [17,22,24]. Specifically, nanoparticles synthesized by microemulsion are known to be more difficult to reduce on oxidic carriers. The control of particle size is another advantage for nanoparticle preparation in microemulsion system. The ratio of water to surfactant is one of the size-determining key parameters [25]. The size of the active metal particles depends on the size of the droplets in the microemulsion. The droplet size will be influenced by the water-to-surfactant ratios. An increase of this ratio at constant concentration of surfactant will increase the average diameter of the droplets. Lisiecki and Pileni reported that the size of Cu nanoparticles prepared in a system consisting of bis(2-ethylhexyl) sulfosuccinate (aerosol-OT) surfactant, cyclohexane, and water increased from 2 to 10 nm as water-to-surfactant ratios changed from 1 to 10 [26].

In our previous works, we used impregnation method to prepare the carbon nanotube-supported MoO_3-K_2O higher alcohol synthesis catalyst [27,28]. We showed that CNTs, when used as HA synthesis catalyst support, allow a better metal dispersion and control and minimize the metal-support interactions (formation of mixed compounds). In the present work, we intend to compare the proposed microemulsion technique for MoO_3-K_2O catalyst preparation with the incipient wetness impregnation method for the control of metal particle size using CNTs as catalyst support. The influences of molybdenum particle size on the catalyst physico-chemical properties, activity, and selectivity were assessed and reported.

Methods

Multiwalled CNTs (purity >95%; characteristics are shown in Table 1) were used as support for the preparation of the catalysts. Prior to catalyst preparation, the support was treated with 30% HNO_3 reflux at 120°C overnight, washed with distilled water several times, and dried at 120°C for 6 h.

For the preparation of the first catalyst, molybdenum and potassium particles were synthesized in a reverse microemulsion using a nonionic surfactant Triton X-100 and *n*-hexanol as the oil phase. The concentrations of molybdenum and potassium were adjusted using aqueous solutions of $(NH_4)6Mo_7O_{24}{\cdot}4H_2O$, (Merck, Whitehouse Station, NJ, USA) and K_2CO_3 as 15 wt.% molybdenum and 8 wt.% potassium. The water-to-surfactant molar ratio (w/s) was selected as 2. After 15 min of vigorous stirring, a microemulsion was obtained. Hydrazine was added to develop nanoparticle formation in the core of the micelles by reducing the metal oxides. Then, the purified carbon nanotubes were added under stirring. Afterward, tetrahydrofurane was added dropwise (1 ml/min). The mixture was left to mature and settle slowly overnight and then decanted. The sample was filtered and washed several times with ethanol. Next, the catalyst was dried at 120°C and calcined under argon flow at 450°C for 3 h and slowly exposed to an oxygen atmosphere during the cooling step. The catalyst prepared by this procedure was signified as C_1.

The second catalyst, with the same metal loadings, was prepared by the incipient wetness impregnation method using aqueous solutions of $(NH_4)_6Mo_7O_{24}{\cdot}4H_2O$ and K_2CO_3. After the impregnation step, the catalyst was dried at 120°C and calcined at 450°C under argon flow for 3 h and slowly exposed to an oxygen atmosphere during the cooling step. The catalyst prepared by this procedure was signified as C_2.

ICP-OES

The metal loadings of the fresh and purified supports and the calcined catalysts were performed using Varian VISTA-MPX inductively coupled plasma-optical emission spectrometry (ICP-OES) instrument (Varian, Palo Alto, CA, USA).

BET surface area measurements/BJH pore size distributions

Surface area, pore volume, and pore average diameter of the supports and calcined catalysts were measured using an ASAP-2010 V2 Micrometrics system (Micrometric Instrument Co., Cleveland, OH, USA). The samples were degassed at 200°C for 4 h under 50-mTorr vacuums, and their Brunauer-Emmett-Teller (BET) area, pore volume, and pore diameter were determined.

X-ray diffraction

The phases and particle sizes of the crystals present in the calcined, reduced, and sulfide catalysts were analyzed by X-ray diffraction (XRD) using a Philips Analytical X-ray diffractometer (X'Pert MPD, X'Pert, PANalytical B.V., Almelo, The Netherlands) with monochromatized Co/Kα radiation, 2θ angles from 20° to 80°. The Debye-Scherer formula ($d = 0.9\lambda \ / \ \beta\cos\theta$, where β is the full width at half maximum; λ, the wavelength of X-ray; and 0.9, a constant) was applied to MoO_3 peaks at $2\theta = 43.3°$ (the most intense peak for MoO_3), in order to calculate the average particle sizes.

Table 1 Textural properties of the fresh and purified carbon nanotubes

Sample	BET surface area (m^2/g)	Total pore volume (ml/g)	Average pore diameter (Å)	% Fe
Fresh CNTs	209	0.48	91.62	0.60
Treated CNTs	253	0.59	94.12	0

Transmission electron microscopy

The morphology of the support and catalysts was studied by transmission electron microscopy (TEM). Sample specimens for TEM studies were prepared by ultrasonic dispersion of the catalysts in ethanol, and the suspensions were dropped onto a carbon-coated copper grid. TEM investigations were carried out using a Philips CM20 (100 kV) transmission electron microscope equipped with a NARON energy-dispersive spectrometer with a germanium detector.

Temperature-programmed reduction

The H_2 temperature-programmed reduction (TPR) profiles of the catalysts were performed in order to study the reducibility of the metal species in the catalysts. The calcined catalyst sample (0.05 g) was first purged in a flow of helium at 140°C to remove traces of water and gases that exist in the catalyst, and then cooled to 40°C. Then, the TPR of each sample was performed using 5% H_2 in argon stream at a flow rate of 40 ml/min at atmospheric pressure using Micromeritics TPD-TPR 2900 analyzer (Norcross, GA, USA) equipped with a thermal conductivity detector (TCD), heating at a linearly programmed rate of 10°C/min up to 850°C.

Hydrogen chemisorption

The amount of chemisorbed hydrogen was measured using the Micromeritics TPD-TPR 2900 analyzer system. The calcined catalyst (0.2 g) was first purged in a flow of argon at 140°C to remove traces of water. The temperature was then raised to 500°C at a linearly programmed rate of 10°C/min, and the catalyst was reduced for 12 h under hydrogen flow and then cooled to 40°C. Then, the flow of hydrogen was switched to argon for 30 min in order to remove the weakly adsorbed hydrogen. Afterwards, the temperature-programmed desorption (TPD) of the samples was performed by increasing the temperature of the samples, with a ramp rate of 40°C/min, to a maximum of 500°C under the argon flow. The temperature-programmed desorption spectrum was used to determine the molybdenum dispersion and its average crystallite size.

Table 2 Textural properties of the catalysts

Sample	BET surface area (m^2/g)	Total pore volume (ml/g)	Average pore diameter (Å)
C_1	157	0.46	127
C_2	136	0.42	129

Results and discussion

Characterization overview

Textural properties of the fresh and purified carbon nanotubes are given in Table 1. The data indicate that in the case of acid-washed support, surface area, total pore volume, and average internal diameter of nanotubes increased significantly which will result in better dispersion of metal as well as enhancement of the catalyst activity. Table 1 also shows the Fe content for the untreated fresh CNTs and acid-treated CNTs (determined by ICP). As shown, the amount of encapsulated metals in the fresh CNTs was 0.6 wt.% which decreased to zero after acid treatment.

Results of surface area measurements for the calcined C_1 and C_2 catalysts are given in Table 2. These results show that the BET surface area of the C_1 is 157 m^2/g. However, the BET surface area for the C_2 catalyst is somewhat lower (136 m^2/g), which indicates some more pore blockage by metal oxide clusters in the catalyst prepared by incipient wetness impregnation method than the catalyst prepared by microemulsion method.

The elemental compositions of the calcined catalysts measured by ICP are given in Table 3. The measured contents of the prepared catalysts are found to be slightly lower compared to the targeted values. The discrepancies may be due to the dilution of Mo because of the hygroscopic nature of the Mo precursor.

Figure 1 presents the TEM image of the purified support. As shown, the treated support is comprised of an interwoven matrix of multiwalled CNTs. TEM images of C_1 catalyst made by the microemulsion preparation route is shown in Figure 2. As shown, the very small particles are dispersed mostly inside the tubes and on the outer surface of the CNT walls. In the case of C_1 catalyst, the size of particles located inside the nanotubes and the size of particles located outside the nanotubes are within the range of 1 to 5 nm and 1 to 7 nm, respectively. The TEM image of C_2 catalyst prepared by impregnation method is shown in Figure 3. This figure shows that the particles are distributed both inside and outside of the nanotube walls.

Table 3 Chemical composition of the catalysts

Catalyst	Targeted composition (wt.%)		Measured composition (wt.%)		Particle diameter (nm) calculated by Debye-Scherer formula
	Mo	K	Mo	K	
C_1	15	8	14.2	7.8	5.1
C_2	15	8	14.4	7.9	7.4

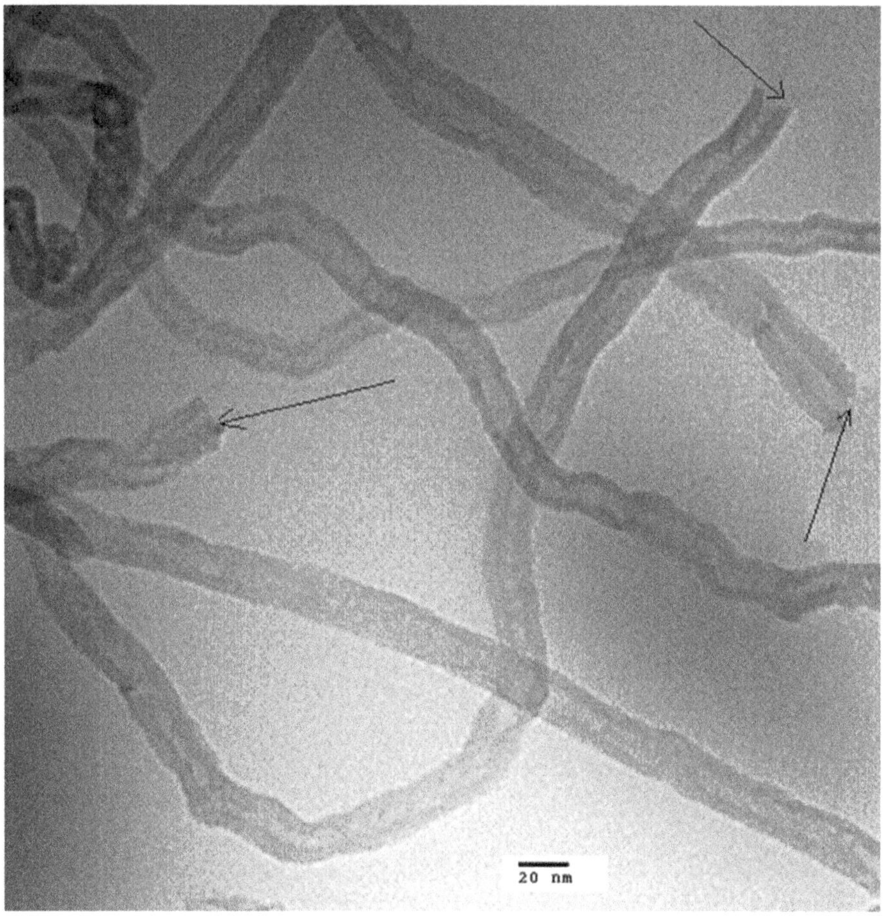

Figure 1 TEM image of purified CNTs. Arrows show open caps.

The size of the particles located inside the nanotubes are in the range of 3 to 9 nm. However, particles located outside the nanotubes have grown to about 16 nm. Indeed, the narrow inner diameter of the CNT channels (8 to 12 nm) restricted the insertion of particles in sizes close to the channel diameter (10 nm).

Figure 4 depicts the size distribution of the particles for C_1 and C_2 catalysts, which is determined using the population of the total particles of the catalyst based on the data taken from five TEM pictures. According to this figure, the average particle size for the C_1 and C_2 catalysts are 4.8 and 7 nm, respectively. It should be noted that using impregnation method to make the catalysts, there is no any factor to control the particle size and metal particles disperse randomly in outer and inner surfaces of the CNTs. However, using microemulsion method, the size of particles is related to the size of water droplets and the water/surfactant ratios. By controlling these parameters, it is possible to direct the size of metal particles.

To determine the crystalline phases, XRD experiments of the calcined catalysts were performed. Figure 5 presents the XRD spectrum of the catalysts with different preparation methods. To detect the phases, we used JCPDS chemical spectra data bank. As shown in this figure, using microemulsion technique decreases the intensity of the peaks corresponding to MoO_3 (peaks at 2θ values of 43.3, 63.2, and 71.9), indicating the smaller particle sizes of MoO_3 in the case of C_1 catalyst. Table 3 shows the average sizes of MoO_3 particles calculated from the XRD patterns, using the Debye-Scherer formula at 2θ value of 43.3. As can be seen, using microemulsion technique decreases the average crystal diameter of MoO_3 from 7.4 to 5.1 nm. Also, the XRD of both catalyst peaks at 2θ values of 25.8, 31.3, 40.4, and 46.3 can correspond to the crystalline structure of K-Mo-O mixed oxide species ($K_2Mo_2O_7$, K_2MoO_4, $K_2Mo_7O_{20}$, KMo_4O_6, and $K_{0.33}MoO_3$) [24-31]. In the case of C_1 catalyst, the corresponding K-Mo-O peaks appeared more significantly. This implies that using microemulsion technique, which significantly decreases the MoO_3 particle sizes, the chemical interactions between K-Mo-O species increase, enhancing the conditions for the formation of alcohols [24-31].

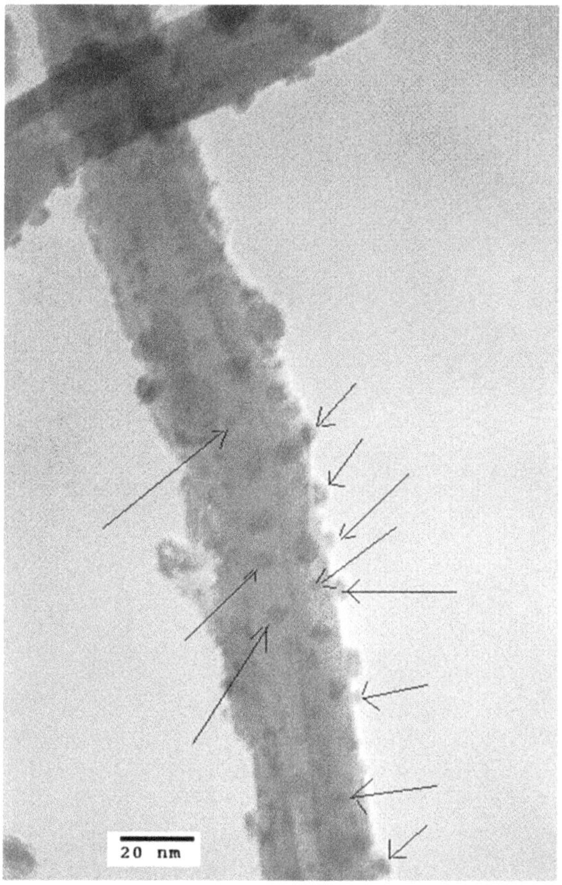

Figure 2 TEM image of the catalyst prepared by microemulsion method. Arrows show metal particles.

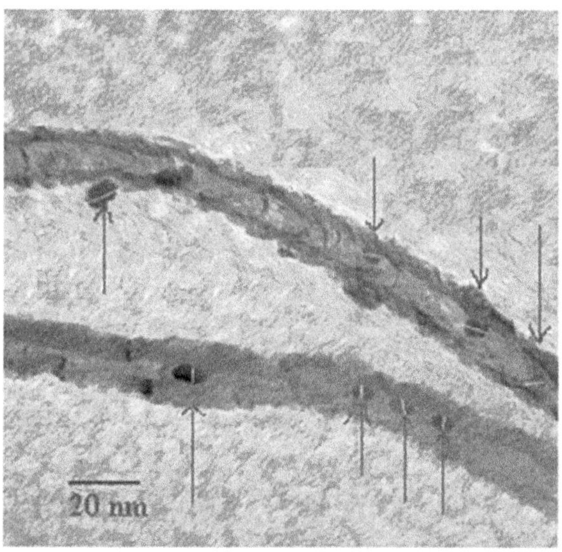

Figure 3 TEM image of the catalyst prepared by impregnation method. Arrows show metal particles.

TPR tests were performed to determine the reducibility of the calcined catalysts. The TPR spectra of the calcined C_1 and C_2 catalysts are shown in Figure 6. In the case of the catalyst prepared by impregnation method (C_2), the low temperature peak (510°C) is typically assigned to the reduction of Mo^{6+} to Mo^{4+}. The second broad peak is generally assigned to the second reduction step, which is mainly the reduction of Mo^{4+} to metallic Mo^0 [24-31]. According to Figure 6, the deposition of Mo nanoparticles synthesized by microemulsion method on the CNTs shift the reduction of the Mo^{6+} to Mo^{4+} (first peak) to a lower temperature compared to the catalyst prepared by impregnation method, indicating higher reducibility of the uniform particles. According to the TEM picture of the catalyst prepared by microemulsion technique, a good number of the small particles are located inside the tubes, which in turn lead to a better interaction with the intern electron-deficient walls of the CNTs and favor the reduction of molybdenum. The tubular morphology of the graphene layers make CNTs different compared to other carbonaceous supports. The deviation of the graphene layers from planarity causes p electron density to shift from the concave inner surface to the convex outer surface, leading to an interior electron-deficient surface and an exterior electron-enriched surface. Thus, the exterior surfaces of the CNTs are electron rich, whereas the interior ones are electron deficient, which could influence metal and metal oxide particles in contact with either surface. It can be concluded that particles located inside the carbon nanotubes are more susceptible for reduction than the particles located on the outside surfaces [24-31].

In addition, it should be noted that using microemulsion method and decreasing the particle sizes promote the formation of Mo-K mixed oxide species (as shown by XRD) that are reducible at lower temperatures. In accordance with the first TPR peak, the second peak temperature for the catalyst prepared by microemulsion method shifted to lower temperatures. The degree of catalyst reduction which is proportional to the areas of the corresponding peaks indicates that using micoemulsion technique resulted in a significant improvement in the reducibility of the catalyst (37% enhancement). In addition to the above-mentioned explanations, this can be due to the significant pore blockage for catalyst prepared by impregnation method.

The results of hydrogen chemisorption tests are given in Table 4. As shown, using microemulsion considerably increased the hydrogen uptake and percentage of dispersion. The particle diameter decreased, which is in agreement with the results of XRD and TEM tests. Higher dispersion and smaller molybdenum cluster sizes in the case of C_1 catalyst will increase the number of sites available for CO conversion and alcohol formation reaction rates.

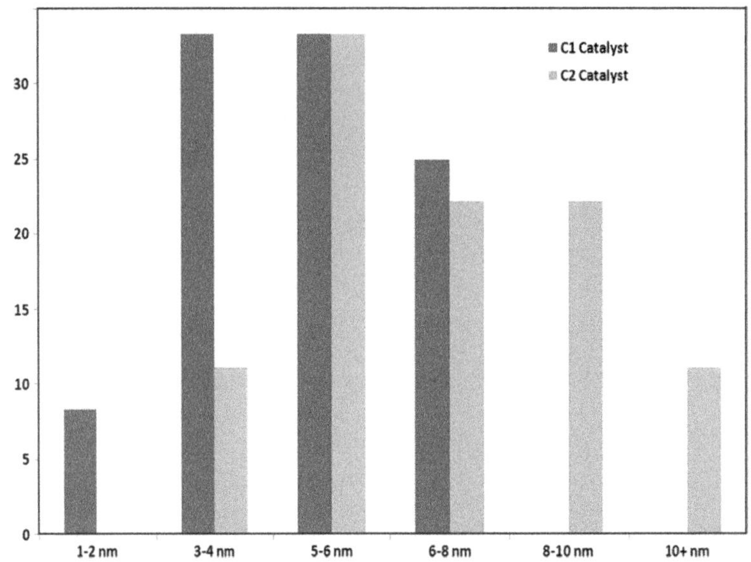

Figure 4 Particle size distribution for C_1 and C_2 catalysts.

Activity and product selectivity results

Synthesis of higher alcohols was carried out after the reduction and sulfidation of catalysts in a fixed bed reactor at 320°C, 70 bars, gas hourly space velocity of 3.6 nLit/g cat./h, and $H_2/CO = 2$. The liquid products were collected at 0°C, and the exit gas was analyzed to measure the CO conversion. The analysis of liquid products indicates that methanol and ethanol are the major products with small amounts of propanol, butanol, and other higher alcohols. The analysis of exit gas indicates that methane is the major component apart from CO, H_2, and CO_2.

Figure 7 presents the percentage of CO and H_2 conversions and different product selectivities during 24 h of continuous synthesis for C_1 and C_2 catalysts. As shown in this figure, for both catalysts, alcohols have the highest product selectivity. It has been suggested that sulfidation of the mixed oxides present in the calcined form of the catalysts and formation of the Mo-K-S species are responsible for the alcohol formation via

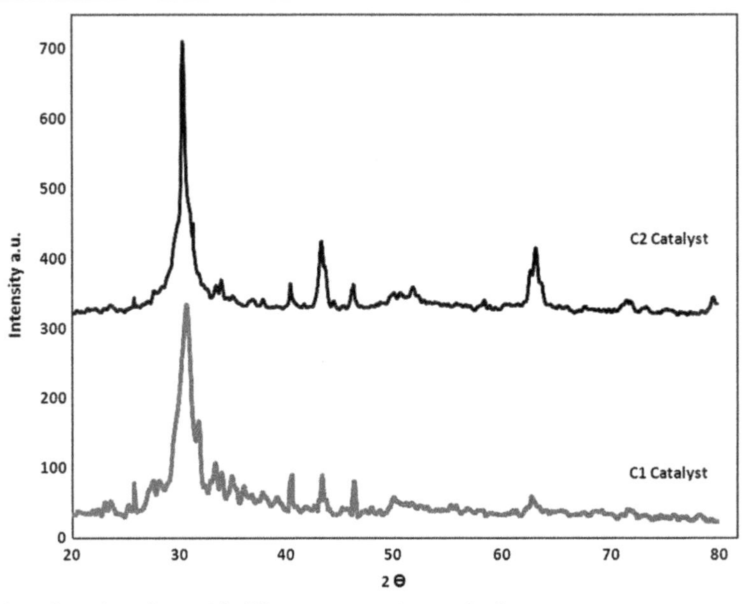

Figure 5 XRD patterns of the calcined catalysts with different preparation methods.

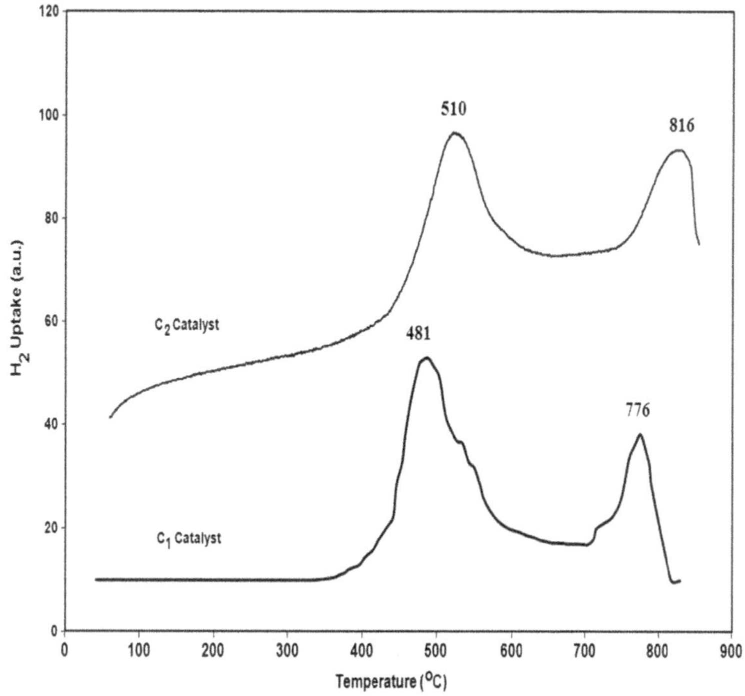

Figure 6 TPR profiles for the catalysts with different preparation methods.

following CO insertion mechanism [28-31].

$$CO + 2H_2 \leftrightarrow CH_3OH \quad (1)$$

$$CH_3OH + H_2 \rightarrow CH_4 + H_2O \quad (2)$$

$$CH_3OH + CO + 2H_2 \rightarrow C_2H_5OH + H_2O \quad (3)$$

$$C_2H_5OH + CO + 2H_2 \rightarrow C_3H_7OH + H_2O \quad (4)$$

$$CO + H_2O \leftrightarrow CO_2 + H_2 \quad (5)$$

In this mechanism, adsorption of CO and hydrogen takes place at different sites to from intermediate surface species. These surface species propagate chain growth through hydrogenation, followed by insertion of molecularly adsorbed CO species to form different long-chain intermediates. The direct hydrogenation of these intermediate hydrocarbon species leads to the formation of hydrocarbon, methanol, and higher alcohols.

Figure 7 shows that using microemulsion technique for the preparation of Mo-K/CNT catalyst increases the CO and H_2 conversions significantly. The percentage of CO conversion and the percentage of H_2 conversion increased from 27.7% to 43.8% (about 58.1% enhancement) and 23.9% to 37.6% (about 57.3% enhancement), respectively. The increase in the percentage of CO and H_2 conversions upon using microemulsion technique can be due to the increase in Mo dispersion (confirmed by XRD, TEM, and H_2 chemisorption tests) and the enhancement in reduction of Mo^{6+} to Mo^{4+} and Mo^{4+} to metallic Mo^0 (confirmed by TPR tests). Increasing dispersion, increases the number of active sites available for reaction, thus increases the activity of the catalyst [24-31].

Interestingly, Figure 7 also shows that using microemulsion technique for the preparation of Mo-K/CNT catalyst increases the selectivity of total alcohols and decreases the selectivity of hydrocarbons significantly. The total alcohol selectivity increased by 35.58%, whereas the selectivity of hydrocarbons decreased by 24.22%. Different causes can explain the significant enhancement in the selectivity of the C_1 catalyst. As shown by XRD tests, using microemulsion technique increased the chemical interactions between K-Mo-O species (the intensity of the peaks corresponding to Mo-K-O mixed oxides increased, i.e., $K_2Mo_2O_7$, K_2MoO_4, $K_2Mo_7O_{20}$, KMo_4O_6, and $K_{0.33}MoO_3$). These mixed oxides will convert to Mo-K-S species like K_2MoS_4 during catalyst reduction and sulfidation processes. The XRD patterns of the reduced and sulfurized C_1 catalyst are shown in Figure 8. As shown in Figure 8, the various diffraction peaks observed in the XRD patterns of oxide samples are completely removed after sulfidation, and the new diffraction

Table 4 H_2 chemisorption results for the catalysts

Catalyst	μmol H_2 desorbed/g cat.	% Dispersion	Particle diameter (nm)
C_1	516	59.25	4.9
C_2	399	51.1	7.1

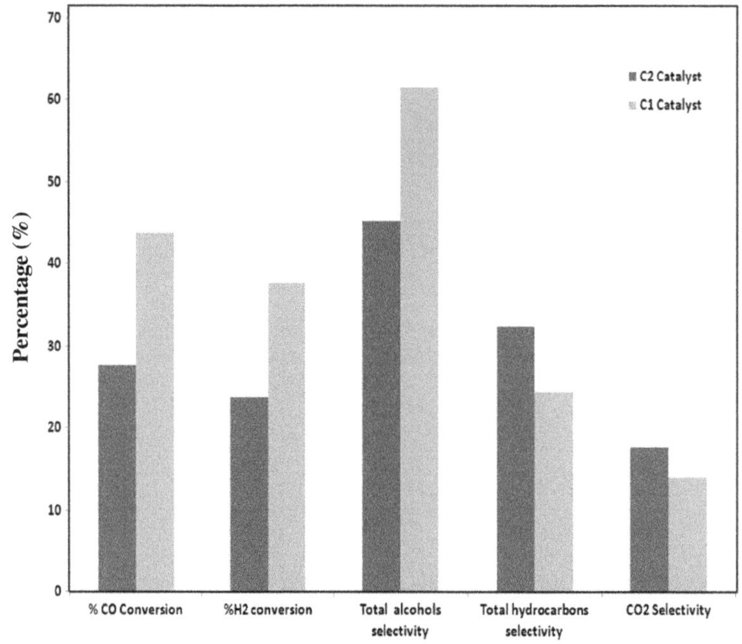

Figure 7 Percentage of CO and H_2 conversions and different product selectivities for C_1 and C_2 catalysts.

peaks with less diffraction intensity representing different sulfide species appeared. This confirmed that newly formed active phases during the sulfidation step are well dispersed on the support. The reflections of MoS_2 crystallites are observed at 2θ values of 17.6°, 38.4°, 44.4°, and 67.9°. The peaks at 2θ values of 15.6°, 28.7° and 61.6° are due to the characteristic reflections of the K-Mo-S species, which represent KMo_3S_3.

Formation of these species can boost the conditions suitable for the formation of alcohols [24]. In addition, the decrease in the selectivity of hydrocarbons can be due to the decreased availability of hydrogenation active centers, i.e. MoS_2 species as also confirmed by XRD results (the intensity of peaks corresponding to MoO_3 decreased in the case of C_1 catalyst).

Figure 9 shows the selectivity trends of individual alcohols for C_1 and C_2 catalysts. As shown, using microemulsion method for the preparation of Mo-K/CNT catalyst dramatically decreases the methanol selectivity and increases ethanol and other higher alcohol selectivities.

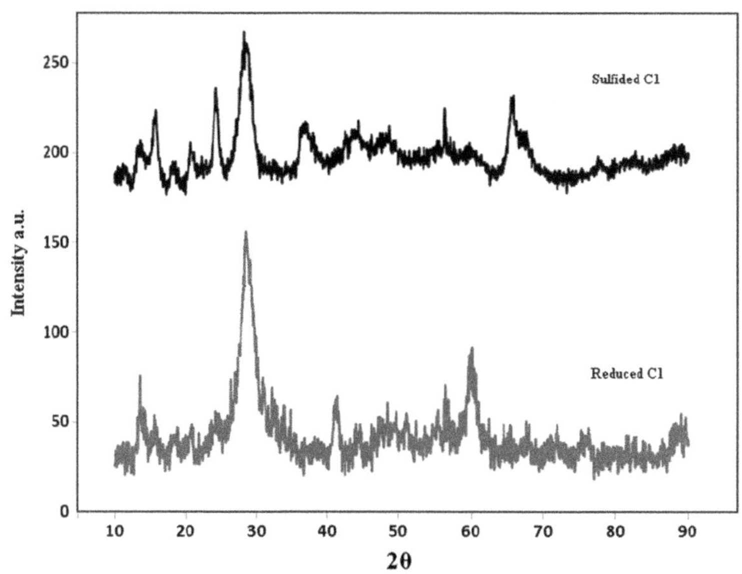

Figure 8 XRD patterns of the reduced and sulfurized C_1 catalyst.

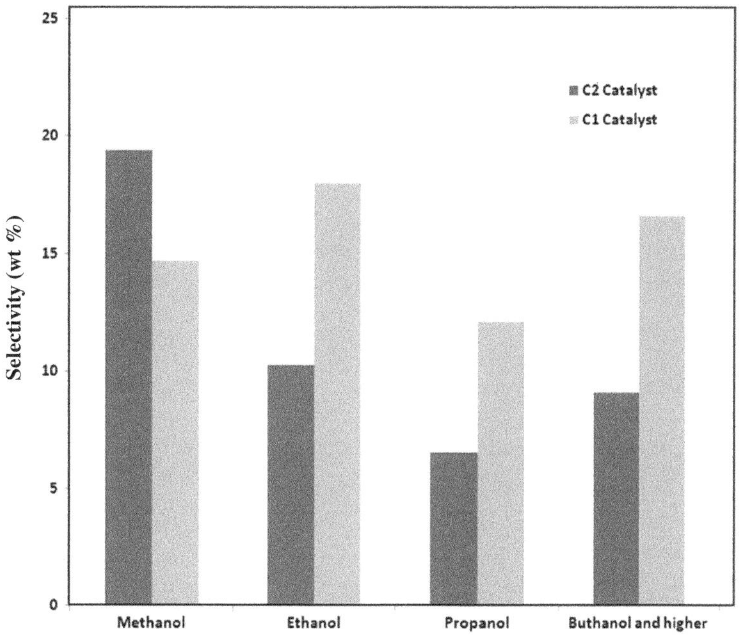

Figure 9 The selectivity trends of individual alcohols for C_1 and C_2 catalysts.

Using microemulsion technique, the methanol selectivity decreased from 19.4% to 14.7%. At the same time, the higher alcohol selectivity (C_2^+OH) increased from 25.9% to 46.7%. In an interesting research, Parmon [32] showed that a decrease in the particle size of the catalyst active phase increases the standard chemical potentials of the components and affects appreciably the adsorption equilibrium. In the case of C_1 catalyst prepared by microemulsion method, the decrease in Mo particle sizes increases the adsorption of CO and decreases the adsorption of H_2. Decreasing the amount of H_2 on the catalyst surface can suppress the formation of methane via reaction 2. Also, increasing the partial pressure of CO on the catalyst surface will express the reactions 3 and 4 which in turn increase the selectivity of ethanol and the selectivity of higher alcohols. Also, the reason may be explained from the XRD and TPD characterizations. It is clear from XRD and TPD that the C_2 catalyst showed larger MoO_3 particle size. MoO_3 will change to MoS_2 during catalyst reduction and calcination, which is responsible for poor activity of this catalyst towards the formation of higher alcohols.

Experimental

Reaction setup and experimental procedure

Higher alcohol synthesis has been performed in a tubular downflow, fixed-bed reactor system. The reactor was made up of Inconel tube of 450-mm length and 22-mm inside diameter. The reactor temperature was controlled via a PID temperature controller. Brooks 5850 mass flow controllers (Hatfield, PA, USA) were used to add H_2 and CO at the desired rate to the reactor. The reactor was packed with 0.5 g of catalyst diluted with 90-mesh-size silicon carbide and housed in a melton salt bath controlled by a temperature controller. The reactor was pressurized to 30 bars with He, and the reduction as well as the sulfurization was carried out for 6 h at 450°C at a heating rate of 2°C/min using a gas mixture containing 5 mol% H_2S in H_2 and a flow rate of 50 ml/min. The temperature was then lowered to the reaction temperature, and the system was then pressurized to the reaction condition. The feed gas mixture (H_2/CO ratio of 2) was passed through mass flow controllers, and the HAS reaction was carried out at steady state under the reaction conditions of 320°C, 70 bars, and gas hourly space velocity of 3.6 nLit/g cat./h over a period of 24 h. The product gas was cooled to 0°C and separated into gas and liquid phases at the reaction pressure. The CO, CO_2, and other gaseous products were monitored with time intervals of 1 h. The products were collected after completion of the run and analyzed by means of three gas chromatographs, a Shimadzu 4C gas chromatograph (Nakagyo-ku, Kyoto, Japan) equipped with two subsequent connected packed columns: Porapak Q and Molecular Sieve 5 Å, and a TCD with argon which was used as a carrier gas for hydrogen analysis. A Varian CP 3800 with a chromosorb column and a TCD were used for CO, CO_2, CH_4, and other non-condensable gases. A Varian CP 3800 with a Petrocol Tm DH100 fused silica capillary column and a flame ionization detector were used for liquid products so that a complete product distribution could be provided.

Conclusions

Carbon nanotube-supported alkalized MoS_2 nanocatalysts prepared by microemulsion and impregnation methods are used to produce higher alcohols from synthesis gas. Using microemulsion technique decreased the average MoO_3 particle sizes from 7 to 4.8 nm, increased the percentage of dispersion from 51 to 59, and increased the percentage of reduction by 37%. In addition, the chemical interactions between K-Mo-O species, enhancing the conditions for the formation of alcohols, are increased. Selectivity and activity were found to be dependent on the catalyst preparation method. Using microemulsion technique increased the percentage of CO conversion from 27.7% to 43.8%. The CNT-supported MoO_3-K_2O nanocatalyst synthesized by microemulsion technique decreased the hydrocarbon selectivity from 32.4% to 24.5% and increased the alcohol selectivity from 45.3% to 61.4%. Finally, using microemulsion method for the preparation of Mo-K/CNT nanocatalyst severely decreased the methanol selectivity and increased ethanol and other higher alcohol selectivities.

Competing interests

The authors declare that they have no competing interests.

Authors' contributions

AT was responsible for the project planning and wrote the manuscript. SK carried out the experiment and collected the data. MS built the setup instrument. All authors discussed the results and read and approved the final manuscript.

Acknowledgments

This work was supported by Iran National Science Foundation (INSF). The authors would like to thank INSF for their support.

References

1. Herman R (1990) Studies in surface and catalyst, Chapter 7. Elsevier, Amsterdam
2. Natta G, Colombo U, Pasquon I (131) In: Emmet PH (ed) Catalysis, vol 5. Reinhold, New York, p 131
3. Morgan GT, Hardy DVN, Procter RA (1932) J Soc Chem Ind Trans 1T:56
4. Smith, K. J.; Anderson, R. B. The Higher Alcohol Synthesis Over Promoted Cu/ZnO Catalysts. Can. J. Chem. Eng. 1983, 61, 40.
5. Smith K. J. and Anderson R. B., "A Chain Growth Scheme for the Higher Alcohols Synthesis," Journal of Catalysis, 1984, 85, 428–436.
6. Quarderer QJ, Cochram GA (1984) Catalytic process for producing mixed alcohols from hydrogen and carbon monoxide. PCT Int. Pat. Publication No. WO84/03696
7. Kinkade NE (1985) Process for producing alcohols from carbon monoxide and hydrogen using an alkali-molybdenum sulfide catalyst. PCT Int. Pat. Publication No. WO 85/03073
8. Jackson GR, Mahajan D (2001) Method for production of mixed alcohols from synthesis gas. U.S. Patent 6,248,796
9. Chaumette P, Courty P, Durand D, Grandvallet P, Travers C (1985) Process for synthesizing a mixture of primary alcohols from a synthesis gas in the presence of a catalyst containing copper, cobalt, zinc, and aluminum. GB Patent, 2,158,730
10. Woo, H. C. Park, K Y Kim, Y G Nam, I.-S.; Chung, J. S.; Lee, J. S. Mixed alcohol synthesis from carbon monoxide and dihydrogen over potassium-promoted molybdenum carbide catalysts. Appl. Catal. 1991, 75, 267-280.
11. Jiang M., Bian G.-Z., and Fu, Y.-L., "Effect of the K-Mo interaction in K-MoO3/γ-Al2O3 catalysts on the properties for alcohol synthesis from syngas," Journal of Catalysis, 1994, 146, 144–154.
12. Lee JS, Kim S, Lee KH, Nam IS, Chung JS, Kim YG, Woo HC. Role of alkali promoters in K/MoS2 catalysts for CO-H2 reactions, Volume 110, Issue 1, 1994, Pages 11–25
13. Tatsumi T., Muramatsu A., Yokota K. and Tominga H., "Mechanistic study on the alcohol synthesis over molybdenum catalysts: Addition of probe molecules to CO—H2", Jounal of Catalysis, 1989, 115, 388-398.
14. Surisetty V. R., Tavasoli A. and Dalai A. K., "Synthesis of higher alcohols from syngas over alkali promoted MoS2 catalysts supported on multi-walled carbon nanotubes," Applied Catalysis A: General, 2009, 365, 243-251.
15. Iranmahboob, J., Toghiani, H., Hill, D.O., 2003. Dispersion of alkali on the surface of Co-MoS2/clay catalyst: a comparison of K and Cs as a promoter for synthesis of alcohol. Applied Catalysis A: General, 247(2):207-218.
16. Iranmahboob J,∗. Hill DO. Alcohol synthesis from syngas over K2CO3/CoS/MoS2 on activated carbon. Catalysis Letters. 78. 1–4(2002)
17. MartŁnez A, Prieto G Breaking the dispersion-reducibility dependence in oxide-supported cobalt nanoparticles. Journal of Catalysis, 245 (2007) 245-470.
18. Kim WY, Haoka T, Kishida M, Wakabayashi K (1997) Appl Catal A 155:283
19. T. Tago; Hanaoka, T; Dhupatemiya, P; Hayashi, H; Kishida, M; Wakabayashi, K, "Effects of Rh content on catalytic behavior in CO hydrogenation with Rh-silica catalysts prepared using microemulsion", CATAL LETT, 64(1), 2000, pp. 27-31
20. Eriksson S, Nylén U, Rojas S, Boutonnet M.∗ Preparation of catalysts from microemulsions and their applications in heterogeneous catalysis, Applied Catalysis A: General 265 (2004) 207–219
21. Hayashi H, Zhe Chen L, Tago T, Kishiba M, Wakabayashi K, Catalytic properties of Fe/SiO2 catalysts prepared using microemulsion for CO hydrogenation. Applied Catalysis A: General Volume 231, Issues 1–2, 2002, Pages 81–89
22. Saib, A.M.; Claeys, M.; van Steen, E. Silica supported cobalt Fischer–Tropsch catalysts: effect of pore diameter of support. Catalysis Today vol. 71 issue 3-4, 2002. p. 395-402
23. Bezemer, G.L.; Radstake, P.B.; Koot, V.; van Dillen, A.J.; Geus, J.W.; de Jong, K.P. Preparation of Fischer-Tropsch cobalt catalysts supported on carbon nanofibers and silica using homogeneous deposition-precipitation Journal of Catalysis, Volume 237, Issue 2, 2006, Pages 291-302
24. Tavasoli A, Sadagiani K, Khorashe F, Seifkordi AA, Rohani AA, Nakhaeipour A, Cobalt Supported on Carbon nanotubes a promissing Novel Fischer-Tropsch synthesis Catalyst, Fuel Process, 2006, 89 (5), 491-498
25. ZHANG X∗, ZHANG F, Synthesis of Pt–Ru–Mo ternary metal nanoparticles by microemulsions, their characterization and electrocatalytic properties, J OF Material Science 39 (2004) 5845–5848
26. Eriksson S, Nylén U, Rojas S, Boutonnet M.∗ Preparation of catalysts from microemulsions and their applications in heterogeneous catalysis, Applied Catalysis A: General 265 (2004) 207–219
27. Tavasoli A, Karimi S, Nikookar H, Molybdenum Loading Effects on the Physico-chemical Properties and Performance of Carbon Nanotubes Supported Alkalized MoS2 Catalysts for Higher Alcohols Synthesis, (2012) Iranian J of Chem & Chem Enj 31:9
28. Tavasoli A, Karimi S, Zolfaghari Z, Taghavi S, Amirfirouzkouhi H, M.A. Babatabar. Cobalt loading effects on the physico-chemical properties and performance of Co promoted alkalized MoS2/CNTs catalysts for higher alcohols synthesis
29. Surisetty V. R., Dalai, A. K. and Kozinski, J., "Intrinsic reaction kinetics of higher alcohols synthesis from synthesis gas over sulfided alkali-promoted Co-Rh-Mo trimetallic catalyst supported on MWCNTs," Energy & Fuels, 2010a, 24, 4130–4137.
30. Surisetty, V. R., Dalai, A. K., and Kozinski J., "Influence of porous characteristics of the carbon support on alkali-modified trimetallic Co–Rh–Mo sulfided catalysts for higher alcohols synthesis from synthesis gas," Applied Catalysis A: General, 2011b, 393, 50-58.

31. Surisetty V. R., Dalai A. K. and Kozinski J., "Effect of Rh promoter on MWCNT-supported alkali-modified MoS2 catalysts for higher alcohols synthesis from CO hydrogenation," Applied Catalysis A: General, 2010c, 381, 282-288.
32. Parmon VN, Thermodynamic Analysis of the Effect of the Nanoparticle Size of the Active Component on the Adsorption Equilibrium and the Rate of Heterogeneous Catalytic Processes. Physical Chemistry, 2007, 413, Part 1, pp. 42–48.

Equilibrium and kinetic studies of cadmium(II) and lead(II) ions biosorption onto *Ficus carcia* leaves

Ahlam M Farhan[1], Ammar H Al-Dujaili[2] and Akl M Awwad[3*]

Abstract

Background: *Ficus carcia* leaves (fig leaves) for the removal of cadmium(II) and lead(II) ions from aqueous solutions have been investigated.

Results: The biosorption of cadmium(II) and lead(II) ions was found to be dependent on the solution pH, initial metal ion concentrations, biosorbent dose, contact time, and temperature. The experimental equilibrium biosorption data were analyzed by two widely used two parameters, Langmuir and Freundlich isotherm models.

Conclusions: Langmuir isotherm model provided a better fit with the experimental data than Freundlich model by high correlation coefficients R^2. Kinetic studies showed that pseudo-second-order described the biosorption experimental data better than the pseudo-first-order kinetic model.

Keywords: Biosorption, *Ficus carcia* leaves, Cadmium(II), Lead(II), Equilibrium isotherms, Kinetic

Background

Cadmium(II) and lead(II) ions have been released to the environment through phosphate fertilizers, electrical wiring, manufacturing batteries, air conditioning tubing, plumbing, electroplating, and pigments. There is a growing demand to find relatively efficient, low-cost, and easily available adsorbents for the adsorption of toxic heavy metals. Different agrowastes have been investigated for the biosorption of Cd(II) and Pb(II) ions such as *Caladium bicolor* [1], neem leaf [2]; heartwood powder of *Areca catechu* [3]; *Ulmus carpinifolia* and *Fraxinus excelsior* tree leaves [4]; green coconut shell powder [5]; mango peel waste [6]; banana, lemon, and orange peel [7]; chemically modified *Moringa oleifera* tree leaves [8]; maize leaf [9]; *Scolymus hispanicus* [10]; processed walnut shell [11]; sugar cane residue or bagasse [12]; pomegranate peel [13], walnut, hazelnut, almond, pistachio, and apricot stone [14]; *Iagenaria vulgaris* shell [15]; grape stalk waste [16]; rice husk [17]; paper mulberry [18]; *Eucalyptus camaldulenis* tree leaves [19]; *Saraca indica* leaf powder [20]; and olive tree pruning waste [21].

* Correspondence: akl.awwad@yahoo.com
[3]Royal Scientific Society, El Hassan Science City, Amman, Jordan
Full list of author information is available at the end of the article

In the present work, we have studied the potential of cadmium(II) and lead(II) biosorption on an agromaterial which *Ficus carcia* leaves (FL) coming from fig fruit tree waste. Results from this study can be used to assess the utility of FL for cadmium(II) and lead(II) removal from aqueous solution.

Methods

Preparation of biosorbent

The raw FL were collected from a local plantation in Jordan. This agricultural waste was thoroughly rinsed with water to remove dust and soluble material. Then it was allowed to dry at room temperature. The dried leaves were grounded to a fine powder in a grinding mill (Retsch RM 100, Thermo Fisher Scientific, NH, USA) and sieved to get size fraction <44 μm and then dried in an oven at 60°C for 4 h.

Preparation of Cd(II) and Pb(II) stock solutions

All the chemicals used were of analytical reagent (AR) grade purchased from Merck, Germany. Stock solutions of 1,000 mg/L of cadmium(II) and lead(II) were prepared from cadmium nitrate [$Cd(NO_3)_2 \cdot 4H_2O$] and lead nitrate [$Pb(NO_3)_2$], respectively, using double distilled water. Desired test solutions of Cd(II) and Pb(II) ions were prepared using appropriate subsequent dilutions of the stock solution. The range of concentrations of Cd(II) and Pb(II) ions prepared from standard solution varies between 10 and 100 mg/L. Before mixing the adsorbent, the pH of each test solution was adjusted to the required value with 0.1 M NaOH or 0.1 M HCl.

Analysis

The concentrations of Cd(II) and Pb(II) ions in the solutions before and after equilibrium were determined by AAS6300 atomic absorption spectrometer (Shimadzu, Japan). The pH of the solution was measured with a WTW pH meter using a combined glass electrode. Fourier transform infrared spectroscopy (FTIR) IR Prestige-21 (Shimadzu, Tokyo, Japan) was used to identify the different chemical functional groups present in the FL. FTIR analyses were also used to determine the functional groups which are responsible for the metal binding with FL. The analysis was carried out using KBr, and the spectral range varies from 4,000 to 400 cm^{-1}.

Biosorption experiments

Batch biosorption experiments were conducted by mixing biosorbent with Cd(II) or Pb(II) ion solutions with desired concentration in 250 mL glass flask. The glass flasks were stoppered during the equilibration period and placed on a temperature-controlled shaker at a speed 120 rpm. The amount of biosorption was calculated based on the difference between the initial (C_o, mg/L) and final concentration (C_e, mg/L) in every flask, as follows:

$$q_e = \frac{C_o - C_e}{M} V \quad (1)$$

where q_e is the metal uptake capacity (mg/g), V the volume of the metal solution in the flask (L) and M is the dry mass of biosorbent (g).

Results and discussion

FTIR analysis

To investigate the functional groups of FL and metal loaded FL, a FTIR study was carried out and the spectra are shown in Figure 1 (a and b). The FL leaves display a number of absorption peaks, reflecting the complex nature of the FL. A strong and broad peak at 3,348 cm^{-1} results due to the stretching of the N-H bond of amino groups, which is indicative of bonded hydroxyl group. A change in peak position to 3,336 cm^{-1} in the spectrum of the metal-loaded FL indicates the binding of cadmium or lead with amino and hydroxyl groups [22-24]. The strong absorption peak at 2,927 cm^{-1} could be assigned to -CH stretching vibrations of $-CH_3$ and $-CH_2$ functional groups. The peak at 1,570 cm^{-1} indicates the fingerprint region of CO, C-O and O-H groups, which exists as functional groups of FL. Shifting of this peak to 1,666 cm^{-1} indicated involvement of these groups in metal binding. The regions between 1,481 and 1,000 cm^{-1} are the fingerprint region, OH and C-H bending vibration, and C-O stretching vibration absorption bands. The absorption peaks at 1,384 and 1,119 cm^{-1} could be attributed to the presence of C-O stretching. These peaks were shifted to 1,477 and 1,014 cm^{-1}, respectively. The intense band at 1,014 cm^{-1} can be assigned to the C-O of alcohols and carboxylic acids. The shift of the peak from 1,084 to 1,014 cm^{-1} also suggests the involvement of the C-O group in binding Cd(II) [22-24].The shifts in the absorption peaks generally observed indicate the existence of a metal binding process taking place on the surface of the *F. carcia* leaves.

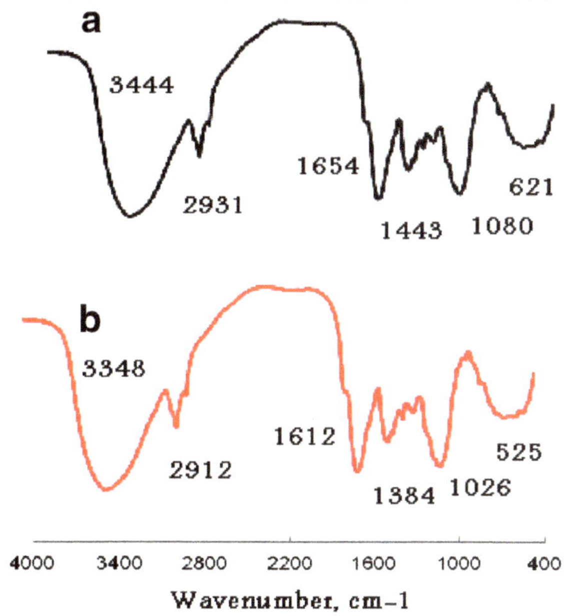

Figure 1 FTIR of (a) FL- and (b) Cd(II)-loaded FL.

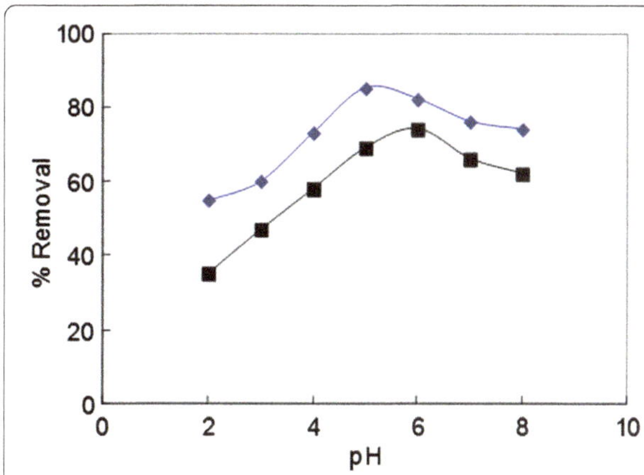

Figure 2 Effect of pH on biosorption of Pb(II) Cd(II) ions. Effect of pH on biosorption of Pb(II) (dark blue diamond) and Cd(II) (black square) ions at 303 K.

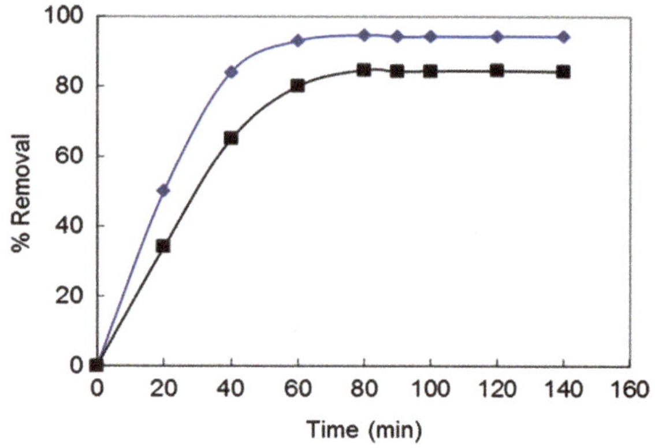

Figure 3 Effect of contact time. Effect of contact time on Pb(II) (dark blue diamond) and Cd(II) ions biosorption (black square) onto FL at 303 K.

Effect of pH

The effect of pH on the biosorption of Cd(II) and Pb(II) ions onto FL was studied at pH 2 to 8 conditions. The maximum biosorption was observed at pH 5.0 for Pb(II) ions and pH 6.0 for Cd(II) ions, as shown in Figure 2. At lower pH values Cd(II) and Pb(II) removal was inhibited, possibly as a result of the competition between hydrogen and metal ions on the sorption sites, with an apparent preponderance of hydrogen ions, which restricts the approach of metal cations as a consequence of the repulsive force. As the pH increased, the active sites in FL would be exposed, increasing the negative charge density on the FL surface, increasing the attraction of metal ions with positive charge, and allowing the biosorption onto the FL surface. In this study, the Cd(II) and Pb(II) ions at pH 5.0 to 6.0 would be expected to interact more strongly with the negatively charged binding sites in the biosorbent, as shown in Figure 2. The decrease in Cd(II) and Pb(II) biosorption above pH 5.0 to 6.0 was probably due to the precipitation of Cd(II) and Pb(II) ions as cadmium and lead hydroxides, but not due to biosorption. As a result, the optimum pH for cadmium and lead ions biosorption was found as 5.0 and 6.0, respectively; and the other biosorption experiments were performed at this pH value.

Effect of contact time

The rate of biosorption is important for designing batch biosorption experiments. Therefore, the effect of contact time on the biosorption of metals was investigated. The biosorption of Cd(II) and Pb(II) ions increased considerably until the contact time reached 80 min at 30°C. Further increase in contact time did not enhance the biosorption process; so, the optimum contact time was selected as 80 min for further experiments (Figure 3).

Effect of the adsorbent concentration

The adsorbent concentration effect on the percentage removal at equilibrium conditions was investigated. It was observed that the amount of Cd(II) and Pb(II) biosorbed varied with varying adsorbent concentration. The amount of Cd(II) and Pb(II) adsorbed increases with an increase in adsorbent concentration from 0.1 to 0.5 g/0.1L. The percentage of metal removal was increased from 58.44% to 98.58% for an increase in FL concentration from 0.1 to 0.5 g/0.1L at initial concentration of 40 mg/L. The increase in the adsorption of the amount of solute is obvious due to the increasing FL surface area.

Biosorption isotherms

An adsorption isotherm describes the fraction of sorbate molecules that are partitioned between liquid and solid phases in equilibrium. The adsorption of Cd(II) and Pb (II) ions onto FL particles was modeled using two adsorption isotherms: Freundlich and Langmuir isotherms.

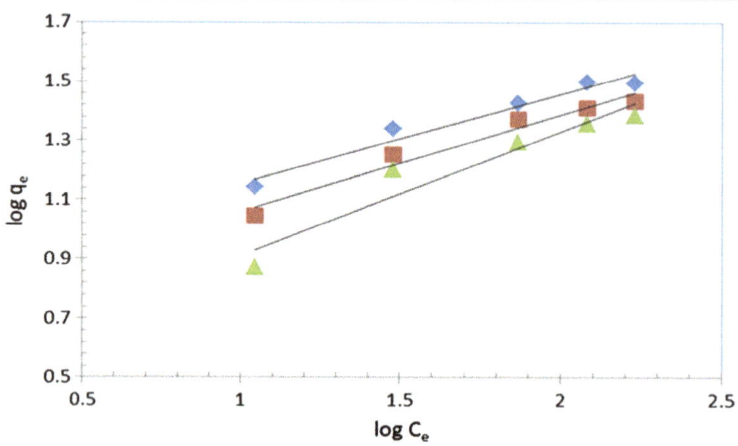

Figure 4 Freundlich adsorption isotherm of Cd(II) ions onto FL. 293 K (blue diamond), 303 K (red square), and 313 K (green triangle).

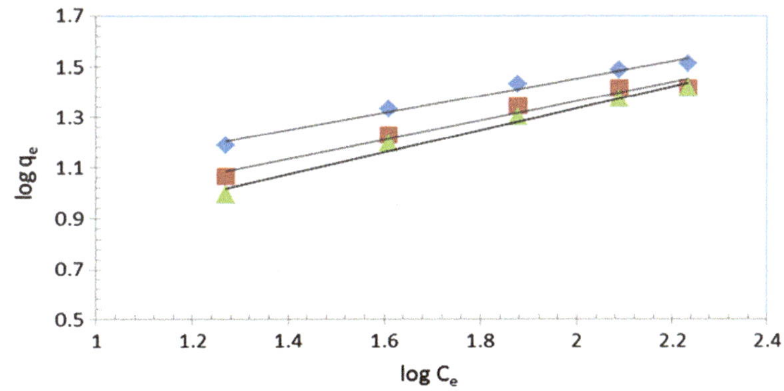

Figure 5 Freundlich adsorption isotherm of Pb(II) ions onto FL. 293 K (blue diamond), 303 K (red square), and 313 K (green triangle).

Freundlich isotherm

The Freundlich isotherm model is the well-known earliest relationship describing the adsorption process. This model applies to adsorption on heterogeneous surfaces with the interaction between adsorbed molecules, and the application of the Freundlich equation also suggests that sorption energy exponentially decreases on completion of the sorption centers of an adsorbent. This isotherm is an empirical equation and can be employed to describe heterogeneous systems and is expressed as follows in linear form [25]:

$$\log q_e = \log K_F + \frac{1}{n} \log C_e, \tag{2}$$

where K_F (mg/g)(L/mg)$^{1/n}$ is the Freundlich constant related to the bonding energy; $1/n$ is the heterogeneity factor and n (L/g) is a measure of the deviation from linearity of adsorption. Freundlich equilibrium constants were determined from the plot of log q_e versus log C_e, Figures 4 and 5, on the basis of the linear of Freundlich equation (Equation 1). The n value indicates the degree of nonlinearity between solution concentration and adsorption as follows: if $n = 1$, then adsorption is linear; if $n < 1$, then adsorption is a chemical process; if $n > 1$, then adsorption is a physical process. The n value in Freundlich equation was found to be 2.39 to 3.96 for FL, Table 1. Since n lies between 1 and 10, this indicates the physical biosorption of cadmium(II) and lead(II) onto FL.

Table 1 Langmuir and Freundlich constants for biosorption of Cd(II) onto FL at different temperatures

T(°C)	Langmuir			Freundlich		
	q_{max}	K_L	R^2	n	K_F	R^2
20	34.13	0.057	0.9994	3.31	7.15	0.9575
30	30.30	0.049	0.9998	3.06	5.43	0.9639
40	28.98	0.030	0.9999	2.39	3.11	0.9234

Langmuir isotherm

The Langmuir isotherm assumes monolayer adsorption on a uniform surface with a finite number of adsorption sites [26]. Once a site is filled, no further sorption can take place at that site. As such, the surface will eventually reach a saturation point where the maximum adsorption of the surface will be achieved. The linear form of the Langmuir isotherm model is described as

$$\frac{C_e}{q_e} = \frac{1}{K_L q_{\max}} + \frac{1}{q_{\max}} Ce, \tag{3}$$

where K_L is the Langmuir constant related to the energy of adsorption and q_{max} is the maximum adsorption capacity (mg/g). Values of Langmuir parameters q_{max} and K_L were calculated from the slope and intercept of the linear plot of C_e/q_e versus C_e as shown in Figures 6 and 7. Values of q_{max}, K_L, and regression coefficient R^2 are listed in Tables 1 and 2. These values for FL biosorbent indicated that Langmuir theory describes the biosorption phenomena as favorable.

The essential characteristics of the Langmuir isotherm parameters can be used to predict the affinity between the sorbate and sorbent using separation factor

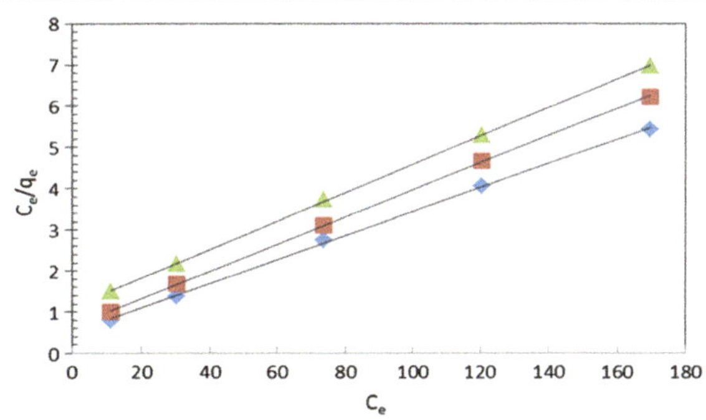

Figure 6 Langmuir adsorption isotherms for Cd(II) ions onto FL. 293 K (blue diamond), 303 K (red square), and 313 K (green triangle).

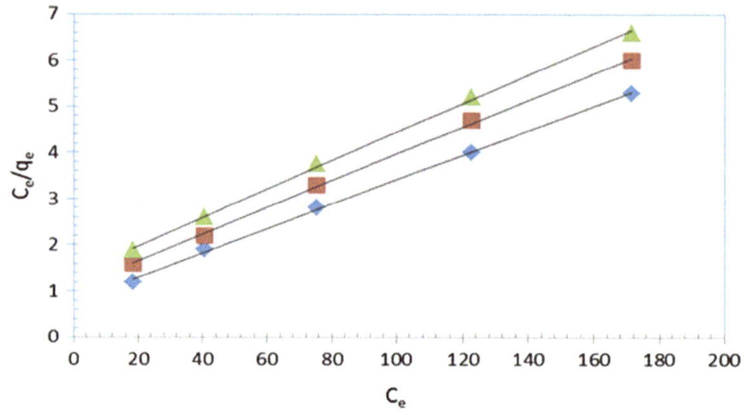

Figure 7 Langmuir adsorption isotherms for Pb(II) ions onto FL. 293 K (blue diamond), 303 K (red square), and 313 K (green triangle).

or dimensionless equilibrium parameter, R_L, expressed as in the following equation:

$$R_L = \frac{1}{(1 + K_L C_o)} \tag{4}$$

where K_L is the Langmuir constant and C_o is the initial concentration of metal ions. The value of separation parameter R_L provides important information about the nature of adsorption. The value of R_L indicated the type of Langmuir isotherm to be irreversible (R_L = 0), favorable (0 < R_L < 1), linear (R_L = 1), or unfavorable (R_L > 1). The R_L was found to be 0.21 to 0.67 for concentrations of 10 to 100 mg/L of Cd(II) and Pb(II). They are in the range of 0 to 1 which indicates the favorable biosorption.

Adsorption kinetic

Parameters from two kinetic models, pseudo-first-order and pseudo-second-order, were fit to experimental data to examine the adsorption kinetics of cadmium(II) and Pb(II) uptake by FL.

Pseudo-first-order kinetics

The pseudo-first-order equation of Lagergren [27] is generally expressed as follows:

$$\frac{dq_t}{dt} = k_1(q_e - q_t), \tag{5}$$

Table 2 Langmuir and Freundlich constants for biosorption of Pb(II) onto FL at different temperatures

T(°C)	Langmuir			Freundlich		
	q_{max}	K_L	R^2	n	K_F	R^2
20	37.74	0.034	0.9992	3.96	5.44	0.9825
30	34.36	0.027	0.9991	2.64	3.82	0.9648
40	32.36	0.022	0.9990	2.41	3.02	0.9678

where q_e and q_t are the sorption capacities at equilibrium and at time t, respectively (mg/g), and k_1 is the rate constant of pseudo-first-order sorption (1/min). After integration and applying boundary conditions, q_t = 0 to q_t = q_t at t = 0 to t = t, the integrated form of Equation 5 becomes

$$\log(q_e - q_t) = \log q_e - \frac{k_1}{2.303}t. \tag{6}$$

The equation applicable to experimental results generally differs from a true first-order equation in two ways: the parameter $k_1(q_e - q_t)$ does not represent the number of available sites, and the parameter log q_e is an adjustable parameter which is often not found equal to the intercept of a plot of log ($q_e - q_t$) against t, whereas in a true first-order sorption reaction, log q_e should be equal to the intercept of log($q_e - q_t$) against t. In order to fit Equation 5 to the experimental data, the equilibrium sorption capacity q_e must be known. In many cases, q_e is unknown and as chemisorption tends to become unmeasurably slow, the amount sorbed is still significantly smaller than the equilibrium amount. In most cases in the literature, the pseudo-first-order equation of Lagergren does not fit well for the whole range of contact time and is generally applicable over the initial 20 to 60 min of the sorption process. Furthermore, one has to find some means of extrapolating the experimental data to $t = \infty$, on treating q_e as an adjustable parameter to be determined by trial and error. For this reason, it is therefore necessary to use trial and error to obtain the equilibrium sorption capacity, in order to analyze the pseudo-first-order model kinetics.

The pseudo-first-order rate constant can be obtained from the slope of plot between log ($q_e - q_t$) against time, t. Figure 8 shows the Lagergren pseudo-first-order kinetic plot for the biosorption of Cd(II) and Pb(II) ions onto FL. The pseudo-first-order rate constant values were

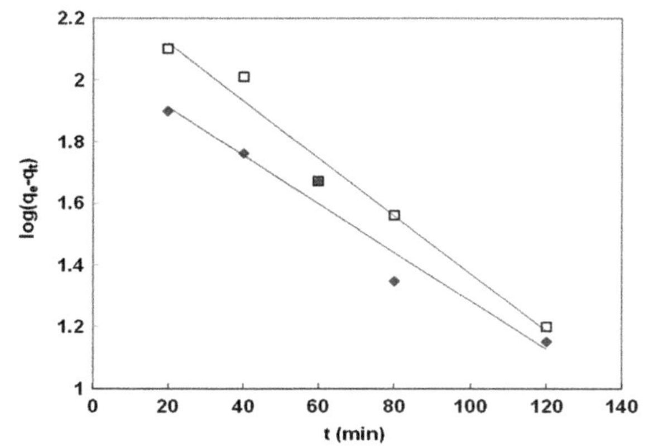

Figure 8 Pseudo-first-order kinetic model for Cd(II) and Pb(II) biosorption. Pseudo-first-order kinetic model for Cd(II) (open square) and Pb(II) (dark blue diamond) biosorption onto FL at 303 K.

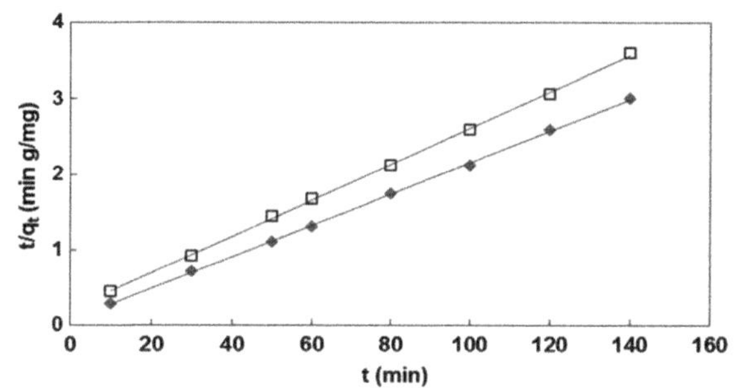

Figure 9 Pseudo-second-order kinetic model for Cd(II) and Pb(II) biosorption. Pseudo-second-order kinetic model for Cd(II) (open square) and Pb(II) (blue diamond) biosorption onto FL at 303 K.

calculated from the slope of Figure 8. The calculated values and their corresponding linear regression correlation coefficient values are shown in Table 3. The linear regression correlation coefficient values show that this model cannot be applied to predict the adsorption kinetic model.

Pseudo-second-order kinetics

The pseudo-second-order rate expression, which has been applied for analyzing chemisorption kinetics rate, is expressed as

$$\frac{dq_t}{dt} = k_2(q_e - q_t)^2, \tag{7}$$

where q_e and q_t are the sorption capacity at equilibrium and at time t, respectively (mg/g); and k_2 is the rate constant of pseudo-second-order sorption, (g/mg/min). For the boundary conditions, $q_t = 0$ to $q_t = q_t$ at $t = 0$ to $t = t$, the integrated form of Equation 7 becomes

$$\frac{1}{(q_e - q_t)} = \frac{1}{q_e}^e + k_2 t. \tag{8}$$

Equation 8 can be rearranged to obtain

$$q_t = \frac{t}{\frac{1}{k_2 q_e^2} + \frac{t}{q_e}}. \tag{9}$$

Table 3 Biosorption kinetic model parameters for Cd(II) and Pb(II) by FL

T(°C)	Pseudo-first order			Pseudo-second order		
	k_1	q_e	R^2	k_2	q_e	R^2
Cd(II)	0.018	117.19	0.9766	0.11	41.67	0.9997
Pb(II)	0.018	201.51	0.9627	0.021	47.85	0.9995

The linear form of Equation 9

$$\frac{t}{q_e} = \frac{1}{k_2 q_e^2} + \frac{1}{q_e} t, \tag{10}$$

where t is the contact time (min), q_e (mg/g) and q_t (mg/g) are the amount of the solute adsorbed at equilibrium and at any time, t. If pseudo-second-order kinetics is applicable, the plot of t/q_t versus t of Equation 9 should give a linear relationship, from which q_e and k can be determined from the slope and intercept of the plot, Figure 9. The pseudo-second-order rate constant k_2, the calculated q_e value, and the corresponding linear regression correlation coefficient value are given in Table 3. At all initial metal concentrations, the linear regression correlation coefficient R^2 values were higher. The higher

Table 4 Biosorption of Pb(II) and Cd(II) ions by different biosorbents

Biosorbent	Pb(II)	Cd(II)	Reference
Pomegranate peel	13.87	-	[13]
Banana cortex	-	67.2	[7]
Orange cortex	-	28.8	[7]
Lemon cortex	-	12	[7]
Processed walnut shell	32	11.6	[11]
Rice husk	0.62	-	[17]
Olive tree pruning waste	33.39	-	[21]
Maize leaf	-	10.18	[9]
Moringa oleifera leaves	209.54	-	[8]
Scolymus hispanicus L	-	54.04	[10]
Ulmus tree leaves	201	80	[4]
Fraxinus tree leaves	172	67.2	[4]
Green coconut shell	-	285.7	[5]
Neem leaf	13.51	1.89	[2]
Heartwood of *Areca catechu*	-	10.66	[3]
Ficus carcia leaves	34.36	30.31	This work

values confirm that the adsorption data are well represented by pseudo-second-order kinetics.

Comparison of *F. carcia* leaves with other biosorbents

A comparative study of the maximum biosorption capacity (q_{max}) of *F. carcia* leaves with those of some other biosorbents reported in literature is given in Table 4. Differences in q_{max} are due to the nature and properties of each biosorbent such as surface area and the main functional groups in the structure of the biosorbent. A comparison with other adsorbents indicated a high Cd(II) and Pb(II) biosorption capacity of the *F. carcia* leaves.

Experimental

All the chemicals used were of analytical reagent (AR) grade purchased from Merck (Germany). All the solutions were prepared with double-distilled water with zero levels of dissolved oxygen. The cadmium(II) and lead(II) stock solutions were prepared from cadmium nitrate [$Cd(NO_3)_2$] and lead nitrate [$Pb(NO_3)_2$], respectively. *F.* carcia leaves were washed with distilled water several times to remove dust and soluble materials and then allowed to dry at room temperature. The dried leaves were grounded to a fine powder and sieved to get size fraction <44 μm and then dried in an oven at 60°C for 4h. Cadmium(II) and lead(II) concentrations were determined by AAS600 atomic spectrophotometer (Shimadzu, Japan). pH measurements were made with a WTW pH meter using a combined glass electrode. Fourier transform infrared spectroscopy IR prestige-21 (Shimadzu, Japan) was used to study the functional groups in biosorbent and biosorbent-loaded metal.

Conclusions

The potential of FL for the removal of Cd(II) and Pb(II) ions from aqueous solutions was dependent on biosorption process such as pH, initial metal ion concentrations, biosorbent dose, contact time, and temperature. The equilibrium data have been analyzed using Langmuir and Freundlich isotherms. The characteristic parameters for each isotherm and related correlation coefficients R^2 were determined. The Langmuir biosorption isotherms were demonstrated to provide the best correlation for the biosorption of Cd(II) and Pb(II) ions onto FL. The maximum monolayer biosorption capacity of FL was found to be 30.31 mg of Cd(II)/g of FL and 34.36 mg of Pb(II)/g of FL. The kinetic results provided the best correlation of the experimental data of biosorption of Cd(II) and Pb(II) onto FL by pseudo-second-order equation. It can be concluded that since the FL is an easily locally available low-cost adsorbent and has a considerable high biosorption capacity, it may be treated as an alternative adsorbent for treatment of wastewater containing cadmium(II) and lead(II) ions.

Competing interests
The authors declare that they have no competing interests.

Authors' contributions
AMA participated in the design of the study, and carried out the experimental work and drafted the manuscript. AHA studied kinetics of the biosorption process. AMF participated in the design of the study and performed the analysis of the samples. All authors read and approved the final manuscript.

Acknowledgment
Authors are thankful to Royal Scientific Society, Amman, Jordan and Baghdad University, Baghdad, Iraq for providing the necessary facilities to carry out this research work.

Author details
[1]Department of Chemistry, Faculty of Science for Women, Baghdad University, Baghdad, Iraq. [2]Department of Chemistry, College of Education, Ibn Al-Haitham e, Baghdad University, Baghdad, Iraq. [3]Royal Scientific Society, El Hassan Science City, Amman, Jordan.

References

1. Horsell M, Spiff AI (2005) Effects of temperature on the sorption of Pb^{2+} and Cd^{2+} from aqueous solution by *Caladium bicolor* (wild cocoyam) biomass. Electron J Biotechnol 8:43–50
2. Babarinde NAA (2011) Kinetic, equilibrium and thermodynamic of the biosorption of Pb(II), Cd(II) and Cr(III) by neem leaf. J Innovative Res Eng Sci 2:291–306
3. Chakravarty P, Sarma NS, Sarma H (2010) Biosorption of cadmium(II) from aqueous solution using heartwood powder of *Areca catechu*. Chem Eng J 162:949–955
4. Sangi MR, Shahmoradi A, Zolgharnein J, Azimi GH, Ghorbandoost M (2008) Removal and recovery of heavy metals from aqueous solution using *Ulmus carpinifolia* and *Fraxinus excelsior* tree leaves. J Hazard Mater 155:513–522
5. Pino GH, Souza de Mesquita LM, Torem ML, Pinto GAS (2006) Biosorption of cadmium by green coconut shell powder. Miner Eng 19:380–385
6. Iqbal M, Saeed A, Zafar SI (2009) FTIR spectrophotometry, kinetic and adsorption isotherms modeling, ion exchange and EDX analysis for understanding the mechanism of Cd2+and Pb2+ removal by mango peel waste. J Hazard Mater 164:161–171
7. Kelly-Vargas K, Cerro-Lopez M, Reyna-Tellez S, Bandala ER, Sanchez Salas JL (2012) Biosorption of heavy metals in polluted water using different waste fruit cortex. Phys Chem Earth 39:26–39
8. Reddy DHK, Harinath Y, Seshaiah K, Reddy AVR (2010) Biosorption of Pb(II) from aqueous solutions using chemically modified *Moringa oleifera* tree leaves. Chem Eng J 162:626–634
9. Babarinde NAA, Babalola JO, Sanni SO (2007) Isotherm and thermodynamic studies of the biosorption of Cd(II) from solution by maize leaf. Inter J Phys Sci 2:207–211
10. Barka N, Abdennouri M, Boussaoud A, El Makhfouk M (2010) Biosorption characteristics of cadmium(II) onto *Scolymus hispanicus* L. as low-cost natural biosorbent. Desalination 258:66–71
11. Almasi A, Omidi M, Khodadadian M, Khamutian R, Gholivand MB (2012) Lead(II) and cadmium(II) removal from aqueous solution using processed walnut shell: kinetic and equilibrium study. Toxicol Environ Chem 94:660–671
12. Sousa FW, Sousa MJ, Isadora RN, Oliveira IRN, Oliveira AG, Cavalcante RM, Fechine PBA, Neto VOS, de Keukeleire D, Nascimento RF (2009) Evaluation of a low-cost adsorbent for removal of toxic metal ions from wastewater of an electroplating factory. J Environ Manag 90:3340–3344
13. El-Ashtoukhy ESZ, Amin NK, Abdelwahab O (2008) Removal of lead(II) and copper(II) from aqueous solution using pomegranate peel as a new adsorbent. Desalination 223:162–173
14. Kazemipour M, Ansari M, Tajrobehkar S, Majdz M, Kermani HR (2008) Removal of lead, cadmium, zinc and copper from industrial carbon developed from walnut, hazelnut, almond, pistachio and apricot stone. J Hazard Mater 150:322–327

15. Mitic-Stojanovic DL, Zarubica A, Purenovic M, Bojic D, Andjelkovic T, Bojic AL (2011) Biosorptive removal of Pb^{2+}, Cd^{2+} and Zn^{2+} ions from water by *lagenaria vulgaris* shell. Water SA 37:303–312
16. Martinez M, Miralles N, Hidalgo S, Foil N, Villaescusal PJ (2006) Removal of lead(II) and cadmium(II) from aqueous solution by grape stalk waste. J Hazard Mater 133:203–211
17. Asrari E, Tavallali H, Hagshenas M (2010) Removal of Zn(II) and Pb(II) ions using rice husk in food industrial wastewater. J Appl Sci Manag 14:159–162
18. Nagpal UMK, Bankar AV, Pawar NJ, Kapadnis P, Zinjarde SS (2011) Equilibrium and kinetic studies on biosorption of heavy metals by leaf powder of paper mulberry (*Broussonetia papyrifera*). Water Air Soil Poll 215:177–188
19. Abdel-Ghani NT, Hefny MM, El-Chaghaby GA (2008) Removal of metal ions from synthetic wastewater by adsorption onto *Eucalyptus camaldulenis* tree leaves. J Chil Chem Soc 53:1585–1587
20. Goyal P, Sharma P, Srivastava S, Srivastava MM (2008) *Saraca indica* leaf powder for decontamination of Pb: removal, recovery, adsorbent characterization and equilibrium modeling. Int J Environ Technol 5:27–34
21. Blazquez G, Martin-Lara MA, Tenorio G, Calero M (2011) Batch biosorption of lead(II) from aqueous solution by olive tree pruning waste: equilibrium, kinetics and thermodynamic study. Chem Eng J 168:170–177
22. Aharoni A, Ungarish M (1977) Kinetics of activated chemisorption part 2. Theoretical models. J Chem Soc Faraday Trans 73:456–464
23. Boparai HK, Joseph M, O'Carroll DM (2011) Kinetics and thermodynamics of cadmium ion removal by adsorption onto nano zerovalent iron particles. J Hazard Mater 186:458–465
24. Ho YS (2006) Review of second order models for adsorption systems. J Hazard Mater 136:681–689
25. Freundlich H, Hellen W (1993) The adsorption of *cis*- and *trans*-azobenzen. J Amer Chem Soc 61:2–28
26. Langmuir I (1918) The adsorption of gases on plane surfaces of glass, mica and platinum. J Amer Chem Soc 40:1361–1403
27. Lagergren S (1898) About the theory of so-called adsorption of soluble substances. K. Sven Vetenskapsakad Handl 24:1–39

Prediction of flooding and pressure drop in a spinning cone column using neural networks

Nasser Saghatoleslami*, Mohammad Amiri and Jamal Rohi Golkhatmi

Abstract

Spinning cone column as a distillation tower has various applications in food industry. It has a complex geometric structure which makes the modeling of liquid and gas regimes in the column rather difficult.In the last decade, artificial neural networks havebeen used in various industries considerably. Unlike empirical correlation, neural network does not requirephysical mechanism which occurs in column. Therefore, in this research the effect of tray speed, pressure drop, cone spacing, and flooding for both small and large scales operation has been examined using artificial neural network. Furthermore, variation of gas and liquid flow rate in an industrial scale spinning cone column has also been evaluated. To obtain this objective, multilayer perceptron structure and Levenberg-Marquart training algorithm has been utilized. The findings of this study reveal that the predictions of this work are much accurate than those obtained from the existing empirical correlation. There also exists a good compatibility between the pressure drop values predicted from the present study and the experimental data in both dry and wet state (normalized bias = −0.00232, mean squared error = 0.0021, and root mean squared error = 0.0021). From the scheme adopted in this work, the spinning cone column capacity at different operating conditions could be estimated more accurately than the exiting correlations.

Keywords: Flooding, Neural networks, Multi layer perceptron, Spinning cone column

Background

Spinning cone column (SCC) as a distillation tower has various applications in the food industry. The spinning cone column is a gas-liquid contacting device consisting of a vertical countercurrent flow system, which contains a succession of alternate rotating and stationary metal cones, whose upper surfaces are wetted with a thin film of liquid. Liquid flows down the upper surfaces of the stationary cones under the influence of gravity and moves up the upper surfaces of the rotating cones in a thin film by the action of the applied centrifugal force. Vapor flows up the column and traverses the successive fixed and rotating cones [1]. Figure 1 shows the liquid and gas flow regimes in a SCC. Flooding limit is an important criterion in the distillation tower, since it is an index of tower's capacity. By knowing the geometric structure of the rotary conical tower, we could determine the effective parameters on the flooding such as gas flow rate, liquid flow rate, cone spacing, and rotation speed of the trays.

In the last decade, artificial neural networks have been used in various industries considerably [2]. Neural networks are nonlinear calculation algorithms for images, signals, and numerical data processing. Regarding some features of neural networks such as internal dynamic of neural networks in prediction, changing information error, unnecessary information in input data, employing neural networks in the engineering field as a tool to control and observe the process performance has been increased substantially. Multilayer perceptron is one of the most frequent structures which are utilized in neural networks [3].

Spinning cone column has a complex geometric structure which makes the modeling of liquid and gas regimes in the column rather difficult. Unlike empirical correlation, neural network does not require physical mechanism which occurs in column. Furthermore, multilayer perceptron structure could predict values outside the training limit. Therefore, in this research, the effect of tray speed, pressure drop, cone spacing, flooding for both small and large scales operation has been examined using artificial neural network. Furthermore, variation of gas and liquid flow rate in an industrial scale spinning cone column has also been evaluated.

Rotatory conical columns

The first research on SCC was carried out in 1937 by Pegram and co-workers which gave the details of the

* Correspondence: slami@um.ac.ir
Department of Chemical Engineering, Ferdowsi University of Mashhad, Mashhad, 9177948974, Iran

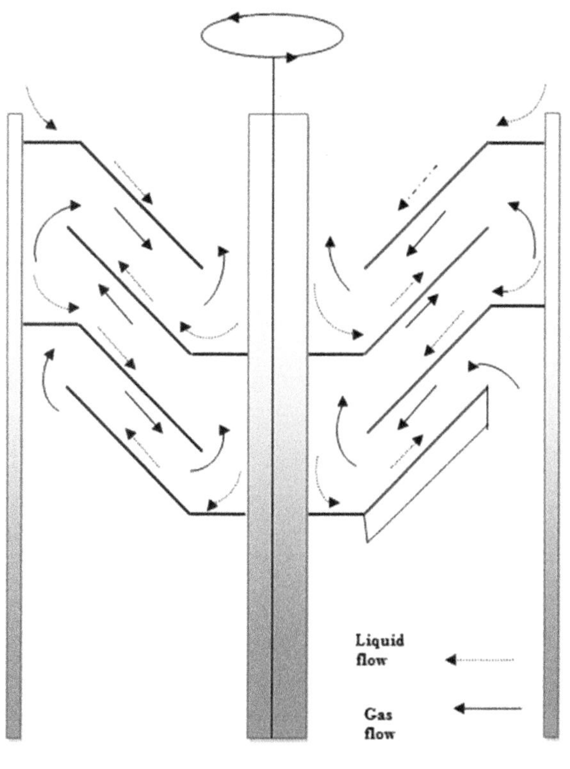

Figure 1 Section of liquid and gas flow path in SCC.

column [4]. Mir and Wilingham in 1939 conducted some experiments on distillation column on a laboratory scale (as cited in Makarytechev et al. [5]). Later on and in 1963, Ziolowski and co-workers performed pressure drop studies on the separation of benzene-tetra-chloride mixture for the first time (as cited in Makarytechev et al. [6]). Liquid film flow on rotatory conical tray was studied by Prince and co-workers in Sydney University, who examined the liquid flow regime in SCC in 1998 and showed that scale up causes a change in liquid flow regime. In a further work, they also examined the speed rate and liquid film thickness on the rotator in 2001 [7]. Macarif and Langrich exhibited a relationship between pressure drop and flooding limit in 2004 and 2005 (as cited in Riley and Sykes [8]).

From 2002, computational fluid dynamics was adopted by some researchers to study on SCC in the absence of liquid phase. Therefore, in this paper, the flooding limits in a rotatory conic distillation column in an industrial scale have been studied in order to improve SCC design and predict tower capacity using a neural network.

Multilayer perceptron structure

The base of multilayer perceptron (MLP) is made up of artificial neurons, which a simple mathematical operation is performed on its inputs, which is shown in Figure 2. The input neuron of the first layer includes P_1, P_2, and P_R, variables and threshold or bias expression. Each of the input variables is multiplied by weight coefficient, and then the grand total is added to bias. Finally, a defined activated function performs a mathematical operation on inputs and results the outputs. MLP structure is a standard combination of inputs, hidden units, and outputs. The outputs of all processing units of each layer are correlated to all units of the next layer. All processing units of the input layer are linear, but in the hidden layer, nonlinear neurons are used. In this structure, sigmoid function in hidden layer and pure line function in output layer have been utilized. The neural networks adopted for training algorithms utilize real data of input and output in order to define a hidden relation between input and output using weight coefficients, bias, and performance functions [9].

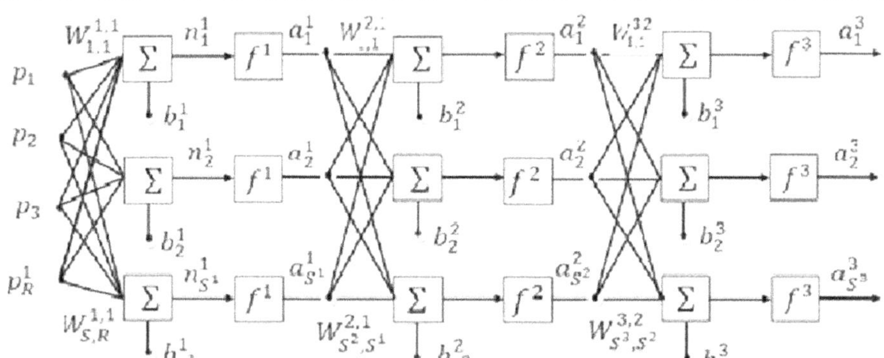

Figure 2 MLP neural network structure.

Methods and experimental

Levenberg Marquarts training algorithms

Although back propagation (BP) algorithm is frequently utilized, however, it has two disadvantages:

- Low speed of convergence
- Static and stoppage of network parameters due to falling in local minimum

Therefore, Levenberg Marquart algorithm which is faster than BP was adopted in the present study [10]. For training purposes, it was necessary to optimize the standard function based upon the networks parameters. Based on the Newton method, variation of parameters in each training step is as follows:

$$\Delta\ \bar{X} = -\left[\nabla_{\bar{x}}^{2}\ J(\bar{X})\right]^{-1}.\nabla\bar{x}\ J\ (\bar{X}) \quad (1)$$

where $\left[\nabla_{\bar{x}}^{2}\ J(\bar{X})\right]$ is a Hessian matrix, and $\nabla\bar{x}\ J\ (\bar{X})$ is a function gradient of J with respect to the neural parameters vector (i.e., $\bar{x}$). As the standard network function is defined as mean sum square errors, therefore we would have:

$$e_v = t_v - a_v \quad (2)$$

$$J\ (\bar{X}) = \frac{1}{2}\sum_{U=1}^{U} e_U^2\ (\bar{X}) = \frac{1}{2}\sum_{p=1}^{p} \{e_U^2\ (\bar{X})\}_p \quad (3)$$

Where $v = p.s_i$ and p is the number of training model and S_L is number of network output.

Levenberg Marquart method is as follows:

$$\Delta\ \bar{X} = -\left[J^T\ (\bar{X})\ J^T\ (\bar{X}) + \eta_m\ J_n\right]^{-1} J^T\ (\bar{X})\ \bar{E}\ (\bar{X}) \quad (4)$$

$$\bar{E} = \left[E_1\ E_2 \ldots E_p\right]^T \quad (5)$$

$$E_p = \left[e_1^p\ e_2^p \ldots e_{S_L}^p\right] \quad (6)$$

$$P = 1,\ 2 \ldots,\ p \quad (7)$$

where μ_m is the Levenberg Marquart learning coefficient and is a nonnegative number [10].

Empirical model

In this paper, experimental data of Langrich and Macariof for industrial scale (i.e., 0.801 m in diameter) and small scale (i.e., 0.148 m in diameter) have been utilized (as cited in Makarytechev et al. [6]). The number of data for industrial and small scale was 40 and 230, respectively. In this research, 40% of data have been used for the network training and the rest for testing. Mean percentage error of the results was about 10%. Table 1 exhibits the details of neural network features adopted in this work.

Table 1 Details of neural network utilized in the present study

Network type	Feed-forward back propagation
Training function	Trainlm (Levenberg Marquardt)
Number of layer	3
Number of data used for network training	132
Number of data used for testing	44
Number of data used for validation	44
Number of neurons in first layer	9
Number of neurons in second layer	1
Transfer function in first layer	Tanhyperbolic
Transfer function in second layer	Purelin
Regression for training	0.998
Regression for validation	0.992
Regression for testing	0.986
Mean squared error	0.0021
Mean absolute error	5
N.B.	−0.00232

The statistical parameters, i.e., normalized bias (NB), mean squared error (MSE), root mean squared error (RMSE), and squared correlation coefficient (R^2), which are described in the following equations, are used for the fitness investigation and error determination of the model. Normalize bias is a parameter which it defines whether the output values has been predicted less or higher than the actual amount.

$$NB = \sum_{1}^{n} \frac{\left(L_{model,i} - L_{exp.,i}\right)/L_{exp.,\ i}}{n}\ x\ 100 \quad (8)$$

$$\mathrm{MSE} = \frac{\sum_{i}^{n}\left(L_{model,i} - L_{exp.,i}\right)^2}{n} \quad (9)$$

$$\mathrm{RMSE} = \sqrt{\frac{\sum_{i}^{n}\left(L_{model,i} - L_{exp.,i}\right)^2}{n}} \quad (10)$$

$$R^2 = \frac{\sum_{i}^{n}\left(L_{exp.,i} - L_{model,mean}\right)^2 - \sum_{i}^{n}\left(L_{model,i} - L_{exp.,i}\right)^2}{\sum_{i}^{n}\left(L_{exp.,i} - L_{model,mean}\right)^2} \quad (11)$$

Results and discussion

Simulation results revealed that neural networks for both cases of industrial and small scale operation must be run separately. This is caused owing to the fact that there would be a change in the liquid flow regime. In this work, effective parameters such as rotation speed, cone

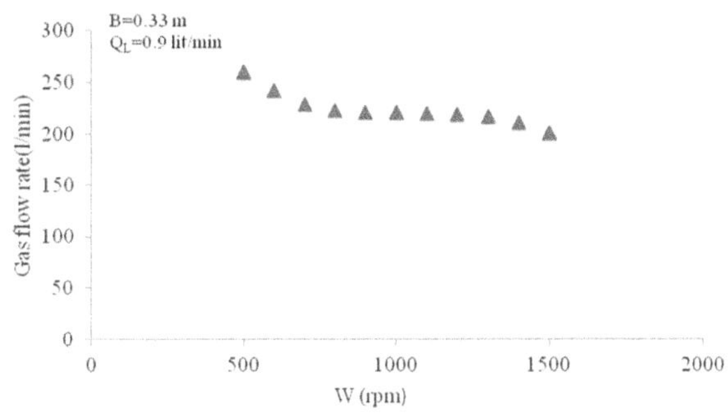

Figure 3 Variations of gas flow rate with respect to the tray speed.

spacing in the beginning of flooding for small scale column, and pressure drop have been examined. Furthermore, the work has been extended to an industrial scale column. Finally, a generalized correlation has been obtained for the flooding limit in industrial scale. These cases have been elucidated in more depth below.

Effect of tray speed in SCC

As the rotation speed increases, it is expected that at low gas flow rate, the flooding in SCC occurs instantaneously. This was verified by the prediction of this work as shown in Figure 3. However, the decline trend has two slopes as shown in Figure 4, and in fact, there exists a point of inflection. Therefore, for an increase in the rotation speed of about 9% (500 rpm) and a reduction of 18%, the gas flow rate decreases accordingly.

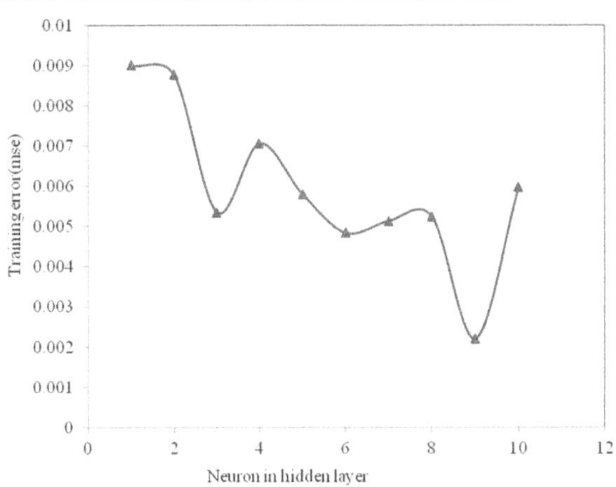

Figure 4 Training error versus the number of neurons in the hidden layer.

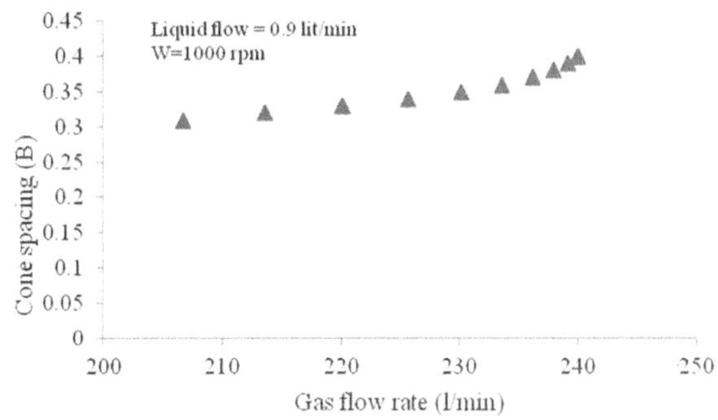

Figure 5 Variations of gas flow rate with respect to the cone spacing.

Effect of cone spacing in SCC

Space between rotary and fixed trays is one of the most important geometric parameters in SCC which is an effective way for flooding to initiate. Figure 5 reveals that as the cone spacing increases (i.e., B), the gas flow rate increases for the formation of flooding which means the column capacity enhances. Figure 5 reveals that a reduction on cone spacing increases the column capacity. On the other hand, as cone spacing increases, the column capacity decreases accordingly. Furthermore, for an increase of about 10% of cone spacing, the capacity of column increases about 8%. However, this reduces to about 2% at the end.

Flooding in SCC

Flooding limit and column capacity in an industrial scale are important criteria in spinning cone column. Operating conditions of the column could be divided into three regions, they are preloading, loading, and flooding and are defined as follows [6]:

$$\mathrm{Fr_{LG}} = \frac{\Delta P_L}{\left(\rho_L - \rho_g\right) g P_C} \tag{12}$$

These limits could be distinguished with $\mathrm{Fr_{LG}}$. Once $0.1 > \mathrm{Fr_{LG}}$, it is defined as preloading, and when $0.1 < \mathrm{Fr_{LG}} < 1$, the regime is called loading, and if $\mathrm{Fr_{LG}} > 1$, the regime is termed flooding. Furthermore, flooding condition could be predicted from the following correlation with a relative error of 20% [6]:

$$1 = 2k_{LL}\sqrt{\frac{\rho_g}{\rho_L}\frac{\omega_{rot}^2 R_{SO}^2 Q_G^6 Q_L^3}{A_{throat}^{9/2} A_{min}^{9/2} P_C^{11/2} g^{11/2}}} \tag{13}$$

Therefore, with the above equation and by knowing the geometric parameter of the column (A_{min}, $A_{throath}$,

P_c, and R_{SO}), we could estimate the gas flow rate at flooding condition for a specified liquid flow rate and rotation speed. A comparative study has been conducted between the predicted results of this study, the experimental data, and the existing correlation as shown in Figure 6.

Pressure drop in SCC

A comparative study between the pressure drop predicted from the present study and the experimental values in both dry and wet state are exhibited in Figures 7 and 8. The results demonstrate a good agreement between them. The mean square errors for predicting pressure drop for this work was 0.0021, and the mean absolute error for all data was about 5%.

Gas and liquid flow rate in industrial scale SCC

Figure 9 shows variations of gas with liquid flow rate for industrial scale spinning cone column at 500 rpm. The results reveal that there is a good agreement between the predicted results of the present study and experimental values. If the rotation rate is different (i.e., different from 500 rpm), the column capacity could be estimated from Figure 4 by increasing the column capacity by 18% for $W = 250$ rpm and decreasing by 9% for $W = 750$ rpm.

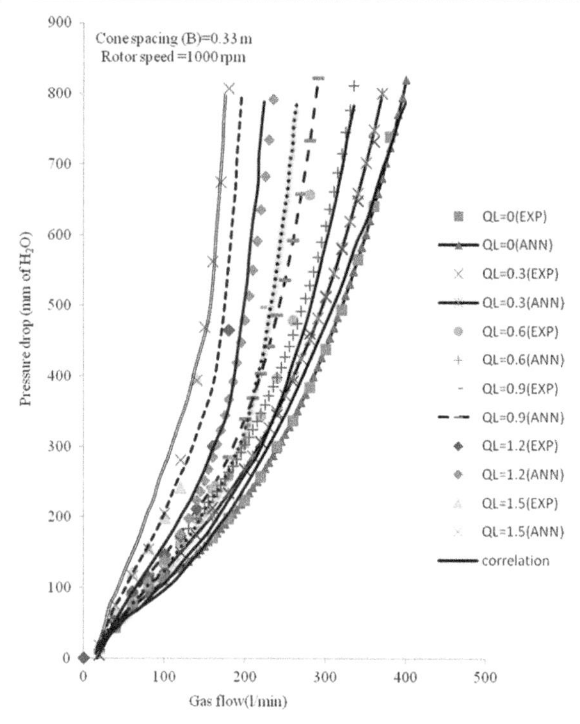

Figure 6 Variations of pressure drop with respect to the gas flow rate.

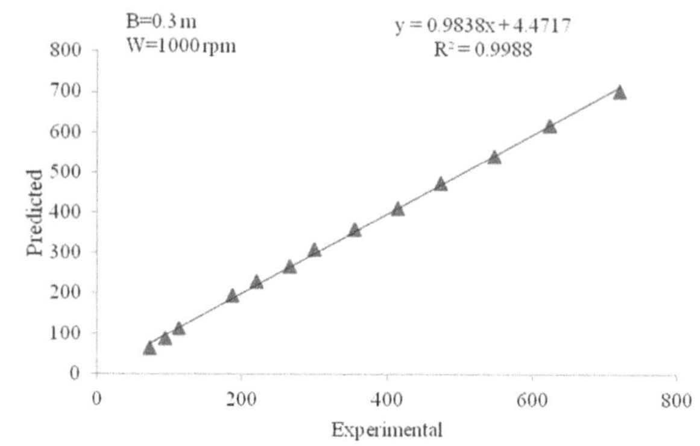

Figure 7 Comparison of pressure drop of the present study and the experimental data in dry state.

Conclusions

Flooding limit and column capacity are important criteria in spinning cone column. In this research and in order to improve spinning cone column design and predict tower capacity, the flooding limits, pressure drop, the effect of tray speed, cone spacing, flooding for both industrial and small scale operation have been assessed using artificial neural network. Furthermore, variation of gas and liquid flow rate in an industrial scale spinning cone column has also been studied. From the findings of this study and from the flooding graph obtained from this work, the spinning cone column capacity in different operating conditions could be estimated. The findings of this study reveal that the predictions of this work are much accurate than those obtained from the existing empirical correlation (about 10%). There also exists a good compatibility between the predicted pressure drop values of the present study and the experimental data in both dry and wet state (MSE = 0.0021, MAE = 5%, and

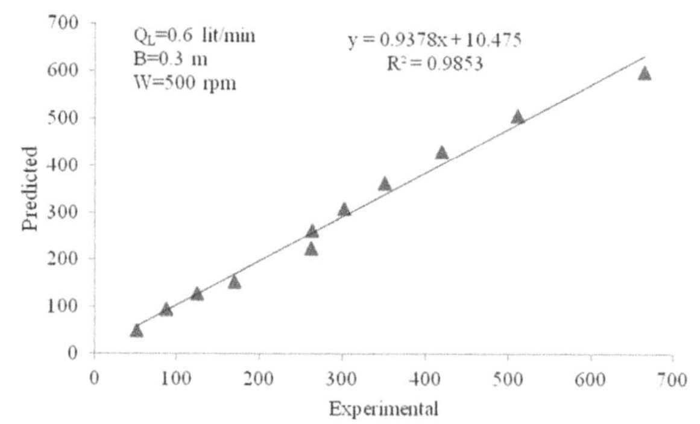

Figure 8 Comparison of pressure drop of the present study and the experimental data in wet state.

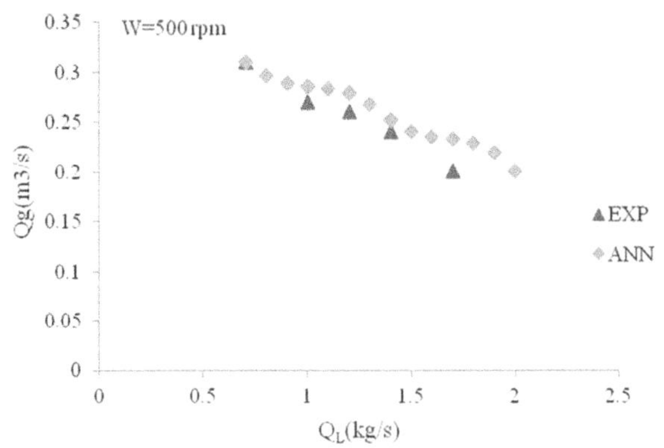

Figure 9 Gas versus the liquid flow rate at 500 rpm.

NB = -0.00232). Therefore, the findings of this study shows that the artificial neural nets technique with multilayer perceptron structure and Levenberg-Marquart training algorithm could be applied as a powerful tool and a cost and time effective way in predicting the spinning cone column capacity and the pressure drop.

Abbreviations

A_{min}: minimum flow area (m^2); A_{throat}: area of the outer throat of the column (m^2); Q_L: liquid flow rate (lit/min); *Qg*: gas flow rate (lit/min); *W*: rotor speed (rpm); *B*: cone spacing (m); *Fr*: Froude number; *g*: acceleration due to gravity (ms^{-2}); k_{LL}: empirical coefficient in the pressure drop correlation; $L_{model,\ i}$: model output value in the test i; $L_{exp,i}$: actual output value in the test i; *n*: number of test samples; P_C: vertical distance between two fixed trays (m); R_{SO}: outer radius of the spinning cone (m); μ_m: Levenberg Marquart learning coefficient; P_G: gas density (kg/m^3); P_L: liquid density (kg/m^3); P: dimensional pressure drop (Pa) subscripts; LG: liquid–gas mixture.

Competing interests

The authors declare that they have no competing interests.

Acknowledgements

Authors wishes to gratefully acknowledge the helpful comments of Dr. Morteza Zivdar and the facilities provided by the Ferdowsi University of Mashhad.

Authors' contributions

Hereby, Nasser Saghatoleslami, Mohammad Amiri, Jamal Rohi Golkhatmi certify that they carried out the work "Prediction of Flooding and Pressure Drop in a Spinning Cone Column Using Neural Networks" and approve the final manuscript.

Authors' information

Nasser Saghatoeslami: Associate of Prof. of Chemical Engineering
Mohammad Amiri and Jamal Rohi Golkhatmi: M.Sc. Research Student

References

1. Zivdar M, Langrish TAG, Prince RGH (2001) Int J of Engi 14:1
2. Chau KW, Cheng CT (2002) Real-time prediction of water stage with artificial neural network approach. Lecture Notes in Artificial Intelligence 2557:715
3. Rumelhart DE, Hinton E, Williams J (1986) Parallel distributed processing 1:318
4. Makarytechev SV, Langrish TAG, Prince RGH (1998) Chem Engin Sci 53:1541
5. Makarytechev SV, Langrish TAG, Prince RGH (2001) Chem Engin Sci 56:77
6. Makarytechev SV, Langrish TAG (2005) Chem Eng Comm 192:445
7. Makarytechev SV, Langrish TAG, Prince RGH (2004) Chem Engin Comm 191:641
8. Riley PC, Sykes SJ (2002) Industrial Applications of spinning cone column technology: a review. Distillation and Absorption 2002 BadenBaden October 2002
9. Saghatoleslami N, Sargolzaei J, Mosavi M, Khoshnoodi M (2006) Int J of Chem and Chem Engin 25:67
10. Saghatoleslami N, Karimi H (2009) Neural networks and its applications in gas, oil and chemical industries. Jahesh Press, Tehran, Iran

17

Liquid-phase microextraction based on the solidification of a floating organic microdrop for trace analysis of BTEX in the aqueous sample

Hakim Faraji[1*] and Ramin Mohammad Ali Tehrani[2]

Abstract

Background: Benzene, toluene, ethylbenzene and xylenes (BTEX) are a group of volatile aromatic compounds. The human exposure to these compounds can comprise serious health consequences. Due to the toxicological properties of BTEX, there is an interest in the chemical control and consequently analytical procedures to measure these compounds. The objective of the present study was to develop a novel liquid-phase microextraction (LPME) based on solidification of floating organic microdrop coupled with gas chromatography flame ionization detection for pre-concentration and determination of BTEX in the aqueous samples.

Results: Microextraction efficiency factors, such as the effects of type and volume of extraction solvent, stirring rate, extraction time, temperature, and ionic strength, have been investigated and optimized: 15.0-μL 1-undecanol microdrop exposed for 15 min floated on the surface of a 10-mL water sample at 45°C, stirred at 1,200 rpm without adding salt. Under the selected conditions, a pre-concentration factor of 211 to 417, limit of detection of 0.07 to 0.18 μg L^{-1} (S/N = 3), limit of quantification of 0.22 to 0.49 μg L^{-1} (S/N = 10), and linearity range of 1.00 to 300 μg L^{-1} for benzene, 0.10 to 300 μg L^{-1} for toluene, 0.05 to 150 μg L^{-1} for ethylbenzene, and 0.02 to 150 μg L^{-1} for xylenes, with coefficient of determination ($0.9979 < r2 < 0.9992$), were observed. Relative standard deviation (RSD %) for analysis of 2.0 μg L^{-1} BTEX compounds in water samples was in the range of 0.9% to 6.4% (n = 5). The relative recovery of different water samples was higher than 79.8%.

Conclusions: The LPME method is a simple, rapid, economic and environmentally friendly technique that can be applied to the trapping of BTEX compounds from aqueous samples.

Keywords: Liquid-phase microextraction, BTEX, Aqueous sample

Background

Benzene, toluene, ethylbenzene and xylenes (BTEX) are a group of volatile aromatic compounds. BTEX are made up of naturally occurring chemicals that are found mainly in petroleum products and are widely utilized in fuels, plastics, paints, pesticides, chemical intermediates, and solvents [1]. Therefore, they can be emitted to the environment from an extensive variety of sources. The human exposure to these compounds can comprise serious health consequences like central nervous system problems, cancer, and damaging effects on the respiratory system, kidney, and heart [2,3]; in order to decrease the human intake of these perilous compounds, a chemical control and, consequently, analytical methods are desirable.

Gas chromatography (GC) is the general technique used for the analysis of BTEX in environmental samples. For the quantification of BTEX in aqueous systems, gas chromatography combined with mass spectrometry (MS) [4-10] and flame ionization detection (FID) [11-15] has been employed in the past. Prior to analysis, for extraction and pre-concentration of target compounds from the aqueous samples, techniques such as liquid-liquid extraction (LLE) and solid-phase extraction have been utilized [16,17]. However, these methods are time-consuming and need large volumes of organic solvents, which are toxic to human health and the environment, and they are also expensive. Solvent-free sample preparation methods or those employing less organic solvent are becoming increasingly

* Correspondence: hakimfaraji@yahoo.com
[1]Department of Chemistry, Varamin-Pishva Branch, Islamic Azad University, Varamin, 33817-74895, Iran
Full list of author information is available at the end of the article

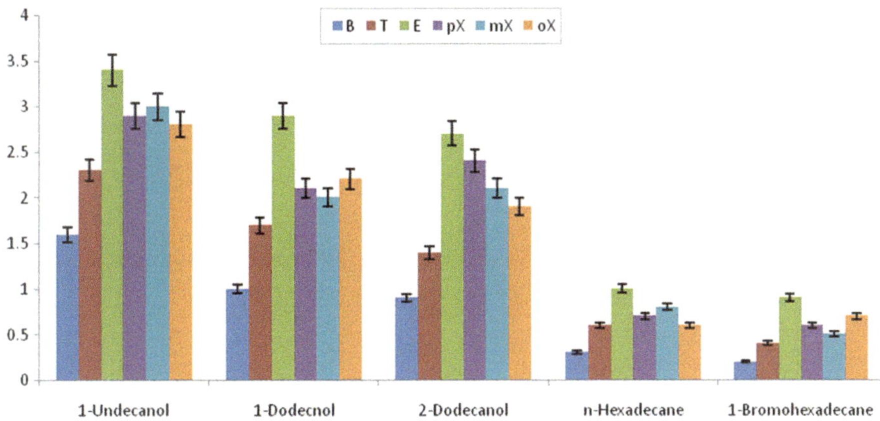

Figure 1 The effect of type of organic solvent volume on the extraction efficiency.

important and have induced a major advance in analytical methodology. Purge and trap [4,7,8], headspace solid-phase dynamic extraction [5], headspace solid-phase microextraction [6,12-14], headspace [9,18], or direct aqueous injection [10,15] have been reported as alternative to traditional methods for sample preparation. Liquid-phase microextraction (LPME) has been reported as a new method in which a low amount of solvent as extractant is used [19,20]. This technique is a simple, low-cost, and rapid procedure requiring lower sample and extractant volumes [21]. Different operational modes have developed from LPME; the most commonly used of which are as follows: single-drop microextraction (SDME) [22], headspace solvent microextraction [11], hollow fiber-based LPME (HF-LPME) [23,24], directly suspended droplet microextraction [25], ionic liquid-based single-drop microextraction [26], and dispersive liquid-liquid microextraction [27].

A modified LPME method [28] has been recently employed for pre-concentration of organic compounds in aqueous samples [29-32]. In this technique, a microdrop of the organic solvent is floated on the surface of an aqueous sample while being agitated by a stirring bar in the bulk of the solution. After the completion of the extraction, the sample vial is inserted into an ice bath to freeze the organic phase. The solidified microdrop is then transferred into a conical vial, where it melts immediately. Finally, the GC can be utilized to determine the species in the extractant.

The objective of the present study was to investigate the feasibility of the novel LPME technique based on solid organic drop combined with gas chromatography with flame ionization detection (GC-FID) for analysis of BTEX in water samples. Effects of various influential parameters, such as the type and volume of extraction solvent, stirring rate, extraction time, sample temperature, and ionic strength, have also been examined. The optimized method, after validation, has been applied to different real water samples.

Results and discussion

The suggested LPME method is established on equilibrium between the analyte concentration in the sample and that in the extractant solvent. The thermodynamic and kinetic

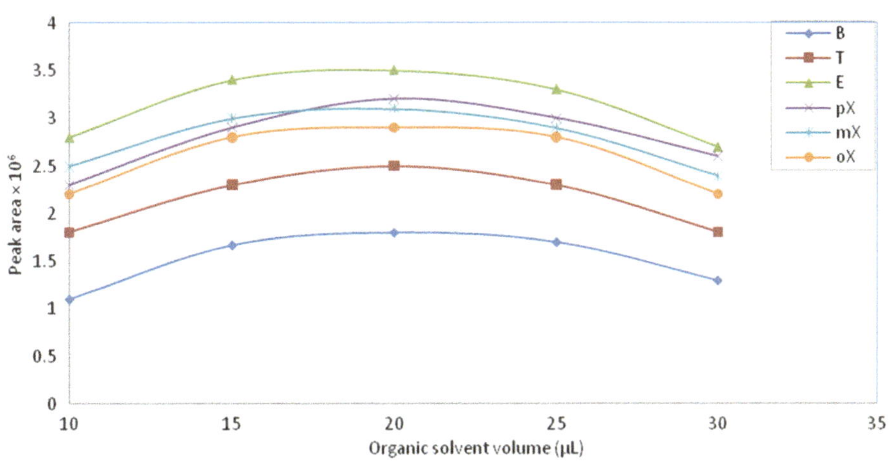

Figure 2 The effect of organic solvent volume on the extraction efficiency.

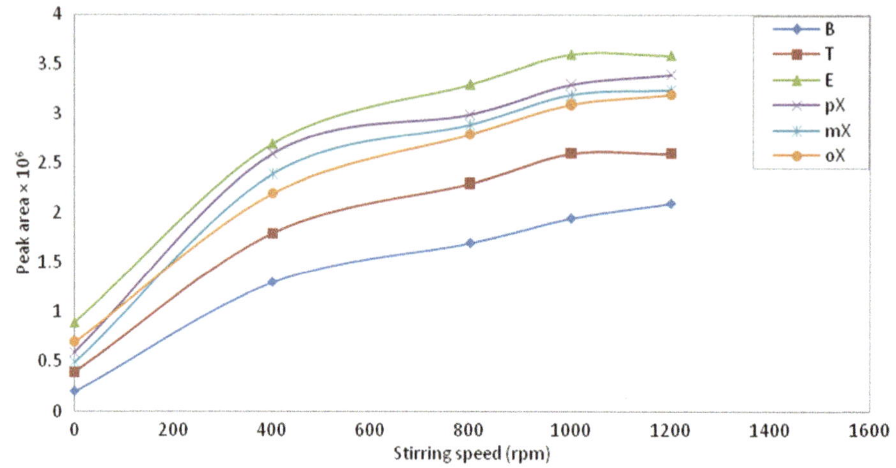

Figure 3 The effect of stirring rate on the extraction efficiency.

equations of LLE are previously demonstrated [33]. To optimize the affecting factors in the analysis of BTEX in aqueous samples by the proposed method, the different parameters such as type and volume of solvent, sample stirring rate, sample solution temperature, extraction time, and salt addition were investigated.

Type of extraction solvent

Selection of an appropriate solvent is a crucial parameter in LPME methods. The range of proper solvents for use with the above mentioned method is limited as the solvents must have low volatility and low solubility in water to be immiscible with water and as the analytes must be soluble in it. The solvent's melting point should be near the room temperature [34]. In this study, five different solvents were evaluated, i.e., 1-undecanol, 1-dodecanol, 2-dodecanol, *n*-hexadecane, and 1-bromohexadecane. Solvent selectivity was evaluated for the extraction of 10 mL of sample containing 2.0 μg L^{-1} of each BTEX. The stirred solution (400 rpm) was kept at 25°C for 20 min using 15 μL of proper organic solvent and without salt addition. The results depicted in Figure 1 reveal that among the solvents studied, 1-undecanol and 1-dodecanol provide higher extraction efficiency for most of the analytes. Because of low melting point, low vapor pressure, and low water solubility, 1-undecanol was selected as the optimized extraction solvent for further investigations. Moreover, its chromatographic peak was easily separated from the analyte peaks. The extraction efficiency was found on the average peak area of each analyte for three replicate analyses.

Volume of extraction solvent

The influence of microdrop volume on the analytical signal was studied in the range of 10.0 to 30.0 μL. Figure 2

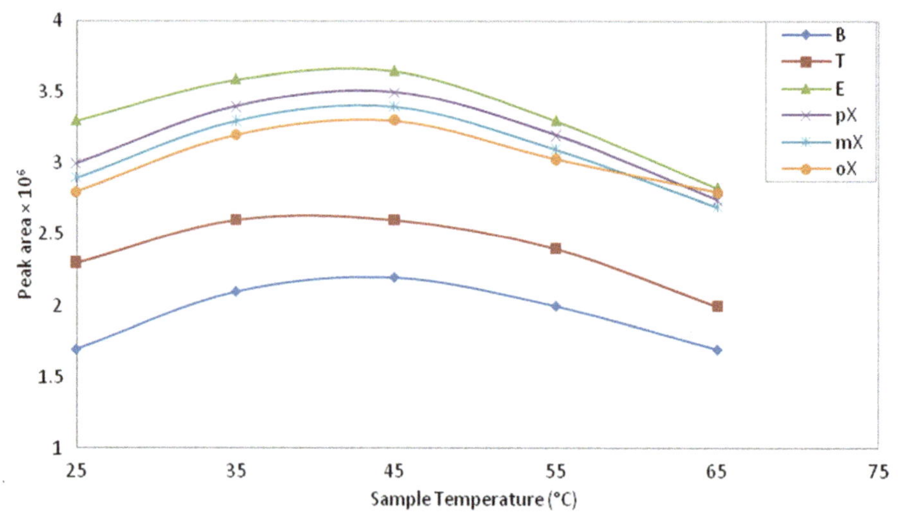

Figure 4 The effect of sample solution temperature on the extraction efficiency.

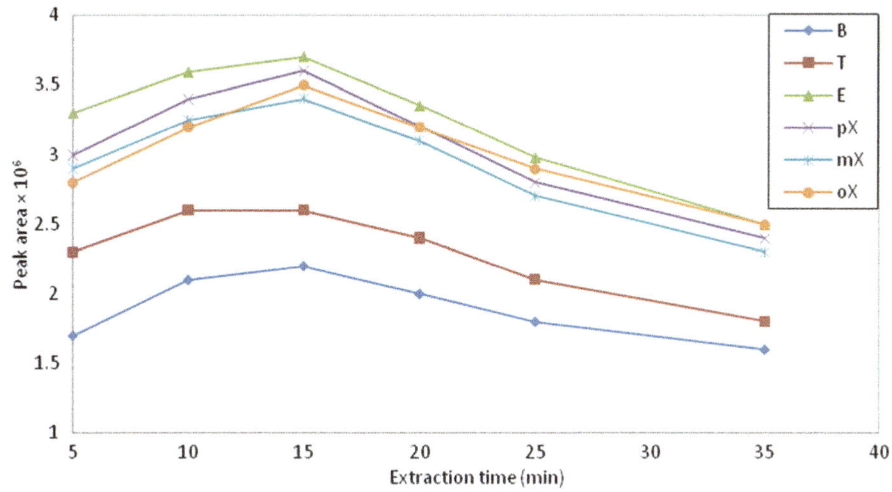

Figure 5 The effect of extraction time on the extraction efficiency.

shows that there was a corresponding increase in the analytical signal of target compounds from 10.0 until 20.0 μL of solvent volumes. Based on LLE equations, rate of the analyte migration into the microdrop is directly related to the surface area between the two liquid phases and is inversely related to the organic-phase volume [35]. Thus, an increase in the drop volume raises the interfacial area following the analytical signals. However, further increase in the microdrop volume results in decrease in the analytical signals [36]. Hence, the volume of 15.0 μL was chosen as the optimal solvent volume.

Stirring rate

Stirring intensity is one of the important parameters that improves the extraction efficiency and lowers extraction time. The agitation can regenerate a new sample solution surface, accelerating the mass transfer from the donor phase to the acceptor phase [37,38]. The optimum stirring rate was evaluated at different stirring rates between 0 and 1,200 rpm. Higher stirring rates were not used because the microdrop was spattered on the vial wall and damaged. Therefore, an optimum stirring rate of 1,200 rpm was selected (Figure 3).

Sample temperature

The effect of sample temperature on extraction efficiency was also investigated between 25°C and 65°C by floating a 1-undecanol microdrop (15.0 μL) for 20 min on the surface of the water samples. It was expected that an increase in temperature leads to an enhancement of the mass transfer coefficients of the goal compounds from the aqueous sample into the microdrop and,

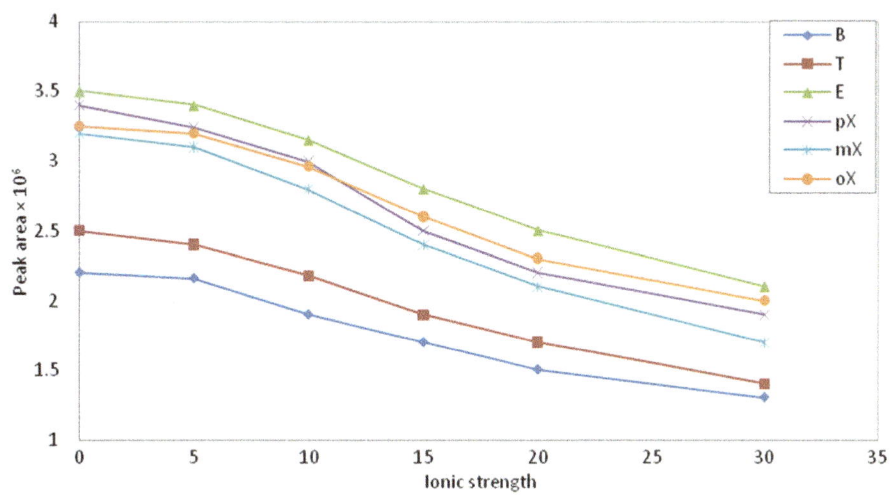

Figure 6 The effect of salt addition on the extraction efficiency.

Table 1 Figures of merit of the proposed method for pretreatment and determination of BTEX

Analytes	LOD (μg L^{-1})	LOQ (μg L^{-1})	LR (μg L^{-1})	r^2	RSD% (n = 5)	PF
Benzene	0.18	0.49	1.00 to 300	0.9979	7.3	211
Toluene	0.12	0.33	0.10 to 300	0.9983	8.6	268
Ethylbenzene	0.07	0.22	0.05 to 150	0.9984	5.7	389
p-Xylene	0.08	0.22	0.02 to 150	0.9992	6.5	417
m-Xylene	0.08	0.22	0.02 to 150	0.9989	3.8	417
o-Xylene	0.08	0.22	0.02 to 150	0.9991	4.2	417

LOD, limit of detection; *LOQ*, limit of quantification; *LR*, linearity range; *RSD*, relative standard deviation; *PF*, pre-concentration factor.

consequently, to higher extraction efficiencies. Nevertheless, the results achieved were in incongruity with this anticipation. In fact, a significant reduction in sensitivity was observed for all the analytes at higher temperature values (>45°C). This behavior, which is disclosed in Figure 4, is likely to be the result of two effects. On one side, it has been reported that high temperatures can decrease the partition coefficients of analytes in the microdrop, resulting in a decrease of the rate of analyte adsorption on the outer surface of the microdrop. On the other hand, the sample temperature affects not only the kinetic and thermodynamic of the analyte extraction, but also those of the water molecules in the sample [39,40]. Thus, optimum extraction temperature was set at 45°C.

Extraction time

As described above, LLE is a type of equilibrium extraction, and the optimal extraction efficiency is obtained when equilibrium is established. Hence, the extraction time performs a very essential role in the whole process. The range of the extraction time investigated was 5 to 35 min with other extraction conditions being constant. Figure 5 shows that there was a corresponding increase in analytical signal from 5 to 15 min, followed by a period of stability. An extraction time of 15 min was selected as a reasonable compromise between enrichment factor and analysis time.

Ionic strength

The ionic strength can be one of the effective parameters on the extraction efficiency. Depending on the target compounds, the presence of salt in aqueous sample can potentially result in contrary effects: it may improve [41], not influence [42], or diminish extraction [24,43]. Apart from the salting-out effect which improves extraction efficiency, it also leads to alter the physical properties of the Nernst diffusion film in the interface of aqueous and organic phases; consequently, the diffusion rate of analytes into the solvent was reduced [44,45]. To investigate this effect on the proposed LPME, the extraction was performed with a 10-mL spiked water sample containing various concentrations of NaCl (0% to 30%, *w*/*v*). The results show that the peak area of BTEX decreased with increasing salt concentration in the aqueous sample (Figure 6). Therefore, it was decided not to alter the salt content of the sample solutions in the following extractions. Hence, salt was not added in this study.

Method validation

Under the above optimum experimental conditions, method validation in terms of linearity, precision, and limits of detection (LOD) and quantifications (LOQ) for each analyte were determined by the accepted procedure of the U.S. Environmental Protection Agency [46]. The results have been presented in Table 1. The calibration curve was found to be linear in the range of 1.00 to 300 μg L^{-1} for benzene, 0.10 to 300 μg L^{-1} for toluene, 0.05 to 150 μg L^{-1} for ethylbenzene, and 0.02 to 150 μg L^{-1} for xylenes with a coefficient of determination (r^2) of >0.9979 (n = 3). Precision was assessed by determination of repeatability, which expresses the precision of the method under the same operating conditions over a short interval of time. It also identified intraday precision and is explained as relative standard deviation (RSD%). The method repeatability was in the range of 3.8% to 8.6% (n = 5). The LOD (S/N = 3) and LOQ (S/N = 10) of target compounds were in the range of 0.07 to 0.18 μg L^{-1}and 0.22 to 0.49 μg L^{-1}, respectively. Furthermore, the pre-concentration factor was

Table 2 The results obtained from analysis of real water samples

Analytes	Recovery of tap water (%)				Recovery of mineral water (%)				Recovery of river water (%)			
	4.0 (μg L^{-1})	(RSD[a])	10 0 (μg L^{-1})	(RSD[a])	4.0 (μg L^{-1})	(RSD[a])	10.0 (μg L^{-1})	(RSD[a])	4.0 (μg L^{-1})	(RSD[a])	10.0 (μg L^{-1})	(RSD[a])
Benzene	87.2	(6.5)	82.5	(8.6)	86.1	(8.4)	88.2	(6.2)	87.5	(9.7)	79.8	(8.4)
Toluene	90.8	(5.1)	83.8	(6.3)	92.7	(6.2)	85.7	(7.5)	91.6	(7.5)	89.5	(9.2)
Ethylbenzene	93.9	(7.2)	88.7	(9.2)	93.1	(9.7)	93.1	(9.7)	93.2	(10.0)	82.3	(10.1)
p-Xylene	93.6	(8.4)	94.3	(5.6)	89.5	(6.8)	90.6	(8.3)	95.8	(6.2)	90.6	(6.8)
m-Xylene	89.8	(6.8)	86.1	(7.4)	97.4	(7.1)	91.8	(4.7)	94.1	(8.7)	86.4	(7.5)
o-Xylene	98.5	(4.5)	99.3	(5.3)	100.4	(7.7)	98.4	(6.5)	101.7	(6.7)	99.1	(9.4)

[a]n = 3.

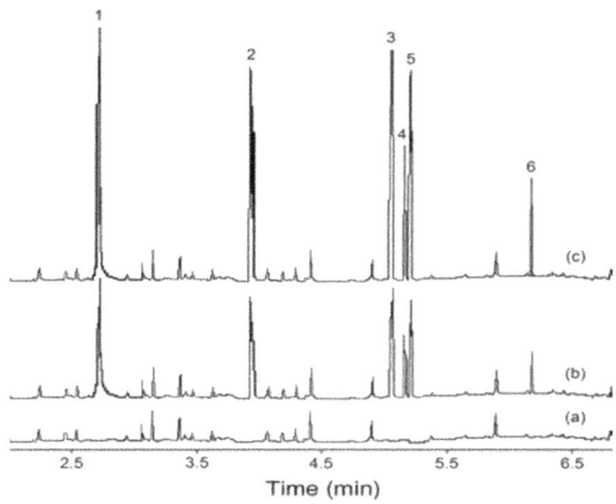

Figure 7 LPME-GC-FID chromatograms from the extraction of BTEX compounds at optimum conditions of the river water sample. a, unspiked; b, spiked by 4.0 μg L^{-1}; and c, spiked by 10.0 μg L^{-1} of the compounds. 1, benzene; 2, toluene; 3, ethylbenzene; 4, *p*-xylene; 5, *m*-xylene; and 6, *o*-xylene.

calculated as the ratio of the peak area after LPME method to the peak area of standard solution at the same concentrations [45,47], which was from 211 to 417 (Table 1).

Application to real samples

Finally, the applicability of the extraction method was evaluated by the analysis of the real samples including tap, mineral, and river water. The results indicated that no BTEX have been detected in analyzed samples using the proposed method. All the real water samples were spiked at two different concentration levels (4.0 and 10.0 μg L^{-1}) to assess the matrix effect. The relative recovery, defined as the peak area ratio of a natural water sample and ultrapure water sample spiked with analytes at the same level, was applied [35]. The relative recoveries of the analytes are given in Table 2, which varies from 79.8% to 101.7%; it shows that the influence of the matrix is not significant on the extraction recoveries. The chromatograms obtained by GC-FID of unspiked river water and those spiked at two concentrations of each analyte after the developed method at optimum conditions are shown in Figure 7.

Conclusion

A comparison of the proposed LLME method and some of the published methods for extraction and determination of BTEX are summarized in Table 3. The recommended LPME provides adequate pre-concentration factor and linear dynamic range, being the least time-consuming method. Precision is similar to that provided by other microextraction techniques reported in the literature. In addition, the LOD is comparable to that reported in most procedures. In conclusion, the LPME method is a simple, rapid, economic, and environmentally friendly technique that can be applied to the trapping of BTEX compounds from aqueous samples.

Experimental

Reagents and chemicals

The BTEX standards (benzene, toluene, ethylbenzene, *p*-xylene, *m*-xylene and *o*-xylene), the organic solvents (HPLC-grade or suprasolv for chromatography), and sodium chloride (analytical reagent grade) were purchased from Merck (Darmstadt, Germany) and Fluka (Buchs, Switzerland). Ultrapure water was prepared by a Milli-Q system (Bedford, MA, USA). An individual stock standard solution of 1,000 mg L^{-1} for each compound was prepared in methanol and stored at 4°C in the darkness. Stock standard mixture of the BTEX compounds (10 mg L^{-1}) was prepared weekly by diluting the stock solutions with methanol. The working standard solutions, with concentration levels ranging from 0.01 to 400 μg L^{-1}, were made daily by diluting the mixed stock solutions with Milli-Q water to the required concentration.

The tap water sample was freshly collected from our laboratory; the bottled mineral water sample was available at the supermarket, packed in polymeric container (Bisheh); and the river water sample was obtained from

Table 3 Comparison of the proposed method with other methods for pretreatment and determination of BTEX

Methods	LOD (μg L^{-1})	LR (μg L^{-1})	r^2	RSD% ($n = 5$)	Recovery (%)	PF	References
SPME	0.01 to 10	0.05 to 2,000	0.9992 to 0.9997	2.96 to 4.16	106 to 113	628 to 733	[48]
HF-LPME	4.8 to 30	5.0 to 300	0.9972 to 0.9988	2.0 to 4.6	89 to 92	41.5 to 128.0	[23]
DI-SDME	5.0 to 10	50 to 20,000	0.9978 to 0.9997	4.72 to 7.74	85.9 to 91.2	43.8 to 64.5	[49]
HS-SDME	0.02 to 0.09	0.01 to 5.0	0.9958 to 0.9997	3.00 to 5.25	88.9 to 103.1	-	[39]
DLLME	0.1 to 0.2	0.2 to 100	0.9991 to 0.9999	0.9 to 6.4	76.0 to 101.0	122 to 311	[27]
DLLME-SFO	0.1 to 0.35	0.25 to 10	0.997 to 0.998	<12	94.6 to 105.9	87 to 290	[50]
Proposed method	0.07 to 0.18	0.02 to 300	0.9979 to 0.9992	3.8 to 8.6	79.8 to 101.7	211 to 417	

SPME, solid-phase microextraction; *HF-LPME*, hollow fiber liquid-phase microextraction; *DI-SDME*, direct-immersion single-drop microextraction; *HS-SDME*, head space single-drop microextraction; *DLLME*, dispersive liquid-liquid microextraction; *DLLME-SFO*, dispersive liquid-liquid microextraction-solidify floating organic drop.

the upstream of Jajrood River (a nonresidential and mountainous area at the northeast of Tehran, Iran) in July 2011. The tap and river water samples were collected in glass bottles. The river water sample was filtered through a filter paper before analysis. The water samples were stored in a refrigerator at 4°C until their analysis.

Apparatus

Analysis was performed using an HP 6890 gas chromatograph (Agilent Technologies, Palo Alto, CA, USA) with FID and equipped with the HP Chemstation software (version A.09.01) used for data analysis. The GC column used for analysis was ZB-WAX, 30 m × 0.32 mm × 0.50 µm (Phenomenex, Torrance, CA, USA). Ultrapure-grade helium (99.9999%, Sabalan Co., Iran) was utilized as the carrier gas at a flow rate of 34 mL/min. Hydrogen for FID was UHP-grade, and air was purified using a Domnick Hunter ZA3500 (EST Analytical, Fairfield, OH, USA) with an additional moisture trap with 13X molecular sieve. The oven program increased from 60°C to 75°C at 15°C/min to 90°C at 3°C/min, then held for 3 min, and extended to 190°C at 25°C/min. Inlet and detector temperatures were 225°C and 300°C, respectively. FID conditions consisted of a hydrogen flow of 40 mL/min, airflow of 450 mL/min, and a helium makeup flow of 40 mL/min. Injection volumes were 2.0 µL with a split of 20:1, and injector temperature was set at 210°C. The solution was stirred with a magnetic heater-stirrer (Heidolph MR 3001 K, Heidolph Instruments, Schwabach, Germany). A simple water bath placed on the heater-stirrer was used for controlling the temperature of the sample solutions.

Methods

An aqueous solution (10 mL) containing a mixture of the BTEX compounds (2.0 µg L^{-1}) was introduced into a vial containing a magnetic stirrer. Then, 15.0 µL of an immiscible extracting solvent (1-undecanol) was dropped onto the surface, and then, the vial was sealed with an aluminum seal with a PTFE septum and stirred for 15 min at 45°C. Subsequently, the sealed vial was placed in an ice bath. After a short time, the organic extractant solidified and was transferred to a clean conical vial where it melted quickly. Finally, 2.0 µL of the extractant was removed for analysis by GC-FID.

Competing interests

The authors declare that they have no competing interests.

Acknowledgements

The authors gratefully acknowledge the financial support provided by the Research Council of the Varamin Branch, Islamic Azad University to carry out this project.

Author details

[1]Department of Chemistry, Varamin-Pishva Branch, Islamic Azad University, Varamin, 33817-74895, Iran. [2]Department of Chemistry, Shahre Rey Branch, Islamic Azad University, Tehran, 1800155144, Iran.

Authors' contributions

HF initiated the study, prepared the design of the study and supervised the acquisition of the data carried out the data analysis, prepared the graphics and prepared the manuscript for submission, review and final acceptance. RMAT carried out the experimental studies and prepared the first draft of the manuscript. All authors read and approved the final manuscript.

References

1. Alberici RM, Zampronio GC, Poppi RJ, Eberlin MN (2002) Analyst 127:230–234
2. Alegretti AP, Thiesen FV, Maciel GP (2004) J Chromatogr B 809:183–187
3. Irwin RJ, Van Mouwerik M, Stevens L, Seese MD, Basham W (1997) Environmental contaminants encyclopedia. National Park Service, Fort Collins
4. Bianchi F, Careri M, Marengo E, Musci M (2002) J Chromatogr A 975:113–121
5. Jochmann MA, Kmiecik MP, Schmidt TC (2006) J Chromatogr A 1115:208–216
6. Nakamura S, Daishima S (2005) Anal Chim Acta 548:79–85
7. Rosell M, Lacorte S, Ginebreda A, Barcel'o D (2003) J Chromatogr A 995:171–184
8. Tanabe A, Tsuchida Y, Ibaraki T, Kawata K, Yasuhara A, Shibamoto T (2005) J Chromatogr A 1066:159–164
9. Lin Z, Wilson JT, Fine DD (2003) Environ Sci Technol 37:4994–5000
10. Zwank L, Schmidt TC, Haderlein SB, Berg M (2002) Environ Sci Technol 36:2054–2059
11. Bahramifar N, Yamini Y, Shariati-Feizabadi S, Shamsipur M (2004) J Chromatogr A 1042:211–217
12. Robert S, Dungan S (2005) Anal Lett 38:2393–2405
13. Dron J, Garcia R, Millan E (2002) J Chromatogr A 963:259–264
14. Ji J, Deng C, Shen W, Zhang X (2006) Talanta 69:894–899
15. Kubinec R, Adamuscin J, Jurdakova H, Foltin M, Ostrovsky I, Kraus A, Sojak L (2005) J Chromatogr A 1084:90–94
16. Beketov VI, Parchinski VZ, Zorov NB (1996) J Chromatogr A 731:65–73
17. Stan HJ, Kirsch NH (1995) Int J Environ Anal Chem 60:33–40
18. Perez Pavon JL, del Nogal Sanchez MD, Garcia Pinto C, Fernandez Laespada ME, Moreno Cordero B (2006) Anal Chem 78:4901–4908
19. Liu H, Dasgupta PK (1996) Anal Chem 68:1817–1821
20. Jeannot MA, Cantwell F (1996) Anal Chem 68:2236–2240
21. Ye C, Zhou Q, Wang X (2007) J Chromatogr A 1139:7–13
22. Psillakis E, Kalogerakis N (2002) Trends Anal Chem 21:54–64
23. Sarafraz-Yazdi A, Amiri AH, Es'haghi Z (2008) Chemosphere 71:671–676
24. Shen G, Lee HK (2002) Anal Chem 74:648–654
25. Sarafraz-Yazdi A, Amiri AH, Es'hagi Z (2009) Talanta 78:936–941
26. Herrador EA, Lucena R, Cardenas S, Valcarce M (2008) J Chromatogr A 1201:106–111
27. Assadi Y, Ahmadi F, Milani Hossieni MR (2010) Chromatographia 71:1137–1141
28. Khalili Zanjani MR, Yamini Y, Shariati S, Jonsson JA (2007) Anal Chim Acta 585:286–293
29. Faraji H, Saber Tehrani M, Husain SW (2009) J Chromatogr A 1216:8569–8574
30. Khalili Zanjani MR, Yamini Y, Yazdanfar N, Shariati S (2008) Anal Chim Acta 606:202–208
31. Farahani H, Norouzi P, Dinarvand R, Ganjali MR (2009) J Sep Sci 32:314–320
32. Dai L, Cheng J, Matsadiq G, Liu L, Li JK (2010) Anal Chim Acta 674:201–205
33. Chen X, Zhang T, Liang P, Li Y (2006) Microchim Acta 155:415–420
34. Leong MI, Huang SD (2008) J Chromatogr A 1211:8–12
35. Zhao L, Lee HK (2001) J Chromatogr A 919:381–388
36. Bagheri H, Khalilian F (2005) Anal Chim Acta 537:81–87
37. Psillakis E, Kalogerakis N (2001) J Chromatogr A 907:211–219
38. Helena P, Locija Zupan IK (1999) Trends Anal Chem 18:272–282
39. Aguilera-Herrador E, Lucena R, Cardenas S, Valcarcel M (2008) J Chromatogr A 1201:106–111
40. Mottola HA (1993) Analyst 118:675–679
41. Kramer KE, Andrews ARJ (2001) J Chromatogr B 760:27–36
42. Ugland HG, Krogh M, Rasmussen KE (2000) J Chromatogr B 749:85–92
43. Psillakis E, Kalogerakis N (2003) J Chromatogr A 999:145–153
44. Bagheri H, Saber A, Mousavi SR (2004) J Chromatogr A 1046:27–33
45. Lambropoulou AD, Albanis TA (2005) J Chromatogr A 1072:55–61

46. U.S. Environmental Protection Agency (1990) Primary drinking-water regulations, maximum contaminant levels. In: Appendix B to part 136, National primary drinking-water regulations. U.S. Code of Federal Regulations, Washington, D.C. Title 40, parts 100–149, revised as of July 1
47. Meng L, Wang B, Luo F, Shen G, Wang Z, Guo M (2011) Forensic Sci Int 209:42–47
48. Sarafraz-Yazdi A, Piri moghadam H, Es'haghi Z, Sepehr S (2010) Anal Methods 2:746–752
49. Sarafraz-Yazdi A, Amiri AH, Es'hagi Z (2010) Intern J Environ Anal Chem 90:1036–1047
50. Vickackaite V, Pusvaskiene E (2009) J Sep Sci 32:3512–3520

The effect of alumina content on the structural properties of ZrO_2-Al_2O_3 unstabilized composite nanopowders

Esmaiel Nouri[1*], Mohammad Shahmiri[1], Hamid Reza Rezaie[1] and Fatemeh Talayian[2]

Abstract

In this research, a sol–gel process was used to synthesize zirconia-alumina (ZrO_2-Al_2O_3) composite nanopowders containing 10, 20, and 50 mol% alumina (Al_2O_3) without any stabilizer. Zirconium acetate hydroxide and aluminum isopropoxide were employed as precursors. The produced composite nanopowders were characterized using a Brunauer-Emmett-Teller (BET) surface area analyzer, Fourier transform infrared (FTIR) spectroscopy, thermogravimetric/differential thermal analysis, X-ray diffraction analysis (XRD), and scanning electron microscopy (SEM). The BET results showed a decrease in surface area and average particle size and an increase in pore size with increasing Al_2O_3 content. FTIR, in corroboration with XRD, confirms the crystallization of corundum (α-Al_2O_3) as one of the alumina phases. The X-ray studies showed that an increase in Al_2O_3 content shifted the tetragonal zirconia (t-ZrO_2)-phase crystallization to higher temperatures. Microstructural characterization by SEM depicted that the particles tend to be more agglomerated with increasing Al_2O_3 content.

Keywords: Sol–gel process, Heat treatment, ZrO_2-Al_2O_3 composite nanopowders, α-Al_2O_3, t-ZrO_2

Background

Zirconia (ZrO_2) has numerous unique properties such as good mechanical strength due to its high toughness, excellent resistance against crack propagation, good thermal resistance, relatively high thermal expansion coefficient, and low thermal conductivity at high temperature [1], which have proven to be superior to other ceramics. Based on these properties, ZrO_2 nanopowders not only have many industrial applications as ceramic bodies, but also can be dispersed as reinforcement in various composite matrixes in order to improve their mechanical properties such as hardness and wear resistance. Studies on the dispersion of ZrO_2 nanopowders, which are well known to be present at a low specific surface area, in various matrixes have been reported in literatures [2-6]. In some cases, wear resistance and also barrier properties of sol–gel-derived coatings were improved by dispersing ZrO_2 nanopowders in the hybrid matrixes [2]. Wang et al. [3] incorporated the ZrO_2 nanopowders into poly methyl methacrylate to overcome some of its drawbacks such as poor heat resistance and weak mechanical surface. Yang et al. [4] reported the highest hardness and the maximum resistance for cracking in Ce-TZP/Al_2O_3 composites (TZP (tetragonal ZrO_2 polycrystal)) containing 20% ZrO_2 nanoparticles. Moreover, several attempts have been made for the combination of hydroxyapatite with yttria-stabilized ZrO_2 nanopowders in order to improve its mechanical properties [5,6].

It is well known that the magnitude of improvement of the mechanical properties depends strongly on the composition, size, and specific surface area of the nanopowders. Therefore, the combination of a hard nanostructured material like Al_2O_3 with a high-toughness ZrO_2 matrix looks as a promising way to produce an excellent composite reinforcement and improve the mechanical properties of the matrix such as hardness, wear resistance, and scratch resistance.

The composite nanopowders have been synthesized via various methods such as chemical vapor synthesis, microwave-hydrothermal route, inert gas condensation,

* Correspondence: enouri@metaleng.iust.ac.ir
[1]Department of Materials and Metallurgical Engineering, Iran University of Science and Technology (IUST), Narmak, Tehran 16846-13114, Iran
Full list of author information is available at the end of the article

laser ablation, and sol–gel processing. The sol–gel method is one of the most promising methods because it offers many advantages for synthesizing high homogeneity and well-controlled properties of nanopowders. Excellent control of the stoichiometry of the precursor solution, customizable microstructure, ease of compositional modification, ease of introducing various function groups of materials, relatively low annealing temperature, and simple and inexpensive equipment are some more advantages of this process [7].

It is also worth mentioning that the ZrO_2 has commonly three polymorphs: monoclinic (m-ZrO_2, below 1,170°C), tetragonal (t-ZrO_2, between 1,170°C and 2,370°C), and cubic (c-ZrO_2, above 2,370°C). The high-temperature cubic and tetragonal phases can be stabilized at room temperature by incorporating some dopants such as Y^{3+} and Ca^{2+} in the lattice [1]. Whereas, according to our previous studies [8,9], tetragonal is the stable crystalline phase, without any stabilizer, at low temperatures in nano and submicron structures of ziconia. Moreover, Garvie et al. [1,10-12] have reported a critical size in ZrO_2 powders in which the tetragonal is the stable crystalline phase below that size.

The main aim of the present research was to gain an understanding of the effect of Al_2O_3 content on the mentioned crystallization and phase transformation in the nanopowders. In addition, the influence of the Al_2O_3 content was investigated on the size and morphology of the composite powders which are determinative factors for particles as reinforcement of the composite matrix. The ZrO_2-Al_2O_3 unstabilized composite nanopowders were synthesized via a sol–gel process and then were characterized using a Brunauer-Emmett-Teller (BET) surface area analyzer, Fourier transform infrared (FTIR), thermogravimetric/differential thermal analysis (TG/DTA), X-ray diffraction analysis (XRD), and scanning electron microscopy (SEM) techniques. An evaluation on the operation of these composite nanopowders as reinforcement in zirconia matrixes produced via a sol–gel process is planned for future work.

Methods

Materials and processing

The ZrO_2 sol was prepared using 3 g of zirconium acetate hydroxide (($(CH_3CO_2)_x$ $Zr(OH)_y$; Sigma-Aldrich Corporation, St. Louis, MO, USA) in 50-ml ethanol (C_2H_5OH) as precursor. In order to prevent the formation of zirconium hydroxide, 2.5 ml of acetone (CH_3COCH_3) was added to the solution. A transparent ZrO_2 sol was obtained after homogenizing the solution with a magnetic stirrer for 1 h at room temperature.

For preparation of the Al_2O_3 sol, 4.39 g of aluminum isopropoxide ($(C_3H_7O)_3Al$; Merck & Co., Inc., Whitehouse Station, NJ, USA) was mixed with 50-ml ethanol at a constant temperature of 80°C, followed by adding 2-ml concentrated nitric acid and 1.5-ml ethyl acetoacetate with stirring for 1 h. The prepared sol also was transparent at room temperature.

The alumina sol was added into the zirconia to prepare ZrO_2-m mol% Al_2O_3 (ZmA hereafter) composite sols, where *m* was equal to 10, 20, and 50. According to the zeta potential changes of the Z20A composite solution versus pH (Figure 1), the pH of the final solution is set to 2.3 by dripping nitric acid. Finally, the composite solutions were homogenized with a magnetic stirrer for 1 h at room temperature. In the end, they were filtered using a 0.22-μm filter.

No precipitate or turbidity was observed after 24 h of aging; then, gelation processes were carried out by instilling commercial ammonia (NH_3) at pH = 4.5 (for the Z20A solution). The obtained gels were dried at 120°C for 24 h. The resulting composite powders were milled in air media in a high-density polythene bottle for 48 h using high-purity zirconia balls with a diameter of 5 mm. The heat treatment procedures were applied with a heating rate of 5°C/min up to 1,100°C and a soaking time of 2 h under an air atmosphere and then cooled over night inside the furnace. The summarized process flowchart of nanopowder preparation is illustrated in Figure 2.

Characterization

Thermal decomposition and structural evolution of the composite nanopowders after annealing were investigated by TG/DTA (PL-STA-1640, Polymer Laboratories, Stanton Redcraft, UK) and FTIR (FTIR-8400 S with IRsolution, Shimadzu Corporation, Kyoto, Japan). Phase

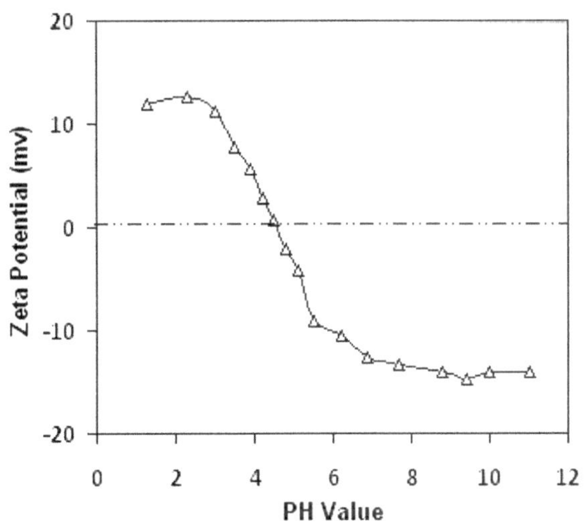

Figure 1 Zeta potential curve of ZrO_2-20Al_2O_3 composite solution.

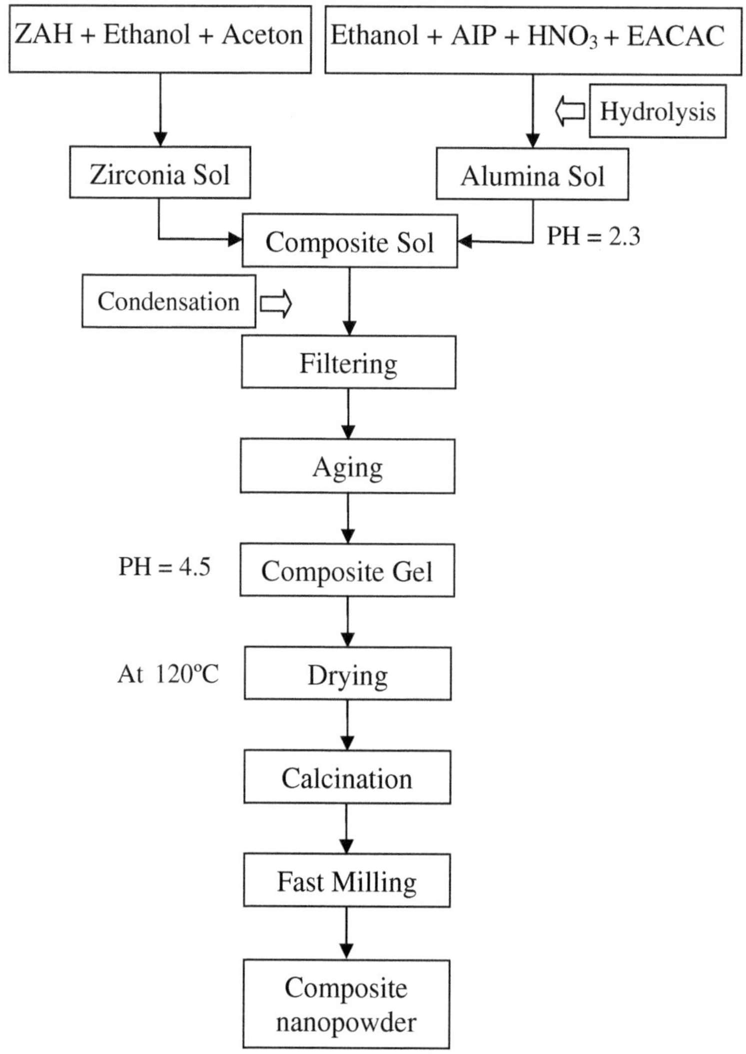

Figure 2 Typical flow chart of the processing of ZrO_2-Al_2O_3 composite nanopowders.

identification and crystallite size of ZmA nanocomposite powders were characterized using a Philips PW1800 X-ray diffractometer (FEI Co., Hillsboro, OR, USA) with nickel-filtered Cu Kα radiation ($\lambda = 1.5406$ Å) at 40 kV and 30 mA. The crystallite size of ZmA nanopowders was determined from the X-ray diffraction data using the following the Debye-Scherrer formula (Equation 1):

$$d = \frac{k\lambda \cos\theta}{\beta}, \tag{1}$$

where d is the crystallite size (in nanometers), $k = 0.9$ is a correction factor to account for particle shapes, λ is the wavelength of Cu target equals 1.54056 Å, β is the full width at half maximum intensity, and θ is the diffraction peak angle (Bragg's angle). Before calculating the crystallite size, line broadening due to the equipment, β_{Equip}, was subtracted from the obtained peak width, β_{Ob}, using Equation 2:

$$\beta^2 = \beta^2_{Ob} - \beta^2_{Equip}. \tag{2}$$

The BET surface area and the pore size distribution of the calcined nanopowders were determined by a BET surface area analyzer (Belsorp-mini 00165 series, BEL Japan, Inc., Osaka, Japan) using N_2 as the adsorptive gas. The morphology changes of the composite powders with different Al_2O_3 contents were examined by a SE-SEM (Wega/Tescan, TESCAN, a.s., Brno, Czech Republic).

Results and discussion

Thermal analysis

The TG-DT analyses for Z20A powders were carried out in air at a temperature range between 25°C and

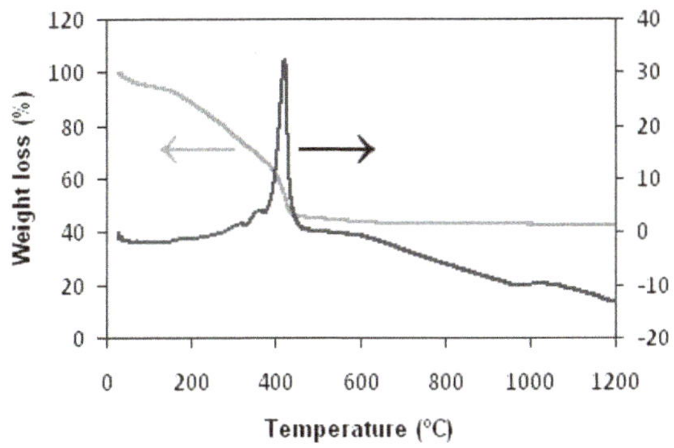

Figure 3 TG/DTA profiles for ZrO_2-$20Al_2O_3$ composite nanopowders at a heating rate of 5°C/min in air.

1,200°C with a heating rate of 5°C/min, and the resulting profiles are shown in Figure 3. The TGA data is compatible with the DTA data. In the first step of the TGA curve, about 8 wt.% is lost at 25°C to 180°C, which was taken to be physically adsorbed H_2O and ethanol evaporation from the powders at the same

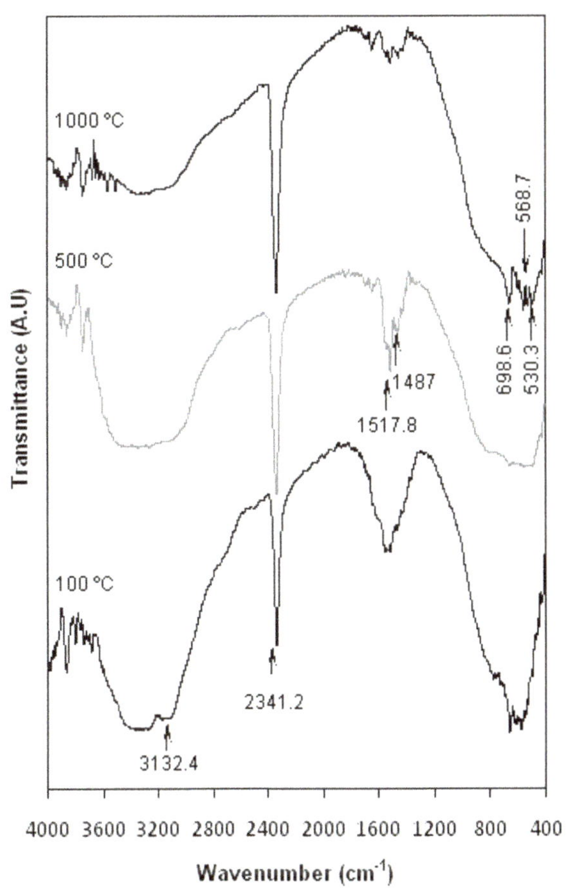

Figure 4 FTIR spectra for ZrO_2-$20Al_2O_3$ composite nanopowders at 100°C, 500°C, and 1,000°C.

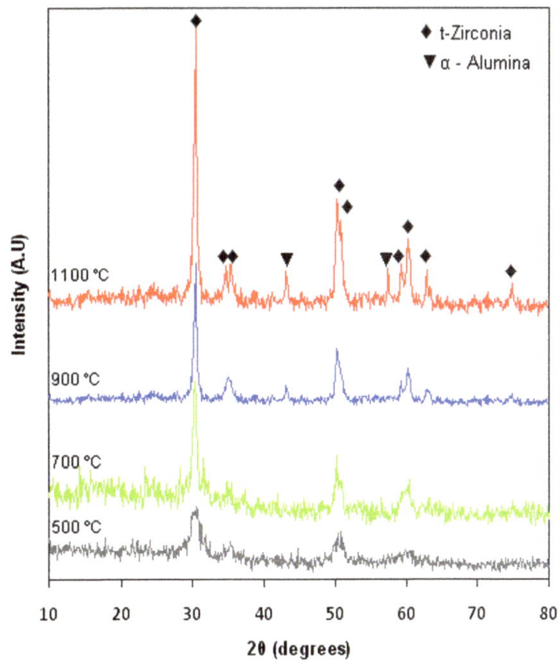

Figure 5 X-ray diffraction patterns of ZrO_2-$10Al_2O_3$ nanopowders synthesized by thermal treatment at 500°C, 700°C, 900°C, and 1,100°C.

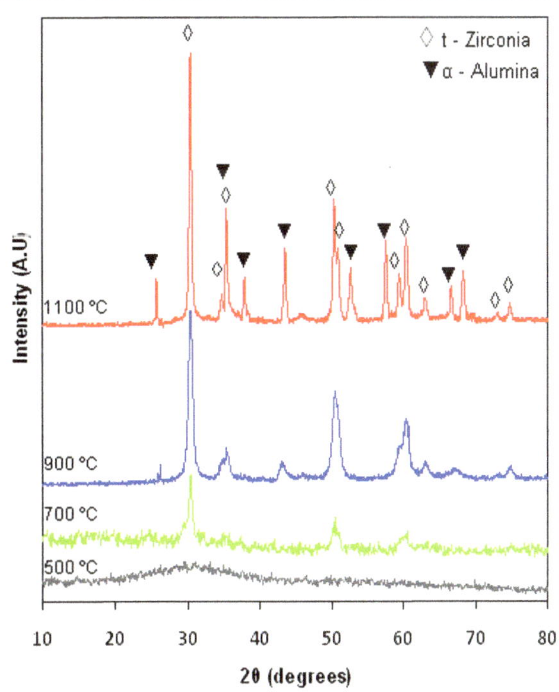

Figure 6 X-ray diffraction patterns of ZrO_2-$20Al_2O_3$ nanopowders synthesized by thermal treatment at 500°C, 700°C, 900°C, and 1,100°C.

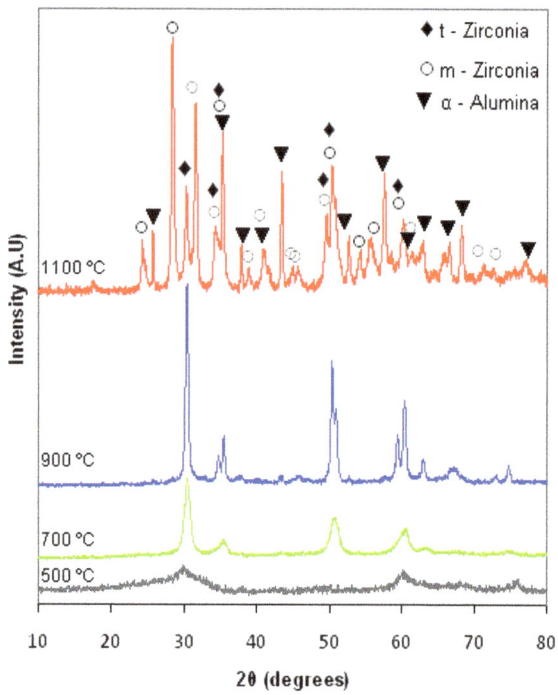

Figure 7 X-ray diffraction patterns of ZrO_2-$50Al_2O_3$ nanopowders synthesized by thermal treatment at 500°C, 700°C, 900°C, and 1,100°C.

time; the DTA curve shows a negative temperature differential, indicating endothermic reaction. The second step of the TGA curve shows about 32 wt.% loss at 180°C to 400°C; that is considered as the debonding of bonded H_2O and the elimination of organic compounds through oxidation. The DTA curve becomes two exothermic reactions at 320°C and 370°C. As the temperature rises, in the last step of the TGA curve, 16 wt.% is lost at 400°C to 460°C. It shows a sharp weight loss accompanied with a significant exothermic peak at 390°C to 470°C in the DTA curve. Possibly, the oxidation of nitro compounds that may be formed by nitration of ethyl acetoacetate and HNO_3 is the cause of this huge exothermic reaction. When the samples were annealed above 500°C, no further weight loss of the organics was observed. In the DTA curve, however, the shift of the base line at about 1,000°C shows that thermal energy was produced during the calcination process, which may correspond to form the crystallized t-ZrO_2 and/or α-Al_2O_3 phases in Z20A composite powders, as will be shown later by XRD analysis.

The FTIR at 4,000 to 400 cm^{-1} for Z20A powders as a function of heat treatment temperature was studied by FTIR spectroscopy, and the results are shown in Figure 4. The broadening absorption peak at 3,450 cm^{-1} belongs to OH-stretching band vibrations. It can be seen that increasing the temperature up to 1,000°C causes a decrease in intensity of this broadening peak, which is due to evaporation of organic compounds and water from the powders. May be due to the moisture adsorbed during the test, the mentioned broad band has appeared even at 1,000°C. The obtained peak at 3,132 cm^{-1} is caused by the C-H vibration of hydrocarbon groups. All of the samples have an intense peak around 2,341 cm^{-1} that can be attributed to the vibrations of O = C = O bonds. Zr-OH-group bending band vibration peaks appeared at 1,517 and 1,487 cm^{-1}. A phenomenon yielded from the repulsion of hydrated groups causes the intensity of these peak vibrations to decrease with increasing thermal treatment temperature. Also, the peaks

Table 1 Characterization of zirconia and alumina in ZmA (where *m* = 10, 20, and 50) nanopowders

Al_2O_3 (mol%)	Temperature (°C)	Identified phases	Crystallite size (nm)		Specific surface area (m^2/g)	Mean pore diameter (nm)
			Zirconia	Alumina		
10	500	t	7	-	164.32	8.55
	700	t	14	-		
	900	t	18	-		
	1,100	t, α	20	11		
20	500	-	-	-	139.46	10.74
	700	t	8	-		
	900	t, α	10	4		
	1,100	t, α	13	8		
50	500	-	-	-	102.78	14.49
	700	t	6	-		
	900	t, α	7	5		
	1,100	t, α, m	9	6		

The nanopowders were calcined at different temperatures for 2 h.

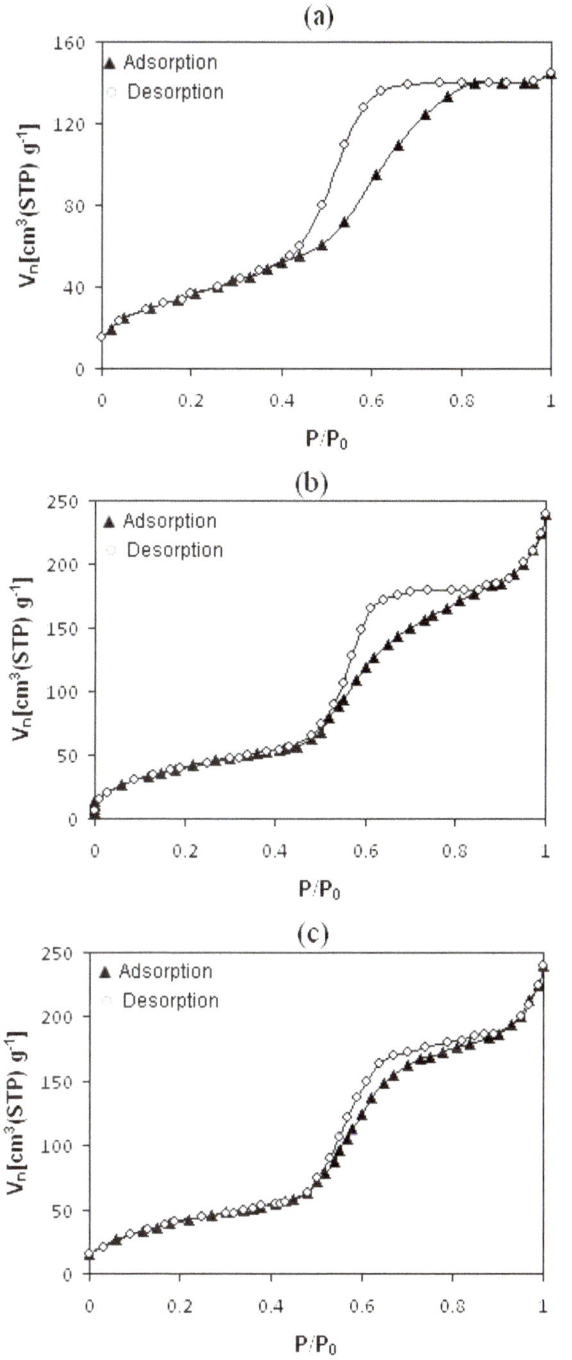

Figure 8 Nitrogen adsorption/desorption isotherm curves of nanopowders calcined at 1,100°C for 2 h and milled. (**a**) 10 mol%, (**b**) 20 mol%, and (**c**) 50 mol% Al_2O_3. P/P_0 is the relation of N_2 pressure in every moment to its saturated vapor pressure; V_n is the volume of nitrogen gas adsorbed at pressure P.

between 1,070 and 612 cm^{-1} correspond to the Al-O vibration [13], whereas the absorption peaks below 670 cm^{-1} indicate Zr-O bond vibrations [14]. Heating the gel at 500°C causes broadening of peaks in the lower frequency range (400 to 1,000 cm^{-1}), owing to the overlapping of Zr-O and Al-O infrared vibrations of alumina and zirconia phases [15]. Thus, these individual peaks could not be detected in the nanopowder calcined at 100°C and 500°C. However, the peaks corresponding to Al-O and Zr-O could be clearly observed in the sample calcined at 1,000°C.

Crystallization

Figures 5, 6, and 7 show XRD patterns of ZrO_2-Al_2O_3 powders containing 10, 20, and 50 mol% Al_2O_3, respectively, that were synthesized at 500°C, 700°C, 900°C, and 1,100°C. Also, the crystallite size of zirconia and alumina phases calculated from the XRD patterns is listed in Table 1.

The XRD patterns indicate that the dried composite gels existed as an amorphous mixture, and heating at temperatures lower than 500°C did not cause crystallization of the gels, except for the Z10A sample, in which some peaks of the t-ZrO_2 phase has appeared (Figure 5). In all samples, the higher intensity and much sharper peak of the t-ZrO_2 peaks were observed with increasing temperature. As shown in Figure 6, the t-ZrO_2 and α-Al_2O_3 crystallized phases appeared in Z20A powders, but in the powders containing 50 mol% Al_2O_3, the t-ZrO_2 phase disappeared, and the m-ZrO_2 phase crystallized beside the α-Al_2O_3 phase when heat treatment process was carried out at 1,100°C (see Figure 7). It seems that the ZrO_2-phase crystallization temperature increases with increasing Al_2O_3 content, which is also reported by other researchers [16-18]. They suggested that the crystallization constraint by the added Al_2O_3 would result in higher activation energy for the crystallization process. The peaks of α-Al_2O_3 were found to appear above 900°C in the Z20A and Z50A powders.

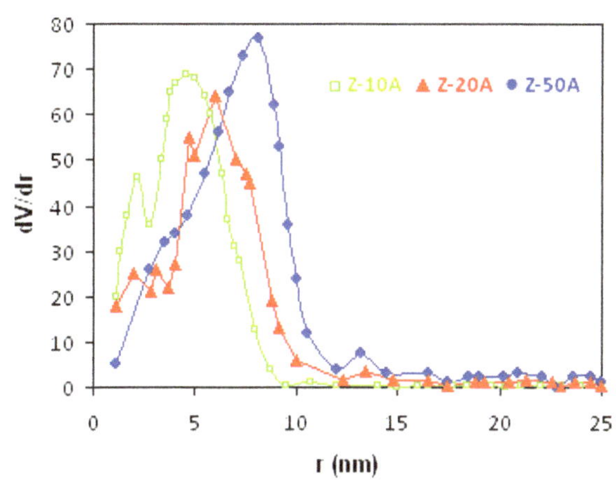

Figure 9 Pore size distribution (dV/dr) of ZrO_2-m mol% Al_2O_3 (where m = 10, 20, and 50) powders. The powders were calcined at 1,100°C for 2 h.

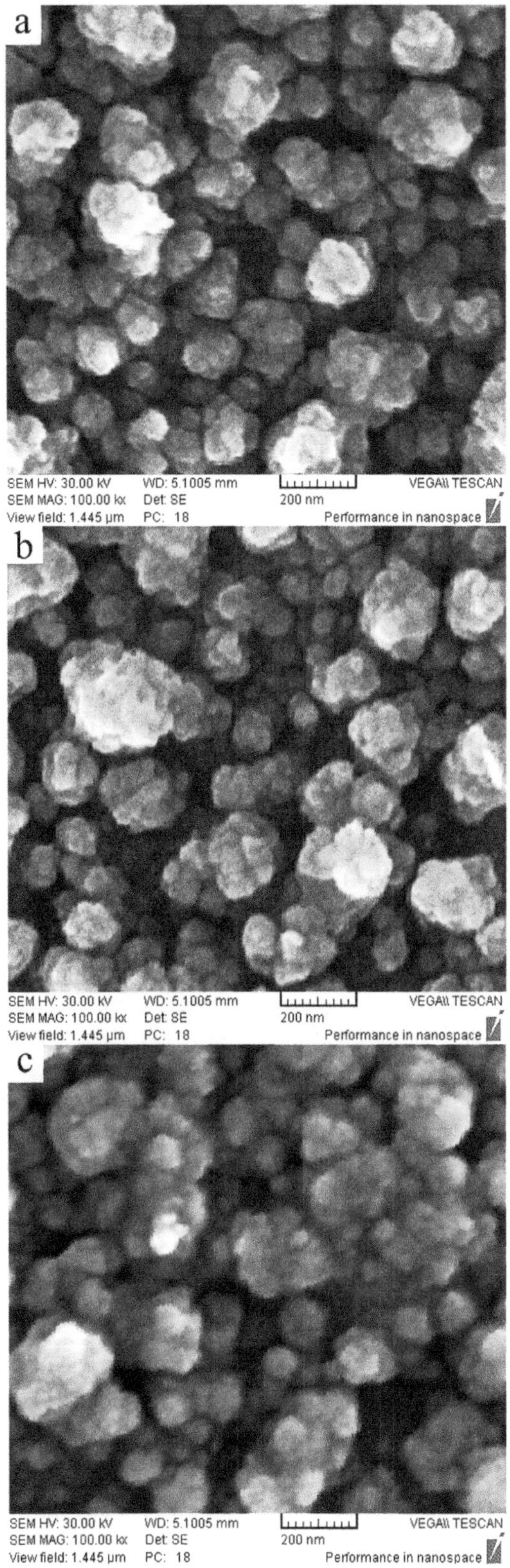

Figure 10 SE-SEM micrographs of the produced nanopowders calcined at 1,100°C for 2 h and milled. (**a**) 10 mol%, (**b**) 20 mol%, and (**c**) 50 mol% Al_2O_3.

However, in these composite powders, the formation of α-Al_2O_3 was obviously detected at a temperature of 1,100°C (Figures 6 and 7), but it was not visible for Z10A (Figure 5). It may suggest the formation of the tetragonal phase as a solid solution in Z10A nanopowders that has been previously reported by Yoshimura et al. [16] and other researchers [17,19].

The nitrogen adsorption and desorption isotherm curves of the synthesized ZmA composite nanopowders are shown in Figure 8a,b,c. These curves show type-IV hysteresis loops that characterize mesoporous adsorbents. The hysteresis loops in these curves result from wedge-shaped capillaries with a closed edge at the narrower side [20]. The decrease in the difference between the desorption and adsorption curves in the hysteresis loops and their shift to higher relative pressures (P/P_0) with increasing amount of alumina may be ascribed to an increase in the pore size, which consequently leads to decreasing the capillary effect.

The specific surface BET areas were obtained from the isotherms for the specimens. Then, the mean pore radii were calculated by the BJH method [21], and the resulting pore size distribution curves are shown in Figure 9.

The results obtained from Figures 5, 6, 7, 8, and 9 are summarized in Table 1. According to the data presented in this table, an increase in alumina content leads to a reduction in the surface area of the nanopowders and a decrease in the average crystallite sizes which have been obtained by substituting their XRD results, in Debye-Scherrer's formula (see Equation 1). The median pore size increases by adding alumina in composite nanopowders, which may be due to a decrease in the t-ZrO_2 content with high surface area.

It should be pointed that according to Garvie's [8-10] studies, the smaller size particles of the t-ZrO_2 phase are more stable at room temperature because of their larger specific surface area. This statement is in a good agreement with the results obtained from Table 1. The large area of particles in the Z50A composite nanopowders has caused the disappearance of the t-ZrO_2 phase and crystallization of m-ZrO_2 beside the α-Al_2O_3 phase.

Microscopic observation

The SEM observations of ZmA (where $m = 10$, 20, and 50) composite nanopowders heat-treated at 1,100°C for 2 h are shown in Figure 10a,b,c, respectively. From these micrographs, it can be seen that most of the particles are spherical in shape. They tend to be larger and agglomerate with increasing Al_2O_3 content that confirms the results obtained from BET analysis. The average agglomerated size of particles in Figure 10c is about 242 nm which is above the critical size for the composite powders, so the monoclinic is the stable phase.

Conclusion

The sol–gel process has been employed to synthesize unstabilized ZmA (where $m = 10$, 20, and 50) composite nanopowders using zirconium acetate hydroxide and aluminum isopropoxide as precursors. The results obtained in the present study are as follows:

1. Increasing the Al_2O_3 content in composite nanopowders shifts the t-ZrO_2-phase crystallization toward higher temperatures. Finally, this phase disappeared in Z50A nanopowders, and the α-Al_2O_3 phase is observed beside the m-ZrO_2.
2. Increasing the Al_2O_3 content results in a decrease in the surface area and crystallite size and an increase in the pore size of nanopowders.
3. The morphology of particles also was affected by Al_2O_3 content, and increasing that tends to form more agglomerated particles.

Competing interests
The authors declare that they have no competing interests.

Authors' contributions
EN carried out the structural studies, participated in the sequence alignment, and drafted the manuscript. MS participated in the sequence alignment and drafted the manuscript. HRR participated in the design of the study and revision of the language of the manuscript. FT conceived the study and participated in its design and coordination. All authors read and approved the final manuscript.

Acknowledgments
The authors wish to thanks the financial support of this work by the Iranian Nanotechnology Initiative Council and Iran Research Institute of Petroleum Industry.

Author details
[1]Department of Materials and Metallurgical Engineering, Iran University of Science and Technology (IUST), Narmak, Tehran 16846-13114, Iran.
[2]Department of Materials Engineering, Sahand University of Technology, Tabriz 11111-53317, Iran.

References
1. Shackelford JF, Doremus RH (ed) (2008) Ceramic and glass materials: structure, properties and processing. Springer, New York
2. Zheludkevich ML, Salvado IM, Ferreira MGS (2005) Sol–gel coatings for corrosion protection of metals. J Mater Chem 15:5099–5111
3. Wang H, XuP ZW, Shen L, Du Q (2005) Transparent poly (methyl methacrylate)/silica/zirconia nanocomposites with excellent thermal stabilities. Polym Degrad Stab 87:319–328
4. Yang G, Li J, Wang G, Yashima M, Min S (2005) Influences of ZrO_2 nanoparticles on the microstructure and mechanical behavior of Ce-TZP/Al_2O_3 nanocomposites. J Mater Sci 40:6087–6090
5. Adolfsson E, Shen Z (2006) Densification of zirconia-hydroxyapatite ceramics without phase changes. Key Eng Mater 309:1141–1144
6. Mobasherpour I, Hashjin MS, Toosi SSR, Kamachali RD (2009) Effect of the addition ZrO_2-Al_2O_3 on nanocrystalline hydroxyapatite bending strength and fracture toughness. Ceram Int 35:1569–1574
7. Sakka S, Kozuka H (2005) Handbook of sol–get science and technology: processing, characterization and applications, vol. I sol–gel processing. Kluwer Academic, New York
8. Nouri E, Shahmiri M, Rezaie HR, Talayian F (2011) Investigation of structural evolution and electrochemical behaviour of zirconia thin films on the 316 L stainless steel substrate formed via sol–gel process. Surf Coat Technol 205:5109–5115
9. Nouri E, Shahmiri M, Rezaie HR, Talayian F (2012) A comparative study of heat treatment temperature influence on the thickness of zirconia sol–gel thin films by three different techniques: SWE, SEM and AFM. Surf CoatTechnol 206:3809–3815
10. Garvie RC (1965) The occurrence of metastable tetragonal zirconia as a crystallite size effect. Phys Chem 69:1238–1243
11. Garvie RC, Goss MF (1986) Intrinsic size dependence of the phase transformation temperature in zirconia microcrystals. J Mater Sci 21:1253–1257
12. Garvie RC (1978) Stabilization of the tetragonal structure in zirconia microcrystals. J Phys Chem 82:218–224
13. Chandradass J, Jun B, Bae DS (2008) Effect of different fuels on the alumina–zirconia nanopowder synthesized by sol–gel autocombustion method. J Non–Crystalline Solids 354:3085–3087
14. Quan ZW, Wang LS, Lin J (2005) Synthesis and characterization of spherical ZrO_2:Eu^{3+} phosphors by spray pyrolysis process. Mater Res Bull 40:810–820
15. Sarkar D, Mohapatra D, Ray S, Bhattacharyya S, Adak S, Mitra N (2007) Synthesis and characterization of sol–gel derived ZrO_2 doped Al_2O_3 nanopowder. Ceram Int 33:1275–1282
16. Yoshimura M, Oh ST, Sando M, Niihara K (1999) Crystallization and microstructural characterization of ZrO_2 (3 mol% Y_2O_3) nano-sized powders with various Al_2O_3 contents. J Alloys Compd 290:284–289
17. Balmer ML, Lange FF, Levi CG (1994) Metastable phase selection and partitioning for $Zr_{(1-x)}Al_xO_{(2-x/2)}$ materials synthesized with liquid precursors. J Am Ceram Soc 77:2069–2075
18. Balmer ML, Lange FF, Jayaram V, Levi CG (1995) Development of nano-composite microstructures in ZrO_2-Al_2O_3 via the solution precursor method. J Am Ceram Soc 78:1489–1494
19. Ferkel H, Naser J, Riehemann W (1997) Laser-induced solid solution of the binary nanoparticle system Al_2O_3-ZrO_2. Nanostruct Mater 8:457–464
20. Jayaseelan DD, Rani DA, Nishikawa T, Awaji H (2000) Powder characteristics, sintering behavior and microstructure of sol–gel derived ZTA composites. J Eur Ceram Soc 20:267–275
21. Gregg SJ, Sing KSW (1982) Adsorption, surface area and porosity. Academic, New York

Electrochemical determination of calcium channel blocker drugs using multiwall carbon nanotube-modified glassy carbon electrode

Abdul Razak Mohamed Sikkander[1,3], Chinnappiyan Vedhi[4] and Paramasivam Manisankar[1,2*]

Abstract

Background: Calcium channel blockers belonging to the dihydropyridine family have been used as a potent arterial vasodilator in the management of angina and cardiovascular diseases. In the design of glassy carbon electrode coatings for the study of the same, multiwall carbon nanotubes (MWCNT) are favored. They are known for the reduction and oxidation of electroactive species towards cathodic and anodic direction with the simultaneous enhancement of the peak current.

Results: The MWCNT-modified glassy carbon electrode exhibited a sharp anodic peak potential at around at 0.5 V for amlodipine (AMLD) andnimodipine (NIMD), and 0.4 V for felodipine (FELD) in the cyclic voltammograms. The antihypertensive drugs were determined using a simple two-step procedure developed which comprised a preconcentration step followed by the differential pulse stripping voltammetric quantification. Concentration calibrations were linear within the range from 0.01 to 0.3 μg/mL for AMLD and FELD, and 0.025 to 0.3 μg/mL for NIMD.

Conclusion: The lower limit detection (LOD) was found to be very low on modified electrode. The LOD is 0.005 μg/mL for AMLD and FELD, and 0.01 μg/mL for NIMD. The prepared electrode showed an excellent electrocatalytic activity towards the oxidation of antihypertensive drugs leading to a remarked improvement in sensitivity. An electrochemical sensor featuring the MWCNT-modified electrode was applied successfully as the result for the calcium channel blocker determination in pharmaceutical samples.

Keywords: Amlodipine, Felodipine, Nimodipine, Electrochemical, Multiwall carbon nanotubes

Background

Calcium channel blocker belonging to the dihydropyridine family has commonly been used as a potent arterial vasodilator in the management of angina and cardiovascular diseases [1]. Electrochemistry plays an important role to study the formation of radical and its reactivity in one-pot systems [2,3]. There are several reports in literature concerning the development of stable carbon nanotube (CNT)-based electrodes for environmental samples, electrochemical sensors, electrocatalysis, and electrochemical estimation of drugs and compounds of biological interest [4,5]. However, simple but effective method for the development of homogeneously and stably assembled CNT-based electrode is particularly desired for electroanalytical determinations. Multiwall carbon nanotube (MWCNT)-based electrodes were prepared generally by casting MWCNT suspension on conventional electrode surface [6,7]. The resulting electrodes have been successfully utilized in the sensitive detection of various biological molecules such as uric acid [4], folic acid [6], and cytochrome c [7]. Generally, MWCNT-based electrodes enhance the detection sensitivity and improve reversibility as they can promote electron transfer [8].

Amlodipine besylate is a dihydropyridine derivative with calcium antagonist activity [9]. The main metabolic pathway is oxidation of dihydropyridine ring to the pyridine analog [10]. Amlodipine (AMLD) has side effects such as peripheral edema (often), dizziness, palpitations,

* Correspondence: pms11@rediffmail.com
[1]Department of Chemistry, Periyar University, Salem, Tamil Nadu 636011, India
[2]Department of Industrial Chemistry, Alagappa University, Karaikudi, Tamil Nadu 630003, India
Full list of author information is available at the end of the article

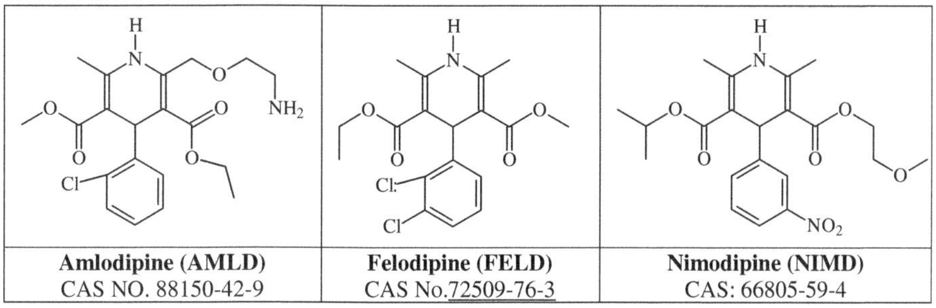

Scheme 1 Calcium channel blocker drugs.

muscle-, stomach- or headache, dyspepsia, nausea (very often), blood disorders, development of breasts in men (gynecomastia), impotence, depression, insomnia, tachycardia, and gingival enlargement (sometimes). The amlodipine besylate in tablets and biological fluids was determined using chromatographic techniques such as gas chromatography (GC) [11], liquid chromatography (LC) [12], and high performance thin layer chromatography [13]. Chromatographic methods offer a high degree of selectivity but need sample clean-up and relatively heavy instrumentation. Few spectrophotometric methods have been reported for amlodipine besylate determination [14]. A differential pulse voltammetric method has been described for the determination of amlodipine besylate in tablets at pH 5.5 [15]. The electrochemical behavior at the dropping mercury and glassy carbon electrodes (GCE) of some calcium antagonist drugs has been reported [16].

Nimodipine (NIMB), a dihydropyridine calcium channel antagonist, has potent vasodilatory effect on cerebral vessels and increases cerebral blood flow. Nimodipine, in high doses (2 to 10 mg/kg), has been shown to improve memory retention and/or memory recall process in aged rats. The Food and Drug Administration has classified the side effects into groups based on dose levels given every 4 h. For the high dosage group (90 mg), less than 1% of the group experienced adverse conditions including itching, gastrointestinal hemorrhage, thrombocytopenia, neurological deterioration, vomiting, diaphoresis, congestive heart failure, hyponatremia, decreasing platelet count, disseminated intravascular coagulation, and deep vein thrombosis. The electrochemical reduction of 4-nitrophenyl-1,4-dihydropyridine compounds such as nifedipine, nitrendipine, nimodipine, nicardipine, and furnidipine in protic media [17] follows the general pattern of nitroaromatic compounds involving a single four-electron step producing the hydroxylamine derivative. However, the electrochemical reduction of nitrendipine, nimodipine, and nifedipine in mixed aqueous-DMF media resulted in the generation of the one-electron reduction product, the nitro radical anion [18].

Cyclic voltammetry was also performed on insulin, valsartan, and felodipine (FELD) separately with the potential window applied from −0.5 V to +1.00 V vs. Ag/AgCl. Infusion of valsartan and felodipine has been reported to lower the striatum glucose level significantly [19]. Stereospecific determination of nimodipine was reported using an ES-OVM column with detection at 230 nm, resulting in a high specific separation of enantiomers with a limit of

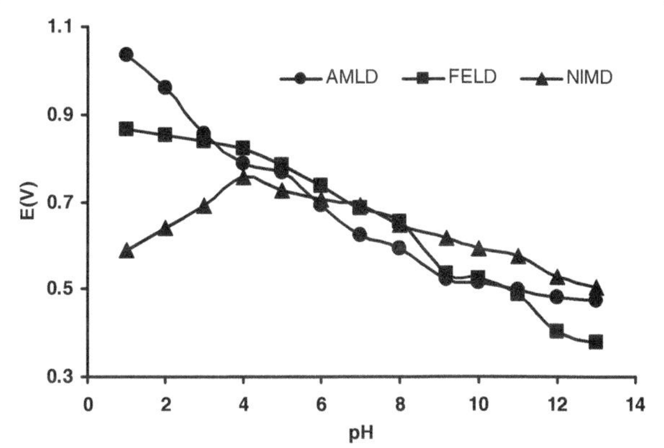

Figure 1 Plot of peak potential vs. pH.

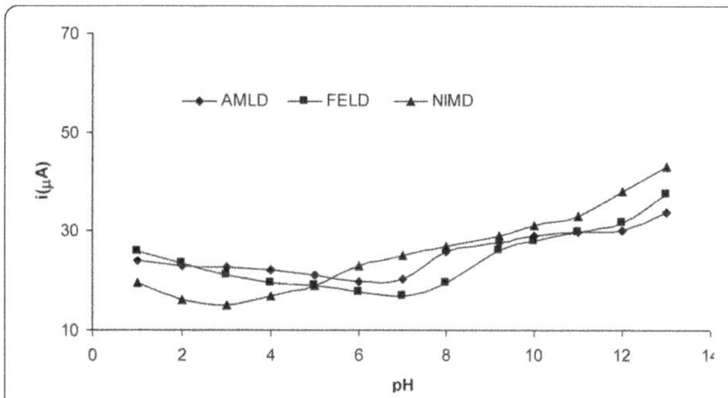

Figure 2 Plot of peak current vs. pH.

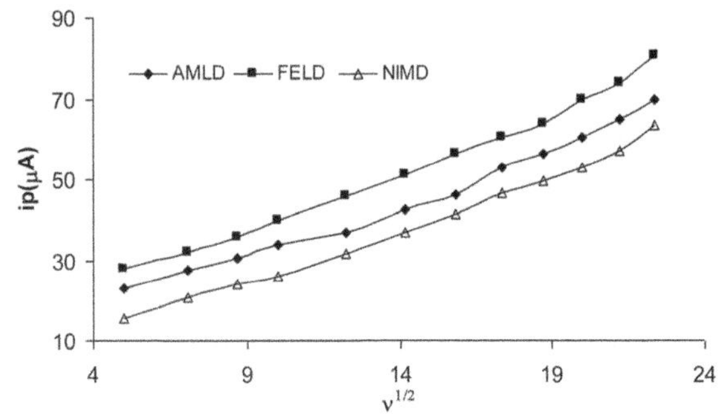

Figure 3 Plot of peak current vs. square root of scan rate.

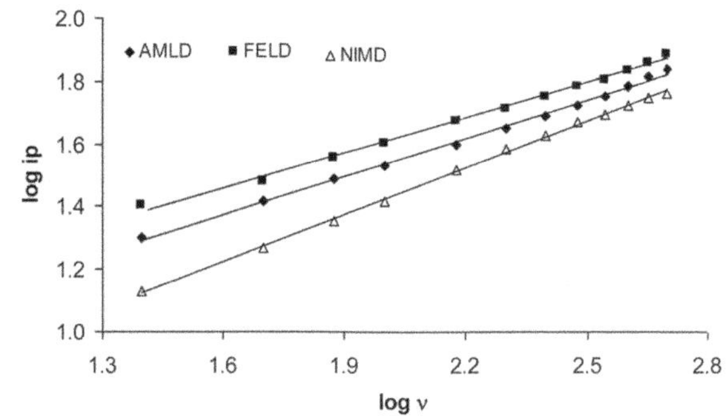

Figure 4 Plot of log peak current vs. log scan rate.

quantitation for each enantiomer of about 1 ng/ml from 1-ml plasma [20]. The stability of the retention volumes of the enantiomers of nimodipine by LC has been checked to guarantee the reproducibility of enantiomer collection to allow automated LC analysis [21]. Introduction of felodipine as an enantioselective determination method for felodipine in human plasma has been described involving off-line detection using capillary GC on a DB-1 column with ECD following chiral separation on a Chiralcel OJ column (Chiral Technologies, Illkirch, France) monitored at 240 nm [22]. Quantitation limits of 0.1 ng/ml were demonstrated for a sample size of 0.5-ml biological fluid, while felodipine [23] and nisoldipine [24] were finally determined by GC-MS in PEI mode [25].

Until now, no publications concerning the electroanalytical determination of antihypertensive drugs in pharmaceutical formulations are available in the literatures. Therefore, the aim of the present investigation is to investigate the electrochemical behavior of antihypertensive drugs on MWCNT-modified GCE and to develop sensitive stripping voltammetric methods for their determination.

Methods

Electrochemical workstation (760C model, CH Instruments, TX, USA) was employed mainly for carrying out electroanalytical studies. The three calcium channel blocker drugs of amlodipine, felodipine, and nimodipine (Scheme 1) were received from CIPLA Ltd, Mumbai, India and used as such.

The stock solutions were made up in methanol/double distilled purified water (TKA-LAB, Thermo Electron LED GmbH, Germany) (80:20). For studies in aqueous methanol media, Britton-Robinson buffers, 4.0, 7.0, 9.2, and 0.1 moldm^{-3} KOH, and 0.1 mol dm^{-3} H_2SO_4 were used as the media for the analysis. Multiwall CNT produced by arc method was purchased from Sigma-Aldrich (MO, USA) and sodium dodecyl sulfate (SDS) from Merck (Merck & Co., Inc, NJ, USA).

Purging of nitrogen was done for analyte solution placed in the electrochemical cell of 15-ml capacity for 25 min under stirring, and then voltammograms were recorded while blanketing nitrogen gas. To get reproducible results, great care was taken in the electrode pretreatment. The GCE was pretreated in two ways as described earlier [26].

Preparation of MWCNT-modified GCE

One milligram MWCNT was dispersed in 1 mL of 0.1 M sodium dodecyl sulfate using an ultrasonicator to give black suspensions [27]. Cast films were prepared by placing 5 μL of the MWCNT/surfactant suspensions on GCE and then evaporating them in an oven at 50°C.

Results and discussion

In the experiments carried out for the effect of pH, pH 13.0 was chosen as the best for further electrochemical studies. In the case of experiments on the influence of modifier, the *i*p and the *E*p values are presented in Figures 1 and 2. As it is observed, the MWCNT-modified electrode gives higher peak intensity values for all the pharmaceuticals. The modifier also influences the *E*p values, although the shifts caused in the peaks of the analytes are not significantly relevant.

In the investigation of cyclic voltammetric behavior of drugs, the peak potentials correlated well with log scan rate and resulted in straight lines. The fractional 'αn' value calculated from the slope confirmed irreversible electron transfer. The peak current showed an increasing trend with sweep rate and straight lines with good correlation coefficients, indicating adsorption. The plots, i_p vs. $\nu^{1/2}$, were curve lines (Figure 3). The log peak current vs. log scan rate showed straight line (Figure 4), and its slope value is above 0.5. Thus, it is further confirmed that adsorption of the substrate on the pores and surface of the electrode and the overall reaction was adsorption-controlled.

In the investigation of differential pulse stripping voltammetric behavior of drugs, maximum peak current was observed for AMLD, FELD, and NIMD at 30, 20, and 10 s respectively. The decreased current above the

Figure 5 SEM photographs. (**a**) AMLD, (**b**) FELD, and (**c**) NIMD on MWCNT-modified GCE.

maximum current signal condition might be due to the saturation of the electrode surface and blocking of the products formed on the surface. The accumulation of the drugs on the modified electrode surface was ascertained by carrying out SEM analysis.

SEM was employed to study the surface morphology of the three accumulated drugs on MWCNT-coated glassy carbon electrode. The stem-like structure of the coating confirmed the presence of MWCNTs on GCE. The average tube size of the material is 50 nm as reported earlier by us [27,28]. The drug AMLD was adsorbed on MWCNT electrode during the accumulation and exhibited granular sponge-like structure (Figure 5a). FELD exhibited broken pitch and sponge-like structure (Figure 5b), and NIMD exhibited tide sponge-like structure (Figure 5c). Different surface morphology confirmed the accumulation of drugs on the MWCNT/GCE.

The initial scan potential was also an important parameter in controlling both peak potential and peak height in the stripping voltammogram. The initial potential was varied between −0.4 to 0.2 V, and an initial scan potential at −0.3 V for AMLD, −0.1 V for FELD, and 0 V for NIMD led to higher peak current response. Pulse height was varied between 0.025 and 0.25 V. This variation had shown a decrease in peak current with increase in applied pulse height after 0.05 V. Hence, pulse height of 0.05 V was chosen due to increased current response for all drugs. The effect of pulse period demonstrated that the stripping peak current increased up to 50 ms and then decreased with an increase in pulse width from 75 to 125 ms for three drugs. The peak current decreased with an increase in pulse width from 25 to 100 ms, and a pulse width of 50 ms was selected. Thus, the maximum peak current conditions were arrived at, and the results are presented in Table 1. These conditions were used to study the effect of concentration.

Analytical characteristics

Typical differential pulse stripping voltammograms for AMLD, FELD, and NIMD obtained under the maximum peak current experimental conditions were presented in Figure 6. As the concentration of the drugs increased, the stripping peak current increased. Calibration plots were made and presented in Figure 7. The limits of concentration were 0.01 to 0.3 μg/mL for AMLD and FELD, and 0.025 to 0.3 μg/mL for NIMD. The LOD is 0.005 μg/mL for AMLD and FELD, and 0.01 μg/mL for NIMD. The precision of the method was ascertained by measuring the peak current of the drugs' response in five standard samples. Ten replicates were analyzed, and standard deviations were calculated. The relative standard deviation was 2.5% for a concentration 50 μg/mL of AMLD, 2.7% for the same concentration of FELD, and 3.1% for 50 μg/mL of NIMD. The low value of standard deviation indicated good reproducibility and feasibility of this method for the determination of drugs.

Table 1 Optimum experimental conditions in DPSV

Variable	Range studied			Optimum value		
	AMLD	FELD	NIMD	AMLD	FELD	NIMD
pH	1 to13	1 to 13	1 to 13	13.0	13.0	13.0
Accumulation potential (V)	0 to 0.6	0 to 0.6	0 to 0.6	0.3	0.4	0.3
Accumulation time (s)	15 to 90	15 to 90	15 to 90	15	30	30
Initial scan potential (V)	−0.2 to 0.3	−0.2 to 0.3	−0.2 to 0.3	0.1	0.1	0
Pulse height (PH; mV)	25 to 150	25 to 150	25 to 150	50	75	50
Pulse width (PW; ms)	25 to 150	25 to 150	25 to 150	50	50	50
Scan increment (SI; mV)	2 to 20	2 to 20	2 to 20	4	6	6
Stirring rate (rpm)	50 to 250	50 to 250	50 to 250	250	250	250
Rest period (s)	2 to 10	2 to 10	2 to 10	5	5	5

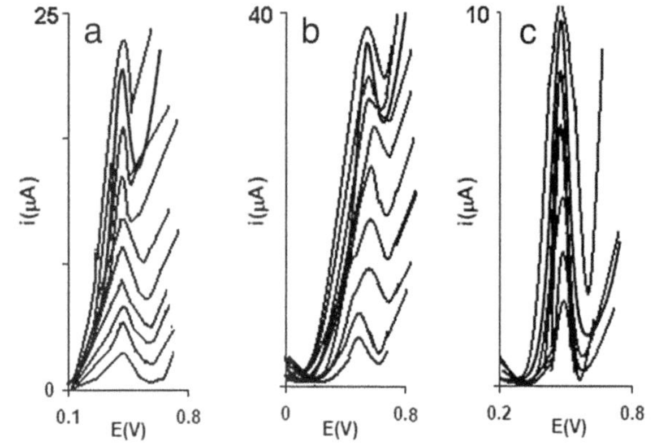

Figure 6 Differential pulse stripping voltammetry. DPSV of (**a**) AMLD (**b**) FELD and (**c**) NIMD under optimum conditions on MWCNT-modified GCE.

In the square wave voltammetric determination of AMLD on SWCNT/EPPGE and MWCNT/EPPGE for 1.0 × 10^{-9} mol L^{-1} and 5.0 × 10^{-9} mol L^{-1} [29], respectively, 0.025 μg mL^{-1} by spectrofluorimetric method was reported [30]. Another compound of FELD was also reported: 0.3 to 1.5 μg cm^{-1} by AAS [31] and range studies of 1.0 to 40.0 μg/mL by spectrophotometric method [32].

Selectivity of electrode is very important. Among the electrode system, the modified MWCNT is very sensitivity for all three drug compounds. Range studies are also very low compared to those reported in the literature since MWCNT was selected for the determination of the drugs.

The reproducibility of electrode was also an important parameter for electrochemical determination technique. The MWCNT modified system is highly stable up to 100 cycles of experimental conditions. After the 100 cycles, the reproducibility decreased slowly. The number of experiments continued from101to 200 cycles, and the 10% reproducibility decreased.

Pharmaceutical sample analysis

In order to evaluate the applicability of the proposed method, three commercial samples in combination or in pure form containing anyone of AMLD, FELD, and NIMD were selected. The pharmaceutical samples were collected from medical shops at Karaikudi, Tamilnadu, India. Various tablets having AMLD, FELD, and NIMD were examined for the estimation of content of drugs. The tablets were dissolved in methanol, and then the filtrate was further evaporated to get the drug in pure form. The residue was dissolved in known quantity of methanol and transferred into a 250-ml calibrated flask and made up to the mark. A 10-ml portion of this solution was transferred into a 50-ml calibrated flask, and 0.1 mM NaOH containing 50% aqueous methanol was used to dilute the contents of the flask to the required volume. The standard addition method was used. An aliquot of 0.05 ml of the 0.1 μg/mL standard stock solution was added to the solution prepared as described above. Differential pulse stripping voltammetric studies under the maximum current signal experimental conditions were carried out and the trace amount of drugs in the sample were determined. A relative standard deviation of 2.9% was obtained for 0.1 μg/mL AMLD for ten identical measurements. The relative standard deviation of 2.9% was obtained for 0.1 μg/mL of FELD and NIMD for ten identical measurements. Thus, the suitability of this method for the determination of AMLD, FELD, and

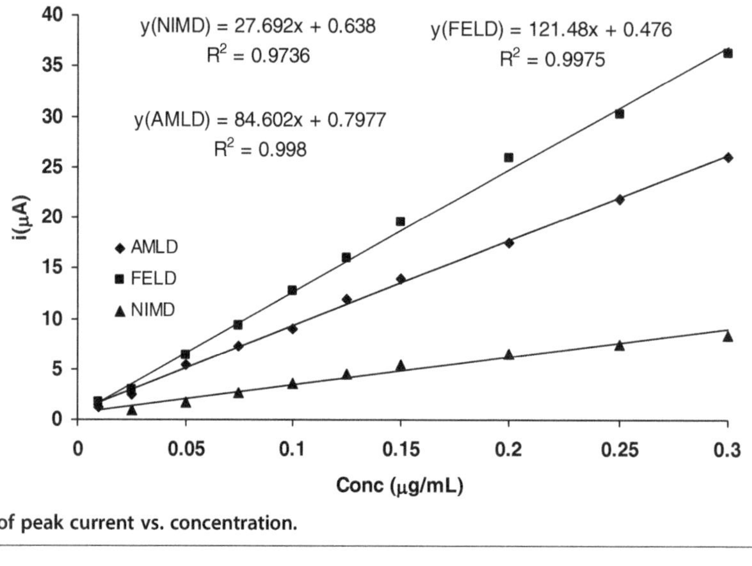

Figure 7 Calibration plot of peak current vs. concentration.

Table 2 Amount of drugs in tablets determined by DPSV in tablets

Brand name	Company name	Tablets (mg)	Experimental value(mg)	% RSD value
Amlodipine				
Amace	Systopic	5	4.80	2.1
Amlodac	ZydusMedica	10	9.96	2.2
Amlopres	Cipla	10	9.85	2.1
Myodura	WOCKHARDT	10	9.90	2.2
Lama	Stadmed	10	9.88	2.5
Card	Jagsonpal	10	9.95	2.3
Calchek	Ipca	10	9.94	2.7
Felodipine				
Felogard	Cipla	10	9.93	2.1
Plendil	Astra Zeneca	10	9.90	2.4
Nimodipine				
Nimodip	USV	30	29.27	2.5

NIMD in real sample was verified. The results are presented in the Table 2.

Experimental

Effect of pH

Britton-Robinson buffer/acid/alkaline solution was selected as the support electrolyte to find the optimal pH values for every analyte. The range of pH investigated was from 1.0 to 13.0. Values outside of this interval did not give either oxidation or reduction waves. Moreover, in some cases, signals were very close to the discharging current of the background, making the quantification very difficult. The pH affects both peak potential (*E*p) and peak intensity (*i*p) values. With respect to the first parameter, *E*p values for the three pharmaceuticals decrease (or increase in absolute value) with pH, being more positive (Figure 1). Figure 2 shows the dependence of *i*p with respect to pH for all the three drugs. From the curves, the optimal pH values for every one of them were deduced. This study was focused in order to find particular zones of potential for every compound that allowed the sequential determination of the three pharmaceuticals in a unique biological sample.

Influence of modifier

*i*p and the *E*p values using unmodified GCE and MWCNT-modified carbon electrode were studied. The modifier is expected to give to higher *i*p value and the longest distance from the discharging current of the background.

Cyclic voltammetric behavior of drugs

Cyclic voltammetric behavior of amlodipine, felodipine, and nimodipine on MWCNT-modified GCE was carried

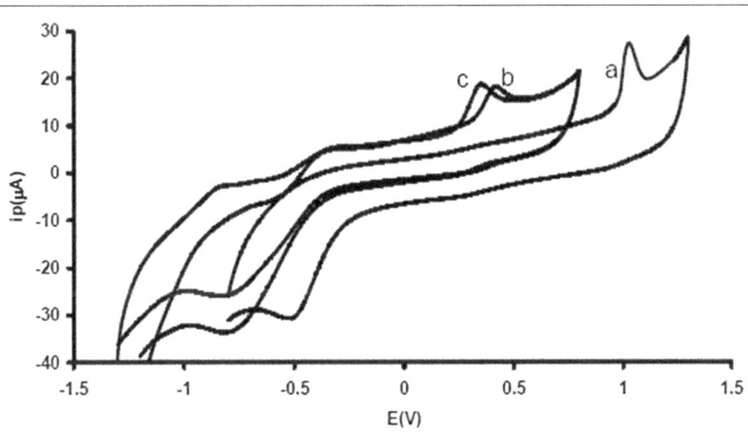

Figure 8 Cyclic voltammogram. (**a**) 305 μg/mL amlodipine, (**b**) 190 μg/mL felodipine, and (**c**) 210 μg/mL nimodipine on GCE at pH 13.0. Scan rate is 100 mV/s.

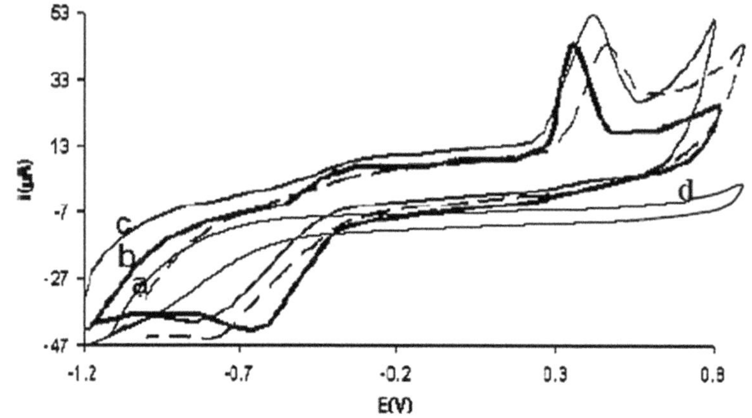

Figure 9 Cyclic voltammogram. (**a**) 305 μg/mL amlodipine, (**b**) 190 μg/mL felodipine,(**c**) 210 μg/mL nimodipine, and (**d**) absence of drugs on MWCNT/GCE at pH 13.0. Scan rate is 100 mV/s.

out in pH 13.0. Figures 8 and 9 represent the cyclic voltammograms recorded for all the three drugs on GCE and modified glassy carbon electrode. They exhibited one oxidation peak with larger current and one reduction peak with lesser current for modified GCE, but bare GCE shows lower current and high potential difference from modified electrode. The anodic peak was taken for further discussion due its analytical characteristic because of larger peak current.

Differential pulse stripping voltammetry

Differential pulse stripping voltammetry (DPSV) experiments were carried out to ascertain the best conditions for the adsorption process. Many preconcentration-stripping experiments were performed for different accumulation potentials and at an accumulation time of 15 s to evaluate the electrostatic attraction/repulsion between electrode surface and the drugs. When accumulation potential changed from −0.1 to +0.5 V, the maximum responses were obtained at 0.1 V for all three drugs. Maximum peak current was found for an accumulation potential in the positive region at 0.1 V because of the electrostatic interaction between the positive nature of electrode at this potential and the electron-rich substrate. After fixing the accumulation potential, the accumulation time was varied between 10 to 60 s.

Conclusion

MWCNT-modified GCE allowed the successful determination of AMLD, FELD, and NIMD drugs with a detection limit of 0.005 μg/mL for AMLD and FELD, and 0.01 μg/mL for NIMD. The anodic peak current varies linearly under optimized conditions in the concentration range from 0.01 to 0.3 μg/mL for AMLD and FELD, and 0.025 to 0.3 μg/mL for NIMD. The results obtained are promising and demonstrate the utility of the developed method for the determination of drugs in pharmaceutical formulations. The specificity of the voltammetric method was also investigated in the presence of substances present in drugs. Thus, the present investigation revealed that the proposed method is simple, specific, sensitive, and effective for the determination of three calcium channel blocker at MWCNT-modified glassy carbon electrode in pharmaceutical formulations.

Competing interests

The authors declare that they have no competing interests.

Authors' contributions

AMS carried out all the experimental components of the study. CV and PM monitored the results and conclusions drawn by AMS. All authors read and approved the final manuscript.

Author' information

AMS is an assistant professor of the Department of Chemistry, Velammal Engineering College, Chennai, India. CV is an assistant professor of the Department of Chemistry, V.O.C College, Tuticorin, India. PM is currently the director of Distance Education and professor and head of the Department of Industrial Chemistry, Alagappa University, Karaikudi, India.

Acknowledgments

We wish to express our deep sense of gratefulness and respectable to Dr. P. Manisankar, professor and head of the Department of Industrial Chemistry, Alagappa University, Karaikudi, for his inspiring guidance, valuable suggestions, and demonstrative encouragement throughout this exploration. We express our colossal thanks to Dr. C. Vedhi, assistant professor of the Department of Chemistry, V.O.C College, Tuticorin for his kind help in each and every possible corners of successful completion of this research work. Our special thanks to Thiru. M.V. Muthuramalingam, Chairman and to Thiru. M.V.M. Velmurugan, chief executive officer, Velammal Educational Trust, Chennai for their encouragement in completing this research work. We also thank Dr. J. Nandagopal and Dr. V.R. Manoj from the Department of Chemistry for their support during the manuscript preparation.

Author details

[1]Department of Chemistry, Periyar University, Salem, Tamil Nadu 636011, India. [2]Department of Industrial Chemistry, Alagappa University, Karaikudi, Tamil Nadu 630003, India. [3]Department of Chemistry, Velammal Engineering College, Chennai, Tamil Nadu 600066, India. [4]Department of Chemistry, V.O. Chidambaram College, Tuticorin, Tamil Nadu 628008, India.

References

1. Senturk Z, Ozkan SA, Ozkan Y (1998) J Pharm Biomed Anal 16:801
2. Squella JA, Bollo S, Nunez-Vergara LJ (2005) Curr Org Chem 9(6):565
3. Arguello Da Silva J, NuñezVergara LJ, Bollo S, Squella JA (2006) J Electroanal Chem 591:99
4. Wang J, Musameh M, Lin Y (2003) J Am Chem Soc 125:2408
5. Goyal RN, Tyagi A, Bachheti N, Bishnoi S (2008) Electrochim Acta 53:2802
6. Wang CH, Li CY, Wang CF (2006) Microchim Acta 152:233
7. Wang J, Li M, Shi Z, Li N, Gu Z (2002) Anal Chem 74:1993
8. Valentini F, Amine A, Orlanducci S, Terranova ML, Palleschi G (2003) Anal Chem 75:5413
9. Abdel-Kader Gazy A (2004) Talanta 62:575
10. Dollery C (1999) Therapeutic drugs, 2nd edition. Churchill Livingstone, UK, p 151
11. Maurer HH, Arit JW (1999) J Anal Toxicol 23:73
12. Dhorda UJ, Shetkar NB (1999) Indian Drugs 36:638
13. Ilango K, Kumar PB, Prasad VRV (1997) Indian J. Pharm Sci 59:336
14. Rahman N, Azmi SNH (2001) IL Farmaco 56:731
15. Altiokka G, Dogrukol AK-D, Tuncel M, Aboul-Enein HY (2002) Arch Pharm 335:104
16. Belal F, Abdine H, Zoman N (2001) J Pharm Biomed Anal 26:585
17. Ellaithy MM, Zuman P (1992) J Pharm Sci 81:191
18. Nunez-Vergara LJ, Bollo S, Alvarez A, Squella JA, Blazquez M (1993) Journal of Electroanalytical Chemistry 345:121
19. Ahmad F, Yusof APM, Bainbridge M, Ghani SA (2008) Biosens Bioelectron 23:1862
20. Wanner-Olsen H, Gaarsker FB, Mikkelsen EO, Jakobsen P, Voldby B (2000) Chirality 12:660
21. Muck W, Bode H (1994) Pharmazie 49:130
22. Soons PA, Roosemalen MCM, Breimer DD (1990) J Chromatogr 528:343
23. Dru JD-Y, Hsieh JY-K, Matuszewski BK, Dobrinska MR (1995) J Chromatogr 666:259
24. Heinig R, Muschalek V, Ahr G (1994) J Chromatogr 655:286
25. Nobuo I, Masahiro N (2002) J Biochem Biophys Methods 54:255
26. Muralidharan B, Gopu G, Vedhi C, Manisankar P (2009) J Appl Electrochem 39:1177
27. Manisankar P, AbiramaSundari PL, Sasikumar R, Palaniappan SP (2008) Talanta 76:1022
28. Manisankar P, AbiramaSundari PL, Sasikumar R, Jestin Roy D (2008) Electroanalysis 20:2076
29. Goyal RN, Bishnoi S (2010) Bioelectrochemistry 79:234
30. Shaalan RA, Belal TS (2010) Drug Test Analy 2(10):489
31. Canlica M, Islimyel S (2005) Turk J Chem 29:141
32. Revanasiddappa HD, Deepakumari HN, Mallegowda SM, Vinay KB (2011) AnaleleUniversităŃii din Bucuresti 20:189

Photocatalytic degradation of tylosin via ultraviolet-activated persulfate in aqueous solution

Mohammad Hossein Rasoulifard[1*], Hamed Majidzadeh[1], Fatemeh Torkian Demneh[1], Elaheh Babaei[1] and Mohammad Hassan Rasoulifard[2]

Abstract
Some of organic materials such as antibiotics are hazardous contaminants in the aquatic environment because of their adverse effects on aquatic life, environment, and humans. In this work, a batch reactor of ultraviolet (UV) light and peroxydisulfate were studied for the degradation of tylosin as a model antibiotic in water. The effect of different parameters such as UV irradiation, peroxydisulfate concentration, antibiotic concentration, and pH on removal of tylosin was investigated carefully. No significant destruction was observed using separate UV illumination. More than 90% removal occurred when peroxydisulfate was added to the solution in optimum levels of tylosin and peroxydisulfate concentration.

Keywords: Peroxydisulfate, Tylosin, Photodegradation, Wastewater treatment

Background

Antibiotics play an important role due to their high consumption rates in both veterinary and human medicine. These materials such as tylosin (Table 1) which is used in veterinary medicine to treat bacterial infections in a wide range of species at low concentrations in the environment the development of antibiotic resistant bacteria [1]. In fact, bacteria have been observed to transfer their resistance in laboratory settings as well as in the natural environment [2]. Furthermore, the presence of antibiotics in wastewaters has increased in recent years, and their abatement will be a challenged in the near future.

Antibiotic wastewater has high chemical oxygen demand (COD) and low biochemical oxygen demand (BOD); hence, biological processes are unsuitable for the wastewater treatment. Advanced oxidation processes (AOPs) have proved to be highly effective in the degradation of most of the pollutants in wastewaters [3]. AOPs are alternative techniques of destruction of many other organics in wastewater and effluents. These processes generally involve UV/H_2O_2, UV/O_3, UV/ $S_2O_8^{2-}$ or UV/Fenton's reagent for the oxidative degradation of contaminants [4-6]. Recently, there were numerous studies on UV/ $S_2O_8^{2-}$ application in wastewater treatment mainly because of high reactivity of UV/ $S_2O_8^{2-}$ process and high solubility of peroxydisulfate [4].

Peroxydisulfate is a strong oxidant ($E_0 = 2.05$ V) which can be purchased in the form of ammonium, sodium or potassium salt. It has been reported that the reaction rate constants of UV/ $S_2O_8^{2-}$ and UV/ H_2O_2 (the most common process) is the same [7]. Moreover, peroxydisulfate benefits UV/H_2O_2 and other similar approaches by the following reasons: (1) peroxydisulfate ions seem to be more promising because of the potential quenching effect of using H_2O_2 when the process is not well controlled (such as overdosing) [7]. (2) Since peroxydisulfate is a solid oxidant, it would be more capable for industrial uses in comparison to liquid oxidants such as H_2O_2. (3) Peroxydisulfate salts are much cheaper than other oxidants like hydrogen peroxide and ozone [8,9].

In this study, degradation of tylosin as an antibiotic pollutant from contaminated water by UV/ $S_2O_8^{2-}$ process was studied. Moreover, effect of different parameters such as UV irradiation, peroxydisulfate concentration, antibiotic concentration, and pH was investigated.

* Correspondence: m_h_rasoulifard@yahoo.com
[1]Water and Wastewater Treatment Research Laboratory, Department of Chemistry, Faculty of Science, University of Zanjan, Zanjan, 45371-38791, Iran
Full list of author information is available at the end of the article

Table 1 Characteristics of tylosin

Structure	$\lambda_{(max)}$ (nm)	Formula	M_w (g mol^{-1})
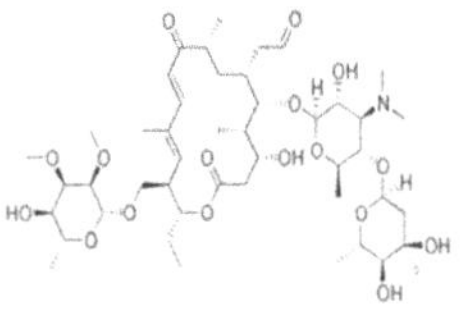	300	$C_{46}H_{77}NO_{17}$	916.1

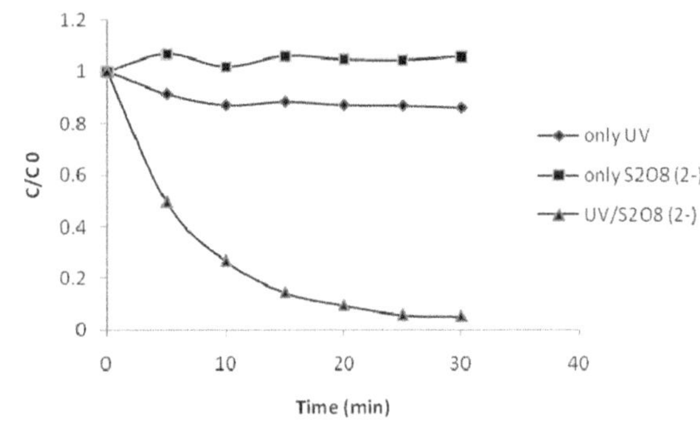

Figure 1 Effect of UV radiation and peroxyldisulfate in oxidative decolorization of tylosin. $[Tylosin]_0 = 40$ ppm, $[S_2O_8^{2-}]_0 = 2$ mM.

Experimental

Tylosin was obtained from Razak Co. (Islamic Republic of Iran, P.O.B 13185-1671, Tehran, Iran) and used without further purification. Characteristics of this antibiotic are shown in Table 1. The antibiotic is determined in an aqueous medium using a scanning UV–vis spectrophotometer (Shimadzu 160, Shimadzu Corporation, Nakagyo-ku, Kyoto, Japan). Ammonium peroxydisulfate was obtained from Merck & Co., Inc. (Whitehouse Station, NJ, USA). Dilute solutions of sodium hydroxide and hydrochloric acid were used for pH adjustment. A 30-W UV lamp was manufactured by Philips (Koninklijke Philips Electronics N.V., St. Vali-asr Ave., Iran). The volume of sample was 100 ml, and all the temperature (25 ± 4°C). Peroxydisulfate concentration varied from 0.5 to 20 mM, while antibiotic concentration varied from 20 to 80 ppm. The effect of pH was investigated in the range of 2.4 to 9.8.

Results and Discussion

Effect of UV irradiation and peroxydisulfate on degradation of tylosin

Degradation of tylosin was investigated with UV irradiation only, $S_2O_8^{2-}$ without UV radiation, and UV radiation plus $S_2O_8^{2-}$ in concentrations of 2 mM of peroxydisulfate, the antibiotic concentration of 40 ppm and natural pH (pH = 6.7). In the first case without any peroxydisulfate, approximately, we had maximum degradation of 11%, while using peroxydisulfate without UV irradiation, we had no degradation. UV plus peroxydisulfate outweighs the two mentioned cases clearly by 80% removal (Figure 1). This effective degradation is due to the formation of hydroxyl and sulfate radicals because of UV illumination. Reactions of peroxydisulfate were slow at normal temperature. Thus, as summarized in Equations 1 to 5,

$$S_2O_8^{2-} + \text{photons or heat} \rightarrow {}_2SO_4^{\bullet -} \quad (1)$$

thermal or photochemical activated decomposition of $S_2O_8^{2-}$ ion to $SO_4^{\bullet -}$ radical is proposed for acceleration of the process [10]:

$$SO_4^{\bullet -} + RH_2 \rightarrow SO_4^{2-} + H^+ + RH^\bullet \quad (2)$$

$$RH^\bullet + S_2O_8^{2-} \rightarrow R + SO_4^{2-} + H^+ + SO_4^{\bullet -} \quad (3)$$

$$SO_4^{\bullet -} + RH \rightarrow R^\bullet + SO_4^{2-} + H^+ \quad (4)$$

$$2R^\bullet \rightarrow RR(\text{dimer}) \quad (5)$$

where R is an organic reagent.

Once $SO_4^{\bullet -}$ is formed, it can produce a rapid attack on any oxidizable agent including organic contaminants (e.g., tylosin) [11]. Also, available oxidants in the solution and their corresponding intermediates are indicated in Equations 6 to 12:

$$SO_4^{\bullet -} + H_2O \rightarrow HSO_4^- + OH^\bullet (k = 500 \pm 60 s^{-1}) \quad (6)$$

$$HSO_4^- \rightarrow H^+ + SO_4^{2-} \quad (7)$$

$$OH^\bullet + S_2O_8^{2-} \rightarrow HSO_4^- + SO_4^{\bullet -} + \frac{1}{2}O_2 \quad (8)$$

$$SO_4^{\bullet -} + OH^\bullet \rightarrow HSO_4^- + \frac{1}{2}O_2 \quad (9)$$

$$2OH^\bullet \rightarrow H_2O_2 \quad (10)$$
(expect in alkaline solution)

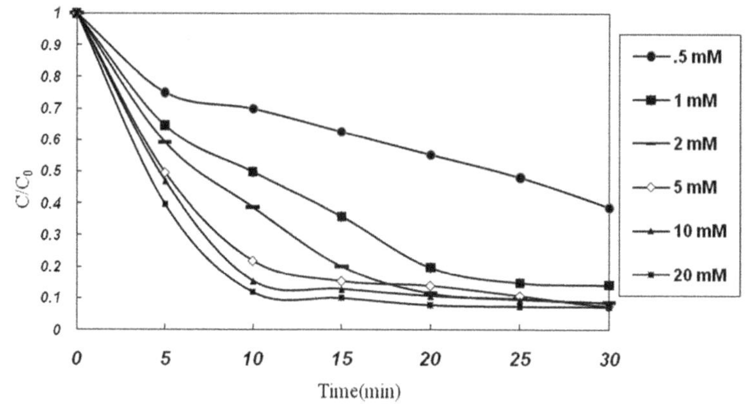

Figure 2 Effect of initial concentration of peroxydisulfate in oxidative degradation of tylosin. $[\text{Tylosin}]_0 = 40$ ppm.

$$H_2O_2 \rightarrow H_2O + \frac{1}{2}O_2 \quad (11)$$

(mostly in acidic solution)

$$S_2O_8^{2-} + H_2O_2 \rightarrow 2H^+ + 2SO_4^{2-} + O_2 \quad (12)$$

Both $SO_4^{\bullet-}$ and $OH^\bullet$ are possibly responsible for the destruction of organic contaminants, and either radical may predominate over the other depending on pH conditions. $SO_4^{\bullet-}$ and $OH^\bullet$ react with organic compounds mainly by three mechanisms: hydrogen abstraction, hydrogen addition, and electron transfer. Sulfate radicals exhibit a higher standard reduction potential than hydroxyl radicals at neutral pH, and both radicals exhibit similar reduction potentials under acidic conditions [8]. In general, $SO_4^{\bullet-}$ is more likely to participate in electron transfer reactions than is $OH^\bullet$ which is more likely to participate in hydrogen abstraction or addition reactions [12].

Effect of initial peroxydisulfate concentration

Initial $S_2O_8^{2-}$ concentration has a promising effect on degradation of tylosin. Investigations were made by varying the concentration of $S_2O_8^{2-}$ from 0.5 to 20 mM at fixed initial antibiotic concentration of 40 ppm, natural pH, and room temperature of 25°C ± 1°C. Studies revealed that increase in amount of $S_2O_8^{2-}$ would enhance degradation of the tylosin respectively [13]. These observations can be explained by the fact that the increase in concentration of peroxydisulfate results in higher generation of hydroxyl and sulfate radicals (Figure 2). It is likely because of excessive generation of hydroxyl radicals (Equations 1 and 6) would be recombined to form less reactive H_2O_2 (Equation 13), which is a known quencher of OH· radical (Equation 14).

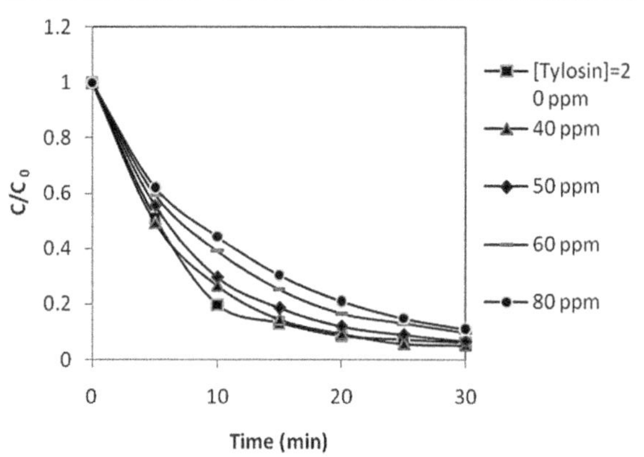

Figure 3 Effect of initial concentration of tylosin in oxidative degradation of antibiotic after 30 min. $[S_2O_8^{2-}]_0 = 2$ mM.

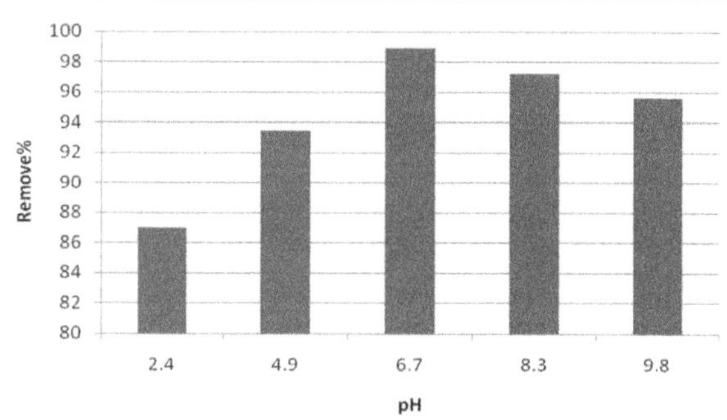

Figure 4 Effect of pH in oxidative decolorization of tylosin. $[\text{Tylosin}]_0 = 20$ ppm, $[S_2O_8^{2-}]_0 = 2$ mM.

Therefore, the destruction of tylosin was slightly slowed down at higher $S_2O_8^{2-}$ dosages. However, such a recombination effect of the radical was likely not very effective due to the low steady-state concentrations of the radicals; higher decay rates of tylosin at higher $S_2O_8^{2-}$ dosages are still expected [5,13]:

$$2OH^{\bullet} \rightarrow H_2O_2 \quad \text{(Only in acidic to neutral pH)} \tag{13}$$

$$OH^{\bullet} + H_2O_2 \rightarrow H_2O + HO_2^{\bullet} \tag{14}$$

Effect of initial tylosin concentration

The initial antibiotic concentration has a remarkable effect on photolytic degradation of tylosin. Tylosin concentration varied from 20 to 80 ppm at constant dosage of peroxydisulfate (2 mM). Degradation of 91% is observed in 20 ppm, while only 78% of degradation is achieved in 80 ppm concentration of tylosin (Figure 3). As shown in Figure 3, the higher concentration of tylosin was, the lower the degradation rate would be. One possible reason may be that the increase in antibiotic concentration decreases the ratio of hydroxyl radical to tylosin and percentage of degradation reduces in result.

Effect of the initial pH

The effect of initial pH was investigated in the range of 2.4 to 9.8 in constant concentration of tylosin (40 ppm), peroxydisulfate (2 mM), and temperature (25°C ± 1°C; Figure 4). pH 6.7 was found to be the most effective level of pH in decolorization of tylosin. However, effective photodegradation of tylosin was observed at all pH levels, which reveals the efficiency of this method in treatment of wastewater in different regions. The destruction performance increased from low to initial neutral pH (pH = 6.7) levels, but it started to reduce when basic pH was performed. The decreasing photodecay at pH ≥ 7 can be explained by the following reasons: (a) the instability of H_2O_2 at high pH level, (b) relatively higher amounts of $OH^{\bullet}$ and $SO_4^{\bullet-}$ were generated catalytically in alkaline conditions, which induced recombination of these two radicals (Equation 9), though this could be minor [14].

Conclusion

The application of peroxydisulfate along with UV irradiation as an advanced oxidation process at laboratory scale introduces an effectual and safe method for degradation of tylosin. Almost no antibiotic removal was achieved using peroxydisulfate alone. On the other hand, destruction percentage of 11% was obtained using UV irradiation alone. Finally, more than 97% of antibiotic content was removed using UV irradiation and peroxydisulfate simultaneously. Degradation rate of tylosin was dependent on the antibiotic concentration, peroxydisulfate concentration, and pH. Increase in antibiotic concentration would decrease the degradation as degradation of 91% was observed in 20 ppm, while only 78% of degradation was achieved in 80 ppm concentration of tylosin. Neutral pH (6.7) was monitored as the optimum pH; however, all levels of pH demonstrate satisfactory removal.

Competing interests

The authors declare that they have no competing interests.

Authors' contributions

AB carried out the molecular genetic studies, participated in the sequence alignment and drafted the manuscript. JY carried out the immunoassays. MT participated in the sequence alignment. ES participated in the design of the study and performed the statistical analysis. FG conceived of the study, and participated in its design and coordination. All authors read and approved the final manuscript. All authors read and approved the final manuscript.

Acknowledgment

The authors are grateful to the University of Zanjan for the financial support.

Author details

[1]Water and Wastewater Treatment Research Laboratory, Department of Chemistry, Faculty of Science, University of Zanjan, Zanjan, 45371-38791, Iran. [2]Department of Vetinerary, Islamic Azad University of Tabriz, Tabriz Branch, Tabriz, 1655, Iran.

References

1. Walter MV, Vennes JV. Occurrence of multiple-antibiotic-resistant enteric bacteria in domestic sewage and oxidation lagoons. *Appl Environ Microbiol.* 1985; 50(4): 930–933.
2. Kanai H. Drug-resistance and distribution of conjugative R plasmids in Escherichia coli strains isolated from healthy adult animals and humans. *Jpn J Vet Sci.* 1983; 45: 171–178.
3. Pera-Titus M, Garcia-Molina V, Baños MA, Gimenez J, Esplugas S. Degradation of chlorophenols by means of advanced oxidation processes: a general review. *Appl Catal B Environ.* 2004; 47: 219–256.
4. Vilhunen SH, Sillanp MET. Ultraviolet light emitting diodes and hydrogen peroxide in the photodegradation of aqueous phenol. *J Hazard Mater.* 2009; 161: 1530.
5. Daneshvar N, Aber S, Seyed Dorraji MS, Khataee AR, Rasoulifard MH. Photocatalytic degradation of the insecticide diazinon in the presence of prepared nanocrystalline ZnO powders under irradiation of UV-C light. *Sep Purif Technol.* 2007; 58(1): 91–98.
6. Daneshvar N, Rasoulifard MH, Khataee AR, Hosseinzadeh F. Removal of C.I. Acid Orange 7 from aqueous solution by UV irradiation in the presence of ZnO nano powder. *J Hazard Mater.* 2007; 143: 95–101.
7. Chu W, Lau TK, Fung SC. Effects of combined and sequential addition of dual oxidants (H2O2/S2O8(2-)) on the aqueous carbofuran photodegradation. *J Agric Food Chem.* 2006; 54(26): 10047–10052.
8. Anipsitakis GP, Dionysiou DD. Radical generation by the interaction of transition metals with common oxidants. *Environ Sci Technol.* 2004; 38(13): 3705–3712.
9. Anipsitakis GP, Dionysiou DD. Degradation of organic contaminants in water with sulfate radicals generated by the conjunction of peroxymonosulfate with cobalt. *Environ Sci Technol.* 2003; 37: 4790–4797.
10. Salari D, Daneshvar N, Niaei A, Aber S, Rasoulifard MH. Immobilization of TiO_2 Nanopowder on glass beads for photocatalytic decolorization of an Azo Dye C. I. Directed Red 23. *J Environ Sci Health Part A.* 2008; 43: 1–7.

11. Liang CJ, Bruell CJ, Marley MC, Sperry KL. Thermally activated persulfate oxidation of trichloroethylene (TCE) and 1,1,1-trichloroethane (TCA) in aqueous systems and soil slurries. *Soil Sedim Contam*. 2003; 12: 207.
12. Minisci F, Citterio A. Free-radical addition to olefins in the presence of redox systems. *Acc Chem Res*. 1983; 16: 27.
13. Rasoulifard MH, Marandi R, Majidzadeh H, Bagheri I. Ultraviolet light-emitting diodes and peroxydisulfate for degradation of basic red 46 from contaminated water. *Environ Eng Sci*. 2011; 28: 229–235.
14. Lau TK, Chu W, Graham NJD. The aqueous degradation of butylated hydroxyanisole by UV/S2O8(2-): study of reaction mechanisms via dimerization and mineralization. *Environ Sci Technol*. 2007; 41(2): 613–9.

Process optimization studies of Congo red dye adsorption onto cashew nut shell using response surface methodology

Senthil Kumar Ponnusamy[1*] and Ramalingam Subramaniam[2*]

Abstract

Background: High cost of commercially available activated carbon increases the treatment cost. Hence, no-cost or low-cost, alternate adsorbents are the current thirst in this area of research. The limitations of conventional methods of experimentation can mislead the optimum operating conditions. The present investigation mainly deals with utilization of the low-cost, agricultural waste adsorbent, cashew nut shell, and the determination of the optimum conditions for Congo red dye removal from an aqueous solution using response surface central composite design methodology.

Results: The adsorbent, cashew nut shell, can remove the Congo red dye completely from the aqueous solution. The parameters pH, adsorbent dose, initial dye concentration, time, and temperature considered for this investigation play an important role in the adsorption studies of Congo red dye removal. The optimum values of pH, adsorbent dose, initial dye concentration, time, and temperature were found to be 3.2, 24.76 g/L, 20 mg/L, 67 min, and 30°C for complete removal of Congo red dye, respectively. The experimental values were in good agreement with predicted values.

Conclusions: Adsorption of Congo red dye from an aqueous solution with a low-cost adsorbent prepared from an agricultural waste was studied to determine the potential use of this adsorbent for complete removal of the dye. Batch studies were performed to address the influence of various parameters such as pH, adsorbent dose, initial dye concentration, time, and temperature. Response surface central composite design methodology was used to find the interaction among the variables and to determine the optimum conditions towards the adsorption of dye from the aqueous solution. The optimum values of pH, adsorbent dose, initial dye concentration, time, and temperature were found to be 3.2, 24.76 g/L, 20 mg/L, 67 min, and 30°C for complete removal of Congo red dye, respectively. The experimental values were in good agreement with predicted values.

Keywords: Cashew nut shell, Congo red dye, Response surface methodology, Statistical analysis, Optimization

Background

The dye effluents are considered to be highly toxic to the aquatic species and are found to affect the symbiotic process by disturbing the natural equilibrium by reducing photosynthetic activity and primary production due to the colorization of the water [1]. Removal of these dyes from effluents in an economic way remains as a major problem for textile industries [2,3]. The most commonly used methods for color removal are biological and chemical precipitation. However, these processes are effective and economic only in cases where solute concentrations are relatively high [4]. Even though many methods were explored to address this environmental and economic concern, there are advantages and disadvantages in each and every method used for dye removal from wastewaters [5]. Many physicochemical methods have been tested, but adsorption is considered to be superior to other techniques. This is attributed to its low cost, easy availability, simplicity of design, high efficiency, easy operation, biodegradability, and ability to treat dyes in more concentrated forms [6,7]. The adsorption technique has been proven to be an excellent way to treat effluents,

* Correspondence: senthilchem8582@gmail.com; raams@louisiana.edu
[1]Department of Chemical Engineering, SSN College of Engineering, Chennai 603 110, India
Full list of author information is available at the end of the article

offering advantages over conventional processes, especially from the environmental point of view [2]. Weber and Morris [2] have identified many advantages of adsorption over several other conventional treatment methods. The adsorption of dyes on various types of materials has been studied in detail. Those include the following: activated carbon [8], peat [9], chitin [10], silica [11], hardwood [12], hardwood sawdust [13], bagasse pith [14], fly ash [15,16], mixture of fly ash and coal [17], chitosan fiber [18], paddy straw [19], rice husk [20], slag [21], chitosan [22], acid-treated spent bleaching earth [23], palm fruit bunch [24], and bone char [25]. Adsorption with activated carbon has been used widely for the removal of dyes from wastewater [26-31]. Activated carbon was being used as a potential adsorbent because of its high efficiency [32]. Commercially available activated carbons are usually derived from natural materials such as wood or coal which are expensive [33,34]. Increase in the price of activated carbon results in increased treatment cost for the developing countries like India. Hence, no-cost or low-cost, alternate adsorbents with equivalent capabilities of activated carbon are the current thirst in this area of research. Consequently, low-cost activated carbons from agricultural solid wastes are being investigated for a long time.

The conventional method of studying a process by changing one variable at a time and keeping the other variables at a constant level does not depict the combined effect of all the factors involved. Moreover, this method is time-consuming and also requires a large number of experiments to determine optimum levels, which may or may not be reliable. These limitations of the conventional method can be eliminated by simultaneously varying all the affecting parameters by using a statistical experimental design such as the response surface methodology (RSM) [35]. RSM, initially described by Box and Wilson [36], is an experimental approach to identify the optimum conditions for a multivariable system. RSM is a collection of statistical and mathematical techniques useful for developing, improving, and optimizing processes that can be used to evaluate the relative significance of all the factors involved in the process even in the presence of complex interactions. The main objective of RSM is to determine the optimum operational conditions for the system or the process [37]. The application of statistical experimental design techniques in adsorption process development can result in improved product yields, reduced process variability, and reduced development time and overall costs [38].

The main objectives of the present investigation are the utilization of the low-cost, agricultural waste adsorbent, cashew nut shell (CNS), and the determination of the optimum conditions for Congo red (CR) dye removal from an aqueous solution using response surface central composite design methodology. In this study, the central composite design was selected to determine the effect of experimental parameters and their interactions over the removal of CR. The interactions between factors that influence the percentage of CR removal were established. The optimum value of the parameters was also determined for removal of CR from the aqueous solution using RSM.

Methods

Cashew nut shells collected from Pudukkottai District, Tamilnadu, India were used as an adsorbent. This natural waste was thoroughly rinsed with water to remove dust and the soluble material and was allowed to dry at room temperature. The dried natural waste was crushed into powder and sieved through the sieves. CNS particles with a mesh size of 200 to 30 were taken for the adsorption studies. The proximate and ultimate analysis of cashew nut shells are shown in Table 1. The specific surface area and the pore structure of the CNS were determined by using a surface area and pore size analyzer (Autosorb-I, Quantachrome Instruments, Boynton Beach, FL, USA) on nitrogen adsorption at 77 K. The specific surface area was calculated using the BET equation [39]. It was found that the BET surface area, pore volume, average pore diameter, and bulk density of the CNS were 395 m^2/g, 0.4732 cm^3/g, 5.89 nm, and 0.415 g/cm^3, respectively.

Congo red (CI = 22120) was supplied by Merck (Mumbai, India). A stock solution of CR dye was prepared (100 mg/L) by dissolving a required amount of dye powder in deionized water. The stock solution was diluted with deionized water to obtain the desired concentration ranging from 20 to 100 mg/L.

The concentration of CR in the experimental solution was determined from the calibration curve prepared by measuring the absorbance of different known concentrations of CR solutions at $\lambda_{max} = 497$ nm using a UV–vis spectrophotometer (Shimadzu, Kyoto, Japan). The pH of the solution was measured with a Hanna pH meter using a combined glass electrode (model HI 9025C, Hanna Instruments, Singapore).

Adsorption experiments were conducted as per the design developed with response surface central composite design methodology. The experiments were carried out in 250-mL Erlenmeyer flasks with the working volume of 100 mL of reaction mixture. The initial pH of the solution was adjusted to the desired value by adding 0.1 M NaOH or HCl. The flasks were shaken for the specified time period in a rotary shaker. The flasks were withdrawn from the shaker after the desired time of reaction. The residual dye concentration in the reaction mixture was analyzed by centrifuging (5,000 rpm, R-24 REMI Centrifuge, Mumbai, India) the reaction mixture and then measuring the absorbance of the supernatant at the wavelength that corresponds to the maximum absorbance of the sample. Dye concentration in the reaction mixture was calculated

Table 1 Properties of the cashew nut shells

Proximate analysis (wt.%)		Ultimate analysis (wt.%)		Ash chemical composition (wt.%)	
Volatile matter	65.21	Carbon	45.21	Silica	64.53
Moisture	9.83	Hydrogen	4.25	Iron oxide	3.27
Ash	2.75	Oxygen	37.75	Aluminum oxide	2.19
Fixed carbon	22.21	Nitrogen	0.21	Calcium oxide	26.89
		Sulfur	Nil	Magnesium oxide	2.49
		Moisture	9.83	Sodium oxide	0.63
		Ash	2.75		

from the calibration curve. The λ_{max} values of the wastewater samples varied by ±10 nm from the λ_{max} values of pure samples. Each determination is repeated three times, and the results given were their average value. The percentage of CR removal was taken as a response (Y) of the experimental design and calculated as follows:

$$\% \text{ CR dye removal} = \left(\frac{C_i - C_f}{C_i}\right) \times 100, \quad (1)$$

where C_i is the initial concentration of dye in the solution and C_f is the final concentration of dye in the solution.

The pH, adsorbent dose, initial dye concentration, time and temperature are chosen as the independent input variables and the efficiency of dye removal as dependent response variable. Independent variables, experimental range and levels for congo red dye removal are given in Table 2. A 2^5 full factorial experimental design [36], with ten replicates at the center point, 10 experiments at axial points and thus a total of 52 experiments are employed in this study. The alpha value is chosen as face centered ($\alpha = 1$). When $\alpha = 1$, the axial points are placed on the 'cube' portion of the design. This is an appropriate choice when the cube points of the design are at the operational limits. The center point replicates are chosen to verify any change in the estimation procedure, as a measure of precision property. The experimental plan showing the coded value and uncoded value of the variables together with the dye removal efficiency for CR is given in Table 3. For statistical calculations, the variables X_i are coded as x_i according to the following relationship:

$$x_i = \left(\frac{X_i - X_0}{\delta_X}\right), \quad (2)$$

where x_i is the dimensionless coded value of the variable X_i, X_0 is the value of the X_i at the center point, and δ_X is the step change.

The behavior of the system is explained by the following quadratic equation:

$$Y = \beta_0 + \sum \beta_i x_i + \sum \beta_{ii} x_i^2 + \sum \beta_{ij} x_i x_j \quad (3)$$

where β_0 is the offset term, β_i is the linear effect, β_{ii} is the squared effect, β_{ij} is the interaction effect, and x_i is the dimensionless coded value of the variable X_i.

The results of the experimental design are studied and interpreted by MINITAB 14 (State College, PA, USA) statistical software to estimate the response of the dependent variable.

Table 2 Experimental range and levels of independent process variables

Independent variable	Range and level		
	−1	0	+1
pH (X_1)	2	6	10
Adsorbent dose (X_2, g/L)	5	17.5	30
Initial dye concentration (X_3, mg/L)	20	60	100
Time (X_4, min)	0	60	120
Temperature (X_5, °C)	30	45	60

Results and discussion

The experiments were performed at the specified combinations of the physical parameters using statistically designed experiments in order to study the combined effects of these factors. The main effects of all the variables taken into consideration in this work are presented in Figure 1. From the figure, it is clear that the variables considered for this investigation play an important role in the adsorption studies of CR dye removal since each variable has a significant contribution on the percentage of dye removal. A main effect occurs when the mean response changes across the levels of a factor. The main effect plots were used to plot data means when multiple factors are involved in the system/process and also to compare magnitudes of marginal means. The points in the plot were the means of the response variables at the various levels of each factor, with a reference line drawn at the grand mean of the response data. Grand mean is the mean of all observations, as opposed to the mean of

Table 3 Central composite design matrix with experimental and predicted values

Coded values					Uncoded values					% Removal	
pH	Adsorbent dose (g/L)	Initial concentration (mg/L)	Time (min)	Temperature (°C)	pH	Adsorbent dose (g/L)	Initial concentration (mg/L)	Time (min)	Temperature (°C)	Experimental	Predicted
−1	1	1	−1	−1	2	30	100	0	30	0	−0.05
−1	−1	1	−1	1	2	5	100	0	60	0	0.04
0	0	0	0	0	6	17.5	60	60	45	81.21	80.99
1	1	1	−1	−1	10	30	100	0	30	0	−0.34
−1	−1	−1	1	−1	2	5	20	120	30	62.33	62.35
0	0	0	0	0	6	17.5	60	60	45	81.21	80.99
−1	−1	−1	1	1	2	5	20	120	60	57.57	57.88
0	0	0	0	0	6	17.5	60	60	45	81.21	80.99
−1	1	1	1	1	2	30	100	120	60	79.34	79.45
1	1	1	−1	1	10	30	100	0	60	0	−0.05
−1	1	−1	−1	−1	2	30	20	0	30	0	0.03
−1	1	−1	1	−1	2	30	20	120	30	98.52	98.45
1	−1	1	−1	−1	10	5	100	0	30	0	−0.01
1	−1	−1	1	−1	10	5	20	120	30	52.89	53.02
1	−1	−1	−1	1	10	5	20	0	60	0	−0.03
1	−1	1	1	−1	10	5	100	120	30	38.54	38.53
1	1	1	1	1	10	30	100	120	60	70.34	70.34
0	1	0	0	0	6	30	60	60	45	86.04	86.43
0	0	0	0	−1	6	17.5	60	60	30	82.08	82.33
0	0	0	0	0	6	17.5	60	60	45	81.06	80.99
−1	1	−1	−1	1	2	30	20	0	60	0	−0.02
1	1	1	1	−1	10	30	100	120	30	73.93	74.23
−1	1	−1	1	1	2	30	20	120	60	94.24	94.22
0	−1	0	0	0	6	5	60	60	45	68.31	68.43
0	0	0	0	0	6	17.5	60	60	45	81.21	80.99
−1	−1	1	1	−1	2	5	100	120	30	48.56	48.15
−1	0	0	0	0	2	17.5	60	60	45	81.67	81.94
0	0	1	0	0	6	17.5	100	60	45	81.39	81.66
1	1	−1	1	1	10	30	20	120	60	85.82	85.40
0	0	0	1	0	6	17.5	60	120	45	66.43	66.67
1	0	0	0	0	10	17.5	60	60	45	77.08	77.33

Table 3 Central composite design matrix with experimental and predicted values *(Continued)*

0	0	0	0	0	6	17.5	60	60	45	81.21	80.99
0	0	0	0	0	6	17.5	60	60	45	81.21	80.99
1	−1	−1	−1	−1	10	5	20	0	30	0	−0.14
1	−1	−1	1	1	10	5	20	120	60	48.94	48.96
−1	−1	−1	−1	−1	2	5	20	0	30	0	−0.03
−1	−1	−1	−1	1	2	5	20	0	60	0	−0.33
1	1	−1	1	−1	10	30	20	120	30	89.29	89.22
−1	−1	1	−1	−1	2	5	100	0	30	0	0.39
−1	−1	1	1	1	2	5	100	120	60	43.68	43.62
0	0	−1	0	0	6	17.5	20	60	45	88.74	88.98
1	−1	1	−1	1	10	5	100	0	60	0	0.04
0	0	0	0	0	6	17.5	60	60	45	81.21	80.99
0	0	0	0	0	6	17.5	60	60	45	81.21	80.99
−1	1	1	−1	1	2	30	100	0	60	0	−0.16
0	0	0	−1	0	6	17.5	60	0	45	0	0.28
1	1	−1	−1	1	10	30	20	0	60	0	0.38
−1	1	1	1	−1	2	30	100	120	30	83.75	83.74
1	−1	1	1	1	10	5	100	120	60	34.47	34.41
0	0	0	0	1	6	17.5	60	60	60	79.98	80.24
0	0	0	0	0	6	17.5	60	60	45	81.21	80.99
1	1	−1	−1	−1	10	30	20	0	30	0	0.02

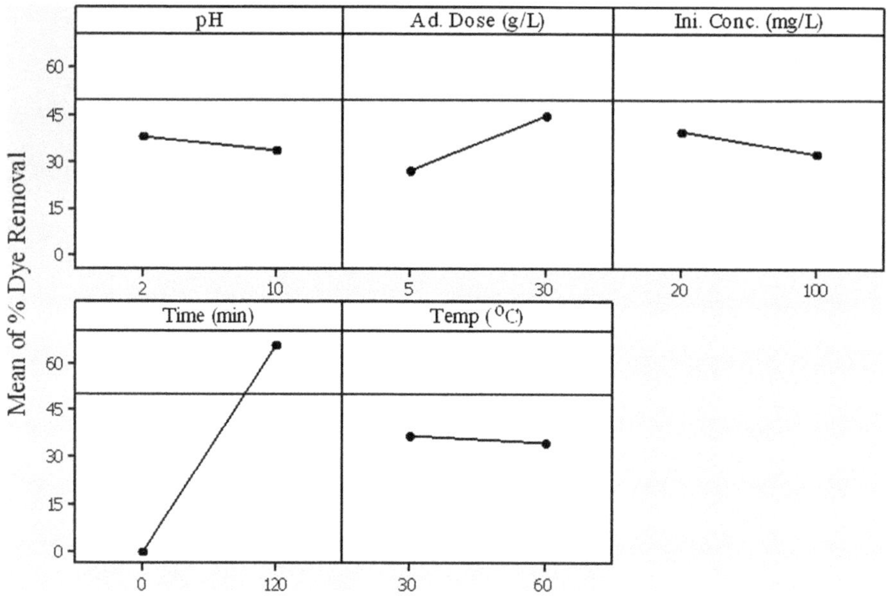

Figure 1 Main effect plot of parameters on CR dye removal.

individual variables. The line drawn across the plot (center line) represents the grand mean. From Figure 1, it is clear that the grand mean of the response falls at around 50% dye removal. Out of the variable considered, adsorbent dose and time are the dominant factors for the percentage of dye removal when compared to other variables.

The percentage of CR dye removal decreases with an increase in initial pH of the system due to its influence on the surface properties of the CNS and ionization/dissociation of the CR molecule. At higher pH, the grand mean drops down to 30%. The maximum removal of dye can be obtained at a low pH. Low pH leads to an increase in H^+ ion concentration in the system, and the surface of the CNS acquires a positive charge by absorbing H^+ ions. As the CNS surface is positively charged at low pH, a significantly strong electrostatic attraction appears between the positively charged CNS surface and anionic dye molecule, leading to maximum adsorption of dye. As the pH of the system increases, the number of negatively charged sites increases and the number of positively charged sites decreases. A negatively charged surface site on the CNS does not favor the adsorption of anionic dye molecules due to the electrostatic repulsion. Furthermore, lower adsorption of the CR dye in alkaline medium is also due the competition from excess OH^- ions with the anionic dye molecule for the adsorption sites.

The percentage of CR dye removal is increased with an increase in adsorbent dose. At higher CNS-to-CR concentration ratios, a very fast superficial sorption onto the CNS surface occurs which produces a lower CR concentration in the solution than when the CNS-to-CR concentration ratio is lower. This is because a fixed dose of CNS can only adsorb a certain amount of dye. Therefore, with higher adsorbent dosage, a larger volume of effluent can be purified with a fixed dosage of CNS. The decrease in the amount of CR adsorbed on to an adsorbent q_e (mg/g) with increasing CNS dose is due to the split in the flux or the concentration gradient between the CR concentration in the solution and the CR concentration in the surface of the CNS. Thus, with increasing CNS dose, the amount of dye adsorbed onto the unit weight of CNS gets reduced, thereby causing a decrease in the q_e value.

The effect of initial dye concentration on adsorption shows that the percentage of CR dye removal decreases with an increase in the initial concentration of CR dye. The initial dye concentration provides the necessary driving force to overcome the resistances to the mass transfer of CR between the aqueous phase and the solid phase. The increase in an initial dye concentration also enhances the interaction between CR and CNS. Therefore, an increase in an initial concentration of CR enhances the adsorption uptake of CR. This is due to an increase in the driving force of the concentration gradient, as an increase in the initial dye concentration.

The effect of contact time on the percentage of CR dye removal increases rapidly with time. This indicates that a longer contact time favors the reaction towards the equilibrium between the dye and adsorbent. It may be due to the fact that the available CR dye molecules are preferentially adsorbed to the most active sites of the CNS. The short contact time between the CNS and CR dye molecules gives the less percentage of CR dye removal and equilibrium adsorption capacity and vice versa. The adsorption of CR dye onto the CNS was investigated as a function of temperature, and the maximum removal of CR dye was obtained at lower temperature. This is mainly due to the decrease in the surface activity, suggesting that adsorption between CR and CNS was an exothermic process.

Response surface methodology has been successfully applied for optimizing conditions for CR dye removal. The experimental results are analyzed through RSM to obtain an empirical model for the best response. The uncoded regression model equation (second-order polynomial) relating the removal efficiency and process parameters is developed and given in Equation 4. The mathematical expressions of the relationship between the independent variables and dependent response are given in terms of uncoded factors. Apart from the linear effect of the parameter for the dye removal, the RSM also gives an insight into the quadratic and interaction effect of the parameters. These analyses are done by means of Fisher's F test and Student's t test. Student's t test was used to determine the significance of the regression coefficients of the parameters. The P values are used as a tool to check the significance of each of the interaction among the variables, which in turn may indicate the patterns of the interactions among the variables. In general, the larger the magnitude of t and smaller the value of P, the more significant is the corresponding coefficient term [40]. The regression coefficient and the t and P values for all the linear, quadratic, and interaction effects of the parameters are given in Table 4. From the high t values and very small P values, it is observed that the coefficients for the linear effect and quadratic effect of all the factors are highly significant. The high P values of 0.593, 0.224, and 0.752 for the interaction effect of pH and adsorbent dose, adsorbent dose and temperature, and initial concentration and temperature did not seem to be significant, respectively. The response surface plots to estimate the removal efficiency over independent variables are shown in Figure 2. The peak of the surface plot shows an optimum value of the percentage of dye removal for the relative effects of variable pH and adsorbent dose (Figure 2a), adsorbent dose and temperature (Figure 2b), initial concentration and time (Figure 2c), and initial concentration and temperature (Figure 2d) by keeping the other three variables at the midpoint values. From Figure 2, it is clear that the interaction effects of adsorbent dose-pH, adsorbent dose-temperature, and initial concentration-temperature are not significant towards the percentage of dye removal. The response surfaces of mutual interactions between the variables other than those discussed above are found to be elliptical which explains the significance of those variables. The significance of these interaction effects between the variables would have been lost if the

Table 4 Estimated regression coefficients and corresponding *t* and *P* values

Term	Coefficient	SE coefficient	*t*	*P*
Constant	2.92257	1.35459	2.158	0.039
pH	0.96179	0.13781	6.979	0
Adsorbent dose (g/L)	0.79432	0.04156	19.112	0
Initial concentration (mg/L)	−0.31653	0.01378	−22.969	0
Time (min)	2.12829	0.00661	321.876	0
Temperature (°C)	−0.13244	0.06976	−1.899	0.067
pH*pH	−0.08497	0.0108	−7.866	0
Adsorbent dose (g/L)*Adsorbent dose (g/L)	−0.02278	0.00111	−20.594	0
Initial concentration (mg/L)*Initial concentration (mg/L)	0.00271	0.00011	25.056	0
Time (min)*Time (min)	−0.0132	0.00005	−274.939	0
Temperature (°C)*Temperature (°C)	0.00131	0.00077	1.71	0.097
pH*Adsorbent dose (g/L)	0.00052	0.00096	0.54	0.593
pH*Initial concentration (mg/L)	−0.00046	0.0003	−1.515	0.14
pH*Time (min)	−0.00961	0.0002	−47.954	0
pH*Temperature (°C)	0.00169	0.0008	2.113	0.043
Adsorbent dose (g/L)*Initial concentration (mg/L)	−0.00025	0.0001	−2.62	0.014
Adsorbent dose (g/L)*Time (min)	0.01201	0.00006	187.376	0
Adsorbent dose (g/L)*Temperature (°C)	0.00032	0.00026	1.242	0.224
Initial concentration (mg/L)*Time (min)	−0.00152	0.00002	−76.049	0
Initial concentration (mg/L)*Temperature (°C)	−0.00003	0.00008	−0.319	0.752
Time (min)*Temperature (°C)	−0.00116	0.00005	−21.718	0

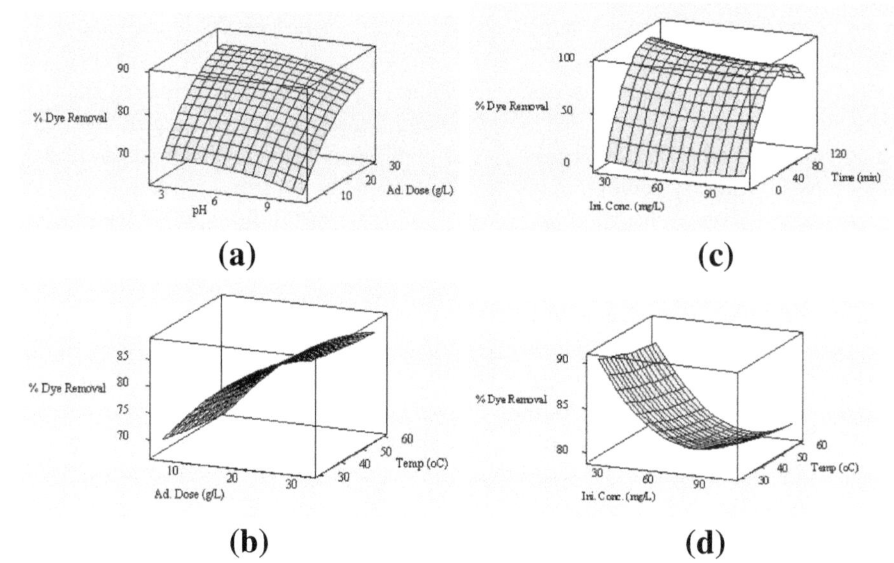

Figure 2 Response surface plot of CR dye removal. (**a**) pH and adsorbent dose, (**b**) adsorbent dose and temperature, (**c**) initial concentration and time, and (**d**) initial concentration and temperature.

experiments are carried out by conventional methods. The experimental results and the results of theoretically predicted responses (using the model equation) are shown in Table 3.

$$\begin{aligned}\eta = {} & (2.9226) + (0.9618\ X_1) + (0.7943\ X_2) \\ & -(0.3165\ X_3) + (2.1283\ X_4) - (0.1324\ X_5) \\ & -\left(0.0850\ X_1^2\right) - \left(0.0228\ X_2^2\right) + \left(0.0027\ X_3^2\right) \\ & -\left(0.0132\ X_4^2\right) + \left(0.0013\ X_5^2\right) + (0.0005\ X_1\ X_2) \\ & -(0.0005\ X_1\ X_3) - (0.0096\ X_1\ X_4) \\ & +\ (0.0017\ X_1\ X_5) - (0.0003\ X_2\ X_3) \\ & +(0.012\ X_2\ X_4) + (0.00032\ X_2\ X_5) \\ & -\ (0.0015\ X_3\ X_4) - (0.00003\ X_3\ X_5) \\ & -(0.0012\ X_4\ X_5),\end{aligned} \tag{4}$$

where X_1 is the pH, X_2 is the adsorbent dose (g/L), X_3 is the initial concentration of adsorbate (mg/L), X_4 is the time (min), and X_5 is the temperature (°C).

The statistical significance of the ratio of the mean square variation due to the regression and mean square residual error is tested using analysis of variance (ANOVA). ANOVA is a statistical technique that subdivides the total variation in a set of data into component parts associated with specific sources of variation for the purpose of testing hypothesis on the parameters of the model [41]. According to the ANOVA (Table 5), the F statistic values for all regressions are higher. The large value of F indicates that most of the variation in the response can be explained by the regression equation. The associated values of P lower than 0.01 indicate that the model is statistically significant [42]. The ANOVA table also shows a term for the residual error, which measures the amount of variation in the response data left unexplained by the model. The form of the model chosen to explain the relationship between the factors and the response is correct and has very good agreement with the experimental value. The high value of R^2 and R^2_{adj} explains that the model is significant.

The $F_{\text{Statistics}}$ value of 48,028 is greater than the tabulated $F_{20,31}$ which indicates that the fitted model exhibits a lack of fit of 0.000 at the 99% confidence level. The ANOVA indicates that the second-order polynomial model Equation 4 is highly significant and adequate to represent the actual relationship between the response (percent removal efficiency) and variables, with P values (0.0000) and a high value of the coefficient of determination (99.7%). This implies that 99.7% of the sample variation is explained by the independent

Table 5 ANOVA results for the quadratic model

Source	*DF*	Seq SS	Adj SS	Adj MS	*F*	*P*
Regression	20	71037.4	71037.4	3551.87	48028	0
Linear	5	40891.6	11392.6	2278.52	30810	0
Square	5	26915.5	26915.5	5383.09	72790	0
Interaction	10	3230.3	3230.3	323.03	4368	0
Residual error	31	2.3	2.3	0.07		
Lack of fit	22	2.3	2.3	0.1	45.91	0
Pure error	9	0	0	0		
Total	51	71039.7				

$R^2 = 0.997$; $R^2_{adj} = 0.994$. DF, degrees of freedom; Seq SS, sequential sums of squares; Adj SS, adjusted sums of squares; Adj MS, adjusted mean squares.

variables. Even though the interaction terms pH and adsorbent dose, adsorbent dose and temperature, and initial concentration and temperature are found to be insignificant, the oval model holds good agreement with the experimental data.

The model Equation 4 is optimized using a multistage Monte-Carlo optimization technique [43]. The optimum values of the process parameters are obtained in uncoded units. The optimum values of the process variables for the maximum removal efficiency are 3.2, 24.76 g/L, 20 mg/L, 67 min, and 30°C for the variables pH, adsorbent dose, initial dye concentration, time, and temperature, respectively. The percentage of dye removal at the optimum conditions simulated from the model is 100%. The experiments were done at optimum conditions, and the percentage of dye removal thus obtained is 99.99% which has a very close agreement with the model. These results closely agree with the optimum values obtained from the response surface methodology as 3.2557, 24.7634 g/L, 20 mg/L, 67.1815 min, and 30°C for the variables pH, adsorbent dose, initial dye concentration, time, and temperature, respectively. These results confirm that the RSM could be effectively used to optimize the process parameters in complex processes using the statistical design of experiments.

Experimental

Adsorption experiments were conducted as per the design developed with response surface central composite design methodology. All the experiments were done in duplicate to minimize the experimental error. The results given in Table 3 are the average experimental values. The detailed experimental procedure was discussed in 'Methods' section.

Conclusion

A novel, low-cost adsorbent prepared from agricultural waste, cashew nut shell, is successfully applied for the removal of CR from its aqueous solution. The specific surface area, pore volume, and average pore diameter calculated by the BET equation states the effectiveness of the CNS adsorbent for the removal of CR from aqueous solution. The effect of pH, adsorbent dose, initial dye concentration, time, and temperature on the adsorption was studied with the 52 experiments designed by the 2^5 full factorial central composite designs. This study clearly shows that the response surface methodology is one of the suitable methods to optimize the best operating conditions to maximize the dye removal. The statistical analysis results proved the significance of the model developed from experimental data to optimize the parameters. A significant interaction was observed in CR removal with the combination effects of pH, adsorbent dose, initial dye concentration, time, and temperature. The optimum values of pH, adsorbent dose, initial dye concentration, time, and temperature were found to be 3.2, 24.76 g/L, 20 mg/L, 67 min, and 30°C for the complete removal of CR dye, respectively. The experimental values were in good agreement with the model predicted values.

Competing interests
The authors declare that they have no competing interests.

Authors' contributions
SP carried out the adsorbent preparation, characterization, and experimentation and wrote the methods and experimental sections. RS carried out the experimental design, data analysis, optimization, and manuscript writing. Both authors read and approved the final manuscript.

Authors' information
RS is working in the Department of Chemical Engineering, University of Louisiana at Lafayette, Lafayette, USA. His areas of research are environmental engineering, bio processing, renewable energy, and process modeling and simulation. SP is working in the Department of Chemical Engineering, SSN College of Engineering Chennai, India. Currently, he is doing research in the fields of waste water treatment and chemical reaction engineering.

Acknowledgments
The authors are grateful for the financial support from the SSN Trust, Chennai, India.

Author details
[1]Department of Chemical Engineering, SSN College of Engineering, Chennai 603 110, India. [2]Department of Chemical Engineering, University of Louisiana at Lafayette, Lafayette, LA 70504, USA.

References
1. Malik A, Taneja U (1994) Utilizing fly ash for color removal of dye effluents. Am Dyestuff Rep 83:20–27
2. Weber WJ, Morris JC (1964) Advances in water pollution research. Pergamon Press, New York
3. Arslan I, Balcioglu IA, Bahnemann DW (2000) Advanced chemical oxidation of reactive dyes in simulated dyehouse effluents by ferrioxalate-Fenton/UVA and TiO_2/UV-A processes. Dyes Pigments 47:207–218
4. Vadivelan V, Kumar KV (2005) Equilibrium, kinetics, mechanism, and process design for the sorption of methylene blue onto rice husk. J Colloid Interface Sci 286:90–100
5. Garg VK, Amita M, Kumar R, Gupta R (2004) Basic dye (methylene blue) removal from simulated wastewater by adsorption using Indian Rosewood sawdust: a timber industry waste. Dyes Pigments 63:243–250
6. Sanghi R, Bhattacharya B (2002) Review on decolorization of aqueous dye solutions by low cost adsorbents. Color Technol 118:256–269
7. Meshko V, Markovska L, Mincheva M, Rodrigues AE (2001) Adsorption of basic dyes on granular activated carbon and natural zeolite. Water Res 35:3357–3366
8. Dedrick RL, Beckmann RB (1967) Kinetics of adsorption by activated carbon from dilute aqueous solution. AIChE J 3:68–75
9. McKay G, Allen SJ, McConvey IF, Otterburn MS (1981) Transport processes in the adsorption of colored ions by peat particles. J Colloid Interface Sci 80:323–339
10. McKay G, Blair HS, Gardner JS (1983) Rate studies for the adsorption of dyestuffs on chitin. J Colloid Interface Sci 95:108–119
11. McKay G (1984) Analytical solution using a pore diffusion model for a pseudo irreversible isotherm for the adsorption of basic dye on silica. AIChE J 30:692–697
12. Asfour HM, Fadeli OA, Nasser MM, El-Geundi MS (1985) Equilibrium studies on adsorption of basic dyes on hardwood. J Chem Technol Biotechnol 35:21–27

13. Asfour HM, Fadeli OA, Nasser MM, El-Geundi MS (1985) Colour removal from textile effluents using hardwood sawdust as adsorbent. J Chem Technol Biotechnol 35:28–35
14. McKay G, Geundi El, Nasser MM (1987) Equilibrium studies during the removal of dyestuff's from aqueous solutions using bagasse pitch. Water Res 21:1513–1518
15. Khare SK, Panday K, Srivastava RM, Singh VN (1987) Removal of Victoria Blue from aqueous solutions by fly ash. J Chem Technol Biotechnol 38:99–104
16. Gupta GS, Prasad G, Singh VH (1988) Removal of chrome dye from aqueous solutions by fly ash. Water Air Soil Pollut 37:13–24
17. Gupta GS, Prasad G, Singh VH (1990) Removal of chrome dye from aqueous solutions by mixed adsorbents: fly ash and coal. Water Res 24:45–50
18. Yoshida H, Fukuda S, Okamota S, Kataoka T (1991) Recovery of direct dye and acid dye by adsorption on chitosan fiber equilibria. Water Sci Technol 23:1667–1676
19. Narmata D, Manzoor A (1993) Dye adsorption by a new low cost material Congo red. Ind J Environ Protec 13:570–576
20. Lee CK, Low KS (1997) Quaternized rice husk as sorbent for reactive dyes. Bioresour Technol 61:121–125
21. Ramakrishna KR, Viraraghavan T (1997) Use of slag for dye removal. Waste Manage 17:483–488
22. Juang RS, Tseng RKL, Wu FC, Lee SH (1997) Adsorption behaviour of reactive dyes from aqueous solution on chitosan. J Chem Technol Biotechnol 70:391–399
23. Lee CK, Low KS (1999) Removal of some organic dyes by acid-treated bleaching earth. Process Biochem 34:451–465
24. Nassar MM (1999) Intraparticle diffusion of basic red and basic yellow dyes on palm fruit bunch. Water Sci Technol 40:133–139
25. Ko DCK, Porter JF, McKay G (2000) Optimized correlations for the fixed bed adsorption of metal ions on bone char. Chem Eng Sci 55:5819–5829
26. Malik PK (2003) Use of activated carbons prepared from sawdust and rice-husk for adsorption of acid dyes: a case study of Acid Yellow 36. Dyes Pigments 56:239–249
27. Goyal M, Singh S, Bansal RC (2004) Equilibrium and dynamic adsorption of methylene blue from aqueous solutions by surface modified activated carbon. Carbon Sci 5:170–179
28. Bangash FK, Manaf A (2004) Kinetics of removal of dye (Basic blue 3) from aqueous solution by activated charcoal from the wood of *Braunsonitia papyrilefra* (Paper Mulberry). J Chem Soc Pak 26:111–115
29. Lorenc-Grabowska E, Gryglewicz G (2006) Adsorption characteristics of Congo red on coal-based mesoporous activated carbon. Dyes Pigments 68:1–7
30. Namasivayam C, Kavitha D (2002) Removal of Congo red from water by adsorption onto activated carbon prepared from coir pith, an agricultural solid waste. Dyes Pigments 54:47–58
31. Teng RL, Wu FC, Juang RS (2003) Liquid-phase adsorption of dyes and phenols using pinewood-based activated carbons. Carbon 41:487–495
32. Khattri SD, Singh MK (1998) Colour removal from aqueous solutions by adsorption. Ind J Chem Technol 5:230–234
33. Singh BK, Rawat NS (1994) Comparative sorption equilibrium studies of toxic phenols on fly ash and impregnated fly ash. J Chem Technol Biotechnol 61:307–317
34. Hameed BH, Ahmad AL, Latiff KNA (2007) Adsorption of basic dye (methylene blue) onto activated carbon prepared from rattan sawdust. Dyes Pig 75:143–149
35. Elibol M (2002) Response surface methodological approach for inclusion of perfluorocarbon in actinorhodin fermentation medium. Process Biochem 38:667–773
36. Box GEP, Wilson KN (1951) On the experimental attainment of optimum conditions. J R Stat Soc Ser B13:1–45
37. Myers RH, Montgomery DC, Anderson-Cook CM (2009) Response surface methodology, process and product optimization using designed experiments 3rd edition. Wiley, Hoboken
38. Annadurai G, Juang RS, Lee DJ (2002) Factorial design analysis of adsorption of activated carbon on activated carbon incorporated with calcium alginate. Adv Inorg Environ Res 6:191–198
39. Brunauer S, Emmett PH, Teller E (1938) Adsorption of gases in multimolecular layers. J Am Chem Soc 60:309–319
40. Montgomery DC (1991) Design and analysis of experiments, 3rd edition. Wiley, New York
41. Segurola J, Allen NS, Edge M, Mahon AM (1999) Design of eutectic photo initiator blends for UV/curable acrylated printing inks and coatings. Prog Org Coat 37:23–37
42. Kim HM, Kim JG, Cho JD, Hong JW (2003) Optimization and characterization of UV-curable adhesives for optical communication by response surface methodology. Polym Test 22:899–906
43. Conley WC (1984) Computer optimization techniques. Petrocelli Books, Princeton

22

Micelles, mixed micelles, and applications of polyoxypropylene (PPO)-polyoxyethylene (PEO)-polyoxypropylene (PPO) triblock polymers

Vijender Singh[1,2], Poonam Khullar[2], Pragnesh N Dave[1*] and Navjot Kaur[3]

Abstract

This review gives a brief outline of various micellar properties of triblock polymers such as critical micellization concentration, critical micellization temperature, and microviscosity. Detailed discussion of the effect of temperature on micellar properties of various triblock polymer mixtures is given. Applications of triblock polymers in solubilization as drug delivery agents, as nano drug, for the synthesis of gold nanoparticles, for cobalt determination, etc. are discussed.

Keywords: Micellization, Critical micelle concentration, Cloud point, Critical micelle temperature

Review

Characteristic features of triblock polymers

Triblock polymers (TBPs) also known as block copolymers belong to a class of nonionic surfactants. They are commercially available under the trade names Pluronics (BASF, Ludwigshafen, Germany), Poloxamers (ICI, London, England), or Synperonics (ICI) and are highly surface-active compounds. The block copolymers consist of a linear hydrophobic polyoxypropylene (PPO) block with hydrophilic polyoxyethylene (PEO) blocks on each side with the structure PEO-PPO-PEO. Their hydrophilic-lipophilic balance (HLB) depends on the PEO-to-PPO mass ratio [1,2]. The rich polymorphic behavior of Pluronics is modulated by their HLB [3]. The PEO-PPO-PEO TBPs are commercially available in a range of molecular weights and PPO-PEO composition [4]. Greater amount of oxyethylene groups, for instance, implies significant aqueous solubility. TBPs are the subjects of fundamental as well as technological research [5-7] that arises from their ability to form micelles, macro- or micro emulsions, and several liquid crystalline phases [8-10]. Because of low toxicity and high biodegradability, these copolymers have extensive industrial applications in detergency, paint industries [11], cosmetics, dispersion stabilization, foaming, lubrication [12,13], and pharmaceutical formulation [14,15]. Other important applications include radiation-damaged cell repair and treatments, controlled drug delivery, bioprocessing [16], nuclear waste processing [17,18], food processing, as well as in agricultural formulations [19-21]. Further applications are emulsion [22] and stabilization in polymerization reaction [23], nanomaterial synthesis [24,25], and coal processing [26].

Fundamentally, they have tremendous advantages than conventional neutral polymers due to the presence of both hydrophilic (i.e., PEO) and hydrophobic predominant (i.e., PPO) moieties in the same polymer molecule. In analogy with the low molecular weight surfactants [27], block copolymers form aggregates of different kinds, depending on the molecular weight, block sizes, solvent composition, and temperature. The low molecular weight Pluronics are viscous oils or pastes, and high molecular weight Pluronics are amorphous solids. The different types of blocks within the copolymer are usually incompatible with one another, and as a consequence, block copolymers self-assemble in melts and in solutions. They also show anomalous behavior over a certain range of temperature, an effect shown, in general, by block copolymers in selective solvents [28,29] which has been found due to the presence of more hydrophobic component present as impurity [30,31].

* Correspondence: pragnesh7@yahoo.com
[1]Department of Chemistry, KSKV Kachchh University, Bhuj, Gujarat 370001, India
Full list of author information is available at the end of the article

Micellization

Critical micelle concentration

The parameter of greatest fundamental value is the critical micellization concentration (CMC), the copolymer concentration at which micelles start forming [32]. TBPs are the nonionic surfactants, which limit the number of techniques available to measure the CMC in comparison to those of ionic surfactants. Also in comparison with conventional low molecular weight surfactants, there is some inherent complexity in the micellization of block copolymers, which depends strongly on their composition [33-42]. The blocks are not completely monodisperse even for a copolymer with a narrow distribution of molecular weight, and accordingly, no sharp CMC/critical micellization temperature (CMT) has been observed for block copolymers. Generally, the CMC spans over a much larger concentration interval than observed with conventional surfactants. The CMC is also sensitive to the temperature which is likely to extend the concentration range over which the CMC occurs [33]. Various techniques such as light scattering [33,43], fluorescence spectroscopy [44,45], NMR [46], specific volume [47,48], and small-angle neutron scattering (SANS) [49,50] have been frequently used to gain new insight into the aggregation behavior of these systems. The process of self-association can be induced by increasing the concentration of TBPs above the CMC and adjusting the temperature to CMT [30,51-56].

Micellization in TBPs is understood to arise due to the following reasons [29,57]. As the temperature of a block copolymer solution is raised, the PPO block progressively loses its hydration sphere, resulting in greater interactions between the PPO blocks. On the other hand, the PEO blocks retain their strong interaction with water; thus as is common for all amphiphilic molecules, the differing phase preferences of the blocks drive the copolymers to form micelles. Structural studies [31,58-63] have shown that the micelles form a hydrophobic core consisting mainly of weakly hydrated PPO blocks, which are surrounded by an outer shell known as corona of almost fully hydrated PEO blocks (Figure 1). There is a broad temperature range above the CMT where the micelles coexist in a solution with unimers. Above the transition region, most of the block copolymer molecules form micelles [64-71]. The reason that a higher temperature is needed to form micelles is that the effective PEO-PEO, PPO-PPO, and PEO-PPO interactions are temperature-dependent. At some temperature, the effective PPO-PPO attraction will dominate over the PEO-PEO repulsion, and micelles will form (Figure 1). CMT of some common triblock polymer is listed in Table 1.

Micelle formation is an extremely temperature-dependent entropy-driven process resulting in a large decrease in CMC upon increasing the temperature. This behavior has led to the wide applicability of CMT as a convenient micellar parameter. Above the CMT, unimers and micelles exist in the state of equilibrium with most of the copolymer molecules in the micellar form. The effects of temperatures on the properties and structure of the PEO-PPO-PEO copolymer solution have been studied extensively [59,72]. An interesting property of the aqueous micellar system is its ability to enhance the solubility in water of otherwise water-insoluble hydrophobic compounds. This occurs because the core of the micelle provides a hydrophobic microenvironment suitable for solubilizing such molecules.

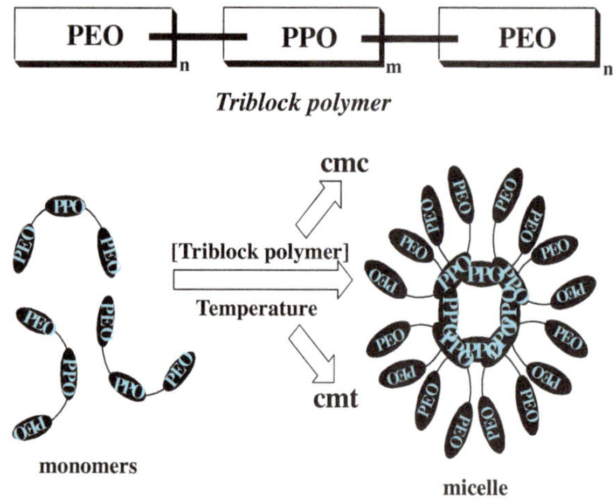

Figure 1 Schematic representation of the general structure of triblock polymer and micelle formation. With respect to the increase in the concentration and temperature.

Many of these copolymers associate in aqueous solution to form spherical micelles [8,59,73], while at higher concentration, block copolymers can also self-assemble into lyotropic liquid crystals [54,74,75]. The progressive growth of the hydrophobic core with increasing temperature due to the increasing dehydration of PEO blocks in the corona induces instability in the spherical micellar dispersion, leading to the formation of rod-like structures. The addition of multivalent salts to aqueous copolymer solutions produces a dramatic effect on the transformation of

Table 1 CMT data 0.1% solution of TBPs

TBPs	Concentration (mM)	CMT (°C)
L64	0.345	39.5
P65	0.294	46
P84	0.238	37
P85	0.217	37.5
F88	0.088	48.5
P103	0.202	24.5
P104	0.169	27.5
P105	0.154	27
F108	0.068	36

the copolymer [76]. At low concentration, spherical micelles form, whereas at high concentration, the formation of hexagonal, cubic, and lamellar liquid crystalline phases takes place [58]. The morphology spectrum runs from micelles (L1), through a cubic array of micelles (I1), hexagonally packed rods (H1), and cubic bicontinuous spheres (V1), to lamellae (L3) where phase inversion takes place, and the inverted morphologies develop: cubic (V2), reversed hexagonal (H2), cubic (I2), and reversed micelle (L2) [77,78]. Regular micelles are formed in polar solvents such as water and alcohols; thus, the corona has a hydrophilic character. Micelles formed in nonpolar solvents such as toluene have a hydrophobic corona and are referred to as 'reverse.' Recently, much attention has been devoted to highly asymmetric block copolymers that form aggregates characterized by a large core and a thin coronal shell of the soluble block: the crew-cut systems [79,80]. 'Crew-cut' micelle-like aggregates represent a new type of aggregate. They are formed via the self-assembly of a highly asymmetric amphiphilic block copolymer, in which the insoluble core-forming blocks are much longer than the soluble corona-forming blocks [81,82]. One of the noteworthy phenomena associated with crew-cut aggregates is the accessibility of a wide range of morphologies [79,81,83,84]. These include spheres, rods, vesicles, lamellae, large compound micelles, large compound vesicles, a hexagonally packed hollow hoop structure (the 'HHH' structure) [85], onions [86], a bowl-shaped structure [87], and several others [88,89].

Cloud point and microviscosity

Another parameter of great practical importance of TBPs in aqueous phase is the cloud point (C_P) [73,90,91]. At higher temperatures well above the CMT, the copolymer solution becomes opaque because the phase separation between the polymer and water occurs. The temperature at which cloudiness appears due to the precipitation is the C_P of TBPs [92,93]. The C_P phenomenon in TBPs is related to the core and corona model of the TBP aggregates. This model suggests that the core of TBP micelles mainly consists of PPO units while the corona occupies PEO units. The presence of ether oxygens both in the PPO as well as PEO units allows some number of water molecules to be even available in the core. An increase in the temperature thus dehydrates the TBP micelles even at optimum temperatures by expelling water molecules, which are weakly associated with ether oxygens through the electrostatic interactions. At C_P, the attractive interactions between PEO blocks and water molecules are sufficiently weak that the PEO blocks become completely dehydrated. It has been observed [94,95] that any factor which would increase the number of water molecules in TBP micelles would result in an increase in C_P and vice versa. The TBPs containing a larger PPO block than that of PEO have higher microviscosity [96-98]. The microviscosity is strongly affected by the PPO block. It appears that the larger its molecular weight, the more viscous the micelle interior. Among the two different polymers, the micelles with a larger PPO block exhibit higher microviscoscity. Cloud point of some common triblock polymer solutions of 1% concentration is cited in Table 2.

Mixed micelles

Like conventional surfactants, TBPs show a clear micelle formation process which can be best demonstrated by the change in I_1/I_3 pyrene ratio versus concentration plots. Since the anterior of the TBP micelle is constituted by the predominantly hydrophobic PPO units surrounded by hydrophilic PEO, therefore pyrene can easily solubilize in the core of the micelle and hence can act as a fine probe for the micelle formation process. The I_1/I_3 ratio is sensitive to the microenvironment of pyrene solubilization, and hence, its variation explains the micelle formation process. In aqueous TBP solutions (Figure 2), it usually starts from approximately 1.75 (value in pure water) and decreases to a constant value around 1.3 where it is solubilized in the TBP micelle. Similar methods can very well be adopted for the measurements of CMC for the mixed components over the whole mole fraction range.

P103 + TBPs mixtures

P103 consists of 60 PO units in comparison to 34 EO units. It makes this TBP predominantly nonpolar. The variation of mixed CMC values for different mixtures of P103 is shown in (Figure 3) along with the ideal mixing. It is usually difficult to determine from the CMC profiles which mixture shows greater attractive or repulsive

Table 2 Cloud point of 1% solution of TBPs

TBPs	Cloud point (°C)
L35	73
F38	>100
L42	37
L43	42
L44	65
L62	32
L63	34
P84	74
P85	85
F127	>100
P123	90
L122	19
L64	58
P103	86

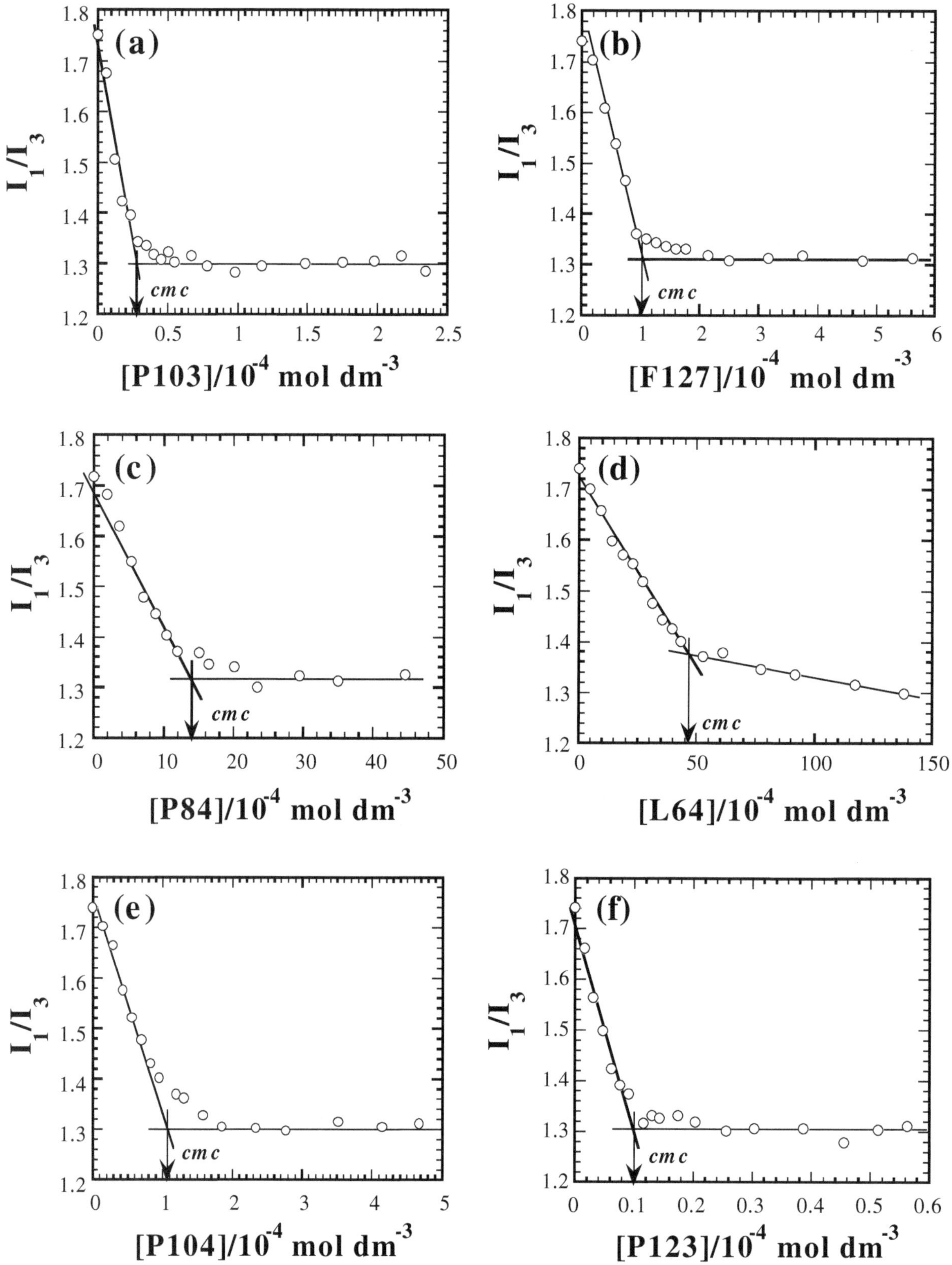

Figure 2 Critical micelle concentration of different TBPs.

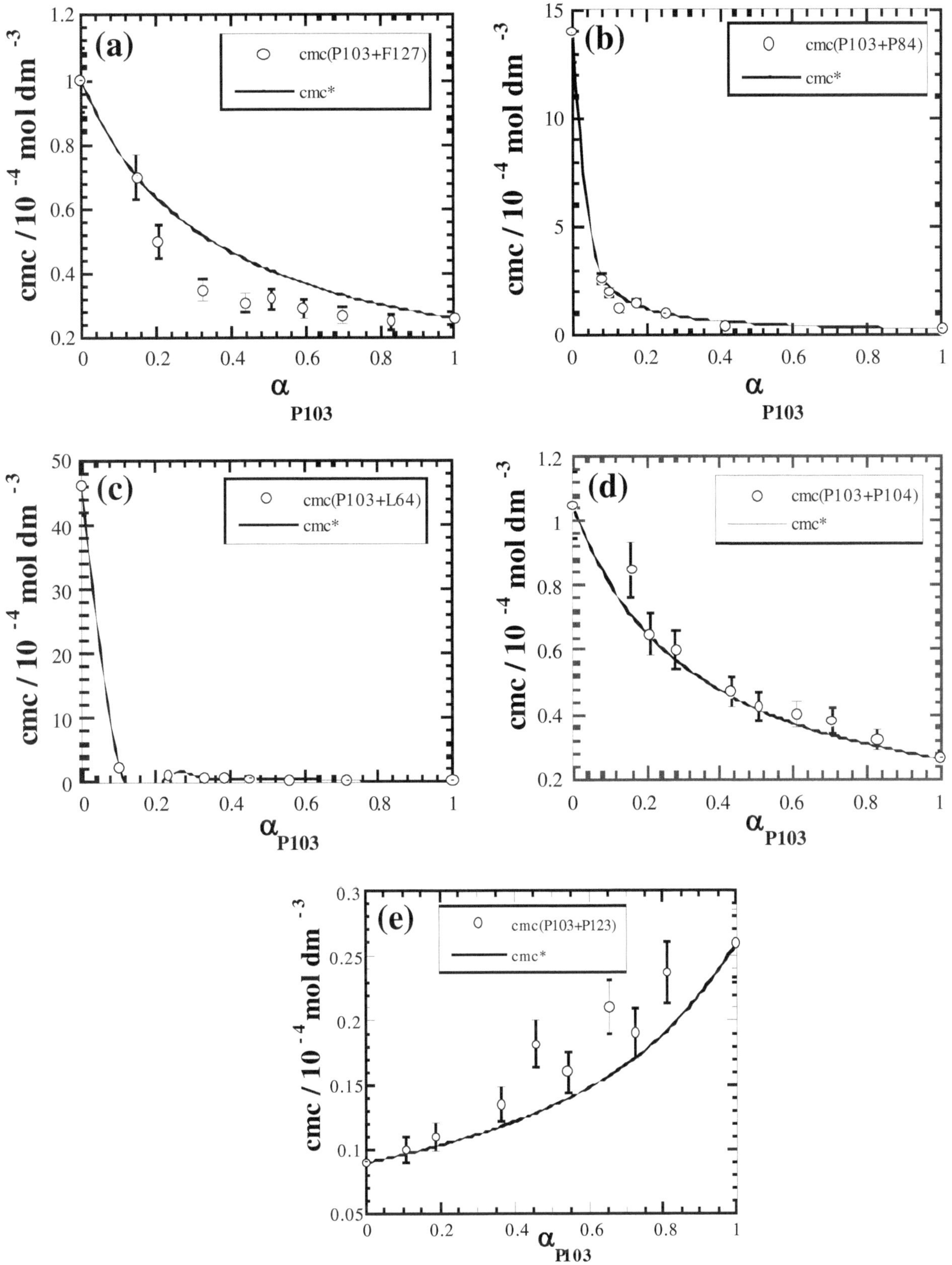

Figure 3 Critical micelle concentration versus mole fraction of P103. For (**a**) P103 + F127, (**b**) P103 + P84, (**c**) P103 + L64, (**d**) P103 + P104, and (**e**) P103 + P123 binary mixtures, respectively.

interactions due to the course of micellization; therefore, the regular solution theory based on the ideal mixing is the best way to establish the non-ideal behavior in terms of parameter 'β.' The average of β values over different mixing ratios for each mixture helps to understand the fact on how one mixture of intermolecular interactions is different from other mixtures. Average β values plotted against the PPO/PEO ratio for different mixtures are shown in (Figure 4). A positive and negative value respectively indicates the operating repulsive and attractive interactions between the components of the mixtures. β_{Avg} becomes more negative as PPO/PEO increases and approaches 1.5. Further increase in the PPO/PEO ratio instantaneously converts β_{Avg} from a negative to a positive value. Components with lower PPO/PEO ratio (i.e., F127, P84, and L64) show attractive interactions with P103, while the component with greater PPO/PEO ratio (i.e., P104 and P123) shows unfavorable mixing with P103 which is clearly related to the delicate balance between the PO and EO units in a mixed state. P103 contains a greater number of PO (70) units in comparison to EO (38) units which makes P103 a predominantly hydrophobic polymer. Thus, in the event of micelle formation with other TBPs, P103 will prefer to have maximum synergistic mixing with similar predominantly hydrophobic TBPs such as L64 rather than P104 and P123 with much higher number of PO units in comparison to EO which will induce steric effects in the core, resulting in the unfavorable mixing.

Temperature effect on P103 + F127/P123 mixtures

Temperature is another very important parameter which can drastically affect the microenvironment of the TBP micelle. PO units show dramatic loss of water or hydration in comparison to EO units as temperature increases. In the first combination, the relative difference between the molecular weights of PPO blocks of P103 and F127 is much less in comparison to that among PEO blocks. In the latter case, this difference is more significant for PPO blocks rather than that of PEO blocks. Hence, we want to see which combination of these two would produce greater non-ideal behavior (favorable or unfavorable) in their mixed state and also under the effect of temperature variation.

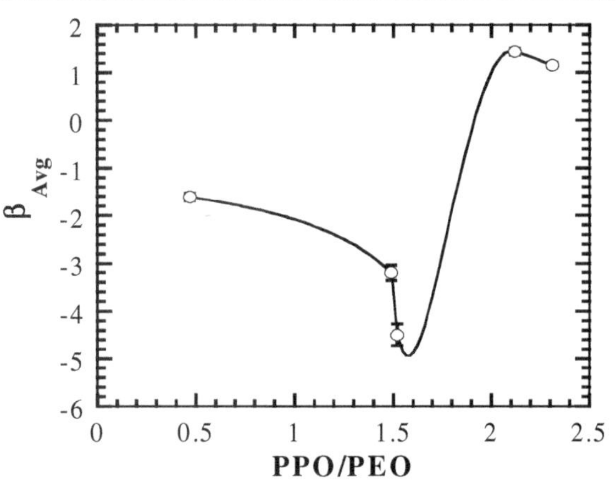

Figure 4 Plot of interaction parameter, β_{Avg}, versus PPO/PEO ratio of the triblock polymers.

The mixed CMC values are shown graphically in Figures 5 and 6 along with the ideal CMC values. The experimental CMC values for the P103 + F127 mixture (Figure 5) are lower than the ideal behavior at 25°C, and this difference decreases as the temperature increases from 25°C to 40°C. In contrast, the experimental values are quite close to the ideal ones for the P103 + P123 mixture (Figure 6) at 25°C but become significantly higher as temperature increases. It means that P103 + F127 shows attractive interactions at 25°C that decrease with the temperature, while P103 + P123, on the other hand, behaves almost ideally at 25°C but becomes increasingly non-ideal as temperatures increases. Since the temperature variation has a dramatic effect on the hydration of mixed TBP micelles, therefore, this effect should reflect from the viscosity measurements. Plots of excess relative viscosity ($\Delta\eta_r$) of both mixtures over the whole mole fraction range (Figure 7) demonstrate negative deviation at 25°C for P103 + F127 that decreases as the temperature increases, while at 40°C, it almost follows the additivity rule. Whereas, $\Delta\eta_r$ for P103 + P123 mixtures shows a comparatively very weak negative deviation at 25°C from an ideal behavior that decreases and reverts to a weak positive deviation as temperature increases.

The negative deviations in $\Delta\eta_r$ values generally arise from the decrease in fluidity in the bulk upon mixing two components. That is happening due to attractive interactions between the micelles of P103 and F127 in order to form mixed micelles which enhance the aggregation, and hence, a greater number of aggregates reduce fluidity. However, when temperature increases, it causes dehydration of PO and EO groups which is relatively rapid for the PO group compared to the EO group and hence leads to instability to the mixed micelles, That, in turn, reduces the aggregation due to the mixed micelle formation, and hence, fluidity increases and approaches the ideal behavior. On the other hand, the weak negative deviation in P103 + P123 mixtures at 25°C is again due to increased fluidity, but that reverts to ideal behavior as temperature increases and even shows positive deviation at higher temperatures. It means that the dehydration in P103 and P123 micelles happens to such an extent that both components start showing unfavorable mixing. It all happens due to the presence of a much larger number of PO units in both P103 and P123 which dehydrate rapidly as temperature increases and even loose solubility in the aqueous phase. This reduces the

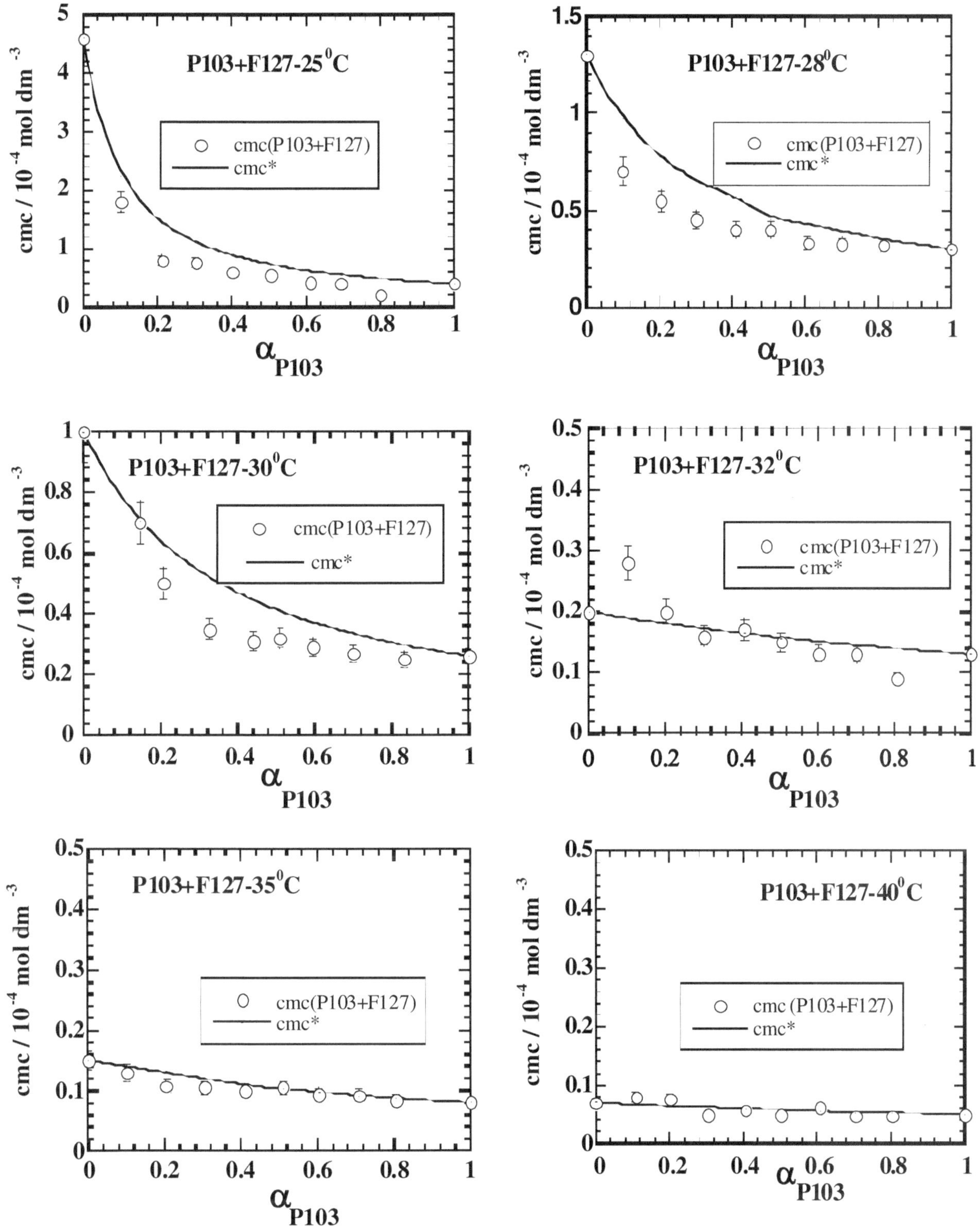

Figure 5 Plot of mixed CMC versus α_{P103} for P103 + F127 mixture at various temperatures. CMC* is the ideal CMC value.

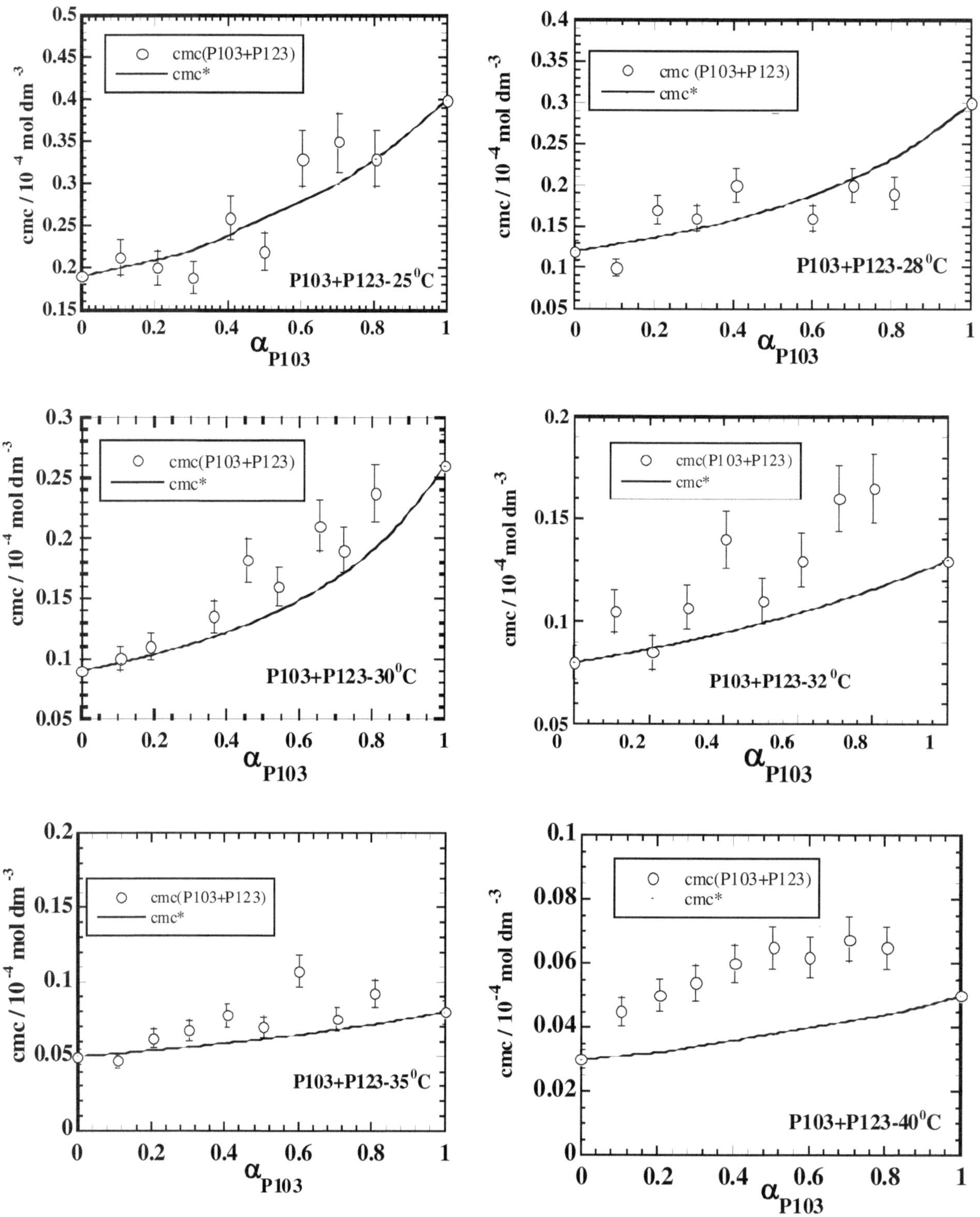

Figure 6 Plot of mixed CMC versus α_{P103} for P103 + P123 mixture at various temperatures. CMC* is the ideal CMC value.

micelle formation in each case, and hence, the system approaches the ideal behavior as temperature starts increasing but eventually reverts to positive deviations at a higher temperature due to decreased solubility.

Temperature effect can further be evaluated by computing the regular solution parameter β for each mixture at $\alpha_1 = 0.5$ ($\beta_{0.5}$) (Figure 8). Both mixtures start with negative β values at 25°C, which become less negative

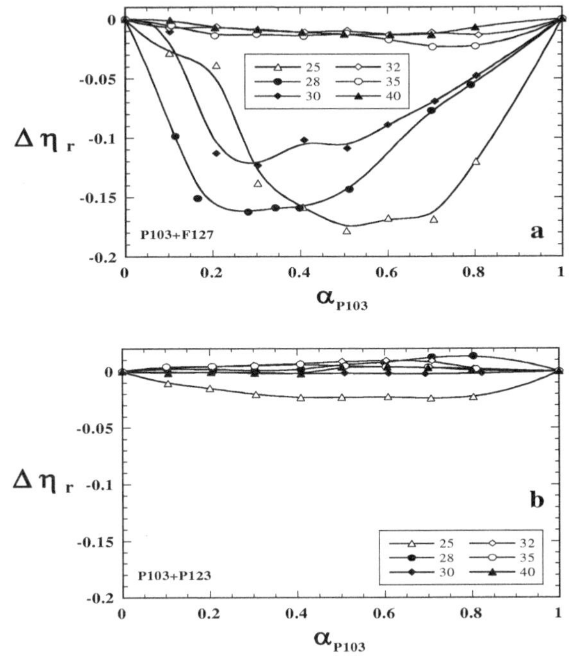

Figure 7 Plot of excess relative viscosity, $\Delta\eta_r$, versus α_{P103}. For (**a**) P103 + F127 and (**b**) P103 + P123 mixtures over the whole mole fraction range at different temperatures.

with the increase in temperature. The values remain negative from 25°C to 40°C for P103 + F127, while those for P103 + P123 change to positive somewhere close to 30°C. Within the framework of the regular solution theory, the negative β value is explained on the basis of attractive interactions between the unlike TBP components, while the positive value can be due to the unfavorable mixing. Thus, an increase in the temperature brings unfavorable mixing among both unlike TBP components in both kinds of mixed micelles. A greater

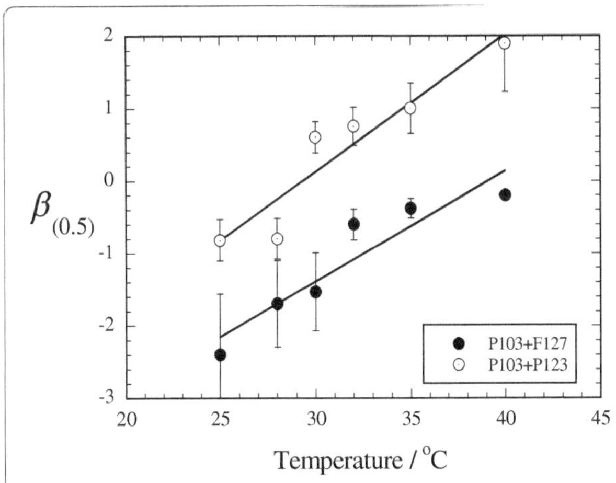

Figure 8 Plot of regular solution interaction parameter, $\beta_{(0.5)}$, versus temperature. For P103 + F127 and P103 + P123 mixtures at α_{P103} = 0.5.

dehydration of PO groups in the case of P103 + P123 mixtures with greater number of PO units shows a stronger temperature effect, and that is why β shifts to a positive value at a much lower temperature around 30°C. Whereas, this is not happing over the whole temperature range studied in the case of P103 + F127 mixtures due to the presence of a relatively lesser number of PO groups. Although mixed micelle formation between two TBPs is governed by the predominantly hydrophobic interactions operating between the predominantly hydrophobic PPO domains of unlike TBPs, an increase in the temperature reduces their aqueous phase solubility and adversely affects the hydrophobic interactions responsible for the mixed micelle formation. That is why this effect is more prominent in the case of P103 + P123 mixtures with a larger PPO domain in comparison to that of P103 + F127.

Critical micelle temperature

TBPs are known to have micelle formation with respect to the variation of temperature, and the temperature where it happens is known as CMT, an analogous term used for this purpose to that of CMC where micelle formation occurs due to a change in concentration. CMT is considered to be having more relevance as far as their shelf life under varying temperatures is concerned because many industrial products in the cosmetic industry consist of more than one TBP component. Although, several studies have reported the CMT, little is known about the mixed CMT behavior. Experimental CMT values for some binary mixtures are shown in (Figure 9) along with the ideal mixing. In most cases, they show negative deviations from ideal behavior except in the case of P104 + P103 where the CMT values mainly lie close to that of the ideal behavior. A negative departure of CMT values indicates that the mixed micellization is taking place at lower temperatures and can be explained on the basis of favorable interactions between the unlike TBP components, and that shifts the mixed CMT to lower temperatures. Greater departure accounts for stronger interactions between the components of P103 + L64/P84 binary mixtures. The close-to-ideal mixing happening in the case of P104 $[(EO)_{18}(PO)_{58}(EO)_{18}]$ and P103 $[(EO)_{17}(PO)_{60}(EO)_{17}]$ is due to the little difference between the number of PO as well as EO units. This allows the unlike polymer macromolecule to accommodate in the mixed state without significant alterations in the overall hydrophilic or hydrophobic environment. On the contrary, this is not the case when L64 $[(EO)_{13}(PO)_{30}(EO)_{13}]$ and P84 $[(EO)_{19}(PO)_{43}(EO)_{19}]$ have been taken. In both these cases, there is a large difference in the PO units of these polymers compared with that of P103. Thus, in the event of mixed micelle formation, a larger PPO block of P103 has to accommodate with the

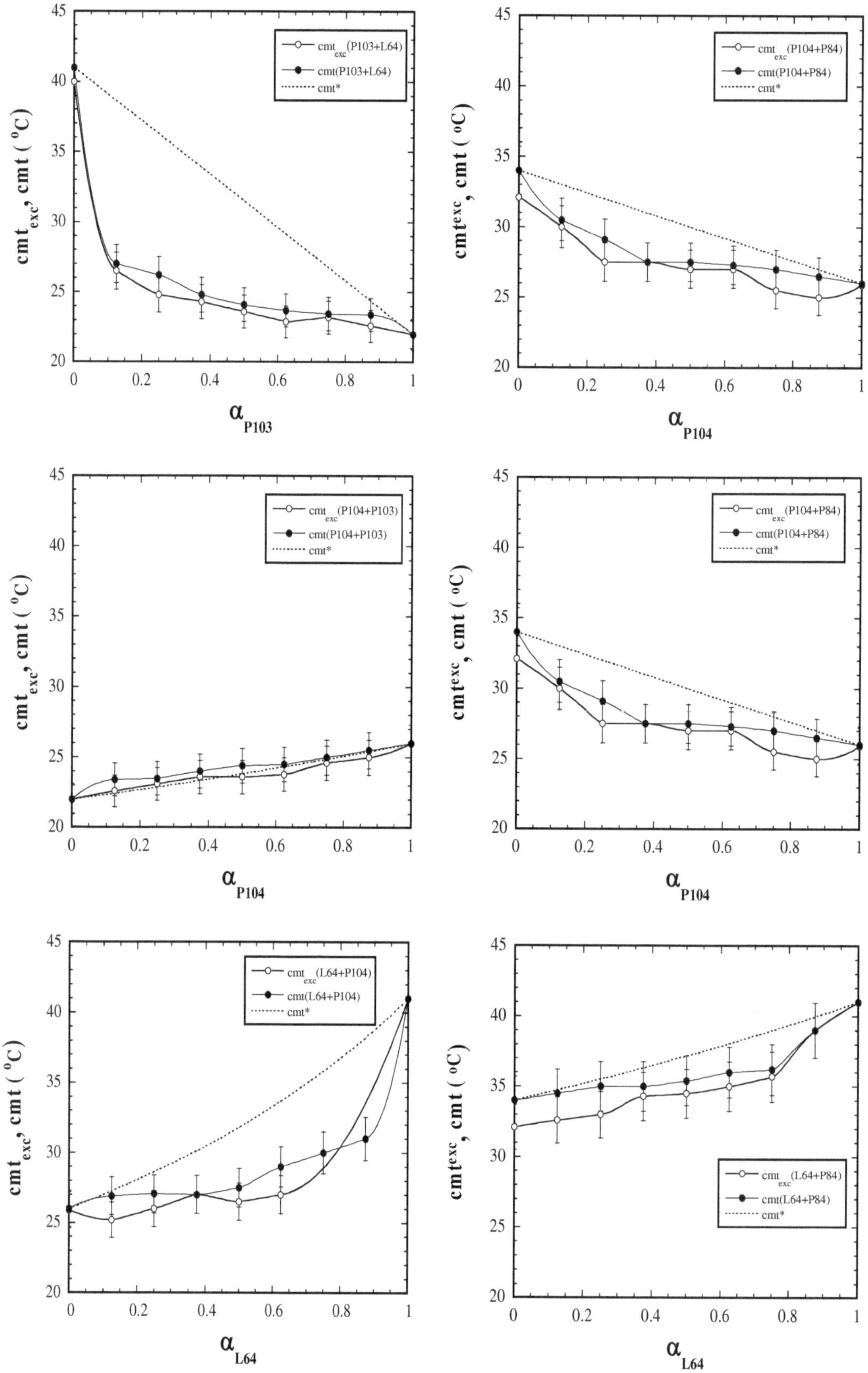

Figure 9 Plot of CMT and CMT$_{exc}$(°C) versus $\alpha_{P103,P104,L64}$. For P103 + L64, P103 + P84, P104 + P103, P104 + P84, L64 + P104, and L64 + P84 binary mixtures.

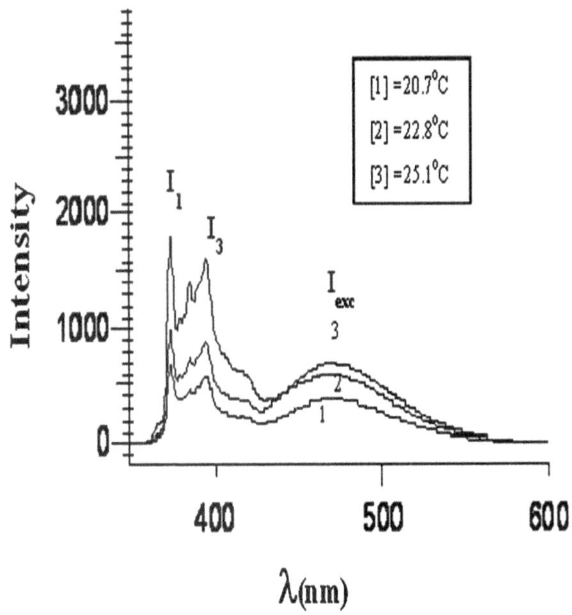

Figure 10 Pyrene emission spectrum at various temperatures of [P103] = 5.0 × 10^{-4} mol dm^{-3}. I_1, I_3, and I_{exc} are the intensities of the first, third, and excimer bands of pyrene.

shorter PPO block of either L64 or L84 in the core of the mixed micelle. This ensures the micelle transitions with greater compatibility among the components in the mixed state (Figure 9).

Quenching of pyrene by a suitable quencher (Q) such as hexadecylpyridinium chloride (HPyCl) under steady-state conditions can also be used to explain these results and assumed that the fluorescence lifetime of pyrene is longer than the residence time of the quencher in a micelle. A suitable [pyrene]/[mixed micelle] and [Q]/[mixed micelle] ratios ensure the Poisson distribution. The fluorescence intensity of the first vibronic band of (Figure 10) pyrene decreases with the increase in [Q] without the appearance of any new band (not shown). A Stern-Volmer relationship is used to calculate the collisional quenching constant, called the Stern-Volmer constant (K_{SV}) [99-102]. These K_{SV} values for various binary mixtures (Figure 11) vary nonlinearly with positive deviations from the ideality, which are predominant in the case of P103 + L64/P84, thus demonstrating that the quenching is facilitated in these mixtures. This can be attributed to the presence of a suitable solubilizing environment provided by the mixed micelles for an effective quenching to take place. The quenching is prominent in the L64/P84-rich region of the mixture which means that the small amount of the induction of P103 generates a favorable environment for the solubilization of both the quencher and pyrene. In the rich region of P103, the much larger micellar core

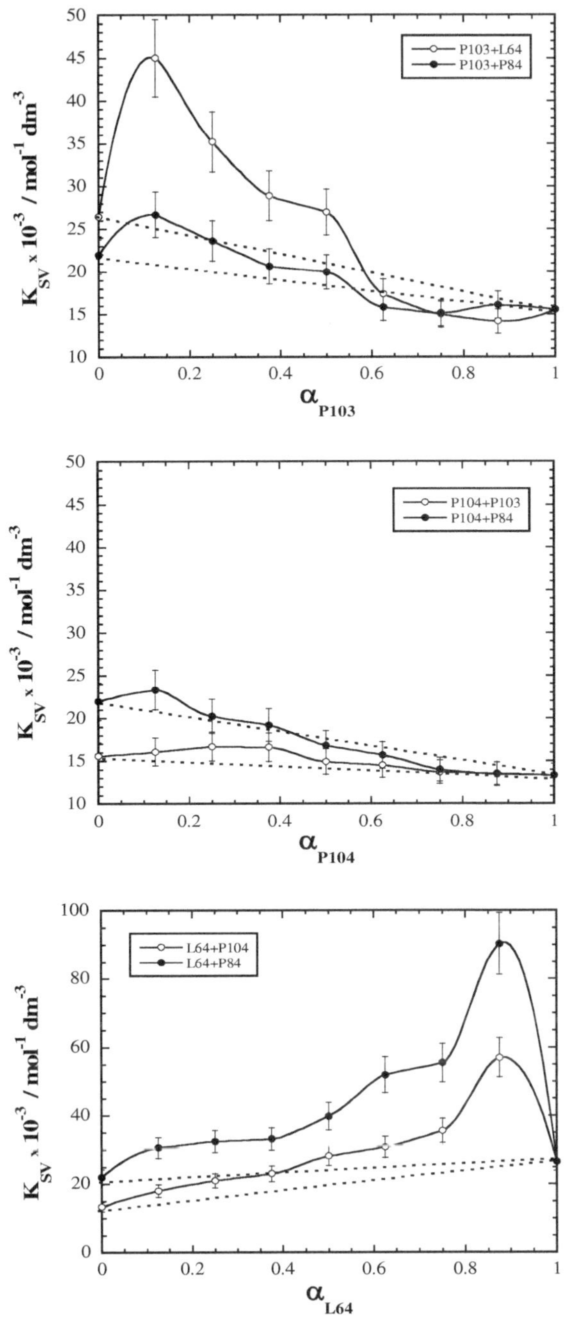

Figure 11 Plot of K_{SV} versus $\alpha_{P103/P104/L64}$. For P103 + L64/P84 binary mixtures, P104 + P103/P84 binary mixtures, and L64 + P104/P84 binary mixtures at 45°C.

with predominantly greater amount of PO units will make the encounters of both the quencher and pyrene difficult; as a result, quenching decreases.

The excimer emission is produced by the collisional quenching between the excited (Py*) and ground state (Py) monomers of the fluorescence probe. Thus, the

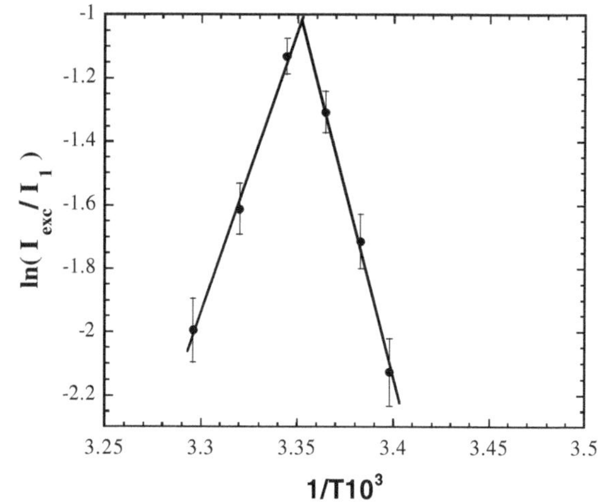

Figure 12 **Plot of ln(I_{exc}/I_1) versus 1/T for pure P104 at [P104] = 5.0 × 10^{-4} mol dm^{-3}.**

mechanism of the excimer dimer (D^*) formation can be written in the following way:

$$Py^* + Py \underset{k_{ED}}{\overset{k_{EF}}{\rightleftharpoons}} D^* \quad (1)$$

$$Py^* \xrightarrow{k_{MF}} Py + hv \quad (2)$$

$$D^* \xrightarrow{k_{DF}} 2Py + hv, \quad (3)$$

where k_{EF}, k_{ED}, k_{MF}, and k_{DF} are the constants of excimer formation, excimer dissociation, monomer fluorescence, and dimer fluorescence, respectively. The kinetics of excimer formation can further be explained under two sets of experimental conditions, i.e., low-temperature and high-temperature behaviors within the temperature range studied herein. At low temperature, I_{exc}/I_1 ratio can be written as:

$$\frac{I_e}{I_1} = \frac{k^1{}_{EF}[Py]}{k_{MF}} \exp\left[{}^-W_{EF}/_{kT}\right], \quad (4)$$

where k^1_{EF} and W_{EF} are the frequency factor (limiting value of k^1_{EF} as T ∞) and activation energy of excimer formation, respectively, and k is Boltzmann's constant. Similarly, at high temperature, the I_{exc}/I_1 ratio is given by:

$$\frac{I_e}{I_1} = \frac{k_{DF}k^1{}_{EF}[Py]}{k^1{}_{ED}k_{MF}} \exp\left[{}^B/_{kT}\right], \quad (5)$$

where B is the excimer binding energy = W_{ED} − W_{EF}, and k^1_{EF} is the frequency factor. Equations 4 and 5 suggest that ln(I_{exc}/I_1) shows an increase and decrease linearly with 1/T, respectively, at fixed [Py]. Figure 12 shows such a variation for P104 at different temperatures. Here, an intersection of two linear lines gives the CMT 26°C. Hence, the kinetics of pyrene solubilization in TBP micelles can be analyzed within two different sets of experimental conditions, one is below and the other above 26°C. The former gives the activation energy for the excimer formation (Figure 13), while the latter gives the binding energy (Figure 14) between Py* and Py.

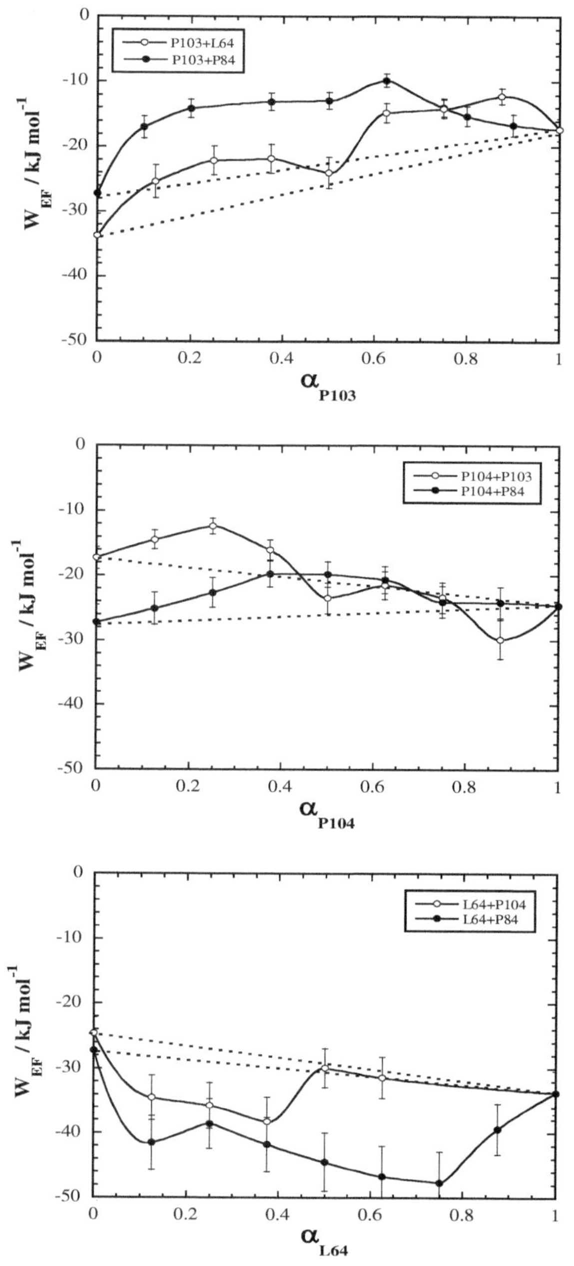

Figure 13 **Plot of energy of activation (W_{EF}) of the pyrene probe for the excimer formation versus $\alpha_{P103/P104/L64}$.** For P103 + L64/P84 binary mixtures, P104 + P103/P84 binary mixture, and L64 + P104/P84 binary mixtures.

Figure 13 shows the variation of W_{EF} over the whole mole fraction range for all mixtures. These values for P103 + L64/P84 (Figure 13, top) show a positive deviation from the ideal mixing (shown by dotted lines),

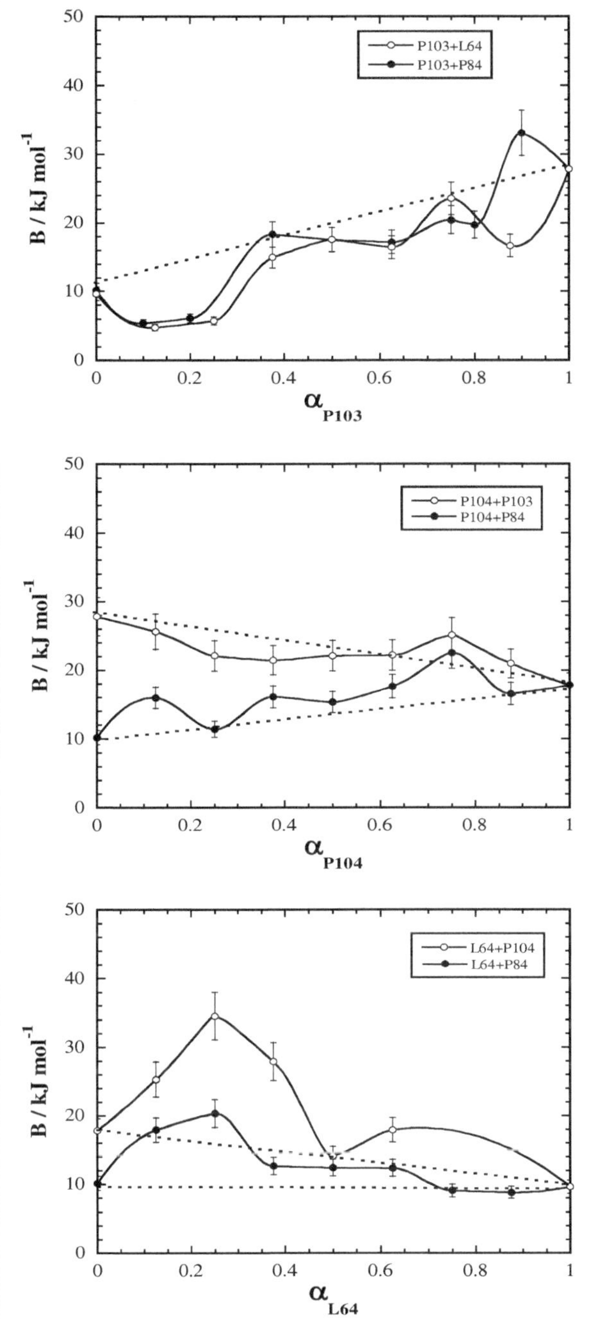

Figure 14 Plot of binding energy (*B*) of the pyrene probe for the excimer formation versus $\alpha_{P103/P104/L64}$. For P103 + L64/P84 binary mixtures, P104 + P103/P84 binary mixture, and L64 + P104/P84 binary mixtures.

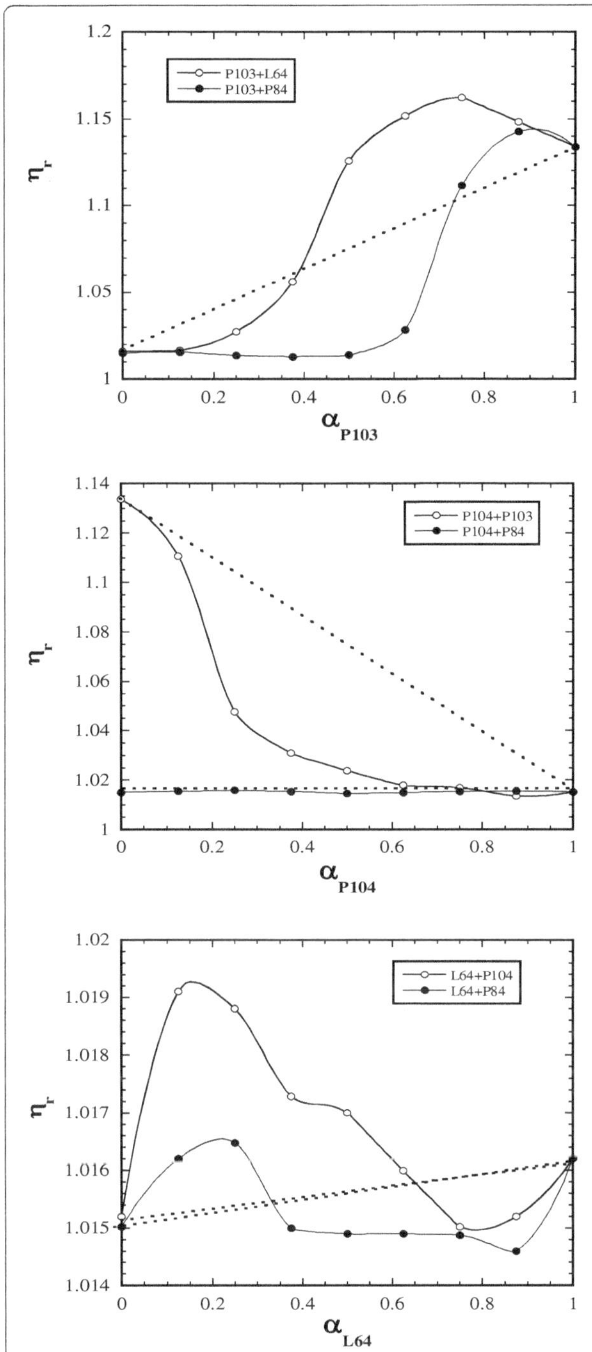

Figure 15 Plot of η_r versus $\alpha_{P103/P104/L64}$. For P103 + L64/P84 binary mixtures, P104 + P103/P84 binary mixtures, and L64 + P104/P84 binary mixtures at 45°C.

while those for P104 + P103/P84 (Figure 13, middle) remain mostly close to the ideal behavior. Figure 13 (bottom) shows negative deviations in W_{EF} values from ideality for L64 + P104/P84 mixtures. A decrease in the W_{EF} values at all mole fractions of P103 + L64/P84 can be attributed to the facilitation of the excimer formation which is also evident from the higher K_{SV} values for these mixtures. These results can also be related to the relative viscosity (η_r) behavior for these mixtures (Figure 15). A smaller η_r value than the ideal behavior especially in the P103-poor region of the mixtures shows the attractive interactions between the components that might be responsible for lower W_{EF} in comparison to their pure states. This is further supported by the negative deviations in B values from the ideal behavior (Figure 14, top). The variation in the B values for these mixtures (Figure 14, middle) fully supports this fact. The η_r of P104 + P84 mixtures (Figure 15, middle) shows clear ideal mixing, but the negative deviation in the case of P104 + P103 is due to unknown reasons. An increase in the W_{EF} value for L64 + P103/P84 mixtures (Figure 13, bottom) from their pure components clearly indicates the reduction in excimer formation. This is again complimentary with the positive deviations in the B values from the corresponding ideal mixing (Figure 14, bottom). Both figures demonstrate that the variation in W_{EF} and B values is mainly predominant in the L64-poor regions of both mixtures. The positive deviations in the η_r (Figure 15, bottom) from the ideal behavior especially in the L64-poor region further confirm these results.

The variation of all the micellar parameters and photophysical properties indicates that the mixed micelles between the components of P103 + L64 and P103 + P84 mixtures form due to attractive interactions. These interactions arise from the mutual compatible arrangement among the unlike TBP monomers in the mixed state in such a way that steric hindrances are minimized. On the other hand, mixtures of P104 + P103 and P104 + P84 prefer to remain ideal in their mixed state, while the mixtures of L64 + P104 and L64 + P84 show mainly unfavorable mixing.

Applications

Solubilization and drug delivery agents

The low solubility in biological fluids displayed by about 50% of the drugs still remains the main limitation in oral, parenteral, and transdermal administration. To overcome these drawbacks, inclusion of hydrophobic drugs into polymeric micelles is one of the most attractive alternatives. Amphiphilic poly(ethylene oxide)-poly(propylene oxide) block copolymers are thermoresponsive materials that display unique aggregation properties in aqueous medium. Due to their ability to form stable micellar systems in water, these materials are broadly studied as hydrosolubilizers for poorly water-soluble drugs. This occurs because the core of the micelle provides a hydrophobic microenvironment, suitable for solubilizing such molecules. The phenomenon of solubilization forms the basis for many practical applications of amphiphiles. The most important applications of PEO-PPO-based copolymers in the pharmaceutical technology field are for attaining improved solubility, stability, release, and bioavailability of drugs [103]. Hydrophobic Pluronic block copolymers (PBC) form lamellar aggregates with a higher solubilization capacity than spherical micelles formed by hydrophilic PBC. However, they also have a larger size and low stability. To overcome these limitations, Kabanov et al. [104] prepared binary mixtures from hydrophobic PBC (L121, L101, L81, and L61) and hydrophilic PBC (F127, P105, F87, P85, and F68). In most cases, PBC mixtures were not stable, revealing formation of large aggregates and phase separation within 1 to 2 days. However, stable aqueous dispersions of the particles were obtained upon (1) sonication of the PBC mixtures for 1 or 2 min or (2) heating at 70°C for 30 min. It was observed that among all combinations, L121/F127 mixtures (1:1% weight ratio) formed stable dispersions with a small particle size. The solubilizing capacity of this system was examined using a model water-insoluble dye, Sudan (III). Mixed L121/F127 aggregates exhibited approximately tenfold higher solubilization capacity compared to that of F127 micelles. Thus, stable aqueous dispersions of nanoscale size were prepared from mixtures of hydrophobic and hydrophilic PBC by using the external input of energy. The prepared mixed aggregates can efficiently incorporate hydrophobic compounds. To enhance stability of micelles in the blood stream upon dilution, Pluronic L121 micelles were cross-linked through their hydrophilic shells [105]. To form the cross-links, the end hydroxyl groups of Pluronic L121 were first chemically converted to aldehydes and then bridged via Schiff bases. This greatly reduced the CMC of the micelles and enhanced the micelle stability. A series of studies used Pluronic P105 micelles for the delivery of Dox into solid tumors in mice [106-109]. In these studies, the localized ultrasonic irradiation of the tumor was applied upon accumulation of the micelles in the tumor interstitium to facilitate the drug release into the tumor cells [108]. Furthermore, the ultrasound-enhanced intracellular uptake of Dox administered with the Pluronic P105 micelles was demonstrated *in vitro* [106]. It was suggested that the enhanced uptake was caused by either micelle disintegration that released free Dox or cell membrane perturbations that facilitated the cellular uptake of the micelles as a whole. Overall, micellar delivery combined with ultrasonic irradiation resulted in a substantial decrease of the tumor growth rates compared to a positive control.

Nano drug delivery

Polymer nanomaterials have sparked a considerable interest as vehicles used for diagnostic and therapeutic agents; research in nanomedicine has not only become a frontier movement but is also a revolutionizing drug delivery field. A common approach for building a drug delivery system is to incorporate the drug within the nanocarrier that results in increased solubility, metabolic stability, and improved circulation time. The recent developments indicate that select polymer nanomaterials can implement more than only inert carrier functions by being biological response modifiers. The Pluronic block copolymers cause various functional alterations in cells. The key attribute for the biological activity of Pluronics is their ability to incorporate into membranes followed by subsequent translocation into the cells and affecting various cellular functions, such as mitochondrial respiration, ATP synthesis, activity of drug efflux transporters, apoptotic signal transduction, and gene expression. As a result, Pluronics cause drastic sensitization of MDR tumors to various anticancer agents, enhance drug transport across the blood-brain and intestinal barriers, and cause transcriptional activation of gene expression both *in vitro* and *in vivo*. Pluronics have a broad spectrum of biological response-modifying activities which make it one of the most potent drug targeting systems available, resulting in a remarkable impact on the emergent field of nanomedicine. Incorporation of low molecular mass drugs into Pluronic micelles can increase drug solubility and drug stability and can improve drug pharmacokinetics and biodistribution. Polymeric micelles were utilized for delivery of CNS drugs across the blood-brain barrier [110,111], oral delivery of drugs [112-114], and tumor-specific delivery of antineoplastic agents [115-117]. For example, neuroleptic drug-loaded Pluronic P85 micelles were targeted to the brain by conjugating the micelles with neurospecific antibodies or insulin as targeting moieties [118]. An improvement of oral bioavailability of a poorly water-soluble phytoestrogen, genistein, was achieved by incorporation of this drug into Pluronic F127 micelles [117]. Pluronic block copolymers were also reported to significantly enhance the bioavailability of various antibacterial and antifungal drugs and to enhance the activity of these drugs with respect to many microorganisms [119-122]. Lee et al. [123] developed a binary mixing system with two Pluronics, L121/P123, as a nanosized drug delivery carrier. The lamellar-forming Pluronic L121 (0.1 wt.%) was incorporated with Pluronic P123 to produce nanosized dispersions (in the case of 0.1 and 0.5 wt.% P123) with high stability due to Pluronic P123 and high solubilization capacity due to Pluronic L121. The binary systems were spherical and less than 200 nm in diameter, with high thermodynamic stability (at least 2 weeks) in aqueous solution. The CMC of the binary system was located in the middle of the CMC of each polymer. In particular, the solubilization capacity of the binary system (0.1/0.1 wt.%) was higher than mono-systems of P123. The main advantage of binary systems is overcoming limitations of mono-systems to allow tailored mixing of block copolymers with different physicochemical characteristics. Kadam et al. [124] investigated the effect of the molecular characteristics of EO-PO triblock copolymers Pluronic P103 ($EO_{17}PO_{60}PEO_{17}$), P123 ($EO_{19}PO_{69}EO_{19}$), and F127 ($EO_{100}PO_{65}EO_{100}$) on micellar behavior and solubilization of a diuretic drug, hydrochlorothiazide (HCT). The CMTs and size for empty as well as drug-loaded micelles are reported. The CMTs and micelle size depended on the hydrophobicity and molecular weight of the copolymer; a decrease in CMT and increase in size were observed on solubilization. The solubilization of the drug HCT in the block copolymer nanoaggregates at different temperatures (28°C, 37°C, 45°C) and pH (3.7, 5.0, 6.7) and in the presence of added salt (NaCl) was monitored by using UV-vis spectroscopy, and solubility data were used to calculate the solubilization characteristics: micelle-water partition coefficient (P) and thermodynamic parameters of solubilization *viz.* Gibbs free energy (ΔG_s°), enthalpy (ΔH_s°), and entropy (ΔS_s°). It is observed that the solubility of the drug in the copolymer increases with the trend: P103 > P123 > F127. The solubilized drug decreased the cloud point (C_P) of copolymers. Results showed that the drug solubility increases in the presence of salt but significantly enhances with the increase in the temperature and at a lower pH in which the drug remains in the nonionized form.

Synthesis of gold nanoparticles

Gold nanoparticles are a great deal of recent interest in the context of emerging nanotechnology applications. At the nanoscale, they exhibit unique quantum and surface properties, different from those of atoms as well as bulk materials [125-128]. Depending on the applications they are being synthesized for, they can be synthesized in many different ways. One of the easiest and convenient ways is the chemical reduction method which involves the use of four basic materials, namely, solvent, metal salt, reducing agent, and stabilizing agent [129-131]. Recently, the use of block copolymers for the synthesis of gold nanoparticles is found to have many advantages; for example, a block copolymer not only plays the dual role of reductant and stabilizer but also provides an economical and environmentally benign way for the synthesis of gold nanoparticles [132-134]. The hydrophobic blocks of the block copolymers (PPO) form the core of these micellar aggregates, whereas the hydrophilic ones (PEO), with the surrounding water molecules, form the corona. The block copolymers can be used to produce metal nanoparticles because of their ability to reduce metal ions. On mixing the aqueous solution of metal (e.g., gold) salt and block copolymers, these polymeric nanostructured matrixes engulf the ionic metal precursors, which after subsequent reduction form nanoparticles. Self-assembly of a block copolymer in this method is utilized to control the synthesis of gold nanoparticles [135]. The formation of gold nanoparticles from $AuCl_4^-$ comprises three main steps: reduction of $AuCl_4^-$ ions by the block copolymers in the solution and formation of gold clusters, adsorption of block copolymers on gold clusters and reduction of $AuCl_4^-$ ions on the surfaces of these gold clusters, and growth of gold particles in steps and finally its stabilization by block copolymers [136]. The role of block copolymers in the synthesis (formation rate, yield, stability, shape, and size of nanoparticles) varies with their molecular weight, PEO/PPO block length, polymer concentration, and temperature [137-139]. The formation of high-concentration gold nanoparticles at room temperature is reported in the block copolymer-mediated synthesis where the nanoparticles have been synthesized from hydrogen tetrachloroaureate (III) hydrate ($HAuCl_4.3H_2O$) using block copolymer P85 ($EO_{26}PO_{39}EO_{26}$) in aqueous solution [140]. The formation of gold nanoparticles in these systems has been characterized using UV-visible spectroscopy and SANS. It showed that the presence of an additional reductant (trisodium citrate) can enhance nanoparticle concentration by manyfold, which does not work in the absence of either of these (additional reductant and block copolymer). Bakshi et al. [141] used aqueous micellar solutions of F68 (PEO_{78}-PPO_{30}-PEO_{78}) and P103 (PEO_{17}-PPO_{60}-PEO_{17}) triblock polymers to synthesize gold (Au) nanoparticles at different temperatures. They observed that all reactions were carried out with the PEO-PPO-PEO micellar surface cavities present at the micelle-solution interface and were precisely controlled by the micellar assemblies. Marked differences were detected when predominantly hydrophilic F68 and hydrophobic P103 micelles were employed to conduct the reactions, and the presence of well-defined predominantly hydrophobic micelles with a compact micelle-solution interfacial arrangement of surface cavities ultimately controlled the reaction.

Mesomorphous behavior

Numerous investigations of the behavior of PEO-PPO-PEO triblock copolymers in aqueous solutions and the adsorption of these copolymers at solid-liquid interfaces were carried out in the past decades [31,52,142]. Also, since the aggregated structure of PEO-PPO-PEO triblock copolymers is controlled depending on temperature, concentration, and the addition of additives, they have been used as structure-directing organic materials for the synthesis of inorganic materials with a controlled size, shape and structure. Bagshaw et al. prepared mesoporous silica molecular sieves using nonionic polyethylene oxide surfactants in a neutral condition [143]. The hydrogen bonding between the hydrophilic part of polymers and the inorganic precursor followed by molecular rearrangement involving the amphiphilic nature of polymers was the key factor for the preparation of this material. Zhao et al. reported on the synthesis of mesoporous silica structures using nonionic alkyl poly(oxyethylene) surfactants and poly(alkylene oxide) block copolymers in an acid media, which included cubic, three-dimensional hexagonal, two-dimensional hexagonal, and lamellar mesostructures [144]. Kim et al. developed an economical and simple method for the preparation of sub-micrometer hematite particles with a narrow size distribution and an isotropic shape. To obtain hematite particles, the ferric ion solution was aged at an elevated temperature in the presence of poly(ethylene oxide)-block-poly(propylene oxide)-block-poly(ethylene oxide) (PEO-PPO-PEO) triblock copolymer EO20PO70EO20 (P123). The resulting particles also show a disordered mesoporous structure and retain their shape after calcinations [145]. Huang et al. synthesized a series of highly ordered mesoporous carbonaceous frameworks with diverse symmetries by using phenolic resols as a carbon precursor and mixed amphiphilic surfactants of poly(ethylene oxide)-*b*-poly(propylene oxide)-*b*-poly(ethylene oxide) (PEO-PPO-PEO) and reverse PPO-PEO-PPO as templates by the strategy of evaporation-induced organic-organic self-assembly [146]. The blends of block copolymers can interact with resol precursors and tend to self-assemble into cross-linking micellar structures during the solvent evaporation process, which provides a suitable template for the construction of mesostructures. An understanding of the organic-organic self-assembly behavior in the mixed amphiphilic surfactant system would pave the way for the synthesis of mesoporous materials with controllable structures.

Other applications

Cobalt determination

da Silva et al. [147] proposed a new method for Co(II) determination based on the use of the triblock copolymer as micellar medium instead of chloroform. The proposed strategy is environmental friendly because the copolymer is biodegradable and nontoxic. The method is based on the formation of a cobalt-1-nitroso-2-naphthol complex in the micellar triblock copolymer compound solution constituted by PEO and PPO. Experimental conditions such as pH, the molecular weight, and the PEO/PPO ratio of the triblock copolymer were optimized. Results obtained for cobalt determination in vitamins with this novel method showed excellent agreement with those obtained using atomic absorption spectrometry.

Boundary lubrication

Lubricants most commonly used in textile manufacturing are composed of fatty acids, mineral oils, ethoxylated acids, and silicones. Poly(oxyethylene)-poly(oxypropylene)-poly(oxyethylene) (PEO-PPO-PEO; Pluronic) triblock copolymers have been of great interest as lubricants due to their numerous advantages, including good solubility in water and organic solvents, compatibility with most surfactants, and availability as electrolyte-free material. The molecular weight and the PEO/PPO ratio of these surfactants can be adjusted to tailor their properties as lubricants, texturizers, softeners, emulsifiers, dispersers, and antistatic and wetting agents. Pluronic copolymers have high wetting and spreading ability, allowing them to form uniform coatings on textiles that can result in low friction, antistatic properties, dye-leveling improvement, and easy cleaning. Li et al. [148] studied lubrication behavior of an aqueous solution of PEO-PPO-PEO symmetric triblock copolymer on thin films of polypropylene (PP), polyethylene (PE), and cellulose. It was observed that the friction coefficient on PP and PE was reduced after adsorption from the PEO-PPO-PEO aqueous solution, while the opposite effect was observed for cellulose surfaces. XPS was used to verify the presence of the lubricant on the polymeric substrates and to evaluate its removal by water washing. The lubricant layer was easily removed with water from the PP and cellulose surfaces, while a durable layer was found on PE.

As membrane material for intermediate-temperature DMFCs

Triblock copolymer/Nafion blend membranes (DuPont, Wilmington, DE, USA) facilitate proton conduction in direct methanol fuel cells (DMFCs) at intermediate temperatures. Hu et al. [149] investigated the interaction between the two polymer components by FT-IR spectroscopy. The blend membranes show higher proton conductivity than recast Nafion under partially anhydrous conditions. Protons can be transported with the assistance of an ether chain under such conditions at elevated temperature. In addition, the membranes exhibit more favorable methanol permeability and selectivity. This kind of blend membrane shows somewhat better performance in DMFC compared to bare recast Nafion at intermediate temperature (≥120°C). This helps to design membrane materials with enhanced proton conductivity under conditions typical of intermediate-temperature DMFCs.

Conclusions

Mixed micelle parameters of different TBPs mixtures are discussed. It has been observed that synergistic mixing is governed by the compatibility between the different blocks of PEO units of different TBPs. A large difference between the PPO and PEO blocks of different polymers leads to the unfavorable mixing. These findings help us to further explore the industrial applications of such TBPs.

Competing interests

The authors declare that they have no competing interests.

Authors' contributions

VS collected the references for this work and did additional studies and calculations. Part of this work is related to the Ph.D. thesis of NK. PK compiled and wrote the review along with PND. All authors read and approved the final manuscript.

Acknowledgments

PK acknowledges the financial assistance from the DST (Ref #SERB/F/0328/2012-13) project.

Author details

[1]Department of Chemistry, KSKV Kachchh University, Bhuj, Gujarat 370001, India. [2]Department of Chemistry, B.B.K. D.A.V. College for Women, Amritsar, Punjab 143005, India. [3]Department of Chemistry, Lyallpur Khalsa College, Jallandhar, (Punjab) 144001, India.

References

1. Schmolka IR (1984) Cosmetics and Toiletries 99:69
2. Bahadur P, Resis G (1991) Tenside Cloud Surf Deter 28:173
3. Alexandridis P, Olsson U, Lindmann B (1998) Langumir 14:2627
4. Parsipany NJ (1989) Pluronic and tetronic surfactants, technical brochure. BASF-Corporation, New Jersey
5. Janert PK, Schick M (1997) Macromolecules 30:137
6. Janert PK, Schick M (1998) Macromolecules 31:1109
7. Alexandridis P (1998) Macromolecules 31:6935
8. Alexandridis P, Hatton TA (1995) Colloids Surf A 96:1
9. Groot RD, Madden TJ (1998) J Chem Phys 108:8713
10. Burger C, Micha MA, Oerstreich S, Foster S, Anntonietti M (1998) Europhys Lett 42:425
11. Liu T, Wu C, Xie Y, Liang D, Zhou S, Nace VM, Chu B (2000) ACS Symp Ser 765:2
12. Edens MW (1996) Surf Sci Series 60:185
13. Alexandridis P (1996) Curr Opin J Colliodal and Interface Sci 1:490
14. Yokoyama M (1992) Crit Rev Therapeutic Drug Carrier Systems 9:213
15. Guzman M, Garica FF, Molpeceres J, Aberturas MR (1992) Int J Pharm 80:119
16. Murhammer DW, Goochee CF (1990) Biotechnol Prog 6:391
17. Harris JM, Zalpisky S (1997) Poly(ethylene glycol): chemistry and biological application. American Chemical Society, Washington, DC
18. Tedder DW, Pohland FG (1990) Emerging technologies in hazardous waste management. American Chemical Society, Washington, DC
19. Rodriguez SC, Singer EJ (1996) Surf Sci Series 20:211
20. Zallipsky S (1995) Bioconjugate Chem 6:150
21. Torchilin VP, Truketskey VS, Whiteman KR, Caliceti P, Ferruti P, Veronesi FM (1995) J Pharm Sci 84:1049
22. Schmolka IR (1997) J Am Oil Chem Soc 54:110
23. Mura JL, Reiss G (1995) Polym Adv Technol 6:497
24. Forster S, Antonietti M (1998) Adv Mater 10:4195
25. Caruso RA, Giersbig M, Willig F, Antonietti M (1998) Ber Bunsen-Ges Phys Chem 102:1540
26. Chander S, Polat H, Mohal B (1995) Trans Soc Min Metall Explor 296:55
27. Lindmann B, Wennerstorm H (1980) Topics of current chemistry. Springer, Berlin, p 87
28. Rices G, Bahadur P, Hurtez G (1985) Block copolymers. In: Mark HF, Kroschwitz J (ed) Encyclopedia on polymer science and engineering, 2nd edition. Wiley, New York, p 314
29. Alexandridis P, Holzwarth JF, Hatton TA (1994) Macromolecules 27:2414
30. Cau F, Lacelle S (1996) Macromolecules 29:170
31. Yang L, Alexandridis P, Steyler DC, Kositza MJ, Holzwarth JF (2000) Langmuir 16:8555
32. Hunter RJ (1987) Foundations of colloid science. Oxford University Press, New York, pp 1–565
33. Thurn T, Couderc S, Sidhu J, Bloor DM, Penfold J, Holzwarth JF, Wyn-Jones E (2002) Langmuir 18:9267
34. Bakshi MS, Kaura A, Kaur G (2006) J Colloid and Interface Science 296:370
35. Bakshi MS, Kaura A, Mahajan RK (2005) Colloids Surf A 262:168
36. Bakshi MS, Sharma P, Kaur G, Sachar S, Banipal TS (2006) Colloids Surf A 278:218
37. Bakshi MS, Sachar S (2006) Colloids Surf A 276:146
38. Bakshi MS, Sachar S (2006) J Colloid Interface Sci 296:309
39. Mahajan RK, Kaur N, Bakshi MS (2005) Colloids Surf A 255:33
40. Bakshi MS, Singh J, Kaur G (2005) Antagonistic mixing behavior of catonic Gemini surfactant and triblock polymers in mixed micelle. J Colloid Interface Sci 285:403
41. Bakshi MS, Kaur N, Mahajan RK (2006) Photochem Photobiol 183:146
42. Mahajan RK, Kaur N, Bakshi MS (2006) Colloids Surf A 276:221
43. Solan SL, Phillips RJ, Cotts PM, Dungan SR (1997) J Colloid Interface Sci 191:291
44. Holland RJ, Parker EJ, Guiney K, Zeld FR (1995) J Phys Chem 99:11981
45. Nakashima K, Anzai T, Fujimoto Y (1994) Langmuir 10:658
46. Almgren M, Bahadur P, Jannson M, Li P, Brown W, Bahadur A (1992) J Colloid Interface Sci 15:1157
47. Williams RK, Simard MA, Jolicoeur C (1985) J Phys Chem 89:178
48. Wen XG, Verrall RE (1997) J Colloid Interface Sci 196:215

49. Goldmints I, Von Gottberg FK, Smith KA, Hatton TA (1997) Langmuir 13:3659
50. Goldmints I, Holzwarth JF, Smith KA, Hatton TA (1997) Langmuir 13:6130
51. Malmsten M, Lindman B (1992) Macromolecules 25:5440
52. Wu C, Liu T, Chu B, Schneider DK, Fraziano V (1997) Macromolecules 30:4574
53. Wanka G, Hoffman H, Ulbricht W (1994) Macromolecules 27:4145
54. Alexandridis P, Nivaggioli T, Hatton TA (1995) Langmuir 11:1468
55. Prud'homme RK, Wu G, Schneider DK (1996) Langmuir 12:4651
56. Alexandridis P, Athanassiou V, Fukuda S, Hatton TA (1994) Langmuir 10:2604
57. Alexandridis P, Holzwarth JF (2000) Curr Opin J Colloid and Interface Sci 5:312
58. Almgren M, Brown W, Hvidt S (1995) Colloid Polym Sci 273:2
59. Mortensen K, Pedersen JS (1993) Macromolecules 26:805
60. Hecht E, Hoffmann H (1995) Colloids Surf A 96:181
61. Goldmints I, Yu GE, Booth C, Smith KA, Hatton TA (1999) Langmuir 15:1651
62. Genz A, Holzwarth JF (1986) Eur Biophys 13:323
63. Kositza MJ, Bohne C, Alexandridis P, Hatton TA, Holzwarth JF (1999) Macromolecules 32:5539
64. Zhou Z, Chu B (1988) J Colloid Interface Sci 126:171
65. Alasden AA, Whatley TL, Florence AT (1982) J Colloid Interface Sci 90:303
66. Attwood D, Collett JH, Tail CJ (1985) Int J Pharm 26:25
67. Zhou Z, Chu B (1988) Macromolecules 21:2548
68. Almgren M, Alisns J, Bahadur P (1991) Langmuir 7:446
69. Tontisakis A, Hilfiker R, Chu B (1992) J Colloid Interface Sci 135:427
70. Wanka G, Hoffman H, Ulbricht W (1990) Colloid Polym Sci 268:101
71. Brown W, Schillen K, Almgren M, Hvidt S, Bahadur P (1991) J Phys Chem 95:1850
72. Chu B, Zhou Z (1996) Surf Sci Ser 60:67
73. Sedev R, Steitz R, Findenegg GH (2002) Physica B 315:267
74. Alexandridis P, Chu D, Khan A (1996) Langmuir 12:2696
75. Alexandridis P, Olsson U, Lindmann B (1997) Langmuir 13:23
76. Mao G, Sukumaran S, Beaucage G, Saboungi ML, Thyagarajein P (2001) Macromolecules 34:552
77. Mitchell DJ, Tiddy GJT, Waring L, Bostock T, MacDonald M (1983) Chem Soc Farady Trans 1 79:975
78. Lawrence MJ (1990) Chem Soc Rev 90:753
79. Zhang L, Eisenberg A (1995) Science (Washington, DC) 268:1728
80. Gao Z, Varshney SK, Wong S, Eisenberg A (1994) Macromolecules 27:7923
81. Shen HW, Zhang LF, Eisenberg A (1999) J Am Chem Soc 121:2728
82. Desbaumes L, Eisenberg A (1999) Langmuir 15:36
83. Zhang LF, Yu K, Eisenberg A (1996) Science 272:1777
84. Zhang LF, Eisenberg A (1996) J Am Chem Soc 118:3168
85. Discher DE, Eisenberg A (2002) Science 297:967
86. Zhang LF, Bartels C, Yu YS, Shen HW, Eisenberg A (1997) Phys Rev Lett 79:5034
87. Talingting MR, Munk P, Webber SE (1999) Macromolecules 32:1593
88. Riegel IC, Eisenberg A (2002) Langmuir 18:3358
89. Yu K, Zhang LF, Eisenberg A (1996) Langmuir 12:5980
90. Cameron NS, Corbierre MK, Eisenberg A (1999) Can J Chem 77:1311
91. Yu G, Eisenberg A (1998) Macromolecules 31:5546
92. Yu YS, Zhang LF, Eisenberg A (1997) Langmuir 13:2578
93. Kindt JT (2002) J Phys Chem B 106:8223
94. van der Schoot P, Wittmer JP (1999) Linear aggregation revisited: rods, rings and worms. Macromol Theory Simul 8:428
95. Wittmer JP, van der Schoot P, Milchev A, Barrat JL (2000) J Chem Phys 113:6992
96. Hu J, Odom TW, Lieber CM (1999) Acc Chem Res 32:435
97. Han YJ, Kim JM, Stucky GD (2000) Chem Mater 12:2068
98. Booth C, Attwood D (2000) Macromol Rapid Commun 21:501
99. Stern O, Volmer M (1990) Phys Z 20:18
100. Wellar A (1976) Prog React Kinet 1:3246
101. Efthik MR, Gheron CA (1976) J Phys Chem 80:486
102. Gorner H, Stammel C, Mattey J (1997) J Photochem Photobiol A: Chem 120:171
103. Chiappetta DA, Sosnik A (2007) Eur J Pharm Biopharm 66:303
104. Oh KT, Bronich TK, Kabanov AV (2004) J Control Release 94:411
105. Yang TF, Chen CN, Chen MC, Lai CH, Liang HF, Sung HW (2007) Biomaterials 28:725
106. Gao Z, Fain HD, Rapoport N (2004) Mol Pharm 1:317
107. Gao ZG, Fain HD, Rapoport N (2005) J Control Release 102:203
108. Rapoport NY, Christensen DA, Fain HD, Barrows L, Gao Z (2004) Ultrasonics 42:943
109. Husseini GA, De la Rosa MA, Diaz T, Gabuji Y, Zeng DA, Christensen WG, Pitt J (2007) Nanosci Nanotechnol 7:1028
110. Kabanov AV, Batrakova EV, Miller DW (2003) Adv Drug Deliv Rev 55:151
111. Spitzenberger TJ, Heilman D, Diekmann C, Batrakova EV, Kabanov AV, Gendelman HE, Elmquist WF, Persidsky YJ (2007) Cereb Blood Flow Metab 27:1033
112. Batrakova E, Han H, Miller D, Kabanov A (1998) Pharm Res 15:1525
113. Batrakova E, Han H, Alakhov V, Miller D, Kabanov A (1998) Pharm Res 15:850
114. Kwon SH, Kim SY, Ha KW, Kang MJ, Huh JS, Im TJ, Kim YM, Park YM, Kang KH, Lee S, Chang JY (2007) Arch Pharm Res 30:1138
115. Alakhov V, Klinski E, Li S, Pietrzynski G, Venne A, Batrakova EV, Bronitch T, Kabanov A (1999) Colloids Surf B Biointerfaces 16:113
116. Kabanov A, Alakhov V (2002) Crit Rev Ther Drug Carrier Syst 19:1
117. Krupka TM, Weinberg BD, Wu H, Ziats NP, Exner AA (2007) Exp Biol Med (Maywood) 232:950
118. Kabanov AV, Chekhonin VP, Alakhov VY, Batrakova EV, Lebedev AS, Melik-Nubarov NS, Arzhakov SA, Levashov AV, Morozov GV, Severin ES, Kabanov VA (1989) FEBS Lett 258:343
119. Saski W, Shah SG (1965) J Pharm Sci 54:277
120. Saski W, Shah SG (1965) J Pharm Sci 54:71
121. Croy SR, Kwon G (2004) J Control Release 95:161
122. Croy SR, Kwon GS (2005) J Pharm Sci 94:2345
123. Lee ES, Oh YT, Seok Youn Yu YS, Nam M, Park B, Yun J, Kim JH, Ho-Taek S (2011) Colloids Surf A Biointerfaces 82:190
124. Kadam Y, Yerramilli U, Bahadur A, Bahadur P (2011) Colloids Surf B Biointerfaces 83:49
125. Sardar R, Funston AM, Mulvaney P, Murray RW (2009) Langmuir 25:13840
126. Giljohann DA, Seferos DS, Daniel WL, Massich MD, Patel PC, Angew CA (2010) Chem, Int Ed 49:3280
127. Homberger M, Simon U (2010) Philos Trans R Soc A 368:1405
128. Sperling RA, Gil PR, Zhang F, Zanella M, Parak W (1896) J Chem Soc Rev 2008:37
129. Jain PK, El-Sayed IH, El-Sayed MA (2007) Nanotoday 2:18
130. Bajpai SK, Mohan YM, Bajpai M, Tankhiwale R, Thomas VJ (2007) Nanosci Nanotechnol 7:2994
131. Goy-Lopez SE, Castro S, Taboada P, Mosquera V (2008) Langmuir 24:13186
132. Rahme K, Oberdisse J, Schweins R, Gaillard C, Marty JD, Mingotaud C, Gauffre F (2008) Chem Phys Chem 9:2230
133. Bakshi MS, Kaura A, Bhandari P, Kaur G, Torigoe K, Esumi KJ (2006) Nanosci Nanotechnol 6:1405
134. Sakai T, Alexandridis P (2006) Chem Mater 18:2577
135. Smart T, Lomas H, Massignani M, Flores-Merino MV, Perez LR, Battaglia G (2008) Nanotoday 13:38
136. Sakai T, Alexandridis P (2005) J Phys Chem B 109:7766
137. Durand-Gasselin C, Capelot M, Sanson N, Lequeux N (2010) Langmuir 26:12321
138. Ray D, Aswal VK, Srivastava DJ (2011) Nanosci Nanotechnol 11:1905
139. Ray D, Aswal VK, Srivastava DJ (2010) Nanosci Nanotechnol 10:6356
140. Ray D, Aswal VK, Kohlbrecher J (2011) Langmuir 27:4048
141. Khullar P, Mahal A, Singh V, Banipal TS, Kaur G, Bakshi MS (2010) Langmuir 26:11363
142. Su YL, Wei XF, Liu HZ, Colloid Interface J (2003) Sci 264:526
143. Bagshaw SA, Prouzet E, Pinnavaia TJ (1995) Science 269:1242
144. Zhao D, Huo Q, Feng J, Chmelka BF, Stucky GD (1998) J Am Chem Soc 120:6024
145. Kim HG, Oh C, Lee YH, Park SH, Oh SG (2007) J Ceramic Process Res 8:177
146. Huang Y, Cai H, Yu T, Sun X, Tu B, Zhao D (2007) Asian J 2:1282
147. da Silva MCH, da Silva LHM, Paggioli FJ (2005) Anal Sci 21:933
148. Li Y, Rojas OJ, Hinestroza JP (2012) Ind Eng Chem Res 51:2931
149. Hu J, Baglio V, Tricoli V, Aricò AS, Antonucci V (2008) J Appl Electrochem 38:543

23

A review on the assessment of polymeric materials used as corrosion inhibitor of metals and alloys

David Ebuka Arthur[1*], Achika Jonathan[1], Paul Ocheje Ameh[1] and Crystal Anya[2]

Abstract

Corrosion control of metals is an important activity of technical, economical, environmental, and aesthetical importance. The use of inhibitors is one of the best options of protecting metals and alloys against corrosion. The toxicity of organic and inorganic corrosion inhibitors to the environment has prompted the search for safer corrosion inhibitors such as green corrosion inhibitors as other more environmental friendly corrosion inhibitors, most of which are biodegradable and do not contain heavy metals or other toxic compounds. Investigations of corrosion-inhibiting abilities of polymeric substances, e.g., plant gums, in addition to being environmentally friendly and ecologically acceptable, have shown that plant products are inexpensive, readily available, and renewable sources of materials. Need for more effective inhibitors has propelled companies such as the Montazhkhimzashchita Trust to develop a pool method of welding sheet vinyl, a method widely employed in gluing on roller sticky bands from thermoplast on a pipe. The article enumerates several kinds of polymeric materials which are suitable for use in combating corrosion, and several which have suitable strength characteristics as to serve in place of scarce expensive metals and alloys.

Keywords: Polymers, Plant gum, Metals, Acid, Alkaline, Corrosion inhibitor

Review

Introduction

Several researches have indicated that some polymers can be used as corrosion inhibitors because, through their functional groups, they form complexes with metal ions and on metal surfaces. These complexes occupy a large surface area, thereby blanketing the surface and protecting the metals from corrosive agents present in the solution. The corrosion inhibition by various cationic polymers such as polyethyleneimine derivative, polyacrylamide derivative, polydicyanodiamide derivative, and anionic polymers such as polymaleic acid derivative, polyacrylic acid derivative, and polyacrylic acid, have been investigated by Sekine et al. [1]. They found that polymers having the COOH group are effective polymer-based systems that can function as corrosion inhibitors. Muller et al. [2] studied the corrosion inhibition of polymethacrylic acid and styrene maleic acid co-polymer on zinc pigments in aqueous alkaline medium.

Abdel Rehim et al. [3] found that the amino polycarboxylic acids, such as diethylenetriaminepentaacetic acid, polyacrylic acid, and polymethacrylic acid, are good inhibitors for metal corrosion. Sedahmed et al. [4] investigated the use of a formulation containing polyethylene oxide, polyacrylamide, and carboxymethyl cellulose (CMC) as corrosion inhibitors for iron in acidic and neutral media using electrochemical methods and found that this formulation is an excellent inhibitor. Khairou and Sayed [5] studied the inhibiting action of polyacrylamide, Polyvinyl alcohol (PVA), sodium polyacrylate, poly(ethylene glycol), pectin, and CMC on the corrosion of Cd in 0.5 M HCl solution. Meena et al. [6] investigated the synergistic effect between CMC and Zn^{2+} on the corrosion inhibition of carbon steel in NaCl solutions and found that the combination produces strong inhibition potential.

* Correspondence: hanslibs@myway.com
[1]Department of Chemistry, Ahmadu Bello University, Zaria, Kaduna State PMB 1044, Nigeria
Full list of author information is available at the end of the article

Studies on the use of some macromolecules (i.e., natural polymers) have also been carried out. For example, Umoren et al. [7,8] studied the potential of gum arabic as a corrosion inhibitor for aluminum in alkaline medium. The inhibition of aluminum corrosion by gum arabic was attributed to the presence of arabinogalactan, oligosaccharides, polysaccharides, and glucoproteins since these compounds contain oxygen and nitrogen atoms which are the centers of adsorption. The inhibitive effect of exudate gum from *Dacroydes edulis* on the corrosion of aluminum in HCl solutions was studied using weight loss and thermometric methods at 30°C to 60°C by Umoren et al. [9,10], and the results revealed that the exudate gum acted as an inhibitor for the corrosion of aluminum in HCl solution. The inhibition efficiency increases with an increase in the concentration of the exudate gum but decreases with increase in temperature.

The effect of naturally occurring exudate gum from *Raphia hookeri* on the corrosion of mild steel in H_2SO_4 between 30°C and 60°C has also been investigated by Umoren et al. [10] using weight loss and hydrogen evolution techniques. Results obtained revealed that the exudate gum is a good inhibitor for the corrosion of mild steel in acidic media. The inhibition efficiency increased with an increase in exudate gum content and decreases with increase in temperature. The adsorption of exudate gum from *R. hookeri* on the mild steel follows Langmuir adsorption isotherm. Guar gum has been shown to be an effective corrosion inhibitor for some metals in aggressive acid environment according to the study by Abdallah [11]. Results obtained show that natural substances act as effective corrosion inhibitors in the different test media. Inhibition efficiency was found to increase with increase in the concentration of the tested material.

It should be pointed out that the potentials of polymers as corrosion inhibitors are dependent on the chemical composition of the polymers, and most of the studies carried out on polymers are done without reference to their chemical properties.

It is well known in surface chemistry that surface reactions are strongly affected by the presence of foreign molecules. Corrosion processes, being surface reactions, can be controlled by compounds known as inhibitors which adsorb on the reacting metal surface.

The term adsorption refers to molecules attached directly to the surface, normally only one molecular layer thick, and not penetrating into the bulk of the metal itself. The technique of adding inhibitors to the environment of a metal is a well known method of controlling corrosion in many branches of technology. A corrosion inhibitor may act in a number of ways: it may restrict the rate of the anodic process or the cathodic process by simply blocking active sites on the metal surface. Alternatively, it may act by increasing the potential of the metal surface so that the metal enters the passivation region where a natural oxide film forms. A further mode of action of some inhibitors is that the inhibiting compound contributes to the formation of a thin layer on the surface which stifles the corrosion process, for example, in order to mitigate aluminum corrosion. The main strategy is to effectively isolate the metal from corrosive agents by the use of corrosion inhibitors. Inorganic substances such as phosphates, chromates, dichromate, and arsenates have been found effective as inhibitors of metal corrosion, but a major disadvantage is their toxicity, and as such, their use has become intolerant for its adverse effect to our societal health in the long run [12].

Corrosion inhibition occurs via adsorption of their molecules on the corroding metal surface, and the efficiency of inhibition depends on the mechanical, structural, and chemical characteristics of the adsorption layers formed under particular condition [13]. Research activities in recent times are geared towards finding alternative corrosion inhibitors to replace the inorganic and organic compounds. Naturally occurring substances have been found to readily satisfy this need. Apart from being readily available, cheap, and a renewable source of materials, naturally occurring substances are eco-friendly and ecologically acceptable. The successful use of naturally occurring substances to inhibit the corrosion of metals in acidic and alkaline environment has been reported by some authors: Ebenso et al. [14], Eddy and Ebenso [15], etc.

Polymers as corrosion inhibitors

In recent years, many branches of national economy for the struggle with corrosion are finding broad application for nonmetallic and chemically stable materials, obtained on the basis of polymers, in form of individual structural materials or linings, protective coatings, coatings, etc. These materials are used for the purpose of economizing on nonferrous metals and highly alloyed steels used by various industries in the role of anticorrosion materials [16]. Several factors such as cost and amount needed, safety to environment and its species, and most importantly, easy availability are needed to be considered when choosing an inhibitor.

An anticorrosion coating for the corrosion protection of metals, liquid Nairit (chloroprene rubber; Yerevan, Armenia), allows us to use these materials widely for the replacement of deficit metals in various branches of national economy. This coating was tested, especially, in ship construction and for the protection of screws (propellers), subjected not only to corrosion but to erosion and cavitations in thermal power equipment and others [16].

They are all, obtainable mainly on the basis of organic products, suitable not only for the preparation of independent structures but also for the lining of devices and equipment, and in the role of compositions and protective coatings. Polymeric materials possessing a practically unlimited chemical stability exceeding even platinum appear to be plastics obtained on the basis of fluorocarbonic compounds, which can be used as independent and lining materials, and in the role of coatings.

The polytetrafluoroethylene or fluoroplast-4 plastics manufactured in the USSR suffer no destruction under the effect of almost all known aggressive media, including such strong oxidizers as fuming nitric acid, aqua regia, at temperatures of up to 2,500°C. Fluoroplast-4 is not void of very essential specific deficiencies: its adhesion to metals and other materials is inconsiderable, which limits its utilization for the lining of devices. Its use for coating is difficult in connection with the insolubility of polytetrafluoroethylene neither in any one of the known organic solvents.

A number of polymers possessing high heat resistance for preservation of their anticorrosion qualities and mechanical strength belong so-called polyorganosiloxanes. Polyorganosiloxane coatings are stable also to the effect of oxygen, ozone, humid atmosphere, ultraviolet rays, etc. and in combination with various fillers (powder aluminum, titanium, boron, etc.) up to 500°C to 550°C and briefly to 700°C to 800°C. Polyorganosiloxane coatings are suitable for the protection against corrosion of smoke pipes, pumps for pumping hot liquids, cracking installations, and other equipment under conditions of high temperatures and under the action of aggressive media [16]. This reputation as well as the presence of heteroatoms, for example O, N, and S, has been found to increase basicity and electron density in polymers and plant extracts, thus intensifying their corrosion potential. O, N, and S are the active centers for the process of adsorption on the metal surface. The inhibition efficiency should follow the sequence O < N < S < P.

Mechanism of action of polymers in a corrosion inhibitor

Many theories to substantiate the mode of action of these green inhibitors have been put forth by several workers. It was suggested that organic substances, which form onium ions in acidic solutions, are adsorbed on the cathodic sites of the metal surface and interfere with the cathodic reaction.

Various mechanisms of action have been postulated for the corrosion-inhibition property of the natural products. The use of organic compounds containing oxygen, sulfur, and especially nitrogen to reduce corrosion attack on steel has been studied in some detail. The existing data show that most organic inhibitors adsorbed on the metal surface by displacing water molecules on the surface and forming a compact barrier. Availability of non-bonded (lone pair) and *p* electrons in inhibitor molecules facilitates electron transfer from the inhibitor to the metal. A coordinate covalent bond involving transfer of electrons from inhibitor to the metal surface may be formed. The strength of the chemisorption bond depends upon the electron density on the donor atom of the functional group and the polarizability of the group. When an H atom attached to the C in the ring is replaced by a substituent group ($-NH_2$, $-NO_2$, $-CHO$, or $-COOH$), the electron density in the metal at the point of attachment changes, resulting in the retardation of the cathodic or anodic reactions. Electrons are consumed at the cathode and are furnished at the anode. Thus, corrosion is retarded. Straight chain amines containing between three and 14 carbons have been examined. Inhibition increases with carbon number in the chain to about 10 carbons, but with higher members, little increase or decrease in the ability to inhibit corrosion occurs. This is attributed to the decreasing solubility in aqueous solution with increasing length of the hydrocarbon chain. However, the presence of a hydrophilic functional group in the molecule would increase the solubility of the inhibitors. The performance of an organic inhibitor is related to the chemical structure and physicochemical properties of the compound-like functional groups, electron density at the donor atom, *p* orbital character, and the electronic structure of the molecule. The inhibition could be due to the following: (1) adsorption of the molecules or its ions on anodic and/or cathodic sites, (2) increase in cathodic and/or anodic over voltage, and (3) the formation of a protective barrier film.

Several researches have indicated that some polymers can be used as corrosion inhibitors because, through their functional groups, they form complexes with metal ions and on the metal surfaces. These complexes occupy a large surface area, thereby blanketing the surface and protecting the metals from corrosive agents present in the solution [17]. The inhibitive power of these polymers is related structurally to the cyclic rings, heteroatom (oxygen and nitrogen), that are the major active centers of adsorption.

Natural polymers, for example gum arabic, were reported by Umoren et al. [8] as potential corrosion inhibitors for aluminum in alkaline medium. The inhibition of aluminum corrosion by gum Arabic was attributed to the presence of arabinogalactan, oligosaccharides, polysaccharides, and glucoproteins since these compounds contain oxygen and nitrogen atoms which are the centers of adsorption.

The inhibitive effect of exudate gum from *D. edulis* in the corrosion of aluminum in HCl solutions was studied using weight loss and thermometric methods at 30°C to

60°C by Umoren et al. [10]. The results revealed that the exudate gum acts as an inhibitor for corrosion of aluminum in HCl solution. The inhibition efficiency increases with an increase in the concentration of the exudate gum but decreases with increase in temperature. The Temkin adsorption isotherm was tested for its fit to the experimental data. The result confirms that the corrosion inhibition of the exudate gum from *D. edulis* was attributed to the adsorption of molecules of phytochemicals present in the exudate gum on the surface of the metal. The free energies and equilibrium constant for the adsorption process were determined. A mechanism of physical adsorption was proposed.

Some natural polymers have also been found to be good corrosion inhibitors. According to Finkenstadt et al. [18], agricultural polymers composed of extracellular polysaccharides secreted by *Leuconostoc mesenteroides* have been shown to inhibit corrosion on corrosion-sensitive metals. The substantially pure exopolysaccharide has a general structure consisting of alpha (1–6)-linked D-glucose backbone and approximately 3% to 4% branching of alpha(1–3) linkages. Corrosion behavior was determined to be strain-dependent. Pore resistance and corrosion rate were calculated from electrochemical impedance spectroscopy (EIS) and polarization studies. The corrosion rate was at least 50% lower, and the pore resistance was twice as high as the control samples, indicating good inhibitory processes. Nanoscale aqueous coatings were examined using AFM and QCM and showed self-repair properties. These materials may be useful in anti-corrosive coating applications.

Natural-occurring polymers

Green corrosion inhibitors are biodegradable and do not contain heavy metals or other toxic compounds. Some research groups have reported the successful use of naturally occurring substances to inhibit the corrosion of metals in acidic and alkaline environments. The effect of addition of halides (KCl, KBr, and KI) was also studied, and the results obtained indicated that the increase in efficiency was due to synergism [10]. Buchweishaija and Mhinzi [19] studied the inhibition efficiency of gum exudates from *Acacia seyal* var. *seyal* using potentiodynamic polarization and EIS technique at a 30°C, and it was found to be a good anodic inhibitor for drinking water with increasing temperature. Umoren et al. [10] also investigated the corrosion properties of *R. hookeri* exudates gum-halide mixtures for aluminum corrosion in acidic medium. *R. hookeri* exudate gum obeys Freundlich, Langmuir, and Temkin adsorption isotherms. Phenomenon of physical adsorption is proposed.

Abdallah [11] also tested the effect of guar gum on carbon steel. It is proposed that it acts as a mixed type inhibitor. The mechanism of action of C steel by guar gum is due to the adsorption at the electrode/solution interface. Guar gum is a polysaccharide compound containing repeated heterocyclic pyrane moiety. The presence of hetero-oxygen atom in the structure makes its adsorption possible by coordinate type linkage through the transfer of lone pairs of electron of oxygen atoms to the steel surface, giving a stable chelate five-membered ring with ferrous ions.

The behavior of corrosion inhibition by various cationic polymers such as polyethyleneimine derivative, polyacrylamide derivative, and polydicyanodiamide derivative, and anionic polymers such as polymaleic-acid derivative, polyacrylic-acid derivative, and polyacrylic acid, was investigated by Sekine et al. [1]. Polymers having the COOH group are one of the effective polymer-based systems. There are several carboxylic acids of low molecular weight, which are used as corrosion inhibitors to many inhibitor formulations. Muller et al. [2] has studied the corrosion inhibition of polymethacrylic acid and styrene maleic acid co-polymer on zinc pigments in aqueous alkaline medium. Abdel Rehim et al. [3] have discussed that the amino polycarboxylic acids, such as diethylenetriaminepentaacetic acid, polyacrylic acid, and polymethacrylic acid, have been used as inhibitors for metal corrosion. Sedahmed et al. [20] have studied the use of the formulation containing polyethylene oxide, polyacrylamide, and CMC as corrosion inhibitors for iron in acidic and neutral media using electrochemical methods. Khairou and Sayed [5] have evaluated the inhibiting action of polyacrylamide, PVA, sodium polyacrylate, poly(ethylene glycol), pectin, and carboxymethyl cellulose (CMC) on the corrosion of Cd in 0.5 M HCl solution. Meena et al. [6] have examined the synergistic effect between CMC and Zn^{2+} on the corrosion inhibition of carbon steel in NaCl solutions.

Synthetic polymers

Synthetic polymers are polymers synthesized in the laboratory to serve as a proficient substitute for the natural ones; they are made mostly to check the demerits and enhance the qualities of natural polymers depending on their use. Plastic materials which are suitable for use in combating corrosion and several which have suitable strength characteristics as to serve in place of scarce expensive metals and alloys include vinyl for piping, sheet graphite for parts of heat exchangers, and liquid Nairit (chloroprene rubber) for anticorrosion protection of ship propellers.

Plastics have also been used successfully to fabricate parts of armatures. Plastics of fluorocarbonic compounds have been found especially chemically stable, resisting ever fuming nitric acid and aqua regia. Antegmit, an anticorrosion and antifriction heat conductive material made from phenolformaldehyde resin and graphite, differs from saturated graphites by impermeability and much lower heat conductivity, but its strength is approximately twice higher than that of saturated graphite.

High chemical stability is possessed by chlorosulfonated polyethylenes, hypalones, used for linings of chemical devices as well as coatings, applied by a brush, by immersion, and dusting. Hypalon-20 at room temperature is affected only by fuming nitric acid, carbon tetrachloride, gasoline, and nitrobenzene. Thermo and fire resistance of Hypalon (DuPont, DE, USA) allow the same role of coating conveyer belts for the transportation of hot materials.

Polypropylene is a highly perspective, polymeric anticorrosion material which is derived from petroleum gasses. Its melting point is 170°C, it is stable against organic solvent as well as in 80% sulfuric acid and caustic sodium, and is frost-resistant. Films from polypropylene possess lower gas permeability than polyethylene films.

Polyorganosiloxanes have proven their value in preventing corrosion. Fiber glass has also found many uses in industrial plant installations. Epoxy resins, epoxy lacquers, and nylon have been found to be of remarkable importance in countering metal corrosion. Umoren et al. [21] studied the corrosion inhibition of mild steel in H_2SO_4 in the presence of gum arabic (naturally occurring polymer) and polyethylene glycol (PEG) (synthetic polymer). It was found that PEG was more effective than gum arabic and acetylthiourea chitosan polymer (ATUCS).Corrosion inhibition study on mild steel using potentiodynamic polarization, EIS measurements, and SEM technique, and ATUCS has shown very good IE in 0.5 M sulfuric acid solution which reaches to 94.5% for 0.76 M concentration [22].

Poly(styrenesulfonic acid)-doped polyaniline has been synthesized, and the influence of this polymeric compound on the inhibition of corrosion of mild steel in 1 M HCl has been investigated using weight loss measurements, galvanostatic polarization studies, electropermeation studies and AC impedance measurements by Manickavasagam et al. [23]. The polymer acts predominantly as an anodic inhibitor.

The inhibitive performance of novel synthesized water soluble triblock copolymers2-(diethylamino)ethyl methacrylate-block-2-(dimethylamino)ethyl methacrylate-block-2-(*N*-morpholino)ethyl methacrylate (PDEA-PDMA-PMEMA) and 2-(diisopropylamino)ethyl methacrylate-block-2-(dimethylamino)ethyl methacrylate -block-2-(*N*-morpholino)ethyl methacrylate (PDPA-PDMA-PMEMA) of two different molecular weights on the corrosion behavior of mild steel in 0.5 M HCl has been reported by Yurt et al. [24].

The inhibitive action of orthomethoxy-substituted polyaniline (poly(*o*-methoxyaniline), a new class of conducting polymer on the corrosion of iron in acidic chloride solution, has been evaluated by EIS, linear polarization resistance, weight loss, and by logarithmic polarization technique. Inhibition efficiencies of nearly 80% to 88% have been observed even at 25-ppm concentration. Double-layer capacitance studies indicate a strong adsorption of the polymer following Temkin adsorption isotherm is largely responsible for its inhibitive action [25]. Volatile corrosion inhibitors (VCIs) are unique. They are organic compounds that protect metal surfaces by emitting a vapor such as an amine-based compound. The nitrogen on the amine has two electrons that are attracted to the polar metal surface.

Once it is attracted to the metal, the rest of the molecule is very hydrophobic and repels water to significantly retard corrosion. Miksic et al. [26], Kuznetsov [27], and Andreev and colleagues [28] studied several amines, their derivatives, and imines used as VCIs. VCIs migrate from the coated area to the unprotected metal area. Metals coated with these VCI paints corrode very little in scribed areas since the films are self healing. The VCI evaporates and redeposits itself where the coating has been scratched. Many VCI coatings also contain other corrosion inhibitors to enforce the corrosion-resistance process [29].

Replacement of nonferrous metals in the manufacture of synthetic fibers with paint and varnish coatings for ferrous metals gave a greater economical effect. Investigations of various coatings obtained on the basis of high molecular epoxy resins with the use of organic amines and mineral acids in the role of hardener have shown that epoxy coatings possess excellent physicomechanical properties and greater acid resistance.

Paint and varnish materials prepared on the basis of epoxy resin differ also in alkaline resistance. Alkaline resistance is characteristic also for perchlorovinyl enamel, which represents a suspension of pigments and plasticizer in a solution of dry perchlorovinyl resin in organic solvents. Enamel obtained from phenolformaldehyde resin is used as benzostable material for the protection of metal surfaces, subjected to corrosion effect of hot lubricating materials. Tests of coatings, prepared on the basis of chlorovinyl copolymers with vinyliden chloride, anticorrosion enamel and non-overgrowing paint, obtained on the basis of a chlorovinyl copolymer with vinylacetate, are suitable for the protection of metal against corrosion and against the effects of sea water.

Advantages and limitations

Polymeric materials are environmentally friendly, non toxic, and relatively less expensive. Most polymers are not easily biodegradable which, in merits, allow for their long time storage and usage on corrosion protection of metals and alloys. To protect against corrosion on most metals, it is necessary to apply a coating with a thickness of 250 to 400 μm. As is known, majority of polymeric

Table 1 Polymeric materials used as corrosion inhibitors

Sample number	Metal/alloy	Inhibitor	Medium	Reference
1	Mild steel	Gum exudate from *Acacia seyal* var. *seyal*	Drinking water	[19]
2	Mild steel	ATUCS	0.5 M H_2SO_4	[22]
3	Aluminum	Polybutadiene acid		[30]
4	Mild steel	*Anogeissus leiocarpus* gum	0.1 M HCl	[31]
5	Carbon steel	Guar gum		[11]
6	Steel	Fluoroplast-3M		[16]
7	Carbon steel used as smoked pipes and pumps for pumping hot water and cracking installations	Polyorganosiloxane coatings		[16]
8	Carbon steel	Polymers or proteins of amino acids: alanine, glycine, and leucine		[32]
9	Aluminum	Polyamino acids: aspartic acid		[3]
10	Carbon steel	Terpolymer (acrylic acid/acrylonitrile)	Water treatment plant	[33]
11	Al and Zn	Saccharides: mannose and fructose		[34]
12	Mild steel	PEG	2 M H_2SO_4	[35]
13	Stainless steel and carbon steel	PVA	0.9% NaCl	[17]
14	Copper corrosion	Extracts of *Medicago sativa*, *Withania somnifera*, *Atropa belladonna* and *Medicago polymorpha*	Saline water (3.4% NaCl)	[36]
15	Stainless steel	*M. polymorpha* Roxb	5% HCl	[37]
16	Carbon steel	2-phosphonobutane-1,2,4-tricarboxylic acid and polyvinylpyrrolidone (PVP)	Cooling water system, H_2SO_4	[38]
17	Aluminum	Hydroxyethyl cellulose	1.0 M HCl	[39]
18	Mild steel	Hydroxyl cellulose	1.0, 1.5 M HCl	[40]
19	Stainless steel	Polyvinylpyrrolidone polymers		[41]
20	Steel used as ship propellers	Liquid 'nairit' (chloroprene rubber)	Sea water	[16]
21	Aluminum alloy	Polyaniline	0.1 M NaCl	[42]
22	Metal concrete	Epoxy resins	Acid and alkali resistance	[16]
23	Aluminum	Gum arabic	0.1 M NaOH in the absence and presence of KI	[21]
24	Mild steel	Chlorosulphonated polyethylenes: hypalones		[16]
25	Aluminum	Exudate gum from *Pachylobus edulis*	Presence of halide ions in HCl	[9]
26	Aluminum	Polyacrylic acid	Weak alkaline solution	[30]
27	Steel plate	Phosphonated copolymers with polyvinylidene fluoride	Phosphonic acid groups	[43]
28	Zinc (Zn)	Polypyrrole (PPy)-coated mica flakes	Alkaline media	[2]
29	Mild steel	Polyaniline		[44]
30	Low carbon steel	2, 6 ionen, 2, 10 ionen, polyvinylbenzyltrimethyl ammonium chloride and latex	HCl solution	[45]
31	Mild steel	Polymer-polymer complex of polymethacrylic acid and polyacrylamide	Cooling water systems	[46]
32	Low carbon steel	PVP and polyethyleneimine (PEI)	Aqueous phosphoric acid solutions	[47]
33	Mild steel	Polyethylene glycol methyl ether	H_2SO_4 solution	[48]

Table 1 Polymeric materials used as corrosion inhibitors *(Continued)*

34	Aluminum anode	Polyvinylbenzyl trimethylammonium chloride and polydiallyldimethyl ammonium chloride (PDDA)	Alkaline electrolyte	[49]
35	Iron (Fe)	*N*-vinyl-2-pyrrolidone and PVP	1 M H_2SO_4	[50]
36	Fe	Poly(*p*-aminobenzoic acid)	1 M HCl	[51]
37	Fe	Poly(diphenylamine)	0.5 M H_2SO_4	[52]
38	Fe	Poly(4-vinylpyridine-poly(3-oxideethylene) tosyle)	H_2SO_4 solution	[53]
39	Fe	Poly(aminoquinone)	0.5 M H_2SO_4	[54]
40	Copper (Cu)	Poly(*o*-anisidine-*co*-*o*-toluidine)	3% NaCl	[55]
41	Cu	Polyvinylimidazole and benzimidazole	Acidic water, 25°C	[56]
42	Cu	Films of PPy, polyindole and PPy	3.5% NaCl	[57]
43	Cu	PPy and Polythiophene	3.5% NaCl	[58]
44	Cu	PVP and PEI	2 M H_2SO_4, 30°C	[59]

materials are suitable to a temperature of not more than 150°C, and many of them up to 40°C to 50°C (vinylplast, polyisobutylene, etc.).

Majority of polymeric materials have low mechanical strength; that is why their use is limited. Although sheet vinyl has high mechanical strength and can withstand a temperature of about 100°C for a while, a clear reason why its need is high in recent times is that it can be used for making faolite columns and pumps as well as lining metal pipes. Warning from plant scientists: If polymeric materials, some of which are directly gotten from a plant source such as plant gums, are excessively used as corrosion inhibitors to prevent the corrosion of metals, the plant kingdom will slowly diminish; metals will be protected at the cost of destruction of plant kingdom.

Coatings applied directly to metals normally use conventional corrosion inhibitor pigments such as zinc, aluminum, zinc oxide (ZnO), modified ZnO, and calcium ion exchanged amorphous silica gel. Using corrosion inhibitor pigments has several disadvantages. Some pigments contain metals that are toxic. Several, including metallic zinc, have high densities and settle. A number of pigments react with the resins in the coating. Additional pigmentation also requires added wetting agents that may affect corrosion resistance. A list of various polymeric materials that have been used as corrosion inhibitors is given in Table 1.

Conclusions

It has been shown that polymers especially the water soluble ones are efficient corrosion inhibitors in different aqueous media. Mechanism of inhibition are mainly attributed to adsorption and depends on the metal, physicochemical properties of the molecule such as functional groups, steric factors, aromaticity at the donor atom and *p* orbital character of donating electrons, as well as the electronic structure of the molecules. In other words, the efficiency of polymers as corrosion inhibitor depends not only on the characteristics of the environment in which it acts, the nature of the metal surface, and electrochemical potential at the interface, but also on the structure of the inhibitor itself, which includes the number of adsorption active centers in the molecule, their charge density, the molecular size, the mode of adsorption, the formation of metallic complexes, and the projected area of the inhibitor on the metallic surface. The results of the series of investigations have revealed that the processes involved in corrosion inhibition are not uniform with respect to all classes of compounds so far investigated and are not even constant or consistent with one inhibitor in a given system. Indeed, the overall process is a function of the metal, corrodent, inhibitor structure, and concentration, as well as temperature.

Competing interests

The authors declare that they have no competing interests.

Authors' contributions

DA, AJ, PA, and CA participated in the design of the study. DA conceived of the study, and participated in its design and coordination. All authors read and approved the final manuscript.

Authors' information

DEA was born in Minna in 1986 and was raised in a Christian home. He is a Physical Chemist and a graduate of Ahmadu Bello University, Zaria, where he obtained his bachelors and masters degrees in chemistry. As a scientist, his desire is to further his carrier by accepting challenging responsibilities in a dynamic progressive course that will create and maintain a friendly working environment and offer opportunity for growth.

Acknowledgments

We would like to acknowledge the editor Joseph Gerard Bautista of the Journal Editorial Office, Springer Open for his help and gift as a brilliant editor as well as his contribution to the final form and beauty of this work.

Author details
[1]Department of Chemistry, Ahmadu Bello University, Zaria, Kaduna State PMB 1044, Nigeria. [2]Department of Chemistry, Federal University of Technology, Minna, Niger State PMB 65, Nigeria.

References

1. Sekine I, Sanbongi M, Hagiuda H, Oshibe T, Yuasa M, Imahama V, Shibata Y, Wake T (1992) J ElectrochemSoc 139(11):3167
2. Muller B, Forster I, Klager W (1997) Corrosion inhibition of zinc pigments in alkaline media by polymers. Progr Org Coating 31(3):229–233
3. Abdel Rehim SS, Tohommy FM, Seleet MM (1984) Effect of some polyaminopolycarboxylic acids on the corrosion of steel in sulphate solutions. Surf Tech 21(2):169–177
4. Sedahmed GH, Nagy Soliman M, El-Kholy NS (1982) J Appl Electrochem 12(4):479
5. Khairou KS, El Sayed A (2003) Inhibition effect of some polymers on the corrosion of Cd in a hydrochloric acid solutions. J Appl Polymer Sci 88(4):866–871
6. Meena BV, Anthony N, Mangayarkarasi K, Jayaram P, Rajendran S (2000) Proc. of Tenth National Congress on Corrosion Control. NCC and CECRI, Karaikudi, pp 241–247
7. Umoren SA, Ogbobe O, Ebenso EE, Ekpe UJ (2006) Effect of halides on the corrosion inhibition of mild steel in acidic medium using polyvinyl alcohol. Pigm Resin Technol 35(5):284–292
8. Umoren SA, Ebenso EE, Okafor PC, Ekpe UJ, Ogbobe O (2006) Effect of halides on the corrosion inhibition of mild steel in alkaline medium using polyvinyl alcohol. J Appl Polymer Sci 103:2810–2816
9. Umoren SA, Obot IB, Ebenso EE (2008) Corrosion inhibition of aluminium using exudate gum from *Pacchylobus edulis* in the presence of halide ions in HCl. Electron J Chem 5(2):355–364
10. Umoren SA, Ogbobe O, Igwe IE, Ebenso EE (2008) Inhibition of mild steel corrosion in acidic medium using synthetic and naturally occurring polymers and synergistic halide additives. Corrosion Sci. doi:10.1016/j.corsci.2008.04.015
11. Abdallah M (2004) Guar gum as corrosion inhibitor for carbon steel in sulphuric acid solutions. Portugaliae Electrochimica Acta 22:161–175
12. Obot IB, Obi-Egbedi NO, Umoren SA, Ebenso EE (2010) Synergistic and antagonistic effects of anions and *Ipomoea invulcrata* as green corrosion inhibitor for aluminium dissolution in acidic medium. Int J Electrochem Sci 5(7):994–1007
13. Eddy NO, Ibok UJ, Ebenso EE (2009) Adsorption, synergistic inhibitive effect and quantum chemical studies on ampicillin and halides for the corrosion of mild Steel. J Appl Electrochem. doi:10.1007/s10800-009-0015-z
14. Ebenso EE, Ekpe UJ, Ibok UJ (1998) Studies on the inhibition of mild steel corrosion by some plant extracts in acidic medium. Discov Innovat 10:52–59
15. Eddy NO, Ebenso EE (2010) Corrosion inhibition and adsorption properties of ethanol extract of *Gongronema latifolium* on mild steel in H_2SO_4. Pigm Resin Technol 39(2):77–83
16. Klinov IY (1962) New polymeric materials in anticorrosion technology. Trudy. Vsesoyuznoy Mezhvuzovskoy Nauchnoy Konferentsii po Voprosam Bor'by s Korroziyey (Translations of the Inter-Higher Educational Institute Scientific Conference on Questions of Fighting Corrosion). Gostoptekhizdat, Moskva, pp 296–303
17. Rajendran S, Sridevi SP, Anthony N, John Amalraj A, Sundearavadivelu M (2005) Corrosion behaviour of carbon steel in polyvinyl alcohol. Anti-Corros Method M 52(2):102–107
18. Finkenstadt VL, Bucur CB, Cote GL, Evans KO (2011) Agricultural polymers as corrosion inhibitors [abstract]. International Chemical Congress of Pacific Basin Societies 8:1
19. Buchweishaija I, Mhinzi GS (2008) Natural products as a source of environmentally friendly corrosion inhibitors: the case of gum exudates from *Acacia seyal* var. *seyal*. Portugaliae. Electrochim Acta 26(3):257–265
20. Sedahmed GH, Soliman M, el Kholy NS N (1980) Electropolishing of vertical copper cylinders in phosphoric acid under natural convention conditions. Surf Technol 11:67–71
21. Umoren SA, Obot IB, Ebenso EE, Okafor PC, Ogbobe O, Oguzie EE (2006) Gum arabic as a potential corrosion inhibitor for aluminium in alkaline medium and its adsorption characteristics. Anti-Corros Method M 53(5):277–282
22. Fekry AM, Riham MR (2009) Acetyl thiourea chitosan as an eco-friendly inhibitor for mild steel in sulphuric acid medium. Electrochim Acta 55(6):1933–1939
23. Manickavasagam R, Jeya K, Paramasivam M, Venkatakrishnalyer S (2002) Poly (styrene sulphonic acid) doped polyaniline as an inhibitor for the corrosion of mild steel in hydrochloric acid. Anti-Corros Method M 49(1):19–26
24. Yurt A, Butun V, Duran B (2007) Effect of the molecular weight and structure of some novel water soluble triblock copolymers on the electrochemical behaviour of mild steel. Mater Chem Phys 105:114–121
25. Sathiyanarayanan S, Balkrishnan K (1994) Prevention of corrosion of iron in acidic media using poly(o-methoxy aniline). Electrochem Acta 39(6):831–836
26. Miksic BA, Tarvin M, Sparrow GR (1989) Surface analytical techniques in evaluation of the effects of VCI organic corrosion inhibitors on the surface chemistry of metals. Corrosion 607:1–10
27. Kuznetsov YI (1996) Inhibiting action and absorption of beta-aminoketones on metals. Zasshchita Metallov 32(5):528–533
28. Andreev NN (1997) To simplified methods of estimation of saturated vapor pressure of volatile corrosion inhibitors.. Eurocorr 97 conference, Trodheim, Norway, pp 22–25. September 1997
29. Prenosil M (1998) Volatile corrosion inhibitor coatings. Cortec Corporation Technical Publications, New Orleans
30. Amin MA, Abd El-Rehim SS, El-Sherbini EEF, Hazzazi OA, Abbas MN (2008) Polyacrylic acid as a corrosion inhibitor for aluminium in weakly alkaline solutions. J Corrosion Sci 51:658–667
31. Nnabuk Okon E, Paul A, Gimba CE, Ebenso EE (2011) GCMS studies On *Anogessus leocarpus* (AI) gum and their corrosion inhibition potential for mild steel in 0.1 M HCl. Int J Electrochem Sci 6:5815–5829
32. Obot IB, Obi-Egbedi NO (2009) Ginseng root: CA new efficient and effective eco-friendly corrosion inhibitor for aluminium alloy of type AA 1060 in hydrochloric acid solution. Int J Electrochem Sci 4(9):1277–1288
33. Kamal Y, Wan Aizan WAR (2003) Effectiveness of acrylic acid based terpolymer as corrosion inhibitor in water treatment. Thesis. Department of Chemical Engineering, University College of Engineering and Technology, Malaysia
34. Berkovic K, Kovac S, Vorkapic-Furac J (2004) Natural compounds as environmentally friendly corrosion inhibitors of aluminium. Acta Alimentaria 33(3):237–247
35. Chetouani A, Medjahed K, Benabadji KE, Hammouti B, Kertit S, Mansri A (2003) Poly (4-vinylpyridine isopentyl bromide) as inhibitor for corrosion of pure iron in molar sulphuric acid. Prog Org Coating 46(4):312–316
36. El-Dahan HA (2006) Natural products as corrosion inhibitors for copper in saline water in presence of sulphide. Egypt J Chem 49(5):589–600
37. Soror TK (2004) New naturally occurring product extract as corrosion inhibitor for 316 stainless steel in 5% HCl. J Mater Sci Tech 20(4):463–466
38. Abdallah A, Khaled M, El Ali B, Emad M (2007) Corrosion inhibition of steel in cooling water system by 2-phosphonobutane-1,2,4-tricaboxylic acid and polyvinyl pyrrolidone (PVP). Arab J Sci Eng 33:1A
39. Arukalam OI, Obidiegwu MU (2011) The inhibition of aluminum corrosion in hydroxyl ethyl cellulose. Acad Res Int 1(3):2223–9553
40. Arukalam OI (2012) The inhibitive effect of hydroxyl cellulose on mild steel corrosion in hydrochloric acid solution. Acad Res Int 2(1):SS35–SS42
41. Mazen MK (2007) The effect of molecular weight on corrosion protection properties of polyvinyl pyrrolidone polymers on stainless steel. Arab J Sci Eng 35(1A):29–39
42. Kinleni KJ, Hayes S, Doering F, Welker S, Koustis G, Hannah B (2005) Electroactive polymers and smart coatings for corrosion inhibition of aluminum alloys. Tri-service corrosion conference., Orlando. 14–18November 2005
43. Bressy-Brondino C, Boutevin B, Hervaud Y, Gaboyard M (2002) Adhesive and anticorrosive properties of poly (vinylidene fluoride) powders blended with phosphonated copolymers on galvanized steel plates. J Appl Polym Sci 83(11):2277–2287
44. Attar MM, Scantlebury JD (2005) Polyaniline as a possible inhibitor for the corrosion of mild steel. J Corrosion Eng 8(1):97–99
45. Bereket GA, Yurt A, Turk H (2003) Inhibition of corrosion of low carbon steel in acidic solution by selected polyelectrolytes and polymers. Anti-Corros Method M 50(6):422–535
46. Morooka M, Sekine I, Tanaki T, Hirosett N, Yuasa M (2001) Effects of polymer-polymer complexes on the corrosion of mild steel in cooling water system (part 2): corrosion investigation in polymethacrylic acid/ polyacrylamide system. Corrossion Eng 50(3):106–114
47. Jianguo J, Lin W, Otieno-Alego V, Schweinsberg DP (1995) Polyvinylpyrrolidone and polyethylenimine as corrosion inhibitors for the

corrosion of a low carbon steel in phosphoric acid. Corrosion Sci 37(6):975–985
48. Dubey AK, Singh G (2007) Corrosion inhibition of mild steel in sulphuric acid solution by using polyethylene glycol methyl ether (PEGME). Port Electrochim Acta 25:221–235
49. Hirai T, Yameki J, Okada T, Yamaji A (1985) Inhibitive effects of Al corrosion by polymer ammonium chloride in alkaline electrolyte. Electrochim Acta 30(1):61–67
50. El-khair A, Mostata B (1983) Inhibition of iron corrosion by soluble monomeric and polymeric vinyl pyrrolidone. Corrosion Prev Control 30(6):14–16
51. Manivel P, Venkatachari G (2006) Inhibitive effect of *p*-aminobenzoic acid and its polymer on corrosion of iron in 1 mol/L HCl solution. J Mater Sci Technol 22(3):301–305
52. Jeyaprabha C, Sathiyanarayanan S, Phani KLN, Venkatachari G (2005) Investigation of the inhibitive effect of poly (diphenylamine) on corrosion of iron in 0.5 M H2SO4 solutions. J Electroanal Chem 585(2):250–255
53. Chetouani A, Medjahed K, Sid-Lakhdar KE, Hammouti B, Benkaddour M, Mansri A (2004) Poly(4-vinylpyridine-poly(3-oxideethylene) tosyle) as an inhibitor for iron in sulphuric acid at 80°C. Corrosion Sci 6(10):2421–2430
54. Sathiyanarayanan SK, Balakrishnan K, Dhawan SK, Trivedi DC (2005) Influence of poly(aminoquinone) on corrosion inhibition of iron in acid media. Applied Surf Sci 252:966–975
55. Pawar P, Gaikwad AB, Patil PP (2007) Corrosion protection aspects of electrochemically synthesized poly (*o*-anisidine-*co*-*o*-toluidine) coatings on copper. Electrochim Acta 52:5958–5967
56. Xue G, Lu Y, Shi G (1994) SERS studies of synergistic effect of corrosion inhibition of Cu by two component inhibitor systems of polyvinyl imidazole and benzimdazole. Appl Surf Sci 74(1):37–41
57. Tuken T, Tansug G, Yazici B, Erbil M (2007) Poly(N-methyl pyrrole) and its copolymer with pyrrole for mils steel protection. Surf Coating Technol 2002:146–154
58. Tuken T, Yazici B, Erbil M (2005) Polypyrrole/polythiophene coating for copper protection. Prog Org Coating 53:38–45
59. Schweinsberg DP, Hope GA, Trueman A, Otieno-Alego V (1996) An electrochemical and SERS study of the action of polyvinylpyrrolidone and polyethylenimine as inhibitor for copper in aerated H_2SO_4. Corrosion Sci 38:587–599

Kinetics of the Fischer-Tropsch synthesis on silica-supported cobalt-cerium catalyst

Mohsen Mansouri[1], Hossein Atashi[1*], Ali Akbar Mirzaei[2] and Reza Jangi[1]

Abstract

Background: The process of converting synthesis gas into liquid fuels (Fischer-Tropsch synthesis) is a well-known technology. Among all Fischer-Tropsch synthesis (FTS) catalysts, Co- and Fe-based ones are applicable for industrial processes due to their low cost and high activity and selectivity. In this experimental study, a kinetic model has been developed for FTS reactions using co-precipitation technique and Co-Ce/SiO_2 as the catalyst.

Results: Rate data have been obtained for CO hydrogenation over a co-precipitated well-characterized Co-Ce/SiO_2 catalyst, studied in a fixed-bed micro-reactor at atmospheric pressure under the conditions of 200°C to 300°C, H_2/CO feed ratio (mol/mol) of 1 and 1.5, and space velocity in the range 2,700 to 5,200 h^{-1}. Characterization of both precursor and calcined catalysts was carried out using powder X-ray diffraction, scanning electron microscopy, and Brunauer-Emmett-Teller surface area measurements.

Conclusions: The kinetic parameters were estimated with nonlinear regression method. The data were best fitted by a Langmuir-Hinshelwood-Hougen-Watson approach. The activation energy for the optimal kinetic model was determined to be 31.57 kJ mol^{-1}.

Keywords: Fischer-Tropsch synthesis, Kinetic modeling, LHHW

Background

Fischer-Tropsch synthesis (FTS), the production of liquid hydrocarbons from synthesis gas (CO and H_2) at atmospheric pressure or above and at a few hundred degrees Celsius, is a promising developing option for the environmentally sound production of chemicals and fuels from coal and natural gas [1,2]. Considerable progress has been made in the past two decades in the development of more active and selective cobalt and iron catalysts and more effective reactor/process technologies [3]. The FTS with cobalt catalysts is not complicated by extensive, irreversible changes in the catalyst. With iron catalysts, however, their physical and chemical nature is considerably altered by wax and carbon deposition and by formation of carbides and magnetite [4]. Cobalt-based catalysts are the preferred catalysts for hydrocarbon synthesis because of their high FTS activity, selectivity for long-chain paraffins, and low activity for the water-gas shift reaction [5]. Supported Co catalysts with high specific rates require the synthesis of small metal crystallites at high local surface densities on support and the use of supports or alloys that increase the rate per surface Co (turnover rate). It has been reported that the cobalt and cerium matrix plays a synergistic role in the chemisorption of reactant molecules, thereby controlling the reactivity of the title system in the FTS reaction [6]. Co-Ce catalysts have been investigated for their selectivity for lower molecular weight olefins [7-9]. Mixed cobalt-cerium oxide catalysts supported by SiO_2 [10,11] and TiO_2 [12] have also been used in FTS for the production of hydrocarbons. Another study [11] on the effects of cerium oxide on the particle size of cobalt, reducibility of cobalt, and catalytic performance of a Co/SiO_2 catalyst shows that the degree of cobalt reduction decreases with the addition of cerium, even in low concentrations.

The kinetic description of the FT reaction is a very important task for industrial practice, being a prerequisite for the industrial process design, optimization, and simulation. The kinetics of cobalt-based FT catalysts has been the subject of many researches for decades. The kinetics of the FTS has been studied extensively to describe the reaction rate using a power law rate equation

* Correspondence: atashi.usb@gmail.com
[1]Department of Chemical Engineering, University of Sistan and Baluchestan, Zahedan 98164-161, Iran
Full list of author information is available at the end of the article

or an equation based on certain mechanistic assumptions. The mechanistic kinetic rate expressions for cobalt catalysts are based on the formation of the monomer species as the rate-determining step in the consumption of synthesis gas. Many kinetic equations have been proposed in the literature for various cobalt catalysts. These equations have been obtained either empirically (using a power law rate equation) or to fit a proposed mechanism [13-18]. Atashi et al. [2], Zennaro et al. [4], Yates and Satterfield [18], and Das et al. [19] have displayed the explicit rate equations based on the Langmuir-Hinshelwood-Hougen-Watson (LHHW) adsorption theory that has been developed for the cobalt-based FT catalysts in tabulated format.

Many studies have been dedicated to the kinetics of FTS catalytic reactions over the mono- and bimetallic catalysts. However, it is particularly important to know that no work on the kinetics of FTS reaction over a Co-Ce catalyst has been performed. Thus, in the present work, for the first time, we attempted to investigate the kinetics and mechanism of the CO hydrogenation on the silica-supported Co-Ce catalyst, which was prepared by co-precipitation procedure. For the kinetic study, a series of statistically representative kinetic data was obtained on a well-characterized silica-supported catalyst over a range of commercial reaction conditions. The kinetics of FT reaction were studied, and the rate expressions were tested against experimental data that was obtained on some selected catalysts. A model was successfully devised, and the kinetic parameters were determined. The catalyst characterization was carried out using different methods including powder X-ray diffraction (XRD), scanning electron microscopy (SEM), and Brunauer-Emmett-Teller (BET) surface area measurements.

Methods

Characterization techniques

Powder XRD measurements were performed using a D8 Advance diffractometer (Bruker AXS, Ettlingen, Germany). Scans were taken with a 2θ step size of 0.02 and a counting time of 1.0 s using a CuK_α radiation source generated at 40 kV and 30 mA. Specimens for XRD were prepared by compaction into a glass-backed aluminum sample holder. Data were collected over a 2θ range from 4° to 70°, and the phases were identified by matching experimental patterns to entries in Diffract^{plus} version 6.0 indexing software. The morphology of the catalysts and their precursors was observed by means of a Cambridge S-360 scanning electron microscope (Cambridge Instruments, Cambridge, England).

BET surface areas, pore volumes, and average pore sizes of the catalyst precursor and calcined samples (before and after the test) were measured by N_2 physisorption using a Quantachrome Nova 2000 automated system (Quantachrome, Boynton Beach, FL, USA). Each catalyst sample was degassed under nitrogen atmosphere at 300°C for 3 h. In order to obtain the BET surface areas, pore volumes, and average pore sizes, different samples were evacuated at −196°C for 66 min.

Fixed-bed reactor system

A schematic representation of the experimental setup is shown in Figure 1. Fischer-Tropsch synthesis was carried out in a fixed-bed micro-reactor made from stainless steel with an inner diameter of 12 mm. Three mass flow controllers (Model 5850E, Brooks Instrument, Hatfield, PA, USA) were used to automatically adjust the flow rate of the inlet gases comprising CO, H_2, and N_2 (purity of 99.99%). The mixture of CO, H_2, and N_2 was subsequently introduced into the reactor, which was placed inside a tubular furnace (Model ATU 150–15, Atbin, Mina Rashid Ave, Dubai). The temperature of the reaction was controlled by a thermocouple inserted into the catalytic bed and visually monitored by a computer. The catalyst was pre-reduced *in situ* at atmospheric pressure under H_2/N_2 flow (N_2/H_2 = 1, flow rate of each gas = 30 mlmin^{-1}) at 400°C for 16 h. In each test, 1.0 g of catalyst was loaded, and the reactor operated for about 12 h to ensure that a steady-state operation was attained.

Catalytic evaluation

The experiments were conducted with mixtures of H_2, CO, and nitrogen in a temperature range from 200°C to 300°C and H_2/CO feed ratios of 1:1 and 3:2 (mol/mol) at atmospheric pressure. The arrangements of the parameters and the related levels are shown in Table 1. In all experiments, the space velocities were between 2,700 and 5,200 h^{-1}.

To avoid the effect of deactivation, fresh catalysts were loaded in each experiment. To achieve the isothermal conditions in a catalytic bed, the catalyst was diluted with an inert material (quartz), and axial temperature distribution was ensured using Mear's criterion [20], that is with $L/d_p > 50$. Moreover, the gas hourly space velocity (GHSV) increased to the value in which the CO conversion was almost the same for a variety of catalyst weight, which indicates that film diffusion can be neglected above this GHSV (in this paper, 2,700 h^{-1}). Therefore, the catalyst pellet and the bed can be considered as isothermal. The temperature was also kept constant to within ±1°C by the heater, and the conversion of the reaction was kept low (below 20%) in order not to generate too much heat in the above exothermic reactions. Also, plug flow was assumed for the gaseous feed. The CO

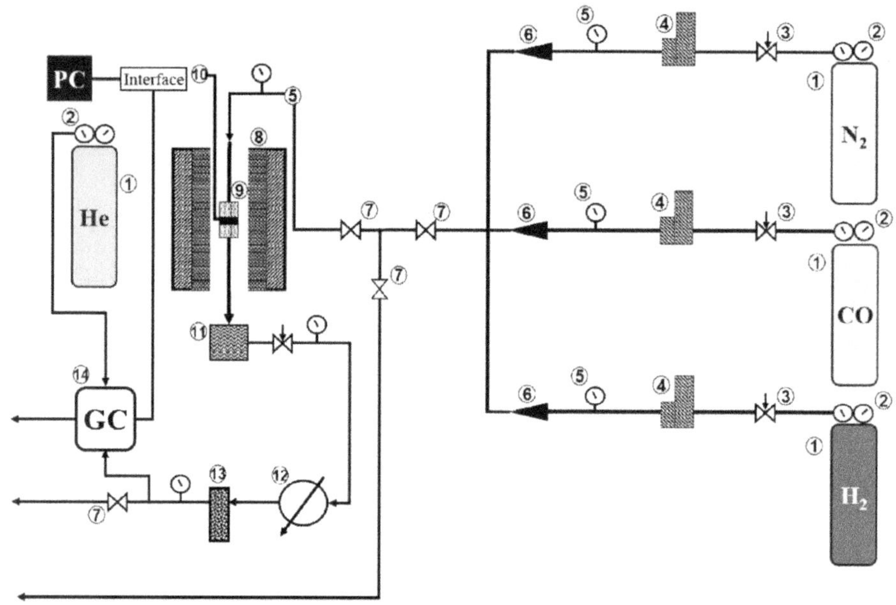

Figure 1 Schematic representation of the reactor used. (1) Gas cylinders, (2) pressure regulators, (3) needle valves, (4) mass flow controllers, (5) monometers, (6) non-return valves, (7) ball valves, (8) tubular furnace, (9) reactor, (10) catalyst bed, (11) trap, (12) condenser, (13) silica gel column, and (14) gas chromatograph (GC).

Table 1 Summary of experimental conditions for the kinetic tests

Number of data	*T* (°C)	P_{H2} (bar)	P_{CO} (bar)	*F/W* (mmol g_{cat}^{-1} min^{-1})	X_{CO} (%)	−*r*CO (mmol g_{cat}^{-1} min^{-1})
1	200	0.333	0.333	0.509	5.83	0.0297
2	210	0.4	0.267	0.498	6.16	0.0306
3	220	0.333	0.333	0.488	8.42	0.0411
4	230	0.4	0.267	0.424	10.21	0.0433
5	240	0.333	0.333	0.469	11.21	0.0526
6	250	0.4	0.267	0.435	12.39	0.0538
7	260	0.333	0.333	0.451	12.40	0.0559
8	270	0.4	0.267	0.443	12.54	0.0555
9	280	0.333	0.333	0.435	12.65	0.055
10	290	0.4	0.267	0.427	12.91	0.0512
11	300	0.333	0.333	0.419	13.19	0.0553
12	200	0.4	0.267	0.407	13.16	0.0535
13	210	0.333	0.333	0.398	13.66	0.0544
14	220	0.4	0.267	0.390	14.46	0.0564
15	230	0.333	0.333	0.450	14.49	0.0652
16	240	0.4	0.267	0.455	16.09	0.0732
17	250	0.333	0.333	0.493	16.28	0.0802
18	260	0.4	0.267	0.282	16.78	0.0474
19	270	0.333	0.333	0.354	17.18	0.0609
20	280	0.4	0.267	0.348	17.94	0.0624
21	290	0.333	0.333	0.342	18.53	0.0633
22	300	0.4	0.267	0.274	19.12	0.0524

P_{tot} = 1 bar after 10 to 15 h of starting the reaction at given conditions.

conversion (%) is calculated according to the normalization method:

$$\text{CO conversion}\ (\%) = \frac{(\text{Moles of } CO_{in}) - (\text{Moles of } CO_{out})}{(\text{Moles of } CO_{in})} \times 100 \quad (1)$$

The experimental reaction rate was determined as follows:

$$\text{Rate of CO conversion} = \frac{(\text{Fractional conversion}) \times (\text{Input flow rate of CO})}{\text{Weight of the catalyst}} \quad (2)$$

Statistical criteria

The unknown parameters for each model were calculated using the experimental data and minimizing the sum of the squares of the differences using the Levenberg-Marquardt algorithm. The experimental reaction rates, the R^2 value (reflects the amount of variance), and the root mean square deviation (RMSD) were used for the evaluation of the fitting.

$$\sigma = \frac{1}{N_{exp}} \sum_{i=1}^{N_{exp}} r_{CO,i}^{exp} \quad (3)$$

$$R^2 = 1 - \left(\frac{\sum_{i=1}^{N_{exp}} \left(r_{CO,i}^{exp} - r_{CO,i}^{cal} \right)^2}{\sum_{i=1}^{N_{exp}} \left(r_{CO,i}^{exp} - \sigma \right)^2} \right)^2 \quad (4)$$

$$\text{RMSD} = \frac{1}{N_{exp}} \left(\sum_{i=1}^{N_{exp}} \left(r_{CO,i}^{exp} - r_{CO,i}^{cal} \right)^2 \right)^2 \quad (5)$$

where $r_{CO,i}^{exp}$ and $r_{CO,i}^{cal}$ denote the experimental and calculated CO conversion rates using each kinetic model, respectively, and N_{exp} represents the number of experimental data with pure variance σ.

To choose the most suitable kinetic expression, different statistical indices can also be used to determine the quality of regression models. In order to find the most appropriate model, the following conditions should be considered [21]: (1) obtained constants must be positive, (2) coefficients of the equation must obey Arrhenius and van't Houff rules, and (3) the optimal model is the one which gives the highest R^2.

Result and discussion

Catalyst screening results

Characterization of this catalyst was carried out using XRD, and the patterns of this catalyst in different stages of precursor, fresh calcined catalyst (before the test), and used calcined catalyst (after the test) at H_2/CO = 2:1 are shown in Figure 2. The catalyst precursor was found to be amorphous. The presence of amorphous phases in the XRD pattern of the precursor makes the other phases undetectable. However, the calcined catalyst before the test showed different phases, and the actual phases identified in this catalyst were $CoSi_2$ (cubic), CeO_2 (cubic), SiO_2 (monoclinic), $CeCoSiO_2$ (tetragonal), Co_2SiO_4 (cubic), and Co_3O_4 (cubic). In order to identify the changes in the calcined catalyst during the reaction and to detect the phases formed, the catalyst after the test was characterized by XRD and its phases were found to be $CoSi_2$ (cubic), CeO_2 (cubic), CoO (cubic), C (hexagonal), Co_2C (orthorhombic), Ce (hexagonal), Co (hexagonal), and CeC_2 (tetragonal). During chemical reaction, some of the oxidic phases in the calcined catalyst before the test transformed into metallic and carbide phases; and in the tested catalyst, there are oxidic and carbide phases, both of which are active phases in the FTS. Carbide phase is active in CO hydrogenation. Formation of this phase with oxidic phases which are active to olefins cause high performance of supported SiO_2 catalyst [22]. Clearly, the XRD study suggests that the precursor undergoes morphological changing during calcination and also FTS chemical reaction. However, the XRD technique may not be sufficiently sensitive to follow the fine details of these changes. Hence, a detailed SEM study of the precursor and calcined catalysts before and after the test at the optimum molar feed ratio of 2:1 was undertaken, and their SEM images on different stages are presented in Figure 3. The SEM observations showed differences in morphology of the precursor and calcined catalysts (before and after the test). The electron micrograph obtained from the catalyst precursor (Figure 3a) was found to be composed of several agglomerates of particles and shows that this material is less dense. After the calcination at 600°C, the morphological features were different from the precursor sample and showed that the agglomerate measure was greatly reduced compared with the precursor (Figure 3b). It may be due to the calcined catalyst surface that is covered with small crystallites of cobalt and cerium oxide, which is in agreement with the XRD results. After FTS chemical reaction, the catalyst texture and its morphology changed (Figure 3c). However, the size of the grains in the tested catalyst grew larger by agglomeration, which may be due to sintering after reactions. The BET surface area, pore volume, and average pore diameter of the catalysts were tabulated in Table 2. As it can be seen, the calcined catalyst before the test has a higher specific surface area (79 m^2g^{-1}) than its precursor (75 m^2g^{-1}), which is in agreement with the SEM results showing that the agglomerate size of the calcined catalyst is less compared with its precursor and therefore leads to an increase in the BET specific surface area of the calcined sample. The high specific surface area of

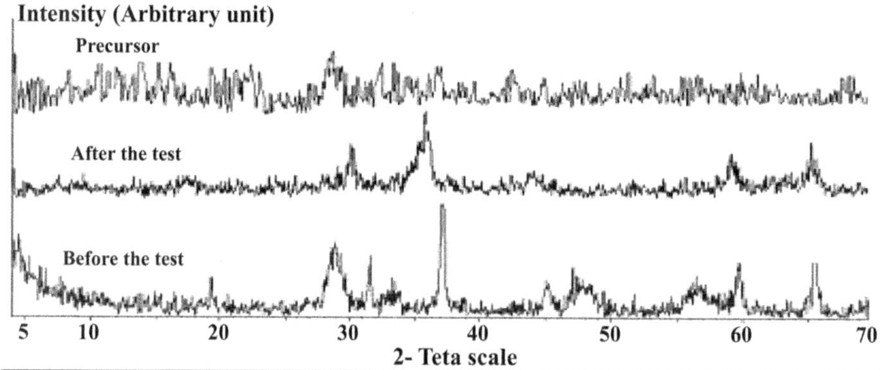

$CoSi_2$	2θ: 29, 48, 57	CeO_2	2θ: 29, 33.5, 48, 57	SiO_2	2θ: 26, 29, 33
CoO	2θ: 37, 42.5	Co_3O_4	2θ: 19, 31, 37, 59, 65.5	Co_2SiO_4	2θ: 19, 37, 45, 49, 59, 65
C	2θ: 41.5, 43, 44	Co_2C	2θ: 37, 39, 59.5, 64.5	Ce	2θ: 32, 36.5, 63
Co	2θ: 41.5, 44.5	CeC_2	2θ: 27, 32.5, 44, 55	$CeCo_2Si_2$	2θ: 37, 46, 59, 60, 66

Figure 2 XRD patterns of precursor and calcined catalysts (before and after the test). Containing 80%Co/20%Ce/15 wt.% SiO_2.

the fresh catalyst before the test allowed a high degree of metal dispersion. The obtained results in Table 2 show that there is no big variation among average pore size diameters of the catalyst in different stages of the precursor and calcined samples (before and after the test).

Development of kinetic equations

CO hydrogenation in FTS process can be analyzed using hydrocarbon chain formation mechanism, which is developed by combining of a number of steps in FTS. The multi-compound product depicts an affluent source of information about the events on the active site by performing high compositional order. Kinetic modeling of the interconnection of surface reaction is an essential step at this point. The mechanism of the hydrocarbon formation during the FTS has been reviewed and discussed by several authors [13,16,23]. On the basis of the nature of CO adsorption and chain initiator intermediates, popular mechanistic proposals include (1) the carbide mechanism [24], wherein CO adsorbs dissociatively and the carbide (Cs) is the chain initiator intermediate, and (2) the enolic mechanism [25], which involves molecular adsorption of CO and the formation of an oxygen intermediate, the enol (HOCs). In the former mechanism, simultaneous dissociative adsorption of CO and H_2 is followed by the hydrogenation of adsorbed carbon by adsorbed hydrogen in a stepwise manner to give methane and higher hydrocarbons. In the latter mechanism, the dissociative adsorption of H_2 and molecular adsorption of CO are followed by the hydrogenation of adsorbed carbon monoxide by adsorbed hydrogen to form an oxygenated intermediate, which reacts with another adsorbed hydrogen to form water and adsorbed carbon, and the reaction of the resulting carbon with adsorbed hydrogen as in the carbide mechanism. For derivation of the LHHW rate expressions, several assumptions were made: (a) The intrinsic reaction rates are proportional to the surface coverage of reactants; (b) CO consumption and -CH_2 preparation reaction have one irreversible limiting step, and the other steps were assumed to be quasi-equilibrium; (c) considering steady-state conditions, mediums do not accumulate on the catalyst surface; (d) due to catalyst dilution with quartz, GHSV > 2,700 h^{-1} utilization, and low

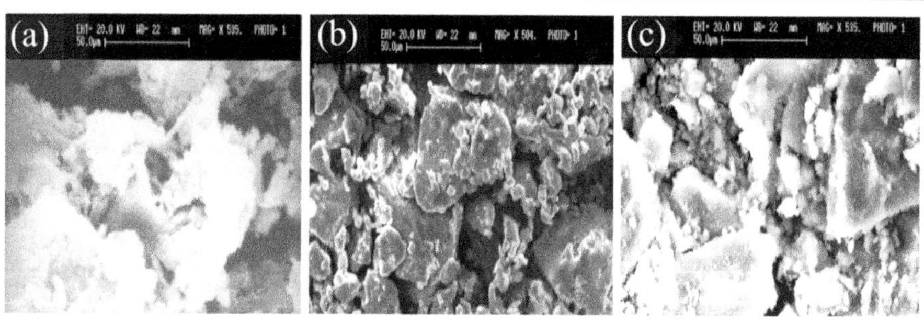

Figure 3 SEM images of the 80%Co/20%Ce/15 wt.% SiO_2catalyst. In the **(a)** precursor,**(b)** catalyst before the test, and **(c)** catalyst after the test at H_2/CO = 2:1.

Table 2 BET results of the precursor and calcined catalysts (before and after the test) containing 80%Co/20%Ce/15wt.%SiO_2

Catalyst	Specific surface area ($m^2\ g^{-1}$)	Pore volume ($cm^3\ g^{-1}$)	Pore size (Å)
Precursor	75.2	2.86×10^{-2}	58.41
Fresh calcined	79.4	6.23×10^{-2}	56.73
Used calcined	53.4	4.63×10^{-2}	60.18

conversion, the operation is assumed to be isothermal; (e) catalysts are fixed and uniformly distributed; (e) the elementary adsorption of hydrogen and carbon monoxide in quasi-equilibrium state within the concentration of gas phase; (f) water is removed irreversibly after CO decomposition; and (g) CO surface concentration is higher compared with hydrogen due to the more difficult adsorption of CO over cobalt catalyst. CO and carbon concentrations are assumed to be dominant concentrations [26-29].

In order to derive rate equations to be adjusted with the data in Table 1, we used the LHHW theory to obtain kinetic models. According to this theory, a reaction mechanism should be adopted. Four mechanisms were offered on the basis of various monomer formations (elementary reactions) and carbon chain distribution pathways. An elementary reaction set on sites for each model is summarized in Table 3.

To derive each kinetic model, one of the elementary reaction (in some case, two or three) steps was assumed as rate-determining step (RDS), and all other steps were considered at equilibrium. Then, all of the models obtained were fitted separately against the experimental data. In the interest of conciseness, only certain selected kinetic models are reported in Table 4.

For example, derivations of the rate equation for FT-IV-2 and FT-I-1 are explained here. The rate expression of the rate-determining step for the FT-IV-2 model, where dissociative H_2 adsorption was the rate-limiting step, can be expressed as irreversible adsorption:

$$-r_{CO} = k_2 P_{H_2} \theta_S{}^2 \tag{6}$$

where $-r_{CO}$ is the rate of disappearance of CO, k_2 is the forward rate constant for hydrogen dissociative adsorption, P_{H_2} is hydrogen partial pressure in the gas, and θ_S is the empty space fraction. The fraction of vacant sites, θ_S, can be calculated from the following balance equation:

$$\theta_S + \theta_{CO} + \theta_H + \theta_{HCO} + \theta_C + \theta_{CH} + \theta_{CH_2} + \theta_O + \theta_{OH} = 1. \tag{7}$$

Hydrogen and carbon monoxide adsorbed dissociatively on cobalt catalyst, but depending on the catalyst

Table 3 Elementary reaction mechanism set for FTS

Model	Number	Elementary reaction
FT-I	1	$CO + 2s \leftrightarrow Cs + Os$
	2	$H_2 + 2s \leftrightarrow 2Hs$
	3	$Cs + Hs \leftrightarrow HCs + s$
	4	$HCs + Hs \leftrightarrow H_2Cs + s$
	5	$Os + Hs \rightarrow HOs + s$
	6	$HOs + Hs \rightarrow H_2O + 2s$
FT-II	1	$CO + s \leftrightarrow COs$
	2	$H_2 + 2s \leftrightarrow 2Hs$
	3	$COs + s \leftrightarrow Cs + Os$
	4	$Cs + H_2 \leftrightarrow CH_2s$
	5	$Os + H_2 \rightarrow H_2Os$
FT-III	1	$CO + s \leftrightarrow COs$
	2	$COs + H_2 \leftrightarrow H_2COs$
	3	$H_2COs + H_2 \leftrightarrow CH_2s + H_2O$
	4	$H_2 + 2s \leftrightarrow 2Hs$
FT-IV	1	$CO + s \leftrightarrow COs$
	2	$H_2 + 2s \leftrightarrow 2Hs$
	3	$COs + Hs \leftrightarrow HCOs + s$
	4	$HCOs + Hs \leftrightarrow Cs + H_2O + s$
	5	$Cs + Hs \leftrightarrow CHs + s$
	6	$CHs + Hs \leftrightarrow CH_2s + s$
	7	$Os + Hs \rightarrow HOs + s$
	8	$HOs + Hs \rightarrow H_2O + 2s$

and reaction condition and the extent of dissociation of CO, their adsorption is varied [30]. However, CO is adsorbed more strongly than hydrogen [31]. In this case, it is assumed that only CO occupies a significant fraction of the total number of sites. Other species were assumed to be negligible in the stoichiometric balance:

$$\theta_S + \theta_{CO} = 1. \tag{8}$$

The surface coverage of carbon monoxide is calculated from the site balance and the preceding reaction steps which are at quasi-equilibrium:

$$CO + S \overset{k_{ads,CO}}{\leftrightarrow} COs \tag{9}$$

$$k_{ads,CO} P_{CO} \theta_S - k_{des,CO} \theta_{CO} = 0 \tag{10}$$

$$\theta_{CO} = K_1 P_{CO} \theta_S \tag{11}$$

$$K_1 = \frac{k_{ads,CO}}{k_{des,CO}}$$

Table 4 Reaction rate expressions for the FTS, r_{FT} ($\text{mmol}\ g_{cat}^{-1}\ \text{min}^{-1}$)

Model of rate controlling	Rate equation
FT-I-1	$\frac{kP_{CO}}{\left(1+aP_{CO}^{1/2}+bP_{H_2}^{1/2}\right)^2}$
FT-I-2	$\frac{kP_{H_2}}{\left(1+aP_{CO}^{1/2}+bP_{H_2}^{1/2}\right)^2}$
FT-I-3	$\frac{kP_{CO}^{1/2}P_{H_2}^{1/2}}{\left(1+aP_{CO}^{1/2}+bP_{H_2}^{1/2}\right)^2}$
FT-I-4	$\frac{kP_{CO}^{1/2}P_{H_2}^{3/4}}{\left(1+aP_{CO}^{1/2}P_{H_2}^{-1/4}+bP_{H_2}^{1/2}\right)^2}$
FT-II-1	$\frac{kP_{CO}}{\left(1+aP_{CO}^{1/2}\right)}$
FT-II-2	$\frac{kP_{H_2}}{\left(1+aP_{CO}^{1/2}\right)^2}$
FT-II-3	$\frac{kP_{CO}}{\left(1+aP_{CO}^{1/2}\right)^2}$
FT-II-4	$\frac{kP_{CO}^{1/2}P_{H_2}}{\left(1+aP_{CO}^{1/2}\right)}$
FT-III-1	$\frac{kP_{CO}}{(1+aP_{CO})}$
FT-III-2	$\frac{kP_{CO}P_{H_2}}{1+aP_{CO}}$
FT-III-3	$\frac{kP_{CO}P_{H_2}^2}{1+aP_{CO}}$
FT-IV-1	$\frac{kP_{CO}}{1+aP_{CO}+bP_{H_2}^{1/2}}$
FT-IV-2	$\frac{kP_{H_2}}{(1+aP_{CO})^2}$
FT-IV-3	$\frac{kP_{CO}P_{H_2}^{1/2}}{\left(1+aP_{CO}+bP_{H_2}^{1/2}\right)^2}$
FT-IV-4	$\frac{kP_{CO}P_{H_2}}{\left(1+aP_{CO}+bP_{CO}P_{H_2}^{1/2}\right)^2}$

where K_1 is the equilibrium constant of the CO adsorption step. Substituting Equation 11 into Equation 8, the ratio of free active site can be expressed as:

$$\theta_S = \frac{1}{1+K_1P_{CO}} \tag{12}$$

Substituting the fraction of vacant sites into Equation 6, the final rate expression is obtained as follows:

$$-r_{CO} = \frac{kP_{H_2}}{(1+aP_{CO})^2}. \tag{13}$$

In the same way, the surface fractions (θ) and the final kinetic models would be obtained briefly.

The rate expression of the RDS for model FT-I-1 can be expressed as follows:

$$-r_{CO} = k_1P_{CO}\theta_S^2 \tag{14}$$

where k_1 is the forward rate constant for carbon monoxide dissociative adsorption, and P_{CO} is carbon monoxide partial pressure in the gas. It is assumed that Cs, Os, and Hs are the abundant surface species, and catalyst site balance is as follows:

$$\theta_S + \theta_C + \theta_O + \theta_H = 1 \tag{15}$$

$$K_1 = \frac{k_1}{k_{-1}} = \frac{\theta_C\theta_O}{P_{CO}\theta_S^2} \tag{16}$$

$$K_2 = \frac{k_2}{k_{-2}} = \frac{\theta_H^2}{P_{H_2}\theta_S^2}. \tag{17}$$

From Equation 16,

$$\theta_O = \frac{K_1P_{CO}\theta_S^2}{\theta_C}, \tag{18}$$

and from Equation 17,

$$\theta_H = K_2^{0.5}P_{H_2}^{0.5}\theta_S. \tag{19}$$

In steady state, the consumption rate of carbon in stage 3 is equal to the elimination rate of oxygen by stage 5, so

$$k_3\theta_C\theta_H = k_5\theta_O\theta_H \Rightarrow \theta_C = \frac{k_5}{k_3}\theta_O. \tag{20}$$

By substituting Equation 18 in Equation 20,

$$\theta_C = \frac{k_5}{k_3}\theta_O = \frac{k_5}{k_3}\left(\frac{K_1P_{CO}\theta_S^2}{\theta_C}\right) \Rightarrow \theta_C = \left(\frac{k_5K_1}{k_3}\right)^{0.5}P_{CO}^{0.5}\theta_S, \tag{21}$$

and by substituting Equation 21 in Equation 18,

$$\theta_O = \frac{K_1P_{CO}\theta_S^2}{\theta_C} = \frac{K_1P_{CO}\theta_S^2}{\left(\frac{k_5K_1}{k_3}\right)^{0.5}P_{CO}^{0.5}\theta_S} \Rightarrow \theta_O = \left(\frac{k_3K_1}{k_5}\right)^{0.5}P_{CO}^{0.5}\theta_S. \tag{22}$$

By substituting Equations 19, 21, and 22 into the site balance equation, the concentration of free active sites will be as follows:

$$\theta_S = \frac{1}{1 + \left(\left(\frac{k_5 K_1}{k_3}\right)^{0.5} + \left(\frac{k_3 K_1}{k_5}\right)^{0.5}\right) P_{CO}^{0.5} + K_2^{0.5} P_{H_2}^{0.5}}, \tag{23}$$

and finally, the rate equation will be as follows:

$$-r_{CO} = \frac{k_1 P_{CO}}{\left(1 + \left(\left(\frac{k_5 K_1}{k_3}\right)^{0.5} + \left(\frac{k_3 K_1}{k_5}\right)^{0.5}\right) P_{CO}^{0.5} + K_2^{0.5} P_{H_2}^{0.5}\right)^2}$$

$$= \frac{k P_{CO}}{\left(1 + a P_{CO}^{0.5} + b P_{H_2}^{0.5}\right)^2} \tag{24}$$

where $k = k_1$, $a = \left(\left(\frac{k_5 K_1}{k_3}\right)^{0.5} + \left(\frac{k_3 K_1}{k_5}\right)^{0.5}\right)$, and $b = K_2^{0.5}$.

Similar derivations were also made for the other models shown in Table 4. It can be seen that the pressure dependency of CO and H_2on the numerator ranges from 1:2 to 1 and 1:2 to 2, respectively. The denominator is quadratic in case of a dual-site elementary reaction, in contrast to a single-site rate-determining step. The denominator consists of the individual contribution of significantly abundant species on the catalyst surface.

Model parameters and model discrimination

CO consumption rate was obtained from the data in Table 1 by using the differential method of data analysis. The kinetic data presented in Table 1 for CO conversion were used for testing the 15 models listed in Table 4. The least square method and nonlinear regression analysis based on the summarized values in Table 1 were used to determine the kinetic model parameters from the experimental data provided in Table 2. Arrhenius and adsorption equations were substituted in kinetic models: Equations 25 and 26 were substituted for k and a, respectively.

$$k = k_0 \exp\left(\frac{-E}{RT}\right) \tag{25}$$

$$a = a_0 \exp\left(\frac{\Delta H}{RT}\right). \tag{26}$$

According to the statistical results obtained by inserting the data and models, the best model can be selected. Table 5 shows the statistical indicators for the FT kinetic models. Based on the statistical criteria and other conditions [21], and also by comparing the values of R^2, RMSD, and variance, it was recognized that FT-IV-2 is the most appropriate model, which was developed based on enolic mechanism. The model FT-IV shows that the dominant mechanism on the catalyst surface is based on the dissociation of hydrogen with associative adsorbed CO and forming methyl monomers. Although in other studies on cobalt catalyst the dominant mechanism was carbide mechanism, it can be concluded that on

Table 5 Estimated values of the statistical indicators for the FT kinetic models

Model	Kinetic parameter						Statistical indicator		
	k_0 (x) (mmol g^{-1} min^{-1} bar^x)	a_0 (x) (bar^x)	ΔH_a (kJ mol^{-1})	b_0 (x) (bar^x)	ΔH_b (kJ mol^{-1})	E (kJ mol^{-1})	R^2	RMSD (10^{-12})	Variance (10^{-7})
FT I 1	0.11 (1)	18.29 (−1/2)	12.25	−91.54 (−1/2)	7.07	−11.27	0.91	5.25	8.328
FT-I-2	2.4E + 03 (−1)	137.35 (−1/2)	7.21	27.05 (−1/2)	7.54	8.49	0.89	5.05	7.578
FT-I-3	0.61 (−1)	−15.41 (−1/2)	4.87	−0.0004 (−1/2)	36.72	2.05	0.78	11.26	38.37
FT-I-4	7.1E + 03 (−5/4)	47.59 (−1/4)	10.02	1.3E + 04 (−1/2)	−18.76	9.86	0.82	10.31	32.13
FT-II-1	9.50 (−1)	1.7E + 04 (−1/2)	40.65			−33.58	0.33	28.06	211.8
FT-II-2	106.97 (−1)	24.43 (−1/2)	1.05			22.22	0.94	5.13	7.071
FT-II-3	765.28 (−1)	1.9E + 03 (−1/2)	5.85			−5.88	0.16	22.34	134.2
FT-II-4	47.34 (−3/2)	86.79 (−1/2)	50.13			−26.36	0.56	16.15	70.11
FT-III-1	12.87 (−1)	6.27 (−1)	33.68			−27.91	0.16	22.34	134.2
FT-III-2	189.18 (−2)	346.87 (−1)	50.04			−26.28	0.56	16.15	70.11
FT-III-3	2.14E + 06 (−3)	1.7E + 03 (−1)	27.64			13.68	0.08	23.35	146.6
FT-IV-1	0.11 (−1)	213.73 (−1)	−13.84	−213.73 (−1/2)	−12.26	25.89	0.93	5.65	9.675
FT-IV-2	8.38 (−1)	15.24 (−1)	−6.55			31.57	0.98	5.05	6.858
FT-IV-3	0.28 (−3/2)	6.42 (−1)	1.24	−18.71 (−1/2)	−5.09	15.94	0.81	5.16	7.726
FT-IV-4	7.68 (−2)	5.85(−1)	10.47	−226.57(−1/2)	−48.91	4.394	0.95	5.28	8.417

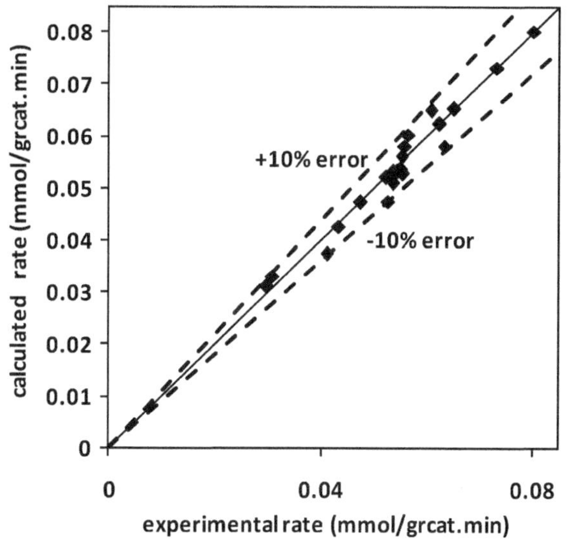

Figure 4 Parity plot of the experimental and calculated CO consumption rates for FT-IV-2.

bimetallic cobalt catalysts, carbon dissociation of CO cannot be done alone. In our previous research on titania-supported Co-Mn catalyst [2], forming the monomer CH_2 was done by reaction of adsorbed CO and hydrogen in two steps, which was assumed as the dominant mechanism. Also, Keyser [32] observed through a study on bimetallic Co-Mn oxide catalyst that a reaction rate equation for the FT reaction based on the enolic mechanism gave results which were marginally better than the results based on the carbide mechanism.

Figure 4 shows a comparison between the experimental data and predicted results of the optimal kinetic model. The solid line in the figure denotes that the calculated $-r_{CO}$ is equal to the experimental one, and the dotted lines over and under the solid line represent 10% deviation. The experimental results were found to be in good agreement with the kinetic model showing about 10% deviation. Therefore, there are appropriate fitting by the LHHW approach in the form of $-r_{CO} = kP_{H_2}/(1 + aP_{CO})^2$. As it has been shown in Table 5 for the best-fitted model (FT-IV-2), the value 31.57 $kJmol^{-1}$ for the activation energy of the Fischer-Tropsch reaction is close to the results reported by Atashi et al. [2].

Experimental

In the present work, we attempted to investigate the kinetics and mechanism of the CO hydrogenation on the silica-supported Co-Ce catalyst, which was prepared by co-precipitation procedure. For the kinetic study, a series of statistically representative kinetic data was obtained on a well-characterized silica-supported catalyst over a range of commercial reaction conditions. The kinetics of Fischer-Tropsch reaction was studied, and the rate expressions were tested against experimental data that were obtained on some selected catalysts.

Materials and processing

The catalyst tested in this study, containing 80%Co/20%Ce/15wt.%SiO_2 which was performed optimally for the conversion of synthesis gas to light olefins [33], was prepared using co-precipitation procedure. Aqueous solutions of $Co(NO_3)_2{\cdot}6H_2O$ (0.5 M) (99%, Merck, White house Station, NJ, USA) and $Ce(NO_3)_3{\cdot}6H_2O$ (0.5 M) (99%, Merck) with different molar ratios were premixed, and the resulting solutions were heated to 70°C in a round-bottomed flask fitted with a condenser. Aqueous Na_2CO_3 (0.5 M) (99.8%, Merck) was added dropwise to the mixed nitrate solution with stirring while the temperature was maintained at 70°C until pH 8 ± 0.1 was achieved. The aged precipitate (120 min) was then filtered and then washed several times with warm distilled water. The precipitate was then dried in an oven at 120°C for 16 h to form a material denoted as the catalyst precursor, which was subsequently calcined in static air in a furnace at 600°C for 6 h to obtain the final catalyst. Then, for the preparation of SiO_2-supported catalyst, the optimal amount of 15 wt.% of SiO_2 based on the total catalyst weight was added to the mixed solution of cobalt and cerium nitrates with the molar ratio of 80%Co/20%Ce and then filtered, washed, dried at 120°C, and calcined at 600°C for 6 h, in the same way as for the unsupported catalyst preparation.

Conclusions

Cobalt-cerium catalyst is an effective catalyst for the hydrogenation of carbon monoxide to light olefins. Experiments for the kinetic of Fischer-Tropsch reaction (hydrocarbon formation) were carried out over the co-precipitated silica-supported Co-Ce catalyst in a fixed-bed micro-reactor over a range of operating conditions. The kinetic models of FTS reaction over a well-characterized 80%Co/20%Ce/15wt.%SiO_2 catalyst, which is used as an optimized catalyst of the process, were studied. The data of this study are best fitted by the simple LHHW approach rate of the form $-r_{CO} = kP_{H_2}/(1 + aP_{CO})^2$ which was developed based on enolic mechanism. The values of the kinetic constants were obtained, and the activation energy was found to be 31.57 $kJmol^{-1}$ for the best model.

Nomenclature

d_p, particle diameter (m); E, activation energy (kJ mol^{-1}); F, molar flow rate (mmol min^{-1}); K, kinetic equation; L, length of the catalytic bed (m); P_i, partial pressure of component i (bar); R, universal gas constant (8.314 J K^{-1} mol^{-1});

RMSD, root mean of standard deviation; R^2, a parameter for goodness of fit; *T*, temperature (°C); *W*, catalyst mass (g); ΔH, heat of adsorption (kJ mol^{-1}); $-r_{CO}$, rate constant of the reaction (mmol g_{cat}^{-1} min).

Greek symbols

θ, surface fraction; σ, variance.

Competing interests

The authors declare that they have no competing interests.

Authors' contributions

MM carried out the experiments, studied the mechanism of CO hydrogenation, and analyzed the data with nonlinear regression method. HA and AAM participated in interpretation of results and drafted the manuscript. RJ helped in the experiments and analysis. All the authors have read and approved the final manuscript.

Authors' information

MM is a PhD student of chemical engineering. HA is a professor of chemical engineering. AAM is a professor of chemistry. RJ has an MSc in Chemical Engineering.

Acknowledgements

The authors wish to gratefully acknowledge the helpful comments of Dr. Kiyanoosh Razzaghi and the facilities provided by the University of Sistan and Baluchestan.

Author details

[1]Department of Chemical Engineering, University of Sistan and Baluchestan, Zahedan 98164-161, Iran. [2]Department of Chemistry, University of Sistan and Baluchestan, Zahedan 98135-674, Iran.

References

1. Bulavchenko OA, Cherepanova SV, Malakhov VV, Dovlitova LS, Ishchenko AV, Tsybulya SV (2009) *In situ* XRD study of nanocrystalline cobalt oxide reduction. Kinet Catal 50:192–198
2. Atashi H, Siami F, Mirzaei AA, Sarkari M (2010) Kinetic study of Fischer–Tropsch process on titania supported cobalt–manganese catalyst. J Ind Eng Chem 16:952–961
3. Pour AN, Shahri SMK, Bozorgzadeh HR, Zamani Y, Tavasoli A, Marvast MA (2008) Effect of Mg, La and Ca promoters on the structure and catalytic behavior of iron-based catalysts in Fischer–Tropsch synthesis. Appl Catal A Gen 348:201 208
4. Zennaro R, Tagliabue M, Bartholomew C (2000) Kinetics of Fischer–Tropsch synthesis on titania-supported cobalt. Catal Today 58:309–319
5. Mirzaei AA, Shirzadi B, Atashi H, Mansouri M (2012) Modeling and operating conditions optimization of Fischer–Tropsch synthesis in a fixed-bed reactor. J Ind Eng Chem 18:1515–1521
6. Spadaro L, Arena F, Granados ML, Ojeda M, Fierro JLG, Frusteri F (2005) Metal–support interactions and reactivity of Co/CeO_2 catalysts in the Fischer–Tropsch synthesis reaction. J Catal 234:451–462
7. Al-Shammary FY, Caga IT, Tata AY, Winterbotton JM (1991) The effect of CeO_2 support upon activity and selectivity of Ru and Co Fischer Tropsch catalysts. Eur J Solid State Inorg Chem 28:453–456
8. Barault J, Guilleminot A, Achard JC, Paul-Boneour V, Percheron GA, Hilaire L, Coulon M (1986) Syngas reaction over lanthanum-cobalt intermetallic catalysts. Appl Catal A Gen 22:273–287
9. Bruce LA, Hoang M, Hughes AE, Turney TW (1993) Ruthenium promotion of Fischer-Tropsch synthesis over coprecipitated cobalt/ceria catalysts. Appl Catal A Gen 100:51–67
10. Dai X, Yu C, Li R, Shi H, Shen S (2006) Role of CeO_2 promoter in Co/SiO_2 catalyst for Fischer-Tropsch synthesis. Chin J Catal 27:904–910
11. Ernst B, Hilaire L, Kiennemann A (1999) Effects of highly dispersed ceria addition on reducibility, activity and hydrocarbon chain growth of a Co/SiO_2 Fischer–Tropsch catalyst. Catal Today 50:413–427
12. Dai X, Yu C, Shen S (2001) Promotion effect of ceria on Fischer–Tropsch synthesis performance over Co/Al_2O_3. Chin J Catal 22:104–108
13. Jager B, Espinoza R (1995) Advances in low temperature Fischer–Tropsch synthesis. Catal Today 23:17–28
14. Ribeiro FH, Von Wittenau AES, Bartholomew CH, Somorjai GA (1997) Reproducibility of turnover rates in heterogeneous metal catalysis: compilation of data and guidelines for data analysis. Catal Rev Sci Eng 39:49–76
15. Sarup B, Wojciechowski BW (1989) Studies of the Fischer-Tropsch synthesis on a cobalt catalyst. II. kinetics of carbon monoxide conversion to methane and to higher hydrocarbons. Can J Chem Eng 67:62–74
16. Wojciechowski BW (1988) The kinetics of the Fischer-Tropsch synthesis. Catal Rev Sci Eng 30:629–702
17. Yang CH, Massoth FE, Oblad AG (1979) Kinetics of CO + H_2 reaction over co-Cu-Al_2O_3 catalyst. Adv Chem Ser 178:35–46
18. Yates IC, Satterfield CN (1991) Intrinsic kinetics of the Fischer-Tropsch synthesis on a cobalt catalyst. Energy Fuel 5:168–173
19. Das TK, Conner WA, Li J, Jacobs G, Dry ME, Davis BH (2005) Fischer-Tropsch synthesis: kinetics and effect of water for a Co/SiO_2 catalyst. Energy Fuel 19:1430–1454
20. Mollavali M, Yaripour F, Atashi H, Sahebdelfar S (2008) Intrinsic kinetics study of dimethyl ether synthesis from methanol on γ-Al_2O_3 catalysts. Ind Eng Chem Res 47:3265–3277
21. Bercic G, Levec J (1992) Intrinsic and global reaction of methanol dehydration over γ-Al_2O_3 pellets. Ind Eng Chem Res 31:1035–1040
22. Zhang HB, Schrader GL (1985) Characterization of a fused iron catalyst for Fischer-Tropsch synthesis by *in situ* laser Raman spectroscopy. J Catal 5:325–332
23. Iglesia E (1997) Design, synthesis, and use of cobalt-based Fischer–Tropsch synthesis catalysts. Appl Catal A Gen 161:59–78
24. Maitlis PM (1989) A new view of the Fischer–Tropsch polymerization reaction. Pure Appl Chem 61:1747–1754
25. Storch HH, Golumbic N, Aderson RB (1951) The Fischer–Tropsch and related synthesis. Wily, New York
26. Botes FG, Breman BB (2006) Development and testing of a new macro kinetic expression for the iron-based. Ind Eng Chem Res 45:7415–7426
27. Van Der Laan GP, Beenackers AACM (2000) Intrinsic kinetics of the gas-solid Fitscher-Tropsch and water gas shift reactions over a precipitated iron catalyst. Appl Catal A Gen 193:39–53
28. Yang J, Liu Y, Chang J, Wang YN, Bai L, Xu YY, Xiang HW, Li YW, Zhong B (2003) Detailed kinetics of Fischer-Tropsch synthesis on an industrial Fe-Mn catalyst. Ind Eng Chem Res 42:5066–5090
29. Zhang R, Chang J, Xu Y, Cao L, Li Y, Zhou J (2009) Kinetic model of product distribution over Fe catalyst for Fischer-Tropsch synthesis. Energy Fuel 23:4740–4747
30. Ponec V, Van Barnevald WA (1979) The role of chemisorption of Fischer-Tropsch synthesis. Ind Eng Chem Prod Res Dev 18:268–271
31. Christman K, Scober O, Ertl G, Neumann M (1974) Adsorption of CO on a Ni (111) surface. J Chem Phys 60:4719–4724
32. Keyser MJ, Everson RC, Espinoza RL (2000) Fischer–Tropsch kinetic studies with cobalt–manganese oxide catalysts. Ind Eng Chem Res 39:48–54
33. Mirzaei AA, Galavy M, Beigbabaei A, Eslamimanesh V (2007) Preparation and operating conditions for cobalt cerium oxide catalysts used in the conversion of synthesis gas into light olefins. J Iran Chem Soc 4:347–463

Modeling of fluidized bed reactor for ethylene polymerization: effect of parameters on the single-pass ethylene conversion

Hassan Farag[1], Mona Ossman[2,3], Moustapha Mansour[1] and Yousra Farid[1*]

Abstract

Background: In this study, we present the developments in modeling gas-phase catalyzed olefin polymerization fluidized bed reactors (FBR) using chromium catalyst technique. The model is based on the two-phase theory of gas-solid fluidization: bubble phase and emulsion phase. The model has proved to be the suitable model in many of past studies. In the proposed model, the bed is divided into several sequential sections. The effect of important reactor parameters such as superficial gas velocity, catalyst injection rate, catalyst particle growth, and minimum fluidization velocity on the dynamic behavior of the FBR has been discussed. The conversion of product in a fluidized bed reactor is investigated and compared with the actual data from the plant site.

Results: A good agreement has been observed between the model predictions and the actual plant data. It has been shown that about 0.28% difference between the calculated and actual conversions has been achieved.

Conclusions: The study showed that the computational model was capable of predicting the hydrodynamic behavior of gas-solid fluidized bed flows with reasonable accuracy.

Keywords: Fluidized bed reactor, Mathematical model, Ethylene conversion, Bed height, Catalyst feed rate

Background

The fluidized bed reactor has a unique physical design, with gas and polymer particles flowing in opposite directions. It consists of metallic catalyst particles that are fluidized by the flow of ethylene gas, and catalyst particles (pre-polymer) are suspended in the ethylene fluid as ethylene gas is pumped from the bottom of the reactor bed to the top. The gas is fed from the base of the reactor and splits into two phases: bubble phase and emulsion phase. Pre-polymer particles are fed in near the top of the reactor, and while the polymerization reaction occurs, the particles grow, increasing in weight and size. Particle segregation occurs in the reactor according to particle weight, so the full-grown polymer particles are removed at the base of the reactor. Non-reacted gases leave the reactor after passing through the disengagement zone [1]. Pre-polymerization is generally carried out in continuous stirred-tank reactors (CSTR) located before the fluidized bed reactor. Catalyst particles are fed into the CSTR along with ethylene and co-monomers to yield pre-polymer. Afterwards, these pre-polymerized catalyst particles are fed into the FBR to complete the ethylene polymerization. A diagram of the industrial fluidized bed reactor system using a CSTR as the pre-polymerization reactor is shown in Figure 1 [2]. In effect, the catalyst is not actually consumed; it is simply incorporated with the polyethylene product as polyethylene molecules remain stuck to the catalyst particle from which they were produced.

The conversion of ethylene is low for a single pass through the reactor, and it is necessary to recycle the unreacted ethylene. Unreacted ethylene gas is removed off the top of the reactor, where it is expanded and decompressed to separate the catalyst and low molecular weight polymer from the gas. After purification, ethylene gas is then recompressed and recycled back into the reactor. Granular polyethylene is gradually removed from the bottom of the reactor as soon as reasonable conversions have been achieved. Typically, a residence time of 3 to 5 h results in a 97% conversion of ethylene.

* Correspondence: Eng_yousra_hamdy@yahoo.com
[1]Chemical Engineering Department, Faculty of Engineering, Alexandria University, Alexandria, Egypt
Full list of author information is available at the end of the article

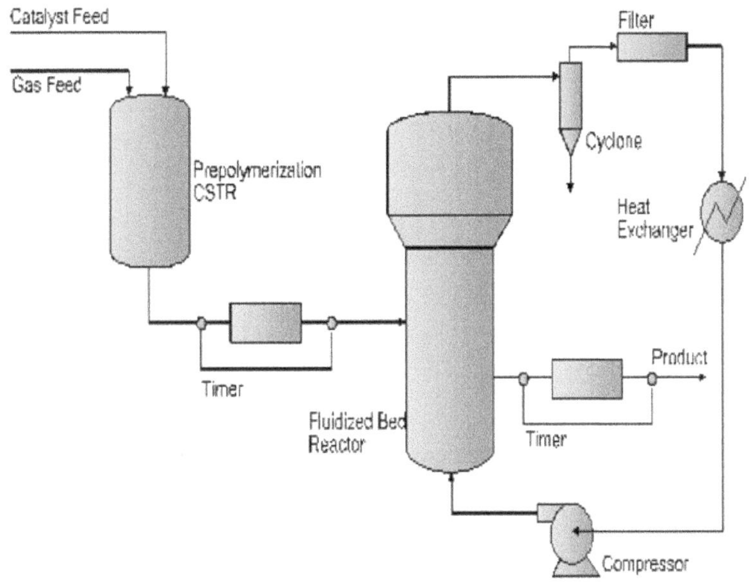

Figure 1 Industrial polyethylene production diagram (BP Chemical Technology, London).

The flow in the fluidized bed reactor can be mathematically modeled using mole balances [1].

An accurate model describing the movement of gas/solid and pressure distribution around a rising bubble was then proposed by Davidson and Harrison [3].This model describes the gas flow through a three dimensional fluidized bed mainly in spherical or semi-spherical shape bubbles through the core; however, depending on the emulsion gas velocity, the region around the bubble may be surrounded by a cloud as a result of emulsion gas circulation between the dense solid phase and the core of the bubble.

Choi and Ray [4] proposed a two-phase model including bubble and emulsion phases with constant bubble size. McAuley et al. [5] proposed a single phase model by modifying Ray's model with additional assumptions. In a comparison between the two models, they have shown that the single phase assumption does not make considerable difference in the results obtained from the models. Hatzantonis et al. [6] in a research work developed the two-phase model by considering the bubble growth effect on hydrodynamic behavior of the reactor and have shown that the developed model has a better agreement with industrial data than single-phase and two-phase models with constant bubble size. Grosso and Chiovetta [7] studied the particle size distribution in the output stream of commercial, fluidized bed reactors for ethylene polymerization using the mathematical model. In another work, Kiashemshaki et al. [8] developed the two-phase model by considering the polymerization reaction not only in the emulsion phase but also in the bubble phase. They have indicated that about 20% of the polymerization reaction occurs in the bubble phase.

Vahidi and Shahrokhi [9] studied the dynamic behavior and control of a fluidized bed reactor for polyethylene production. It has been shown that the control system has satisfactory performances either for set point tracking or load rejection. Hamzehei [10] studied the heat transfer of a poly ethylene fluidized bed reactor without reaction experimentally and computationally at different superficial gas velocities.

The advantage of the present work over the previous ones is that it considered major factors that affect the fluidized bed reactor model, like the bubble size, particle size distribution, and catalyst properties in dynamic simulation of the reactor in order to obtain better understanding of the reactor performance. In the present study, a FBR model is developed to be able to predict the main factors affecting the ethylene conversion. Simulation results indicate that this strategy promotes the control performance considerably.

Methods

Model assumptions

The model assumes the following:

1 The reactor operates at a steady state, under uniform temperature.
2 The reactor is in isothermal operation.
3 The gas flows in plug flow, with gas interchange between phases.

Table 1 Operating conditions and simulation (Sidpec Petrochemical Company)

Parameter	Unit	Value/type
Monomer properties		
Ethylene feed concentration	mol m^{-3}	4×10^{-4}
Hydrogen feed concentration	mol m^{-3}	9.6×10^{-5}
Actual ethylene conversion per pass	%	3.33
Overall ethylene conversion	%	100
Catalyst properties		
Type of catalyst		Silica base chromium catalyst with Al and Ti modifiers
Cr consumption	ppm	5 ppm/1 ton residual polymer
Prepolymer properties		
Prepolymer density	Kg m^{-3}	2,380
Prepolymer particle size	m	190×10^{-6}
Prepolymer feed rate	Kg s^{-1}	0.013
Polymer properties		
Polymer density	Kg m^{-3}	955
Polymer average particle size	m	1.166×10^{-3}
Flow rate of polymer product	Kg s^{-1}	4.16
Minimum particle size	m	1.070×10^{-3}
Maximum particle size	m	1.215×10^{-3}
Density function(PSD)	1/m	0.6489
Particle size distribution	1/m	Range from 190 to 1,215 μ
Reactor parameter		
Temperature	°C	108
Pressure	bar	21.5
Diameter	m	5
Fluidizing velocity	M s^{-1}	0.56
Bed height	m	16
Maximum reactor height	m	22
Bed voidage ε_{mf}	-	0.5
Inert gas properties		
Gas density	Kg m^{-3}	17.05
Composition (% volume)		13.02
Ethylene		14.7
Hydrogen		47.2
Nitrogen		19.06
Ethane		6
Butane		1.08×10^{-2}
Butene-1		

4 The bed is axially divided into N compartments, each one consisting of the bubble, cloud, and emulsion phases.
5 Bubbles grow continuously along the reactor height from their initial diameter until they reach the maximum stable bubble diameter.
6 The bed volume fraction occupied by bubbles and emulsion depends on the flow regime and changes along the reactor height.
7 Solid entrainment and carryover are taken into account by the model. However, particle agglomeration and breakage are not considered.

8 Bed voidage is constant from the distributor until H_{mf} and is equal to 0.5.

Model development

Minimum fluidizing velocity

The superficial gas velocity at minimum fluidization conditions is calculated from the correlations gives by Kunii and Levenspiel [11]:

$$U_{mf} = \frac{Re_{mf}\mu_g}{\rho_g d_p} \tag{1}$$

$$Ar = \frac{g d_p^3 \rho_g \left(\rho_p - \rho_g\right)}{\mu_g^2} \tag{1a}$$

$$Re_{mf} = \left[(33.7)^2 + 0.0408 Ar\right]^{0.5} - 33.7, \tag{1b}$$

where viscosity of the gas mixture is given by the following relation [12]:

$$\mu_{g,\mathrm{mix}} = \frac{\sum y_j \mu_j M_j^{1/2}}{\sum y_j M_j^{1/2}}. \tag{1c}$$

Bubble growth equations

Bubble growth diameter at corresponding bed height is calculated by the correlation given by Wen and Fan [13].

$$\mathrm{b_{di}} = 0.14 \mathrm{d_p} \rho_\mathrm{p} \left(\frac{\mathrm{U_g}}{\mathrm{U_{mf}}}\right) \mathrm{h_i} + \mathrm{d_{bo}}. \tag{2}$$

Initial bubble diameter forms near the distributor calculated from the correlations given by Kunii and Levenspiel [11]:

$$d_{bo} = \frac{2.78\left(U_g - U_{mf}\right)^2}{g}. \tag{2a}$$

The Mori and Wen [14] relation is used to obtain the maximum bubble diameter:

$$d_{bm} = 0.65\left(S\left(U_g - U_{mf}\right)\right)^{0.4} \mathrm{cm}. \tag{2b}$$

Kunii and Levenspiel [11] correlation is used to calculate the bubble rising velocity:

$$U_{bi} = U_g - U_{mf} + 0.711\left(\mathrm{g d}_{bi}\right)^{1/2}. \tag{2c}$$

The relative fraction of the bubble phase for the ith compartment is given by Kunii and Levenspiel [11]:

$$\delta_i = \frac{\left(U_g - U_{mf}\right)}{\left(U_{bi} - U_{mf}\right)}. \tag{2d}$$

The average bubble fraction calculated from the correlations given by Cui et al. (2000) [15]:

$$\delta = 0.534\left[\frac{1 - \exp\left(U_g - U_{mf}\right)}{0.413}\right]. \tag{2e}$$

Kobayashi et al. [16] correlation is used to calculate the number of bubbles:

$$N_i = \frac{6S(H - H_{mf})}{\pi H \Delta h_i^2}, \tag{2f}$$

where the U_{br} is given by the correlation of Grace [17]:

$$U_{br,i} = 0.711(g d_{bi})^{1/2}. \tag{2g}$$

Compartment height

Consider the following equations for the compartment height:

$$\Delta h_i = 2 d_{bo} \frac{(2+m)^{i-1}}{(2-m)^i} \tag{3}$$

$$\Delta h_1 = d_{bo} + \left(1 - \frac{m}{2}\right), \tag{3a}$$

where

$$m = 1.4 d_p \rho_p \frac{U_g}{U_{mf}}. \tag{3b}$$

The total compartment height is calculated using the following equation:

$$H_n = \sum_{i=1}^{n} \Delta h_i \tag{3c}$$

Height at minimum fluidizing conditions

Height at minimum fluidizing conditions is calculated by the correlation given by McAuley et al. [5].

$$H_{mf} = H(1 - \delta). \tag{4}$$

Maximum bubble diameter

Haider and Levenspiel [18] correlation is used to calculate the terminal falling velocity:

$$d_{b,max} = \frac{2U_t^2}{g} \tag{5}$$

$$d_p^{'} = 2.7 d_p \tag{5a}$$

$$d_p^{*} = d_p^{'}\left[\mu_g^{-2}\rho_g\left(\rho_p - \rho_g\right)g\right]^{1/3} \tag{5b}$$

$$U_t^{*} = \left[18 d_p^{*-2} + (2.335 - 1.744\Phi_s) d_p^{*-1/2}\right]^{-1} \tag{5c}$$

$$U_t = U_t^{*}\left[\mu_g \rho_g^{-2}\left(\rho_p - \rho_g\right)g\right]^{1/3}. \tag{5d}$$

Diffusion of ethylene

Specchia et al. [19] correlation is used to calculate the self diffusion of ethylene:

$$D_A = s\frac{U_g d_p}{8.65\left(1 + 1.94\left(\frac{d_p}{D}\right)^2\right)}. \tag{6}$$

Mass transfer coefficient

Mass transfer coefficient is calculated by the correlation gives by Kunii and Levenspiel [20].

$$K_{\text{be},i} = \left[\left(\frac{1}{K_{\text{bc},i}}\right) + \left(\frac{1}{K_{\text{ce},i}}\right)\right]^{-1} \tag{7}$$

$$K_{b,i} = 4.5\frac{U_{\text{mf}}}{d_{b,i}} + \frac{5.85\left(D_A^{1/2} g^{1/4}\right)}{d_{bi}^{\frac{5}{4}}} \tag{7a}$$

$$K_{e,i} = 6.78\left(\frac{\varepsilon_{\text{mf}} D_A U_{b,i}}{d_{bi}^3}\right)^{1/2} \tag{7b}$$

Bed voidage

Bed voidage is calculated by the correlation given by Wen and Fan [13]:

$$(1 - \varepsilon) = \frac{H_{mf}\left(1 - \varepsilon_{mf}\right)}{H} \tag{8}$$

for $h \leq H_{\text{mf}}$

$$(1 - \varepsilon_i) = \frac{H_{\text{mf}}}{H(1 - \varepsilon_{\text{mf}})} - \frac{H_{\text{mf}}(1 - \varepsilon_{\text{mf}})(h - H_{\text{mf}})}{2H(H - H_{\text{mf}})} \tag{8a}$$

for $H_{\text{mf}} \leq h \leq [H_{\text{mf}} + 2(H - H_{\text{mf}})]$.

Particle size distribution of polymer

Grosso and Chiovetta [7] give a correlation to calculate the particle size distribution of the polymer and for the catalyst feed particles:

$$P_b(d_p) = \frac{\left(3\, d_p^5\, \rho_p{}^2 \chi_{\text{cat}}^2\right)}{d_{\text{po}}^6 \rho_{\text{cat}}^2 (1 - \chi_{\text{cat}})} * \frac{\left\{1 + \left(\frac{d_p^3}{d_{\text{po}}^3}\right)\left[\left(\frac{\rho_{\text{cat}}}{\rho_{\text{cat}}} - 1\right)\right]\right\}\exp\left[\rho_p \chi_{\text{cat}}\left(d_{p^2} - dp_o^3\right)\right]}{\rho_{\text{cat}} dp^3 (1 - \chi_{\text{cat}})} \tag{9}$$

$$P_o(r) = \int_{d_{\text{po}}}^{d_p} P_b(d_p) d_p\, d(d_p) \left\{\int_{d_{\text{po}}}^{dp} \left[P_b(d_p) d(d_p)\right]\right\}^{-1}, \tag{9a}$$

For $H_{\text{mf}} \leq h \leq [H_{\text{mf}} + 2(H - H_{\text{mf}})]$.

The flow rate of solids of all sizes removed from the bed

The flow rate of solids of all sizes removed from the bed by the gas flow is calculated by Geldart [21]:

$$F_2 = S\int_{d_{pmin}}^{d_{\text{pmax}}} K^{\bullet}(d_p)\, P_b(d_p)\, d(d_p), \tag{10}$$

where the elutriation constant is calculated as following:

$$K^{\bullet}(d_p) = 23.7\rho_g U_g \exp\left(\frac{-5.4 U_t(d_p)}{U_g}\right) \tag{10a}$$

$$K(d_p) = K^{\bullet}(d_p)\frac{S}{W} \tag{10b}$$

Kunii and Levenspiel [11] correlation is used to calculate the weight of solids in the bed:

$$W = SH(1 - \varepsilon)\rho_{\text{cat}} \tag{10c}$$

The ethylene polymerization rate

The ethylene polymerization rate was adopted based on the model used by Choi and Ray [4] and McAuley et al. [5]:

$$r_{\text{pi}} = K_p C_{ei} X_{cat} \rho_p V_{ei} (1 - \varepsilon_i) M_e \tag{11}$$

$$R_{pi} = r_{pi}\int_{d_{\text{pmin}}}^{d_{\text{pmax}}} P_b(dp)\, d(dp) \tag{11a}$$

$$K_p = k_{\text{po}}\exp\text{-}\left(\frac{E_a}{RT}\right). \tag{11b}$$

Material balance around the Nth compartment

Tanaka [22] has given a correlation for the material balance around the *n*th compartment:

$$k' = \frac{K_{be}\left(U_b - \frac{U_{mf}}{\varepsilon_{mf}}\right)}{\left(U_b + \frac{2U_{mf}}{\varepsilon_{mf}}\right)} \quad (12)$$

$$SU_g C_{b,i-1} = \left(k' V_{bi}\left(C_{b,i} - C_{e,i}\right)\right) + SU_g C_{b,i} \quad (12a)$$

$$K' V_{bi}\left(C_{b,i} - C_{e,i}\right) = R_{pi} V_{ei} \quad (12b)$$

McAuley et al. [5] have given a correlation to calculate the volume of the compartment phases:

$$V_{bi} = S\Delta h_i \delta_i \quad (12c)$$

$$V_{ei} = S\Delta h_i (1 - \delta_i) \quad (12d)$$

The population balance

The population balance is based on a mass balance for particles of size between d_p and $d_p + d(d_p)$, as presented by Kunii and Levenspiel [11,20].

$$F_1 + F_2 + F_o - \sum_{i=1}^{n} R_{pi} = 0 \quad (13)$$

Yong et al. [23] have given a correlation for rate of the increase in particle diameter:

$$(dp) = d\left(\frac{dp}{dt}\right) = \frac{d_{po}\rho_{cat}R_p}{3(1-\varepsilon)d_p^2\rho_p} \quad (13a)$$

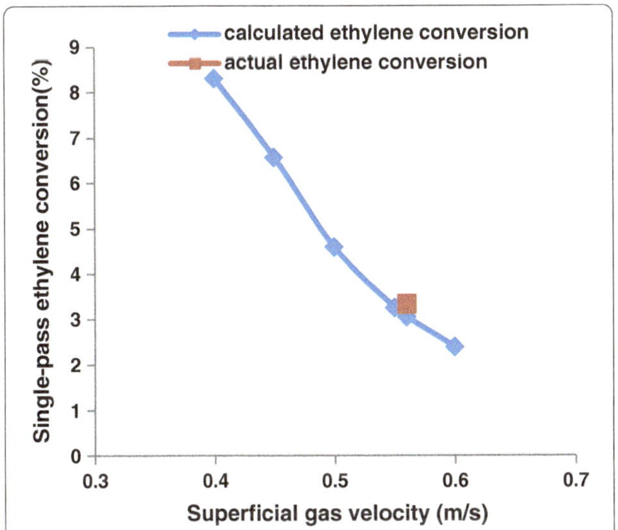

Figure 2 Effect of superficial gas velocity on single-pass ethylene conversion.

The overall mass balance is calculated as:

$$F_o P(d_p) - F_1 P_b(d_p) - WK(r)P_b(d_p) - W\frac{d(dp)P_b(d_p)}{d(dp)} + \frac{3W(dp)(P_b(d_p))}{(d_p)} = 0 \quad (13b)$$

Conversion of ethylene

Kunii and Levenspiel [11] correlation is used to calculate the conversion of ethylene:

$$X_A = \frac{22.4\sum R_p T * 1000}{U_g * 3600 * \left(\frac{yC2}{100}\right) * (1.013 * P) * S * 273 * 28} \quad (14a)$$

$$X_A = \frac{C_{Ain} - C_{Aout}}{C_{Ain}} \quad (14b)$$

Results and discussion

Validation analysis

The validation was carried out using the process reported in BP Chemicals [24].The data used in the validation from SidiKerir Petrochemicals Company for the product HD 5502 GA (Alexandria, Egypt) [25] are shown in Table 1. The results of the validation simulations are shown in the data analysis.

Data analysis

The various states of the effect of different variables on the ethylene single-pass conversion have been studied for the following characteristics:

The effects of superficial gas velocity

Industrial fluidized bed polyethylene reactors are operated at superficial gas velocities ranging from three to six times the minimum fluidization velocity [26]. The major effect of an increase in the superficial velocity, U_g, is a reduction in the time required for a given quantity of gas to pass through the bed.

Figure 2 shows the effect of the superficial gas velocity on the percent ethylene conversion. It shows that high gas velocities reduce the conversion of monomer per pass through the reactor and can lead to a greater elutriation of small particles from the bed. Also, the actual ethylene conversion is in the same trend with the chart. Elutriated particles are prevented from passing out from the reactor and into the gas recycle system using a velocity reduction zone at the top of the reactor, where entrained particles are given the opportunity to drop

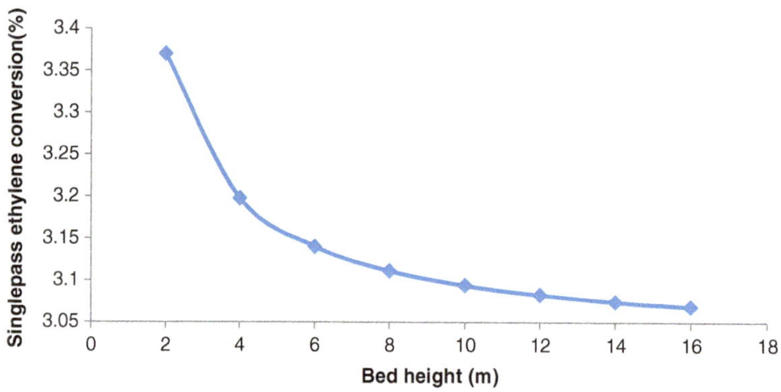

Figure 3 Effect of bed height on ethylene single-pass conversion.

back into the bed. Particle return may also be aided by a cyclone [26].

The effect of bed height

The simulation shows that there is a highly active reaction zone in the top of the reactor; the ethylene conversion decreases by increasing the bed height. It is clear from Figure 3 that single-pass ethylene conversion decreased with increasing bed height. This trend is similar to that found by Fernando and Lona [2].

According to the parametric study of the system, this situation can be avoided by controlling the gas feed velocity. In terms of polyethylene production, it can be enhanced by decreasing the gas feed velocity.

The effect of feed catalyst properties

Figure 4 shows the effect of the catalyst feed rate on the ethylene conversion. It indicates that as the catalyst feed rate increases, the ethylene conversion and particle distribution increase and lead to higher polymerization rates [23].

Figure 5 shows that as Ug decreases from $6U_{mf}$ to $3U_{mf}$, the safe regime, in which the reactor can be operated without a danger of particle melting, decreases accordingly. High gas velocities are required to reduce the risk of particle melting, agglomeration, and subsequent reactor shutdown. However, high gas velocities reduce the conversion of monomer per pass through the reactor and can lead to greater elutriation of small particles from the bed. Elutriated particles are prevented from passing out from the reactor and into the gas recycle system using a velocity reduction zone at the top of the reactor where entrained particles are given the opportunity to drop back into the bed. Particle return may also be aided by a cyclone [26].

Figure 6 shows the effects of catalyst feed rate and bubble size on the single-pass conversion of ethylene. As expected, higher catalyst feed rates lead to higher polymerization rates and to higher conversion. The effects of bubble size on conversion are less obvious. Smaller bubbles have a lower velocity through the bed compared with larger bubbles. As a result, smaller bubbles lead to a larger bubble fraction, δ, in the bed

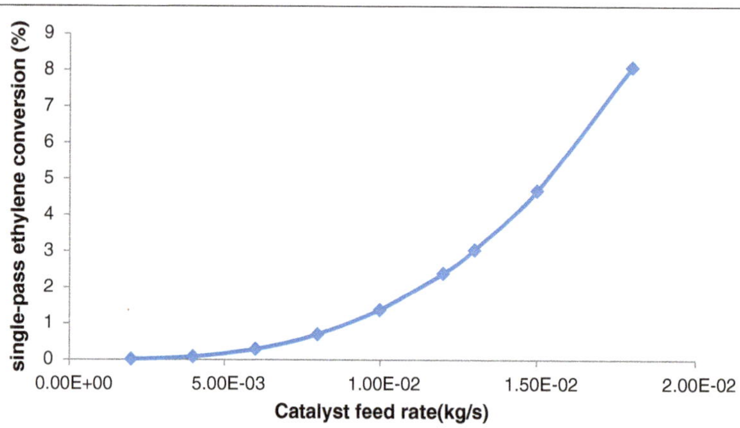

Figure 4 Effect of initial catalyst size on the particle size distribution in the bed.

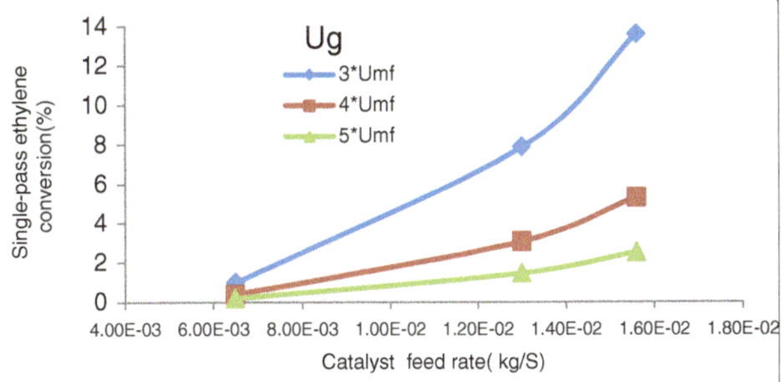

Figure 5 Effect of superficial gas velocity (3 U_{mf} to 6 U_{mf}) and catalyst feed rate on single-pass ethylene conversion.

and to a reduction in the volume of the emulsion phase for a given expanded bed height. Therefore, the residence time for both the solid phase and the catalyst decreases with decreasing bubble size, reducing the quantity of catalyst in the reactor for a given catalyst feed rate. This reduction in the quantity of catalyst tends to reduce R_p, the rate of polymerization, thereby reducing the rate of heat generation in the emulsion phase.

Ethylene conversion

Figure 7 shows a comparison between the calculated conversion and the actual conversion. Good agreement has been observed between the model predictions and the actual plant data. It has been shown that about 0.28% difference between the calculated and actual conversion has been achieved.

Operating variables for sensitivity analysis

Figure 8 shows the sensitivity analysis of the mathematical model using multiplication factors of (0.5, 1, and 1.5) for the variables affected the model. Regarding to catalyst particle diameter, decreasing its value will increase tremendously the single-pass ethylene conversion (by about 4,000 fold); on the other hand, increasing catalyst particle diameter at 50% will decrease the single-pass ethylene conversion to about 0.5% of its original value. Regarding the superficial gas velocity on increasing this value, the single-pass ethylene conversion will decrease. This is understandable since the contact time decreases. From the result shown, it is clear that the single-pass ethylene conversion is not sensitive to change in superficial gas velocity but highly sensitive to catalyst particle diameter. Also, regarding the effect of the total number of compartment, it is clear that increasing total number of compartment in the bed leads to the increase of single pass ethylene conversion. This can be attributed to the increase in the number of compartment in the bed which results in better segregation of the polymer product. It is in accordance with the study results of McAuley et al. [5]. On the other hand, increasing the bed voidage increases the single-pass ethylene conversion. This can be attributed to the increase in bed voidage which would increase the contact surface.

Experimental

The model was solved using Microsoft Excel. The model equations were for the operating conditions shown in Table 1, and the fluidized bed reactor was divided into ten compartments. Each compartment consists of two phases: bubble phase and emulsion phase.

Conclusions

The mathematical model based on the two-phase theory of gas phase fluidization has been developed. The results show that high gas velocities reduce the conversion of monomer per pass through the reactor. The single-pass ethylene conversion decreased with increasing bed height. The ethylene conversion increases as the catalyst feed rate and particle distribution increase. The model has been validated and has shown a good agreement

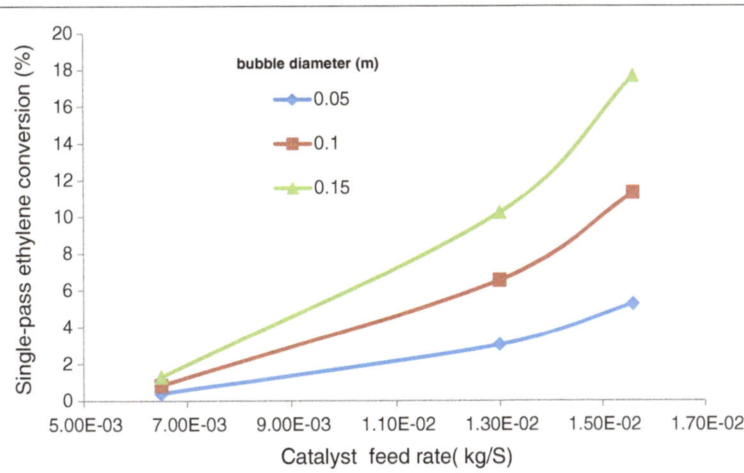

Figure 6 Effect of initial bubble (0.05 to 0.15 m) and catalyst feed rate on single pass ethylene conversion.

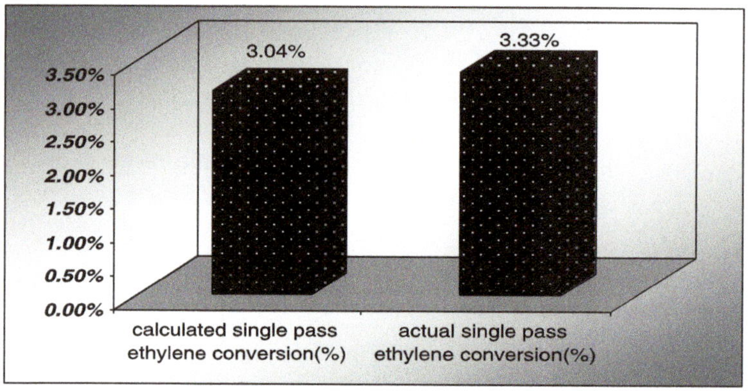

Figure 7 Comparison between actual and calculated ethylene conversions for single pass.

with the actual plant data. The model indicates that the single-pass ethylene conversion is sensitive to catalyst particle diameter, number of compartment, and bed voidage, but change in superficial gas velocity has no significant effect on the model.

Abbreviations

Ar: Archimedes number; C_{Aout}: Ethylene outlet concentration (mol m^{-3}); C_{Ain}: Ethylene inlet concentration (mol m^{-3}); C_{ei}: Monomer concentration in emulsion phase *i*th compartment (mol m^{-3}); d_p: Polymer average particle size(m); d_p^*: Dimensionless particle size; dp_{min}: Minimum particle diameter (m); dp_{max}: Maximum particle diameter (m); d_{po}: Catalyst particle size (m); d_{bo}: Bubble diameter near distributor(m); $d_{b,max}$: Maximum stable bubble diameter(m); $d_{b,i}$: Bubble diameter at (h) height *i*th compartment (m); D_A: Diffusion coefficient for ethylene (m^2s^{-2}); *D*: Reactor diameter (m); E_a: Activation energy (J mol^{-1}); F_o: Catalyst (prepolymer) feed rate (kg s^{-1}); F_1: Flow rate of polymer product (kg s^{-1}); F_2: Flow rate of solids removed by elutriation (kgs^{-1}); *g*: Gravitational acceleration(ms^{-2}); *h*: Bed axial coordinate; *H*: Bed height (m); H_{mf}: Bed height at minimum fluidizing conditions (m); h_n: Distance from the distributor to *n* compartment (m); Δh_1: Initial compartment height (m); Δh_i: Height of *i*th compartment (m); *i*th: Compartment number; K_{bi}: Coefficient of gas interchange in bubble phase *i*th compartment (s^{-1}); K_{ei}: Coefficient of gas interchange in emulsion phase *i*th compartment(s^{-1}); $K_{b,e,i}$: Coefficient of gas interchange between bubble and emulsion phases, overall mass transfer coefficient *i*th compartment (s^{-1}); $K^*(r)$: Elutriation constant as a function of *r* (kg m^{-2} s^{-1}); $K(r)$: Elutriation constant (s^{-1}); K_p: Reaction rate constant (m^3 kg_{cat}^{-1} s^{-1}); k_{po}: Pre-exponential factor (m^3 kg_{cat}^{-1} s^{-1}); M_j: Molecular weight of component *j* (kg mol^{-1}); *m*: Proportionality constant relating the bubble diameter; Re_{mf}: Reynolds number at minimum fluidizing condition; N_i: Number of bubbles in *i*th compartment; *n*: Total compartment number; P_b (*d*): particle size distribution in reactor as a function of *r* (m^{-1}); $P_o(r)$: Particle size distribution of feed catalyst (m^{-1}); r_{pi}: Reaction rate in *i*th compartment (kg s^{-1}); *R*: Gas constant (J mol^{-1} K^{-1}); R_{pi}: Reaction rate for whole solid phase *i*th compartment (kgs – ^{1}); r_{cat}: Catalyst particle radius(m); *r*: Polymer particle radius (m); *S*: Cross-sectional area ($m^2 S = \pi D^2/4$); *T*: Temperature (K); $U_{br,i}$: Velocity *i*th compartment (ms^{-1}); U_t: Terminal velocity of falling particles (m s^{-1}); U_t^*: Dimensionless terminal falling velocity coefficient; $U_t(d_p)$: Terminal velocity of falling particles as function of particle diameter(s^{-1}); U_{bi}: Velocity of bubble rising up through the bed *i*th compartment (m s^{-1}); U_{mf}: Minimum fluidizing velocity (m s^{-1}); U_g: Fluidizing gas velocity (m s^{-1}); V_{bi}: Volume of bubble phase *i*th compartment (m^3); V_{ei}: Volume of emulsion phase *i*th compartment (m^3); *W*: Weight of solids in the bed (kg); X_A: Ethylene conversion; χ_{cat}: Catalyst mass fraction in polymer particle; y_j: Mole fraction of component j.

Greek letters

ρ_p: Polymer density (kg m^{-3}); ρ_g: Gas density (kg m^{-3}); ρ_{cat}: Catalyst density (kg m^{-3}); ρ_j: Density of component j (kg m^{-3}); R (*dp*): Rate of increase in particle diameter (ms^{-1}); μ_g: Gas viscosity (kg m^{-1} s^{-1}); μ_j: Viscosity of component j (kg m^{-1} s^{-1}); ε_i: Bed voidage of the *i*th compartment; ε: Bed voidage; ε_{mf}: Bed voidage at minimum fluidizing conditions; δi: Bubble fraction for *i*th compartment; Φ_s: Sphericity for sphere particles.

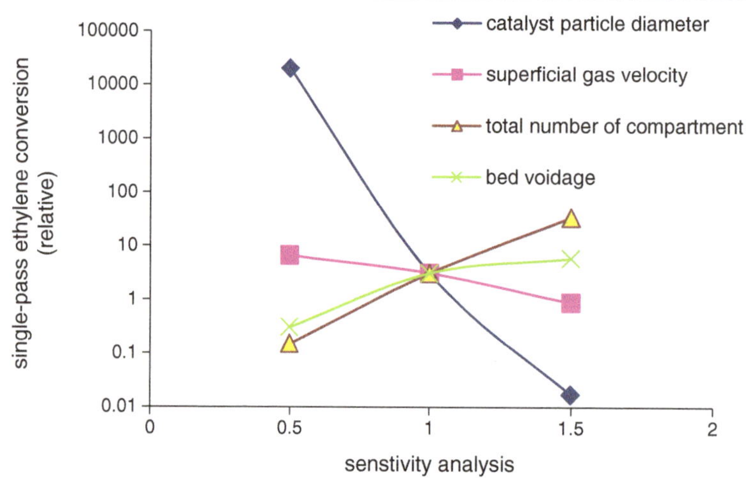

Figure 8 Sensitivity analysis for ethylene conversion.

Competing interest
The authors declare that they have no competing interests.

Authors' contributions
HF and MM drafted the manuscript. MO conceived the study and participated in its design and coordination. YF participated in the design of the study and performed the statistical analysis. All authors read and approved the final manuscript.

Acknowledgment
The authors gratefully acknowledge the faithful support of Eng. Ihab Yahia Abdel-Kader Ahmed (Marketing Research Department Head- SidiKerir Petrochemical Company) for his efforts.

Author details
[1]Chemical Engineering Department, Faculty of Engineering, Alexandria University, Alexandria, Egypt. [2]Petrochemical Engineering Department, Faculty of Engineering, Pharos University, Alexandria, Egypt. [3]City for Scientific Research CSAT, Borg Elarab, Alexandria, Egypt.

References
1. University of Michigan (2008) Diffusion and reaction in porous catalysts. http://www.engin.umich.edu/~cre/12chap/html/12prof2a.htm. Accessed 16 January 2012
2. Fernandes F, Lona L (2003) The fluidized bed reactor with a prepolymerization system and its influence on polymer physicochemical characteristic. Braz J Chem Eng 20(2):171
3. Davidson J, Harrison D (1963) Fluidized particles. Cambridge University Press, England
4. Choi KY, Ray WH (1985) The dynamic behaviour of fluidized bed reactors for solid catalyzed gas phase olefin polymerization. Chem Eng Sci 40(12):2261–2279
5. McAuley KB, Talbot JP, Harris TJ (1994) A comparison of two-phase and well-mixed models for fluidized-bed polyethylene reactors. Chem Eng Sci 49(13):2035–2045
6. Hatzantonis H, Goulas A, Kiparissides C (1998) A comprehensive model for the prediction of particle-size distribution in catalyzed olefin polymerization fluidized-bed reactors. Chem Eng Sci 53(18):3251–3267
7. Grosso WE, Chiovetta MG (2005) Modeling a fluidized-bed reactor for the catalytic polymerization of ethylene: particle size distribution effect. Universidad Nacional Del Litoral, Santa Fe, Argentina
8. Kiashemshaki A, Mostoufi N, Sotudeh R (2006) Two-phase modeling of a gas phase polyethylene fluidized bed reactor. Chem Eng Sci 61:399
9. Vahidi O, Shahrokhi M (2008) Control of a fluidized bed polyethylene reactor. Iran J Chem Chem Engn 27(3):87–101
10. Hamzehei M (2011) Study of heat transfer in the poly ethylene fluidized bed reactor numerically and experimentally. WASET 78:522–530
11. Kunii D, Levenspiel O (1991) Fluidization engineering, 2nd edition. Reed Publishing Inc., USA
12. Reid RC, Pransuitz JM, Poling BE (1987) The properties of gases and liquids. McGraw-Hill, New York
13. Wen CY, Fan LT (1975) Models for flow system and chemical reactors. Marcel Dekker, New York
14. Mori S, Wen CY (1975) Estimation of bubble diameter in gaseous fluidized beds. AIChE 21:109–115
15. Cui HP, Moustoufi N, Chaouki J (2000) Characterization of dynamic gas-solid distribution in the fluidized beds. Chem Eng J 79:135–143
16. Kobayashi H, Arai F, Sunagawa T (1967) Chem Eng Japan 31:239
17. Grace JR (1986) Fluid beds as chemical reactors. In: Geldart D (ed) Gas fluidization technology. Wiley, Chichester
18. Haiderand A, Levenspiel O (1989) Drag coefficient and terminal velocity of spherical and nonspherical particles. Powder Technol 58:63–70
19. Specchia V, Baldi G, Sicardi S (1980) Heat transfer in packed bed reactors with one phase flow. Chem Eng Commun 4:361–380
20. Kunii D, Levenspiel O (1969) Fluidization engineering. Wiley, New York
21. Geldart D (1986) Gas fluidization technology. Wiley, Chichester
22. Tanaka T (1967) MSc Thesis. Hokkaido University, Japan
23. Yong KJ, Choi Y, Kyu (2001) Modeling of particle segregation phenomena in a gas phase fluidized bed olefin polymerization reactor. Chem Eng Sci 56:4069–4083
24. BP Chemicals (1996) The Innovene Process- Chromium catalyst technology manual-(Revision 4).:
25. SidiKerir Petrochemicals, Co. (SIDPEC) (2012) DCS PI system. http://www.sidpec.com/PageDetails.aspx?MenuID=3. Accessed 3 January 2013
26. Wagner BE, Goeke GL, Karol FJ (1981) Process for the preparation of high density ethylene polymers in fluid bed reactor. US Patent 4303771:

Removal of anionic dye from aqueous solution using a cationic carrier

Govindaraju Muthuraman* and Mohamedjakkaria Mohamed Ibrahim

Abstract

Background: The transport of Cibacron Red FN-R (Sigma-Aldrich, MO, USA) anionic dye through a methylene chloride (Merck, India) bulk liquid membrane containing a tetrabutylammonium bromide (TBAB; Merck, India) as an excellent carrier was studied. The extracted dye simultaneously stripped into aqueous stripping solution.

Results: The basic fundamental parameters were determined. These include pH of the feed, nitric acid concentration in the strip solution, TBAB concentration in the membrane phase, initial dye concentration in the feed phase and rate of stirring. The transport efficiency of the dye increased with increasing carrier concentration from $6.06 \pm 0.1 \times 10^{-6}$ to $3.85 \pm 0.1 \times 10^{-4}$ M. The transport study was highly selective for anionic dye, and the efficiency of transport was not affected by the presence of electrolytes such as sodium chloride and sodium sulphates in feed solution.

Conclusions: The percentage of dye transported across the liquid membrane after 70 min was 95 ± 0.1%. When the optimum parameters were used for real textile dyeing effluent for recovery of dyes, the results were found to be satisfactory.

Keywords: Transport, Anionic dye, Tetrabutylammonium bromide, Bulk liquid membrane, Recovery, Efficiency

Background

Among different pollutants of aquatic ecosystem, dyes are of a large and important group of chemicals. They are widely used in industries such as textile, paper, rubber, plastic, cosmetic etc., to colour the products. These dyes are invariably left as industrial waste and consequently discharged mostly to surface water resources. Dyes, even in low concentration, are visually detected and can affect the aquatic life and food web. These coloured compounds are not only aesthetically displeasing but also inhibiting sunlight into the stream, thus affecting the photosynthetic reaction [1]. The removal of colour synthetic organic dyestuff from waste effluents becomes environmentally important. It is rather difficult to treat dye effluents because of their synthetic origins and their mainly aromatic structures, which are biologically non-degradable [2]. Therefore, the decolorization of wastewaters containing these dyes prior to discharge is mandatory by environmental regulations in most countries [3].

* Correspondence: raman.gm@gmail.com
Department of Chemistry, Presidency College, University of Madras, Chennai, India

There are several methods for dye removal such as biological treatment [4-6] coagulation/flocculation [4,7], chemical oxidation and photocatalytic processes [8,9] ozone treatment [10], membrane processes [11-14], adsorption [15-20] and liquid-liquid extraction [21-25]. Among these, liquid membrane is an alternative method over conventional treatment methods for industrial wastewater. This method is low cost, simple, with high efficiency and energy saving in comparison to other processes. In the recent years, a remarkable increase of the application of liquid membranes in separation processes has been observed. These membranes include bulk liquid membrane (BLM), emulsion liquid membrane and supported liquid membrane [26,27]. Transport of textile anionic dyes using cationic carrier by bulk liquid membrane has been studied and found that tetrabutylammonium bromide (TBAB) was an excellent carrier for selective and efficient transport of dyes. At a stirring speed of 250 rpm, 97% dye has been transported from feed to strip within 70 min. The percentage recovery of the dye was not found to decrease in the presence of anions (chloride and sulphates) in the feed solution [28].

Urtiaga et al. [29] investigated the effect of phenolic concentration on the partition coefficient in Cyanex 923 kerosene/water system (Cytec Industries, Inc., Woodland Park, USA) at six different concentrations. They reported that the equilibrium concentration of phenol had a strong effect on the partition coefficient for phenol between trialkylphosphine oxides (Cyanex 923) in kerosene and water. Ma et al. [30] reported the transport of amino acid through di-(2-ethylhexyl) phosphoric acid (D_2EHPA) as carrier in kerosene. The liquid membrane system exhibited excellent transport selectivity for amino acids containing a side chain of high lipophilicity constant, like L-phenylamine, L-tryptophan, L-leucine and L-isoleucine. As a target amino acid with bulky aromatic side chain of L-phenylamine, the influence of its concentration and the concentration of D_2EHPA on the interfacial tension and the effect of temperature on its transport through the BLM were investigated. The pseudo-first-order apparent rate constants of the interfacial transport of L-phenylalanine for the extraction reaction and stripping reaction were estimated.

A kinetics of L-isoleucine transport through a liquid membrane containing D_2EHPA in kerosene was also studied [31]. The influence of pH in the aqueous feed solution, D_2EHPA concentration in the organic phase, the stripping solution composition and H_2SO_4 concentration in the stripping solution were investigated, and the effects of stirring speed and temperature on the transport of L-isoleucine through a BLM were studied. The pseudo-first-order apparent rate constants ($k1$, $k - 1$, $k2$) and the apparent activation energy values of the interfacial transport of extraction, back extraction and stripping have been determined.

In the present work, transport of Cibacron Red FN-R through a methylene chloride bulk liquid membrane containing a TBAB as a carrier (Figure 1) was studied. Different experimental conditions, such as the effect of TBAB concentration in the membrane, the effect of stripping agent, the effect of pH in feed phase, the rate of stirring and the effect of time were also investigated.

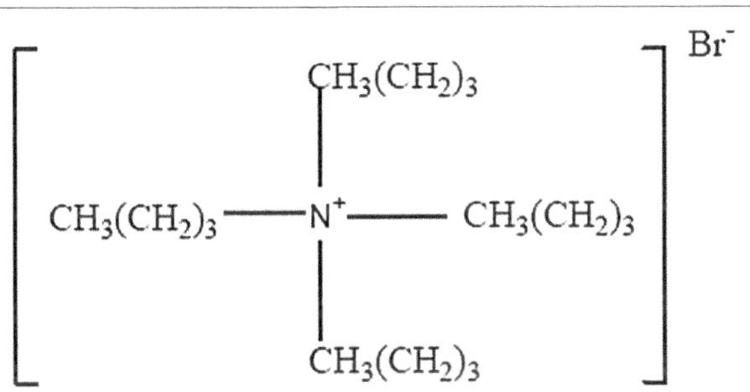

Figure 1 Chemical structure of TBAB as the carrier used to extract dye.

Methods

Liquid-liquid extraction

Solvent extraction experiments were undertaken at 27 ± 0.5°C. Diluted H_2SO_4 and NaOH were used to adjust the pH of the aqueous solutions. A feed phase containing dye (100 mg L^{-1}) and organic phase were introduced in a separating funnel. The phases were mixed gently for known time and then left to separate. The raffinate was collected for the measurement of the remaining dyes in the solution. The distribution ratio and the percentage of extraction were calculated.

Bulk liquid membrane

The BLM apparatus (U-type) used for the study is shown in Figure 2. The inner dimension of the transport cell was 70 mm diameter × 250 mm depth. Experiments were carried out at ambient temperature (27 ± 0.5°C), and the aqueous source (feed) phase (100 mL) dye and strip phase (100 mL) nitric acid were taken in the BLM apparatus. The dichloromethane solution (50 mL) contained TBAB. The organic layer was stirred by Teflon coated magnetic bar (2.0 cm) (DuPont, Haryana, India), and both aqueous layers (feed and strip) were stirred using the stirrer. In transport, experiment samples were taken out from the feed and strip phases, and the dye concentration was measured using spectrophotometer (λ_{max} = 542.5 nm) with good reproducibility.

Kinetic procedure

The variation of the dye concentration with time was directly determined on both donor (feed) (C_d in M) and acceptor (strip) (C_a in M) phases using spectrophotometer at regular intervals for 70 min. The corresponding change of the dye concentration in the membrane phase was determined from the material balance between the

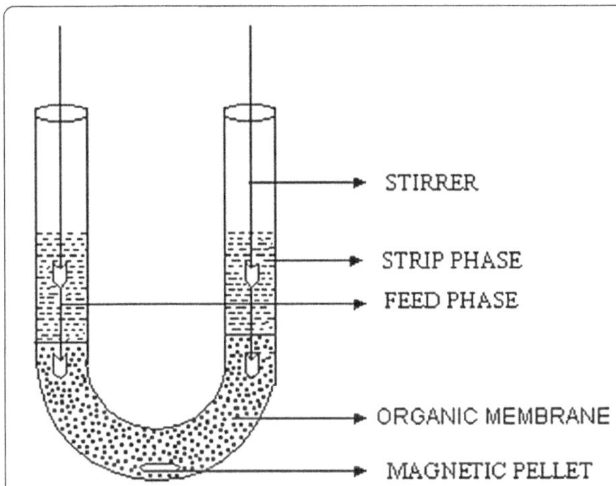

Figure 2 Schematic experimental setup for bulk liquid membrane for transport of dye from aqueous solution.

phases. For practical reasons, dimensionless reduced concentration of dyes in the donor (R_d), membrane (R_m) and acceptor (R_a) were used [32]:

$$R_d = C_d/C_{do},\ R_m = C_m/C_{do},\ R_m = C_a/C_{do}, \quad (1)$$

where C_{do} is the initial dye concentration in the donor phase, while C_d, C_m and C_a represent the dye concentrations in donor, membrane and acceptor phases, respectively. From this expression, the material balance can be established as $R_d + R_m + R_a = 1$. When R_d, R_m and R_a values are inspected, the results suggest that the dye transport obeys the kinetics of the two consecutive irreversible first-order reactions according to the kinetics scheme [33]:

$$C_d \xrightarrow{k_1} C_m \xrightarrow{k_2} \equiv C_a, \quad (2)$$

where k_1 and k_2 are pseudo first-order apparent rate constants of the extraction and the back extraction, respectively. The changes of carrier concentration and other factors affect the dye concentration, donor, membrane and acceptor phases. As a result of this effect, k_1 and k_2 values were changed. The kinetic scheme, Equation (2), for consecutive irreversible reactions can be described by the following rate equations [34]:

$$dR_d/dt = k_1R_d \equiv J_d \quad (3)$$

$$dR/dt = -kR - kR \quad (4)$$

$$dR_a/dt = k_2R_m \equiv J_a, \quad (5)$$

where J stands for flux. When $k_1 \neq k_2$, the integrating Equations (3) to (5) give:

$$R_d = \exp(-k_1t) \quad (6)$$

$$R_m = k_1/k_2 - k_1[\exp(-k_1t) - \exp(-k_2t)] \quad (7)$$

$$R_a = 1 - 1/k_2 - k_1[k_2\exp(-k_1t) - k_1\exp(-k_2t)]. \quad (8)$$

It can be seen that R_d versus t yields a decreasing mono-exponential curve, whereas the time variation of both R_m and R_a was bi-exponential. The actual numerical analysis was carried out by non-linear regression. The variation of R_d, R_m and R_a with time through the liquid membrane is shown in Figure 3.

The maximum value of R_m (when $dR_m/dt = 0$) and the corresponding t_{max} can be written as follows:

$$R_m^{max} = (k_1/k_2)^{-k_2/(k_1-k_2)} \quad (9)$$

$$t_{max} = \ln(k_1/k_2)(1/k_1 - k_2) \quad (10)$$

By considering the first-order time differentiation of Equations (6) to (8) at $t = t_{max}$, one obtains the following equations:

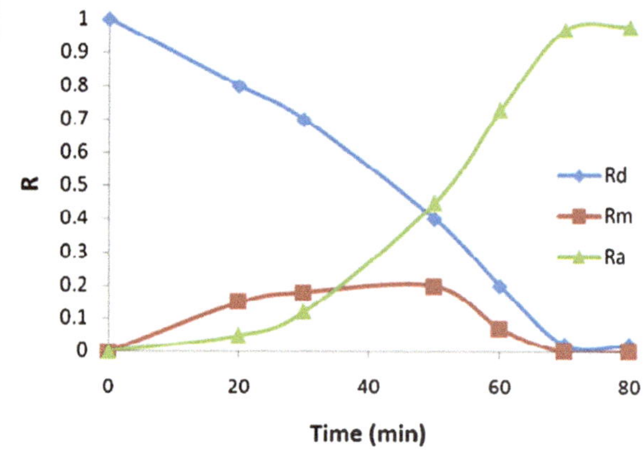

Figure 3 Time dependence of R_d, R_m *and* R_a for transport of dye theoretical curves calculated from Equations (6) to (8). Experimental conditions: donor phase: 1.06×10^{-5} mol L^{-1} dye at pH 2.0 ± 0.1; acceptor phase: 1.0 N HNO_3 solution; membrane phase: 1.55×10^{-3} mol L^{-1} carrier in CH_2Cl_2; stirring speed is 400 rpm.

$$[dR_d/dt]_{max} = k_1(k_1/k_2)^{-k_1/(k_1-k_2)} \equiv J_d^{max} \quad (11)$$

$$[dR_m/dt]_{max} = 0 \quad (12)$$

$$[dR_a/dt] = k_2(k_1/k_2)^{-k_2/(k_1-k_2)} \equiv J_a^{max}. \quad (13)$$

The system is at steady state, so the dye concentration in the membrane ($t = t_{max}$) does not vary with time (Equation (12)).

Results and discussion

Liquid-liquid extraction experiments

Effect of TBAB concentration in the organic phase and pH of the feed phase

The effect of TBAB concentration on the percentage extraction of the dye was investigated in the range of 6.06×10^{-6} to 3.10×10^{-3} mol L^{-1} of TBAB concentration (Table 1). The concentration of the dye used in the part of the study was 100 mg L^{-1}. The extraction efficiency of the dye increases at lower pH, then decreases at higher pH. There was a significant increase in the percentage of extraction when the concentration of TBAB was between 4.85×10^{-5} and 1.55×10^{-3} mol L^{-1}. At 1.55×10^{-3} mol L^{-1} TBAB concentration, the percentage of the extraction was 95.1%. Beyond this concentration, the percentage of the extraction increased very slowly. Sarkar and Dhadke [35] also found that increasing the carrier concentration had no considerable effect on the transport of the bismuth. Hence, the subsequent extraction studies were conducted using 1.55×10^{-3} mol L^{-1} of extractant.

Table 1 Effect of carrier concentration and pH of cibacron FN-R

Sample number	TBAB concentration (mol L^{-1})	pH	Percentage of dye extraction
1	6.06×10^{-6}	2.0	65.1
2	6.06×10^{-6}	4.0	70.1
3	6.06×10^{-6}	6.0	73.1
4	6.06×10^{-6}	10.0	63.1
5	6.06×10^{-6}	12.0	50.1
6	4.85×10^{-5}	2.0	85.1.
7	4.85×10^{-5}	4.0	85.1
8	4.85×10^{-5}	6.0	75.1
9	4.85×10^{-5}	10.0	66.1
10	4.85×10^{-5}	12.0	52.1
11	1.55×10^{-3}	2.0	95.1
12	1.55×10^{-3}	4.0	95.1
13	1.55×10^{-3}	6.0	94.1
14	1.55×10^{-3}	10.0	60.1
15	1.55×10^{-3}	12.0	45.1
16	3.10×10^{-3}	2.0	99.1
17	3.10×10^{-3}	4.0	99.1
18	3.10×10^{-3}	6.0	98.1
19	3.10×10^{-3}	10.0	60.1
20	3.10×10^{-3}	12.0	45.1

Effect of stripping solution

In any extraction process, it is very imperative to back extract the extracted dye from the organic phase and to allow recycling of the organic phase without loss of efficiency. Various inorganic acids such as HCl, HNO_3 and H_2SO_4 were used as stripping agents in this study. Nitric acid stripped the dye very well from the organic phase (Table 2) compared to the other two inorganic acids. Hence, 1N HNO_3 was found to be suitable for stripping the extracted dye. It was found that the maximum amount of dye (96.2%) was stripped within 10 min. Further increase in time did not improve the stripping efficiency.

Effect of stripping reagent concentration

The effect of HNO_3 concentration on the stripping efficiency was studied for the range of 0.1 to 1N, and the result of the percentage stripping is shown in Figure 4. The percentage of stripping was 96% with 1N HNO_3 and 88.97% with 0.6 N HNO_3. The result shows that the higher the concentration of HNO_3, the higher will be the percentage of stripping. This was due to the higher amount of H^+ ion responsible in carrying the anionic dye complex from the organic phase to the aqueous phase. Thus, HNO_3 of 1N concentration was chosen for subsequent studies.

Table 2 Effect of stripping agent concentration

Sample number	1N acid	Percentage of dye stripped		
		1.06×10^{-5} mol L^{-1}	2.12×10^{-4} mol L^{-1}	5.30×10^{-4} mol L^{-1}
1	HCl	90.5	89.2	87.6
1	H_2SO_4	92.1	92.0	90.2
3	HNO_3	96.2	95.2	94.5

Bulk liquid membrane experiments

Effect of stirring speed

The transport experiments were performed at six different stirring speeds, 50, 100, 200, 300, 400 and 500 rpm. It indicates that the stirring speed affects the transport rates of dye through the liquid membrane. Figure 5 shows the flux values at different stirring speeds, and the flux increases almost linearly with the increase of stirring speed below 400 rpm. It also reveals that diffusion was the rate-limiting step in the feed-membrane transport of dye from the feed solution to the membrane phase. The flux increases at lower stirring speeds (50 to 300 rpm), and then is nearly constant at higher stirring speeds (400 to 500 rpm). Yang and Fane [36] also found similar effect at higher stirring speeds. A 'blank' experiment was performed in which the membrane contained no carrier. No detectable transport of dye through the liquid membrane was found, suggesting that the transport of dye through the liquid membrane was fulfilled by the carrier.

Effect of temperature

Figure 6 presents the percentage of the dye stripped from the feed phase to the acceptor/receiving phase (RP) as a function of temperature. It is obvious that the

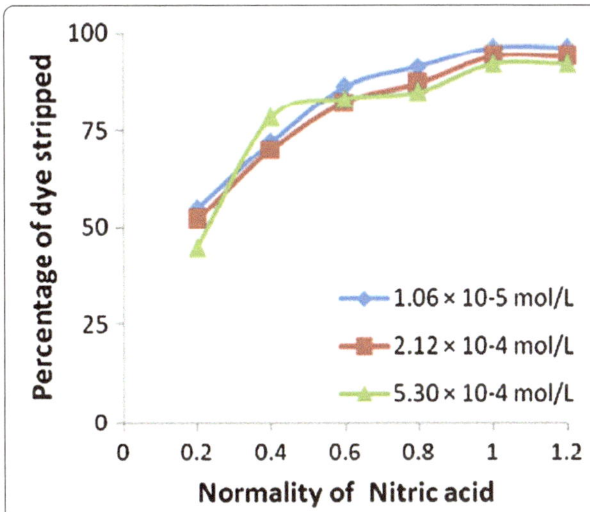

Figure 4 Effect of stripping reagent concentration. Experimental conditions: feed: different concentrations of dye solution; strip: different normality of HNO_3.

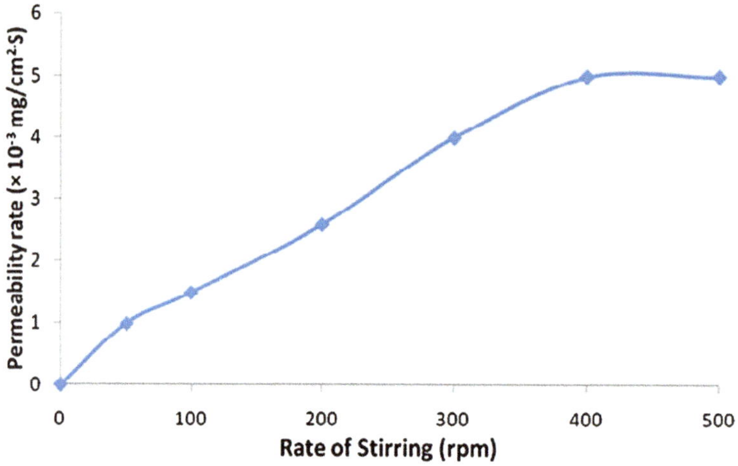

Figure 5 Effect of stirring. Experimental conditions: donor phase 1.06×10^{-5} mol L^{-1} dye at pH 2.0 ± 0.1; acceptor phase: 1.0 N HNO_3 solution; membrane phase: 1.55×10^{-3} mol L^{-1} carrier in CH_2Cl_2; transport time is 70 min.

efficiency of the uptake from the source phase (feed) as well as the release of dye into the receiving (strip) solution enhances with temperature in the range of 10°C to 30°C. The transport was stripping-controlled, so the effects may not be pronounced especially at higher temperature. The results can be interpreted by considering several effects. Among them is the influence of the temperature on the complexation reaction with other ions by the extractant carrier at the donor/membrane interface and on the decomplexation of the dye at the second interface (i.e. membrane/acceptor phase). The effect of temperature on the diffusion of the transported species through liquid membrane, due to the temperature dependency of the membrane viscosity, is another effect of temperature on the transport process. However, the dependency of the distribution of the carrier on temperature should also be taken into account in describing the experimental results.

Effect of time

The time dependence of dye transport through the liquid membrane under the optimum experimental condition was studied. Figure 7 shows a rapid rise in dye concentration in the receiving phase as well as a sharp decrease in the dye concentration in the source phase (SP) during the first 70 min of transport. After 70 min, 95% of the dye was transported into the RP, 3% remained in SP, and about 2% of the dye remained in the organic phase. After 80 min, no significant transport of the dye was noticed. Hence, the time 70 min was recommended.

UV analysis

The UV absorption spectrum of Cibacron Red FN-R in aqueous and organic solutions exhibits an absorption maximum at 585 and 590 nm. The absorption maximum did not change in the stripped dye. This means that

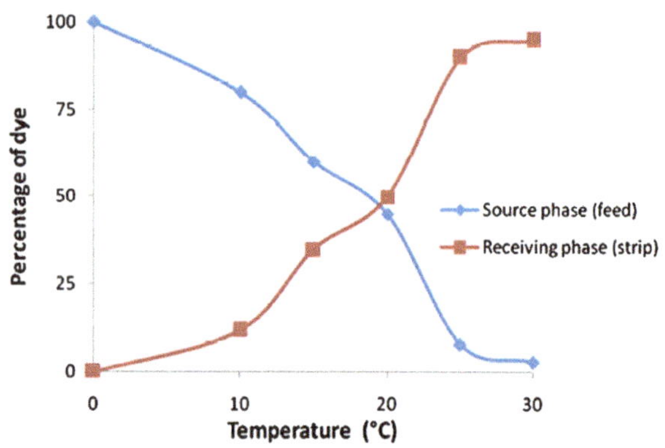

Figure 6 Effect of temperature. Experimental conditions: donor phase 1.06×10^{-5} mol L^{-1} dye at pH 2.0 ± 0.1; acceptor phase: 1.0 N HNO_3 solution;, membrane phase: 1.55×10^{-3} mol L^{-1} carrier in CH_2Cl_2; transport time is 70 min.

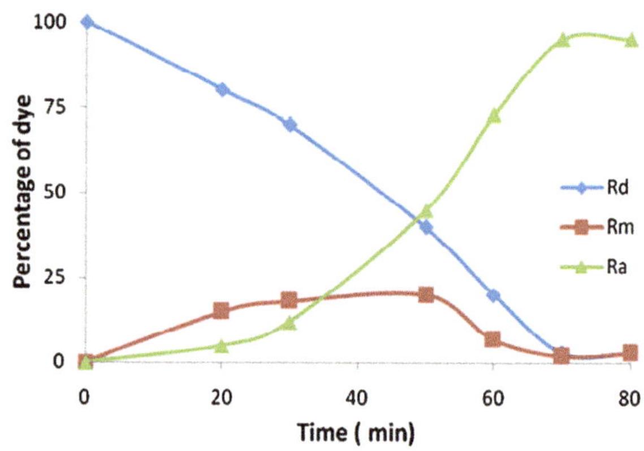

Figure 7 Effect of time. Experimental conditions: donor phase 1.06 $\times 10^{-5}$ mol L^{-1} dye at pH 2.0 ± 0.1; acceptor phase 1.0 N HNO_3 solution; membrane phase 1.55×10^{-3} mol L^{-1} carrier in CH_2Cl_2.

Table 3 Effect of kinetic parameters for transport of dye at different carrier concentration

Sample number	Concentration ($mol\ L^{-1}$)	k_1 (min^{-1})	k_2 (min^{-1})	t_{max} (min)	R_m^{max} (mL)	J_d^{max} (min^{-1})	J_a^{max} (min^{-1})
1	6.06×10^{-6}	0.82×10^3	0.81×10^3	1,852.11	0.32	-0.31×10^3	0.30×10^3
2	4.85×10^{-5}	1.64×10^3	1.52×10^3	502.40	0.35	-0.51×10^3	0.56×10^3
3	3.85×10^{-4}	3.21×10^3	2.91×10^3	208.10	0.37	$-1.1. \times 10^3$	$1.1. \times 10^3$
4	1.55×10^{-3}	3.7×10^3	3.4×10^3	157.54	0.38	-1.30×10^3	1.20×10^3

none of the functional groups were changed at the time of stripping. An enhancement in the absorbance with bathochromic spectral shift was observed in the organic layer. This might be due to the influence of methylene chloride in the organic layer. This spectral study shows that the recovered dye can be reused.

Kinetic parameters

The kinetic parameters for transport of dye are presented in Table 3. As it was seen, the transport rate, k_1, k_2 and R_m^{max} increase, and t_{max} decreases with the increase in carrier concentration. The k_2 values increase with the increase in carrier concentration in the organic phase from 6.06×10^{-6}, 4.85×10^{-5} and 3.85×10^{-4} mol L^{-1}. From the results, 3.85×10^{-4} mol L^{-1} carrier concentration was employed as the optimum carrier concentration.

Application of the developed BLM for textile wastewater

The wastewater from a local textile industry was selected for testing the applicability of the developed bulk liquid membrane system, which was alkaline in nature (pH 12.5 ± 0.1). Under optimised condition (feed phase 100 mL of 1.06×10^{-5} mol L^{-1}, 50 mL of 1.55×10^{-3} mol L^{-1} carrier, acceptor phase 100 mL of 1.0 N HNO_3, equilibration time = 70 min), the textile dye waste was extracted, and the extracted dye was stripped into nitric acid. It was noticed that the extraction and stripping were not affected by the various types of salts present in the textile wastewater.

Experimental

Reagents and instruments

The following inorganic salts, acids and organic solvents were used in the experiments without further purification: NaCl (99.8%), NaOH (>96.0%), tetrabutylammonium bromide (99.5%), Cibacron Red (99.8%), HNO_3 (65%) and dichloromethane (99.5%). The chemicals were obtained from Merck Sharp & Dohme Corp. (Whitehouse Station, USA) and *Thermo Fisher Scientific, Inc.* (Waltham, USA) suppliers. A UV-visible spectrophotometer (Spekol 1200, Analytical Jena, Germany) was used to measure the absorbance of the dye and its concentration; the pH of the aqueous solution was measured by a pH metre (WTW Wissenschaftlich-Technische Werkstätten GmbH, Weilheim, Germany). For agitation of solution, stirrer was used (IKD-KS 50, Germany). Deionized water was used for preparing all the aqueous solutions.

Conclusion

Cibacron Red FN-R was efficiently transported by means of a liquid membrane containing TBAB-CH_2Cl_2. The efficiency of the method depends on the important parameters such as the pH of the feed, dye concentration, carrier concentration, stirring speed and temperature. Flux increases almost linearly with the increase of stirring speed. After 70 min, there was no significant transport of dye noticed. In optimised conditions, the transport efficiency of Cibacron Red FN-R was after 70 min with 95.1 ± 1.1%. The proposed method was applied to real textile effluent and found that the transport efficiency after 70 min was 95%.

Competing interest

The authors declare that they have no competing interests.

Authors' contributions

GM carried out the bulk liquid membrane studies, and MMI participated in the liquid-liquid extraction study. All authors read and approved the final manuscript.

Authors' information

Authors are working in the Department of Chemistry, Presidency College, Chennai, India. Currently, they are doing research in the field of turbidity removal from drinking water using natural coagulants.

Acknowledgements

Authors thank the UGC, New Delhi, India, for supporting this work.

References

1. Malik PK (2003) Dyes Pigments 56:239
2. Dogan M, Ozdemir Y, Alkan M (2007) Dyes Pigments 75:701
3. Chen CH, Chang CF, Ho CH, Tsai TL, Liu SM (2008) Chemosphere 72:1712
4. Allegre C, Moulin P, Maisseu M, Charbit F (2006) J Membr Sci 269:15
5. Pala A, Tokat E (2002) Water Res 36:2920
6. Kapdan IK, Kargi F (2002) Process Biochem 37:973
7. Cooper P (1993) J Soc Dyers Colourists 109:97
8. Liu HL, Chiou YR (2006) J Chin Inst Chem Eng 37:289
9. Clark T, Bruce M, Anderson S (1994) Water Sci Technol 29:421
10. Solozhenko EG, Soboleva NM, Goncharuk W (1995) Water Res 29:2206
11. Buckley CA (1992) Water Sci Technol 25:203
12. Jiraratananon R, Sungpet A, Luangsowan P (2000) Desalination 130:177
13. Koyuncu I (2002) Desalination 143:243
14. Treffry-Goatley K, Buckley CA, Groves GR (1983) Desalination 47:313
15. Karcher S, Kornmuller A, Jekel M (2001) Dyes Pigments 51:111
16. Netpradit S, Thiravetyan P, Towprayoon S (2003) Water Res 37:763

17. Gupta VK, Srivastava SK, Mohan D (1997) Ind Eng Chem Res 36:2207
18. Gupta VK, Suhas JS, Ali I, Saini VK (2004) Ind Eng Chem Res 43:1740
19. Gupta VK, Ali I, Saini VK, Van Gerven T, Van der Bruggen B, Vandecasteele C (2005) Ind Eng Chem Res 44:3655
20. Gupta VK, Mittal A, Gajbe V (2006) J Ind Eng Chem Res 45:1446
21. Muthuraman G, Teng TT (2009) Pro Natu Sci 19:1215
22. Muthuraman G, Teng TT (2009) Ind J Eng Chem 15:841
23. Muthuraman G, Teng TT, Leh CP, Norli I (2009) J Hazard Mater 163:363
24. Muthuraman G, Palanivelu K (2005) Dyes Pigments 64:251
25. Muthuraman G, Palanivelu K (2004) Ind J Chem Technol 11:166
26. Ca K, Walkowiak W (2005) J Membr Sci 266:143
27. Yamini Y, Chaloosi M, Ebrahimzadeh H (2002) Sep Purif Technol 28:43
28. Muthuraman G, Palanivelu K (2005) J Sci Ind Res 64:529
29. Urtiaga A, Ortiz M, Inmaculada N (1997) Sep Sci Technol 32:1157
30. Ma M, Chen B, Luo X, Tan H, He D, Xie Q, Yao S (2004) J Mem Sci 234:101
31. Ma M, Chen B, Luo X, Tan H, He D, Xie Q, Yao S (2002) Ana Chim Acta 456:157
32. Safi AO, Alpaydin S, Sirit A (2006) J Membr Sci 283:448
33. He D, Ma M, Zhao Z (2000) J Membr Sci 169:53
34. Szpakowska M, Naggy OB (1991) J Membr Sci 64:129
35. Sarkar SG, Dhadke PM (1999) Sep Purif Technol 15:131
36. Yang XJ, Fane Sep AG (1999) Puri Technol 34:1873

27

Synthesis and characterization of emulsion polymers using isopropenyl acetate

Nilesh Shinde, Mukesh Kathalewar and Anagha Sabnis*

Abstract

Background: Isopropenyl acetate (IPA) is one of the monomers from acetate family having an unsaturation with a methyl group attached to the ethylenic carbon. The unsaturation present in the monomer can be utilized for free radical polymerization. Irrespective of this, IPA is not studied much in emulsion polymer synthesis. In the current study, we report the preparation of emulsion polymers using IPA and other acrylic monomers. Terpolymers of butyl acrylate (BA), methyl methacrylate, and IPA were prepared via seeded emulsion polymerization with varying IPA proportions on molar basis. Potassium persulfate was employed as an initiator along with Dowfax 2A1 as surfactant (anionic).

Results: The emulsions prepared thereof were characterized for particle size, zeta potential and molecular weight using gel permeation chromatography. The emulsion films were evaluated using Fourier transform infrared spectroscopy, as well as their thermal (differential scanning calorimetry, thermo gravimetric analysis) and mechanical properties to study the effect of isopropenyl acetate on their overall properties.

Conclusions: With increasing proportion of IPA in the formulation, lower conversion was observed which could be due to the lower reactivity of IPA for radical polymerization in the presence of these comonomers. More study for the utilization of IPA in polymer synthesis is needed which, on successful optimization, would give a cost-effective alternative to various commercial acrylic monomers.

Keywords: Isopropenyl acetate, Seeded emulsion polymerization, Terpolymer

Background

Growing awareness and environmental legislations are imposing restrictions on the emission of volatile organic compounds (VOCs). These restrictions have forced the coating manufacturer to formulate coatings comprising low or zero VOCs content. Emulsion polymerization is an important polymerization technique, as it yields high molecular weight polymers with low viscosity. Moreover, polymers can be tailor-made in order to exhibit desired composition and morphology [1-4].

Isopropenyl acetate (IPA) is one of the monomers from acetate family which has an unsaturation with a methyl group attached to the ethylenic carbon. It can be prepared by various methods using synthetic chemicals, but the synthesis from green resources such as waste molasses is a new and attractive field of monomer preparation. This monomer is therefore an eco-friendly and cost-effective option to the acrylic monomers which are widely used. The typical structure of IPA is shown in Figure 1. The unsaturation present in the monomer structure can be utilized for free radical polymerization by various techniques.

IPA was first polymerized via peroxide catalyzed polymerization reaction in 1950 [5]. Until to date, copolymers of isopropenyl acetate with maleic anhydride by both bulk and solution polymerization technique have been reported. Its copolymers with vinyl chloride were also synthesized by Unruh et al. and Basche et al. [6,7]. Copolymer of isopropenyl acetate with indene monomer prepared by radical polymerization at high pressure was also prepared by De et al. to study the relative reactivity ratios [8]. The poly (styrene-*co*-isopropenyl acetate) copolymer prepared by free radical polymerization at 70°C was used as macroinitiator for cationic polymerization of isobutylene to synthesize graft copolymers of styrene-*co*-isopropenyl acetate-*graft*-polyisobutylene [9]. However, its slightly lower reactivity with other acrylic and

* Correspondence: as.sabnis@ictmumbai.edu.in
Department of Polymer and Surface Engineering, Institute of Chemical Technology, N M Parekh Marg, Matunga (East), Mumbai 400019, India

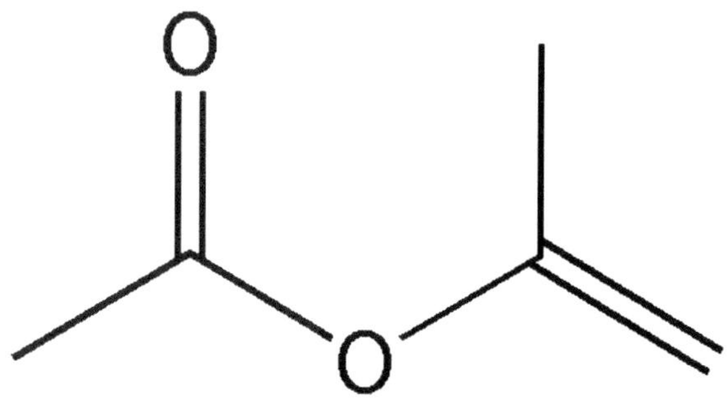

Figure 1 Structure of isopropenyl acetate.

vinyl monomers is a concern for preparing its copolymers with high conversion [10].

IPA has not been studied much via emulsion polymer technique. In the current work, we report the emulsion polymerization using IPA along with other acrylic monomers to study the feasibility of using IPA as a co-monomer for polymer synthesis.

Methods

Materials

Isopropenyl acetate was supplied by Godavari Bio refineries Ltd, Mumbai. BA and MMA were procured from SD fine Chemical Ltd. and were purified by using 10% of NaOH solution followed by washing with deionized water to remove inhibitor and were dried over sodium sulfate. Sodium carbonate, potassium persulfate (KPS), and *tert*-butyl hydroperoxide (TBHP) were lab grade chemicals purchased from S.D. Fine Chemicals Ltd (Worli Road, Mumbai, India). Dowfax 2A1 obtained from The Dow Chemical Company (Mumbai, India) was used as surfactant. Deionized water was used throughout the experiment.

Table 1 Typical formulation for emulsion BMI_{100} (with 1:1:1 mole ratio of IPA/BA/MMA)

Raw material	Quantity (gm)
BA	15.6
MMA	12.2
IPA	12.2
Dowfax 2A1	1.6
KPS	0.2
SBC	0.5
TBHP	1.0
DI water	56.7
Total	100.0

Preparation of emulsion polymers

To study the emulsion polymerization of IPA along with BA and MMA, emulsions with variable molar proportions

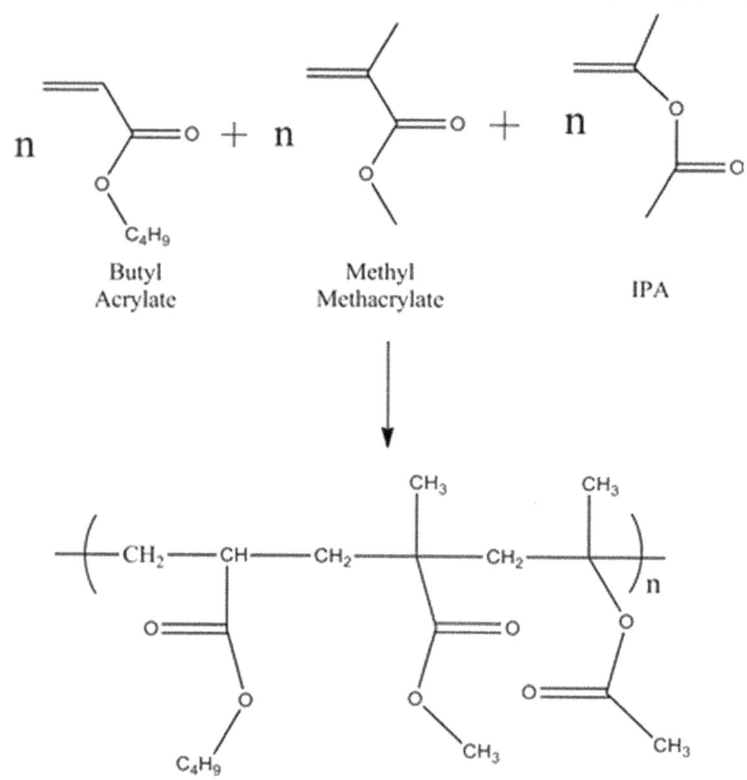

Figure 2 Structure of BA/MMA/IPA terpolymer.

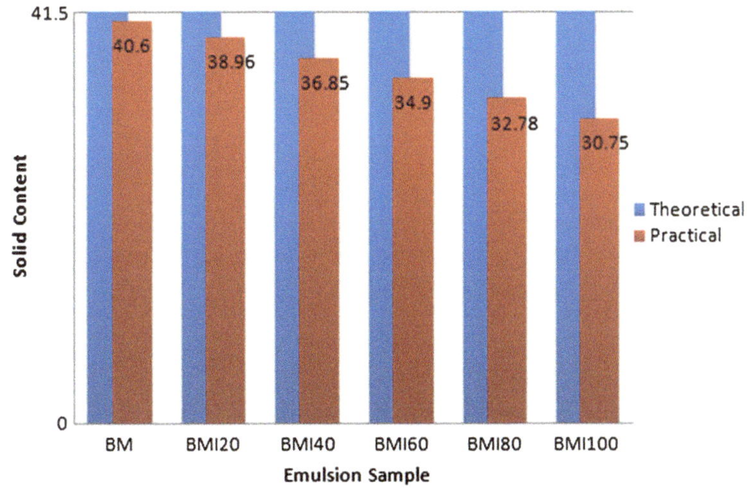

Figure 3 Solid content of respective emulsion.

of IPA were prepared. IPA content was varied from 0.2 to 1.0 mol with respect to each mole of BA and MMA. The emulsion formulations prepared also contained 0.5% radical initiator KPS (of total monomer on weight basis), anionic surfactant Dowfax 2A1 and sodium bicarbonate as a buffer. All the emulsions were formulated for 41.5% solid content. The emulsions with molar concentrations 0.2%, 0.4%, 0.6%, 0.8% and 1.0% of IPA were prepared. These are hereafter denoted as BMI_{20}, BMI_{40}, BMI_{60}, BMI_{80} and BMI_{100} respectively; while the emulsion with 0% IPA is denoted as BM. The typical formulation for emulsion BMI_{100} is summarized in Table 1. Figure 2 is as schematic representation of terpolymer synthesis.

Characterization

The percentage conversion of monomers and nonvolatile matter (NVM) of emulsion polymers were determined by gravimetric method. Malvern Zetasizer instrument (Malvern Instruments Ltd., Malvern, Worcestershire, UK) was used to measure zeta potential (ζ-potential) of emulsions. Particle size of emulsions was determined by Malvern Instruments Ltd. (Mastersizer 2000 Ver. 5.30.010). Viscosities of the emulsions were measured by using Brookfield viscometer (Antonpar, GMBH, Austria) using spindle no.1 at 20 rpm. Molecular weight determination of emulsions was done using GPC by Waters 2414-refractive index detector (Waters

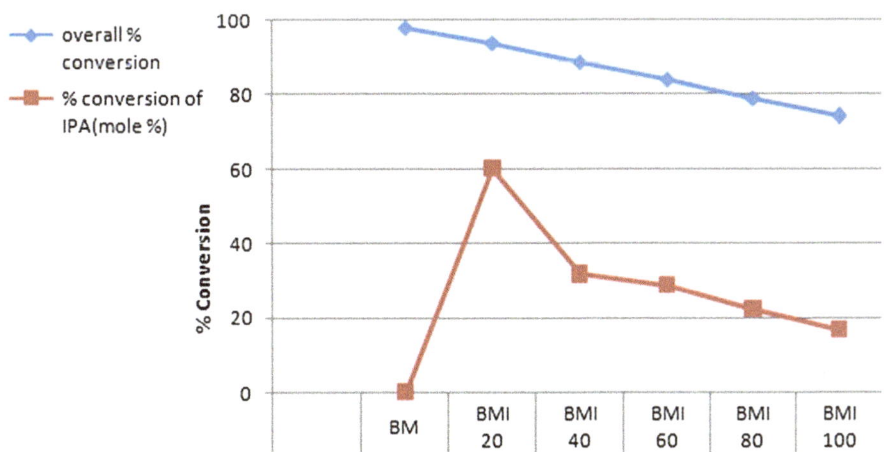

	BM	BMI 20	BMI 40	BMI 60	BMI 80	BMI 100
overall % conversion	97.83	93.87	88.7	84.1	78.98	74.09
% conversion of IPA(mole %)	0	60.06	31.62	28.6	22.32	16.63

Emulsion Sample

Figure 4 Polymerization conversion and conversion of IPA.

Figure 5 Resonance stabilization of monomer free radical.

Corporation, Milford, MA, USA) with flow rate of 1 ml/min in THF as mobile phase.

Fourier transform infrared (FTIR) spectrometric analysis of polymer films was done using Perkin Elmer instrument (PerkinElmer Inc., Waltham, MA, USA) to analyze the chemical structures of the emulsion polymer. Differential scanning calorimetry (DSC) was carried out to determine glass transition temperatures of the emulsion polymer films using TA instrument Q100 DSC (TA Instruments, New Castle, DE, USA). All the samples were analyzed for temperature range of −40°C to 50°C at the heating rate of 10°C/min under nitrogen as a purge gas with a flow rate of 40 ml/min. TGA was conducted to the thermal changes in the emulsion polymer films using Q100 DSC. The films were heated from 25°C to 500°C at the rate of 10°C/min under 40 ml/min nitrogen flow. The tensile strength and percent elongation of the emulsion films were evaluated by using Universal Testing Machine LR-50 K, Lloyd Instrument, West Sussex, UK with 500 N load cell. For measurement of tensile strength, the emulsion films were casted on Teflon sheet and were allowed to dry at room temperature. The films were cut (1 × 10 cm dimension) and conditioned prior to evaluation.

Results and discussion

The aim of this work was to study the feasibility of using isopropenyl acetate as a monomer for the preparation of emulsion polymers. Emulsion polymers were prepared from IPA, butyl acrylate (BA), and methyl methacrylate (MMA). IPA was charged at variable molar concentration with respect to that of BA and MMA. The synthesized emulsion polymers were further evaluated for incorporation of IPA in the polymer chain and to study its effect on the emulsion properties.

Polymerization conversion

Figures 3 and 4 show nonvolatile matter of emulsion polymers and polymerization conversion. It was observed that the percentage conversion of the prepared emulsions and conversion of IPA decreases as the proportion of IPA in the monomer feed is increased from 0.2 to 1 mol. Conversion of IPA was calculated by excluding proportion of BA and MMA from total polymer content by assuming 98% conversion of these monomers based on the conversion in BM polymer. The overall conversion was also observed to decrease in the same order. For every 20% increase in the molar concentration of IPA, the polymer conversion was observed to decrease by up to 5% from polymer BMI_{20} to polymer BMI_{100}. This could be due to the lower reactivity of IPA during polymerization. IPA is reported to show degradative chain-transfer during polymerization. During polymerization process, IPA forms a free radical after elimination of hydrogen atom. The free radical formed is

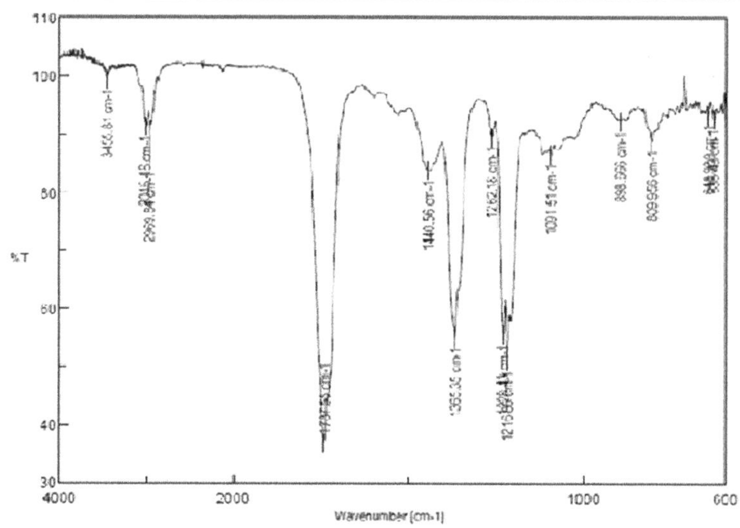

Figure 6 FTIR spectrum of BA/MMA/IPA terpolymer.

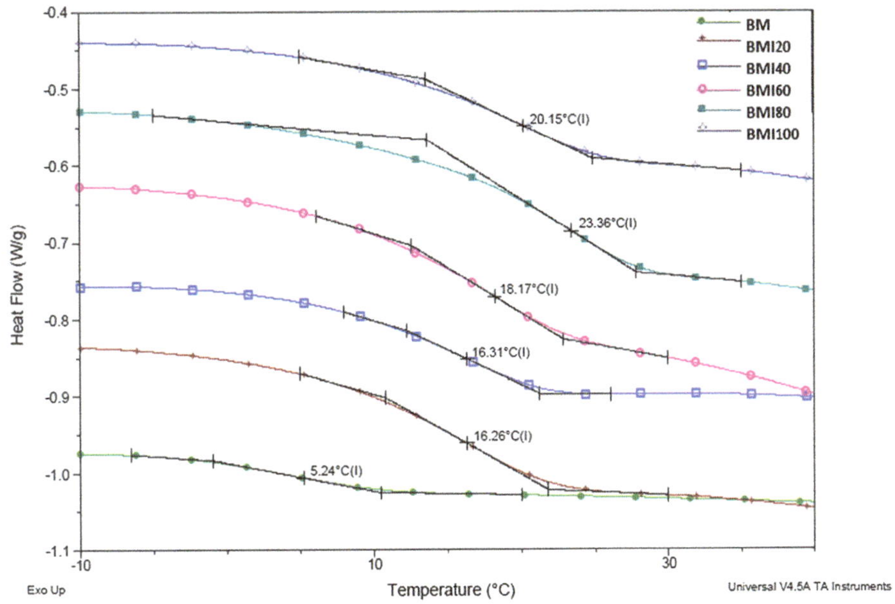

Figure 7 DSC curves of the emulsion films.

further stabilized by resonance (Figure 5) which results in its limited or no polymerization [11].

Particle size and zeta potential

The average particle size for all emulsions was found to be in the range of 0.165 to 0.168 μm. The stabilization of these systems was provided by an anionic surfactant (Dowfax 2A1). The electrostatic stability provided by the surfactant is related to the ζ-potential. Emulsions with the ζ-potential values lower than −35 mV are considered as optimum for negatively charged latex and higher than +35 mV for positively charged system [8,9]. The present emulsions having negatively charged latex exhibiting ζ-potential values less than −35 mV were observed to be stable indicating that the electrostatic interactions were enough to stabilize the emulsions.

Viscosity

All the emulsions were evaluated for viscosity by Brookfield viscometer using spindle no.1 at the speed of 20 rpm and the viscosities for all the samples were observed to be in the range of 0.93 to 0.95 cP at 25°C.

Molecular weight

The emulsions were also characterized by gel permeation chromatography (GPC) to evaluate their molecular weight (Mn). It was observed that the molecular weight of all the emulsion polymers was more than 200,000, and polydispersity was in the range of 1.01 to 1.03.

Fourier transform infrared spectroscopy

The prepared emulsion films were further analyzed by Fourier transform infrared spectroscopy (FTIR). Figure 6 shows FTIR spectrum of polymer BMI_{60}. The transmission band at 3,015 and 2,969 cm^{-1} are the characteristic stretching peaks of C-H (CH_3, CH_2); band at 1,737 cm^{-1} represents stretching vibration peak of C=O; 1,440 cm^{-1} represents distortion vibration of -COO-; 1,365 cm^{-1} shows rocking vibration of CH_3. The stretching vibration of C=C disappeared within the range 1,640to 1,700 cm^{-1} which confirms the conversion of monomers into polymer.

Differential scanning calorimetry

Figure 7 shows the glass transition temperature (Tg) of emulsion polymers. The Tg value for the terpolymer of

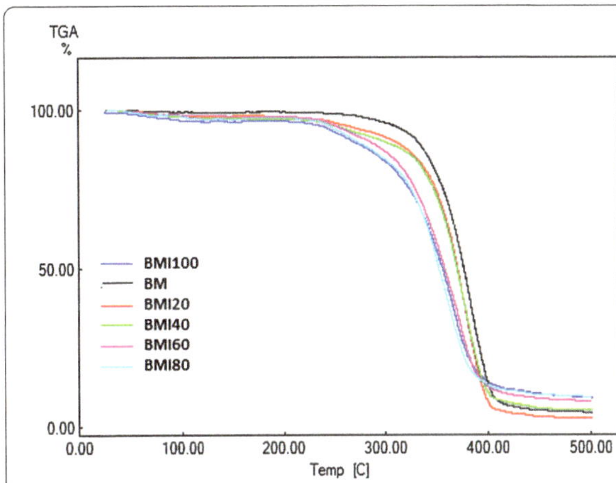

Figure 8 TGA curves of the emulsion films.

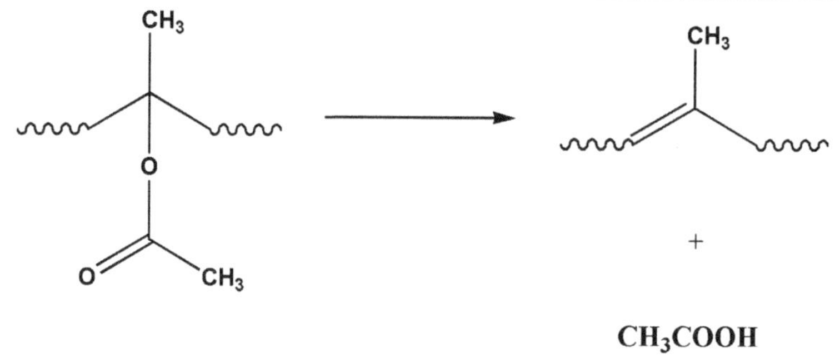

Figure 9 Degradation of IPA in polymer.

IPA, BA, and MMA was observed to be higher than that of copolymer of BA and MMA in all cases. The Tg of homopolymer of IPA (PIPA) is reported to be 45°C [10] which is higher than the Tg of homopolymer of BA (−56°C) and lower than that of PMMA (105°C). The glass transition temperature for these terpolymers suggests that IPA was incorporated in the polymer backbone resulting in higher Tg than that of copolymer due to hard segment of IPA. However, there was no significant change in Tg values of terpolymers to confirm the actual proportion of IPA converted in the polymer. The observations suggest that beyond certain proportion of IPA in the monomer feed, the capacity to undergo polymerization reduces drastically for the IPA monomer and gives limited participation in polymerization process.

Thermo gravimetric analysis

Thermo-gravimetric analysis (TGA) was carried out to investigate the thermal stability of the polymer film. The effect of IPA on the thermal stability of the emulsion polymers was evaluated by comparing the thermal stability of copolymer with that of terpolymers. The TGA graphs for all the samples are as shown in Figure 8. The thermal stability of terpolymers was observed to decrease significantly, as the amount of IPA in the monomer feed was increased. This could be due to the fact that IPA undergoes two-stage decomposition under programmed heating. In the first stage, acetic acid is eliminated via a zipper deactivation mechanism as shown in Figure 9. The second stage generally occurs at higher temperatures which involves fragmentation of the polymer backbone and the formation of cold ring fraction. Figure 10 shows typical degradation of copolymer of IPA/MMA at higher temperature [12,13]. This justifies the discussion made about the decreasing trend observed in the thermal stability of terpolymer.

It was observed that in case of polymer BM, 10% degradation occurred at 332°C whereas polymer BMI_{20} is degraded by 10% at 309°C. Furthermore, it was observed that the increase in the concentration of IPA in the emulsion feed resulted in faster degradation of the polymer as shown in Table 2. From the results, polymer BM was observed to be thermally more stable than terpolymers, and the thermal stability of emulsion terpolymer decreased with the increasing proportion of IPA in polymer.

Tensile strength and percent elongation

The mechanical strength of polymer films was evaluated by measuring tensile strength and percent elongation using Universal Testing Machine. Table 3 shows results of mechanical strength. It was observed that the terpolymer emulsion films showed higher tensile strength and less percent elongation than that of the copolymer films. This could be due to incorporation of harder segment of IPA in backbone of terpolymer as against BA and MMA in the copolymer. While in the case of terpolymer, tensile strength was observed to increase up to 0.6 M of IPA in polymer BMI_{60} which further decreased in

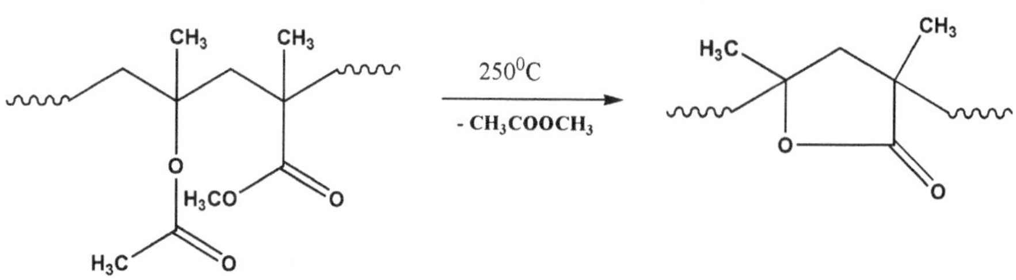

Figure 10 Typical degradation of IPA/MMA copolymer.

Table 2 Temperature at which polymers degrade by initial 10%

Polymer name	10% Degradation temperature (°C)
BM	332
BMI_{20}	309
BMI_{40}	298
BMI_{60}	284
BMI_{80}	272
BMI_{100}	272

polymer BMI_{80} and polymer BMI_{100}. This could be due to the lower reactivity of IPA after 60 mol% concentration in the monomer feed, resulting in higher proportion of softer segments of BA compared to that in polymer BMI_{20}, BMI_{40} and BMI_{60} and thus increasing the overall flexibility and elongation of the films. The similar trend was observed in percentage elongation of the films. It decreased up to 0.1 M of IPA in polymer BMI_{60} and again increased in polymer BMI_{80} and polymer BMI_{100} with increasing concentration of IPA in the polymer backbone.

Experimental

Preparation of pre-emulsion

Pre-emulsion was prepared using calculated amount of BA and MMA with 40% of required KPS, 75% of required surfactant and 60% of required water. The above solution was mixed under magnetic stirrer until a stable pre-emulsion was formed. Ten percent of this pre-emulsion was taken in a four-necked flask to start the polymerization. The calculated amount of IPA was added in the remaining pre-emulsion which was then added to the reaction flask dropwise for duration of 3.5 h. The same procedure was followed for all the emulsion formulations.

Emulsion polymerization

All the emulsions were prepared by seeded emulsion polymerization technique. The polymerization was carried out in a 250-ml four-necked flask equipped with reflux condenser, mechanical stirrer, dropping funnels, and temperature controller. Nitrogen atmosphere was maintained throughout the reaction. The flask was heated over a water bath. The remaining quantities of initiator, surfactant, buffer, and water were directly added into the reactor along with 10% of total pre-emulsion prepared. The reactor was heated to constant temperature of 70°C with a stirring rate of 300 rpm until slight blue color appeared. The reaction temperature was then raised to 78-80°C and the remaining pre-emulsion was fed drop by drop into the flask for a period of 3.5 h. The reaction was further continued for 30 min after the addition of pre-emulsion was over. A digestive catalyst TBHP was added to take care of the unreacted monomer as it blocks the double bond present on the unreacted monomer so that it cannot lead to any further side reaction. The reaction was continued for another 10 min. The reaction temperature was then reduced to room temperature and the emulsion was filtered. All the emulsion polymers were synthesized following the aforesaid procedure under identical conditions.

Table 3 Results of mechanical properties

Emulsion sample	Tensile strength (MPa)	% Elongation
BM	0.62	576
BMI_{20}	1.26	462
BMI_{40}	1.51	417
BMI_{60}	2.7	388
BMI_{80}	1.74	390
BMI_{100}	1.72	412

Conclusions

Emulsion polymers using isopropenyl acetate as a comonomer were successfully prepared by seeded emulsion polymerization. The reactivity of IPA towards copolymerization was observed to reduce with increasing proportion of IPA in the monomer feed. This also affected the overall polymer conversion which could be due to the hindrance provided by the stabilized IPA radicals. The consistency and particle size of emulsions were unaltered in all cases. Incorporation of IPA in the terpolymer showed higher glass transition temperature than those of BA-MMA copolymer, suggesting the presence of adequate quantity of IPA in the polymer backbone. However, the terpolymer emulsion showed lower thermal stability than those of BA-MMA copolymer emulsion due to possible two-stage degradation of IPA at higher temperature. Terpolymer showed higher tensile strength and lower percentage elongation than that of copolymer; but in the case of terpolymer, increase in tensile strength was observed only up to polymer BHI_{60} but beyond that, it was observed to decrease in polymer BHI_{80} and polymer BHI_{100}, exhibiting 0.6 M of IPA in the monomer feed as the optimized one under the scope of this work.

In summary, IPA is a low-cost radical monomer and could be a partial replacement for various acrylic monomers after successful study for the improvement in its polymerization mechanisms, reaction parameters, kinetics of the reaction and conversion.

Competing interests

The authors declare that they have no competing interests.

Authors' contributions

NS carried out all the synthesis work and evaluated film properties. MK and AS participated in interpretation of results and drafted the manuscript. All authors read and approved the final manuscript.

Acknowledgments

We would like to thank Professor VC Malshe for his technical guidance throughout the project. We would also like to thank Godavari Biorefineries Ltd., Mumbai for providing the sample of isopropenyl acetate.

References

1. Chern CS (2006) Emulsion polymerization mechanism and kinetics. Prog Polym Sci 31:443–486
2. Tigli RS, Evren V (2005) Synthesis and characterization of pure poly(acrylate) latex. Prog Org Coat 52:144–150
3. Naghash HJ, Mallakpour S, Mokhtarian N (2006) Synthesis and characterization of silicone-modified vinyl acetate-acrylic emulsion copolymers. Prog Org Coat 55:375–381
4. El Aasser MS, Sudol ED (1997) Features of emulsion polymerization. In: Lovell PA, El-Aasser MS (ed) Emulsion polymerization and emulsion polymers, 2nd edition. Wiley, Chichester
5. Gaylord NG, Eirich FR (1950) Peroxide-catalyzed polymerization of isopropenyl acetate. J Polymer Sci A: Polymer Chem 5(6):743–744
6. Unruh CC, Kenyon WO (1948) Heteropolymer of isopropenyl acetate and maleic anhydride. US Patent 2,452,165, 10 Oct 1948
7. Hasche RL, McMahon EM (1947) Copolymer of isopropenyl acetate and vinyl chloride. US Patent 2,453,317, 3 Sept 1947
8. Priyadarsi D, Sathyanarayana DN (2002) Free radical oxidative copolymerization of indene with vinyl acetate and isopropenyl acetate: synthesis and characterization. J Appl Polym Sci 86:639–646
9. Ma W-Y, Wu Y-X, Li F, Xu R-W (2012) Synthesis of poly(styrene-*co*-isopropenyl acetate)-*g*-polyisobutylene *graft* copolymers *via* combination of radical polymerization with cationic polymerization. Polymer 53:3185–3193
10. Reding FP, Wise EW, Sullivan EM (1967) Solid poly(isopropenyl acetate), US Patent 3,328,369, 27 June 1967
11. Kuwae Y, Kamachi M, Nozakura S-I (1986) Kinetic and electron spin resonance studies on radical polymerization of isopropenyl acetate. Macromolecules 19:2912–2915
12. McNeill IC, Ahemed S, Rendall S, Gorman JG (1998) Thermal degradation behaviour of poly(isopropenyl acetate). Polym Degrad Stab 60:43–52
13. McNeill IC, Ahmed L, Memetea L, Mohammed MH, Zaikov GE, Polishchuk AT (1996) Thermal degradation behaviour of some alternating copolymers. Polym Degrad Stab 52:171–181

$FeCl_3$-mediated efficient method for the synthesis of tetrahydropyran derivatives via cross-cyclization of epoxides and homoallylic alcohols

Nagavani Sunkaraneni, Chandra Mouleswar Rao Jillepalli and Madhukar Jeripothula*

Abstract

Background: An efficient methodology for the synthesis of tetrahydropyran derivatives has been developed under mild reaction conditions. The reactions were carried out using epoxides and homoallylic alcohols as reactants and ferric chloride as catalyst. All the reactions were done at room temperature in methylene dichloride.

Results: In all cases, the reactions proceeded efficiently at ambient temperature under mild conditions showing the generality of the reaction. The reaction also proceeded well with the cyclic epoxide 1,2-dihydronaphthalene oxide to give the corresponding spirotetrahydropyrans in good yield. The proton on the carbon bearing the halide group (4-H) (δ 3.80, J = 4.46 and 11.80 Hz) shows NOE with the proton on the carbons bearing the benzyl group (2-H) (δ 3.46, J = 11.80 Hz) and the proton on the carbon bearing the methyl group (6-H) (δ 3.40, J = 11.80 Hz). This confirms that the protons 2-H, 4-H, and 6-H are on the same side and occupy the axial position of a chair conformation.

Conclusion: The attractive features of this process are mild reaction conditions, which are environmentally friendly, inexpensive reagents, with short reaction times, and cleaner reactions with improved yields, which make it a useful process for the synthesis of tetrahydropyran derivatives.

Keywords: Epoxides, Homoallylic alcohols, Ferric chloride, Tetrahydropyran

Background

Substituted tetrahydropyrans are the common structural motif of many natural products [1,2] such as avermectins, aplysiatoxin, oscillatoxins, latrunculins, talaromycins, acutiphycins,and apicularens. Despite their potential importance to construct structurally complex molecules, the synthesis of dihydropyrans remains underutilized. Although several methods are reported [3-19], many of these procedures involve extended reaction timings and the use of expensive reagents. So, the development of efficient and versatile catalytic methods would be a preferable approach especially that low cost, environmental friendly and mild conditions are in high demand. In this respect, ferric chloride has been found as potential Lewis acid in various organic reactions [16,17,20-23]. Because of the numerous advantages, we undertook a study of the utility of the ferric chloride for the tetrahydropyran synthesis. Epoxides are the most convenient starting materials for the preparation of various compounds because of their ease of formation, wide reactivity, and ability to undergo regioselective ring opening reactions which contribute largely to their synthetic value [24-29]. In this report, we describe a simple and efficient protocol for the cyclization reactions of commercially available ferric-mediated reaction of epoxides and homoallylic alcohols to produce tetrahydropyrans.

Methods

All reactions were carried out under an inert atmosphere of nitrogen in oven-dried glassware with magnetic stirring, unless otherwise noted. DCM was distilled from calcium hydride prior to use. Ferric chloride was purchased from a commercial supplier. Reactions were monitored by thin-layer chromatography (TLC) carried out on 0.25 mm coated commercial silica gel plates (F254 pre-coated glass plates) using UV light as

* Correspondence: madhuj.biochem@gmail.com
Department of Biochemistry, Kakatiya University, Warangal, Andhra Pradesh 506 009, India

visualizing agent and p-Anisaldehyde and heat as a developing agent. Flash chromatography was performed on silica gel (100–200 mesh). ^{1}H and ^{13}C NMR spectra were obtained at 100, 200 or 400 MHz at 298 K, unless otherwise indicated. Abbreviations for multiplicity are as follows: d indicates doublet, t indicates triplet, q indicates quartet, m indicates multiplet, dd indicates doublet of doublet, and dt indicates doublet of triplet. Chemical shifts are reported in ppm referenced to the internal solvent residual of $CDCl_3$ at 7.27 ppm for 1H NMR and 77.1 ppm for ^{13}C NMR, respectively. IR spectra were obtained on an FT-IR spectrophotometer using NaCl plates. Mass spectrometry data were obtained by ESI mass spectrometer.

Results and discussions

In order to delineate the standard operating conditions, a mixture of styrene oxide and 3-buten-1-ol was treated with ferric chloride in dry methylene dichloride. The mixture was stirred at room temperature for 1 h and after work-up; the crude product was purified over silica gel to provide the product in 70% yield. By spectroscopic analysis, the product was confirmed as 3a by comparing with the literature data (Scheme 1).

This methodology has been generalized by reacting a series of epoxides with homoallylic alcohols to give the corresponding tetrahydropyran derivatives in excellent yields ranging from 80% to 95% as illustrated in Figure 1. The corresponding dihydropyran derivatives in good yields range from 65% to 80% (Figure 1). In all cases, the reactions proceeded efficiently at ambient temperature under mild conditions showing the generality of the reaction. The reaction also proceeded well with the cyclic epoxide 1,2-dihydronaphthalene oxide to give the corresponding spirotetrahydropyrans in good yield (Figure 1, entries 7 to 9).

The stereochemistry assignment of 3b was based on the^{1}H NMR spectrum and NOE experiment. From the NOE, it is seen that the proton on the carbon bearing the halide group (4-H) (δ 3.80, J = 4.46 and 11.80 Hz) shows NOE with the proton on the carbons bearing the benzyl group (2-H) (δ 3.46, J = 11.80 Hz) and the proton on the carbon bearing the methyl group (6-H) (δ 3.40, J = 11.80 Hz). This confirms that the protons 2-H, 4-H, and 6-H are on the same side and occupy the axial position of a chair conformation. This observation is consistent with vicinal ^3J couplings and unequivocally confirms a 'R' configuration at C-4 position. The predominant formation of a single stereoisomer is probably due to thermodynamic factors (Figure 2).

Experimental

General procedure

To a stirred solution of 3-buten-1-ol (290 mg, 4 mmol) and styrene oxide (720 mg, 6 mmol) in dry methylene dichloride (20 mL), anhydrous ferric chloride (1.3 g, 8 mmol) was added at room temperature. The mixture was stirred under a nitrogen atmosphere for 1 h. After completion of the reaction as indicated by TLC, the reaction mixture was quenched by adding crushed ice and extracted with methylene dichloride (2 × 25 mL). The combined organic layer was washed with brine, dried over Na_2SO_4 and concentrated under reduced pressure. The obtained crude product was purified by column chromatography over silica gel (ethyl acetate-hexane, 3:7). Both diastereomers of 6a were obtained in a 3:1 ratio. All the products were characterized by their spectral data.

(2R, 4R)-2-benzyl-4-chlorotetrahydro-2H-pyran (cis) (3a)

IR (neat): υ 2960, 2843, 1495, 1453, 1103, 1074, 764, 741 cm−1;1H NMR (400 MHz, $CDCl_3$): δ 7.24 to 7.40 (m, 5H), 3.95 to 4.10 (m, 2H), 3.45 to 3.55 (m, 1H), 3.42 (dt, 1H, J = 2.0, 12.1 Hz), 3.0 (dd, 1H, J = 6.8, 13.8 Hz), 2.75 (dd, 1H, J = 6.5, 13.8 Hz), 2.10 to 2.20 (m, 2H), 1.92 (dt, 1H, J = 4.7, 12.0 Hz), 1.65 (q, 1H, J = 1.5 Hz); ^{13}C NMR (100 MHz, $CDCl_3$): δ 137.8, 129.5, 128.1, 126.7, 78.4, 67.3, 55.6, 42.1, 42.6, 37.5. Analysis calculated for C12H15ClO: C, 68.41; H, 7.18. Found: C, 67.97; H, 7.71.

(Trans) (3a)

IR (neat): υ 3025, 2948, 2846, 1494, 1451, 762, 753, 691 cm−^{1};^{1}H NMR (400 MHz, CDCl3): δ 7.20 to 7.30 (m, 5H), 4.52 to 4.59 (m, 1H), 4.02 to 4.10 (m, 1H), 3.95 (dt, 1H, J = 2.0, 11.7 Hz), 3.86 (dd, 1H J = 5.0, 12.0 Hz), 2.90 (dd, 1H, J = 7.4, 13.8 Hz), 2.70 (dd, 1H, J = 5.9, 13.8 Hz), 2.02 to 2.10 (m, 1H), 1.75 to 2.02 (m, 3H);^{13}C NMR (100 MHz, CDCl3): δ 138.1, 129.3, 128.3, 126.5, 72.2, 62.6, 56.3, 42.2, 39.0, 33.8.

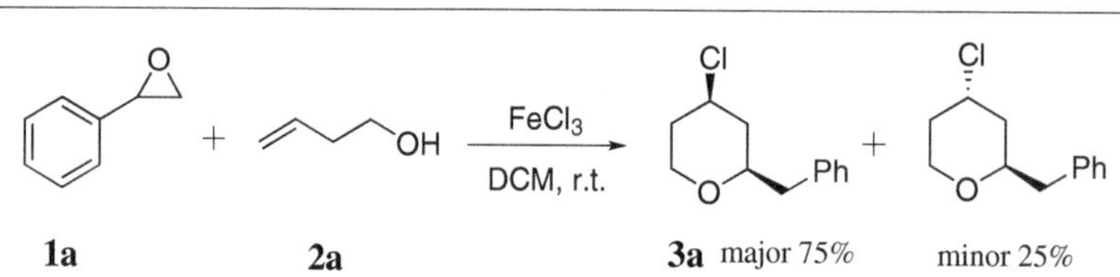

Scheme 1 (2*R*, 4*R*)-2-benzyl-4-chlorotetrahydro-2*H*-pyran (*cis*) (3a).

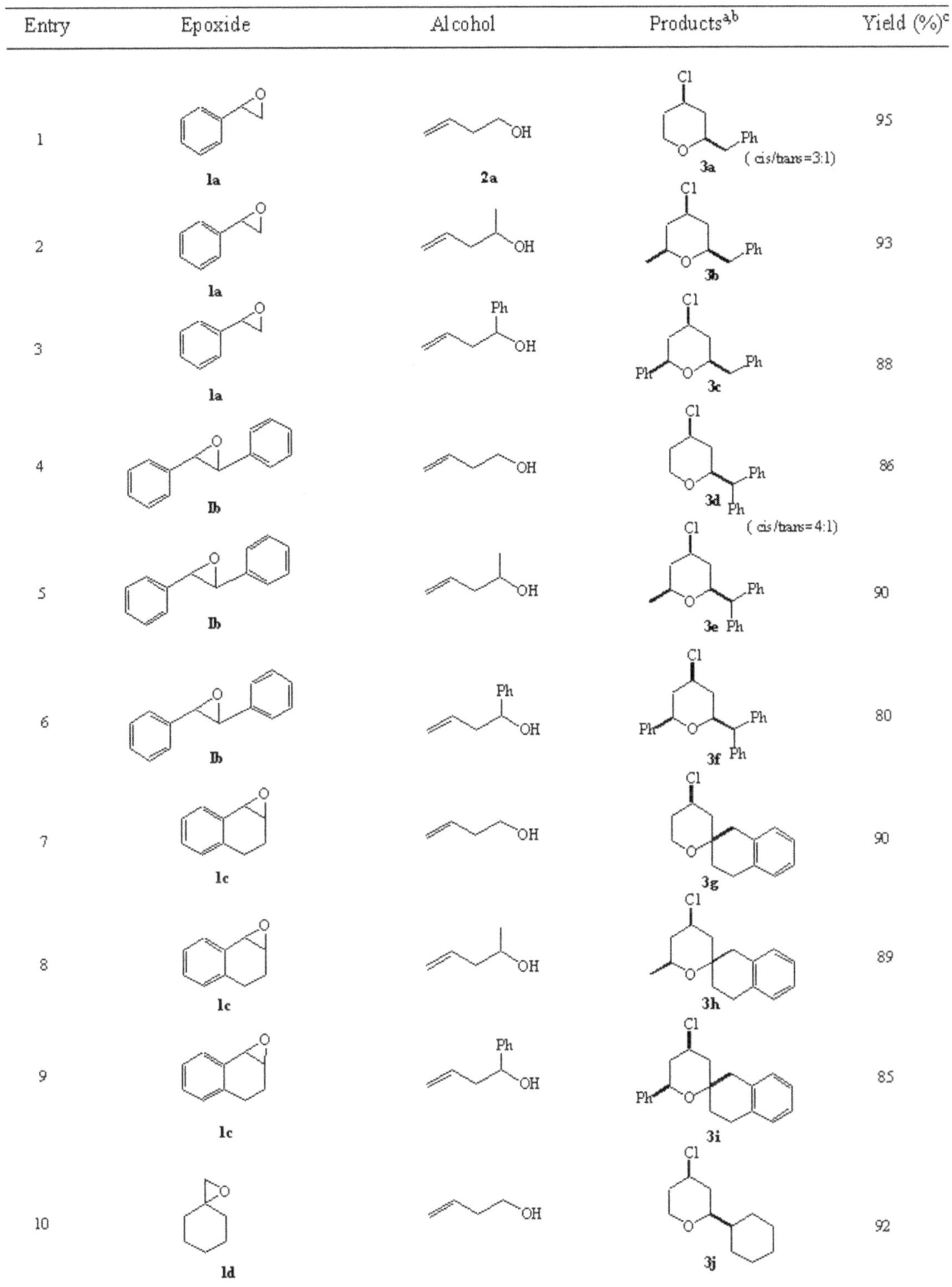

a: All the products were characterized by 1H NMR, IR and mass spectroscopy
b: The minor isomers of THP products were not isolated
c: Yields are Isolated after column chromatography

Figure 1 Ferric chloride-meditated cross-cyclization of epixodes and homoallylic alcohols.

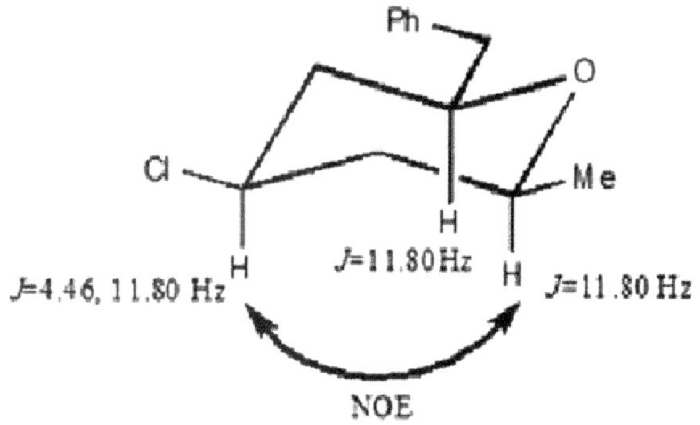

Figure 2 (2*R*, 4*R*, 6*S*)-2-Bbenzyl-4-chloro-6-methyltetrahydro-2*H*-pyran (3b). ^{1}H NMR (200 MHz, CDCl3): δ 7.10 to7.30 (m, 5H), 3.80 to 4.00 (m, 1H), 3.30 to 3.50 (m, 2H), 2.95 (dd, 1H, *J* = 5.9, 13.3 Hz), 2.65 (dd, 1H, *J* = 6.6, 13.3 Hz), 2.05 (dt, 2H, *J* = 2.2, 12.6 Hz), 1.40 to 1.60 (m, 2H), 1.20 (d, 3H, *J* = 5.9 Hz); EIMS *m/z*: 224 (10) M^+, 150 (15), 133 (100), 97 (30), 69 (85), 41 (35). Analysis calculated for $C_{13}H_{17}ClO$ (224.73): C, 69.48; H, 7.62%. Found: C, 69.56; H, 7.87%.

(2R, 4R, 6S)-2-Bbenzyl-4-chloro-6-methyltetrahydro-2H-pyran (3b)

1H NMR (200 MHz, CDCl3): δ 7.10 to 7.30 (m, 5H), 3.80to 4.00 (m, 1H), 3.30 to 3.50 (m, 2H), 2.95 (dd, 1H, *J* = 5.9, 13.3 Hz), 2.65 (dd, 1H, *J* = 6.6, 13.3 Hz), 2.05 (dt, 2H, *J* = 2.2, 12.6 Hz), 1.40 to 1.60 (m, 2H), 1.20 (d, 3H, *J* = 5.9 Hz); EIMS *m/z*: 224 (10) M^+, 150 (15), 133 (100), 97 (30), 69 (85), 41 (35). Analysis calculated for C13H17ClO (224.73): C, 69.48; H, 7.62%. Found: C, 69.56; H, 7.87%.

(2R, 4S, 6R)-2-benzyl-4-chloro-6-phenyltetrahydro-2H-pyran (cis) (3d)

IR (neat): υ 3021, 2956, 2852, 1596, 1492, 1445, 746, 695 cm−1;1H NMR (200 MHz, $CDCl_3$): δ 7.20 to 7.36 (m, 10H), 3.95 to 4.10 (m, 4H), 3.46 (dt, 1H, *J* = 2.0, 8.4 Hz), 2.05 to 2.20 (m, 2H), 1.80 to 1.90 (m, 1H), 1.55 to 1.65 (m, 1H);^{13}C NMR (100 MHz, $CDCl_3$): δ 141.8, 141.7, 128.6, 128.6, 128.4, 128.3, 126.7, 126.4, 78.9, 67.2, 57.4, 56.1, 41.5, 36.9. EIMS *m/z*: 286 M^+.

(Trans) (3d)

IR (neat): υ 3085, 3059, 2961, 2805, 1594, 1491, 750, 698 cm−1;1H NMR (400 MHz, $CDCl_3$): δ 7.10 to 7.30 (m, 10H), 4.52 (dt, 1H, *J* = 8.0 Hz), 4.53 (singlet (s), 1H), 3.92 (triplet (t), 1H, *J* = 12.0 Hz), 3.75to 3.85 (m, 2H), 1.92 to 1.99 (m, 1H), 1.65to 1.75 (m, 3H);^{13}C NMR (100 MHz, CDCl3): δ 142.1, 124.0, 128.6, 128.4, 128.3, 126.5, 126.3, 73.3, 62.5, 57.1, 56.6, 38.2, 33.6. EIMS *m/z*: 286 M^+. Analysis calculated for $C_{18}H_{19}ClO$: C, 75.38; H, 6.68. Found: C, 75.62; H, 7.09.

(2S, 4R)-4-chloro-2-(diphenylmethyl)-tetrahydro-2H-pyran (3e)

1H NMR (200 MHz, $CDCl_3$): δ 7.10 to 7.30 (m, 10H), 3.85 to 4.00 (m, 3H), 3.40to 3.50 (m, 1H), 2.10 (dt, 1H, *J* = 1.4, 2.2 Hz), 1.92 (dt, 1H, *J* = 1.4, 2.2 Hz), 1.40to 1.56 (m, 2H), 1.18 (d, 3H, *J* = 6.6 Hz); EIMS*m/z*: 300 (10) M^+, 166 (45), 134 (100), 98 (25), 69 (70), 41 (30). Analysis calculated for $C_{19}H_{21}ClO$ (300.82): C, 75.86; H, 7.04%. Found: C, 76.15; H, 7.23%.

(2S, 4S, 6R)-4-chloro-2-(diphenylmethyl)-6-phenyltetrahydro-2H-pyran (3f)

1H NMR (200 MHz, $CDCl_3$): δ 7.15to 7.45 (m, 15H), 4.50 (dd, 1H, *J* = 11.2, 12.8 Hz), 4.15 to 4.30 (m, 2H), 4.10 (d, 1H, *J* = 7.2 Hz), 2.40 to 2.55 (m, 1H), 1.60 to 2.10 (m, 3H);^{13}C NMR (200 MHz, CDCl3): δ 40.8, 43.8, 44.3, 56.1, 77.0, 78.7, 125.5, 125.8, 126.4, 126.6, 127.4, 127.7, 128.1, 128.3, 128.5, 128.6, 129.1; EIMS *m/z*: 362 (10) M^+, 271 (10), 236 (10), 165(40), 130 (40), 116 (30), 103 (100), 90 (30), 66 (20), 50 (15). Analysis calculated for C24H23ClO (362.89): C, 79.43; H, 6.39. Found: C, 79.69; H, 6.62.

(2R, 4'R)-4'-chloro-3,3',4,4',5',6'-hexahydro-1H-spiro-[naphthalene-2,2'-pyran] (3g)

1H NMR (200 MHz, $CDCl_3$): δ 6.88 to 7.02 (m, 4H), 4.12 to 4.25 (m, 1H), 3.80 (dd, 1H, *J* = 2.0, 2.7 Hz), 3.65 (dd, 1H, *J* = 2.0, 2.7 Hz), 2.84 to 2.97 (m, 1H), 2.68 to 2.76 (s, 2H), 2.55 to 2.64 (m, 1H), 1.96 to 2.16 (m, 2H), 1.56 to 1.92 (m, 4H); EIMS *m/z*: 236 (25) M^+, 201 (80), 129 (55), 104 (100), 91 (35), 55 (70); Analysis calculated for $C_{14}H_{17}ClO$ (236.74): C, 71.03; H, 7.24%. Found: C, 71.46; H, 7.32%.

(2R,4'R,6'S)-4'-chloro-6'-methyl-3,3',4,4',5',6'-hexahydro-1H-spiro-[naphthalene-2,2'-pyran] (3h)

1H NMR (200 MHz, $CDCl_3$): δ 6.90 to 7.05 (m, 4H), 4.15 to 4.25 (m, 1H), 3.65 to3.85 (m, 1H), 2.50 to 3.02 (m, 4H), 2.01 to 2.12 (m, 2H), 1.45 to 1.90 (m, 4H), 1.17 (d, 3H, *J* = 5.9 Hz). EIMS *m/z*: 250 (15) M^+, 216 (100), 215 (20), 129 (25), 104 (50), 91 (10), 42 (10). Analysis calculated for $C_{15}H_{19}ClO$ (250.76): C, 71.84; H, 7.64%. Found: C, 71.95; H, 7.85%.

Conclusion

In summary, we have described a simple and highly efficient protocol for the preparation of tetrahydropyran derivatives through the reaction between epoxides and homoallylic alcohols using ferric chloride. The attractive features of this process are mild reaction conditions, which are environmentally friendly, inexpensive reagents, with short reaction times, and cleaner reactions with improved yields, which make it a useful process for the synthesis of tetrahydropyran derivatives.

Competing interests
The authors declare that they have no competing interests.

Authors' contributions
NS, carried out synthesis of tetrahydropyran derivatives, CMRJ, participated in the design of the study and performed the spectral analysis. MJ, conceived of the study and participated in its design and coordination. All authors read and approved the final manuscript.

Authors' information
NS and CMRJ are research scholars and MJ is professor from the Department of Biochemistry, Faculty of Science, Kakatiya University, Warangal Andra Pradesh, 506 009, India.

Acknowledgments
Department of Biochemistry, Faculty of Science, Kakatiya University, Warangal.

References
1. Kunze B, Jansen R, Sasse F, Hofle G, Reichenbach H (1998) J Antibiot 51:1075
2. Nicolaou KC, Sorensen EJ (1996) Classics in total synthesis. Weinheim, VCH
3. Li J, Li CJ (2001) Tetrahedron Lett 42:793
4. Yadav JS, Rajasekhar K, Murty MSR (2005) Tetrahedron Lett 46:2311
5. Yadav JS, Rajasekhar K, Murty MSR (2005) Synlett 763(12):1945–1947. doi:10.1055/s-2005-871574. Art ID:D08005ST
6. Yadav JS, Reddy BVS, Gopal CV, Srinivas R, Lingam TR (2002) Synth Commun 32:1803
7. Yadav JS, Reddy BVS, Reddy MS, Niranjan N, Prasad AR (2003) Eur J Org Chem 2003(9):1779–1783. doi:10.1002/ejoc.200210638
8. Ranu BC, Jana U (1998) J Org Chem 63:8212
9. Viswanathan GS, Yang J, Li CJ (1999) Org Lett 1:993
10. Zhang WC, Viswanathan GS, Li CJ (1999) Chem Commun 1999:291. doi:10.1039/A808960D
11. Zhang WC, Li CJ (2000) Tetrahedron 56:2403
12. Yadav JS, Reddy BVS, Kumar GM, Murthy CVSR (2001) Tetrahedron Lett 42:89
13. Yadav JS, Reddy BVS, Sekhar KC, Gunasekar D (2001) Synthesis 6:885
14. Keh CCK, Namboodiri VV, Varma RS, Li CJ (2002) Tetrahedron Lett 43:4993
15. Yadav JS, Reddy BVS, Reddy MS, Niranjan N (2004) J Mol Catal A 210:99
16. Roy UK, Jana PK, Roy S (2007) Tetrahedron Lett 48(7):1183–1186
17. Lee CHA, Loh TP (2006) Tetrahedron Lett 47(10):1641–1644
18. Dobbs AP, Pivnevi L, Penny MJ, Martinović S, Iley JN, Stephenson PT (2006) Chem Commun 29:3134–3136
19. Bothwell JM, Krabbez SW, Mohan RS (2011) Chem Soc Rev 40:4649–4707. doi:10.1039/C0CS00206B
20. Nagano T, Hayashi T (2004) Org Lett 6:1297
21. Bedford RB, Bruce DW, Frost RM, Goodby JW, Hird M (2004) Chem Commun 2004:2822
22. Vidari G, Beszant S, Merabet JE, Bovolenta M, Zanoni G (2002) Tetrahedron Lett 43:2687
23. Iranpoor N, Salehi P (1994) Synthesis 1994(11):1152–1154. doi:10.1055/s-1994-25661
24. Laszlo P, Montaufier MT, Randriamahefa SL (1990) Tetrahedron Lett 31:4867
25. Pelzer S, Kauf T, van Wüllen C, Christoffers J (2003) Journal of Organometallic Chemistry 684:308–314. doi:10.1016/s0022-328X(03)00765-4
26. Nakamura M, Matsuo K, Ito S, Nakamura E (2004) J Am Chem Soc 126:3686
27. Nakamura M, Hirai A, Nakamura E (2000) J Am Chem Soc 122:978
28. Nakamura M, Matsuo K, Inoue T, Nakamura E (2003) Org Lett 5:1373
29. Martin R, Furstner A (2004) Angew Chem Int Ed 43:3955

Experimental evaluation of engine oil properties containing copper oxide nanoparticles as a nanoadditive

Ehsan-o-Ilah Ettefaghi[1], Hojjat Ahmadi[1*], Alimorad Rashidi[2], Seyed Saeid Mohtasebi[1] and Mahshad Alaei[2]

Abstract

Background: The properties of lubricants are mainly the result of adding a material for improving or producing the required properties. Today, different materials with various nanostructures are used as new additives which, because of their unique properties, are used for improving the lubricant's properties. The purpose of the present research is to add copper oxide (CuO) nanoparticles to engine oil and evaluate the produced changes in some of its properties. Also, viscosity, pour point, and flash point of nanolubricants, which are made at different concentrations (0.1, 0.2, and 0.5 wt.%), and also their thermal conductivity coefficient as four quality parameters which are effective in the functionality of engine oil are evaluated.

Results: Upon the obtained results, thermal conductivity coefficient and flash point of nanolubricants with 0.1 wt.% concentration, respectively, had 3% and 7.9% improvement with respect to the base oil. Also, the oil's viscosity, as a very important factor in lubricating issue, at a lower concentration with respect to the engine oil without CuO had no appreciable change.

Conclusion: Among the different methods which have been used for dispersing nanoparticles inside the base oil, using planetary ball mill was determined as the most important method for stabilization of nanoparticles inside SAE 20W50 engine oil. Also, the physical properties of nanolubricants were measured based on the American Society for Testing and Materials standard methods.

Keywords: Copper oxide nanoparticles, Engine oil, Viscosity, Thermal conductivity coefficient

Background

Recent advances and success in chemistry and technology of nano-organic particles with nanosize have provided the possibility of making different metallic oxides. During the past few years, the attention for producing and application of colloid nano-organic particles due to their unique properties such as optical, catalytic, semiconductor, magnetic, and anti-friction properties is continuously developed [1]. Copper oxide nanoparticles are one of the particles with their tribological and anti-wear properties being investigated by some researchers [2,3]. Thottackkad et al. have studied and reviewed the tribological properties of coconut oil, which contains copper oxide nanoparticles, because of very great interest of researchers to plant oil as a lubricant which is used in automobile and also for the purpose of improving its properties. Upon their results, the natural wear and friction coefficient improved and the viscosity of base oil increased with increasing concentration of nanoparticles [3].

With rapid development of nanomaterial technology and stable growth of difficulties which exist for sources of energy, researchers have started to apply nanomaterial technology in the field of increasing heat transfer for the purpose of developing homogenous and stable heat transfer fluids. Nanofluids are produced by dispersing nanoparticles inside typical heat transfer fluids such as water, ethylene glycol, and oil [4]. Recently, an increase in the thermal conductivity coefficient of nanofluids which contain a little amount of metallic particles like Cu or nonmetallic particles like ZnO, Al_2O_3, CuO, and SiC had also been reported. The natural potential of nanomaterials which are dispersed in base fluids for heat

* Correspondence: hjahmadi@ut.ac.ir
[1]Department of Mechanical Engineering of Agricultural Machinery, University of Tehran, Mesbah Cross, Karaj 4111, Iran
Full list of author information is available at the end of the article

transfer also has an important effect on the rate of increase in thermal conductivity coefficient of nanofluids. For example, carbon nanotubes, due to their high thermal conductivity coefficient, are known as an ideal material for making nanofluids [5]. Hwang et al. have studied the heat transfer properties of nanofluids in water base and ethylene glycol base which contain CuO nanoparticles and multiwalled carbon nanotubes. They used transient hot wire method for measuring the thermal conductivity coefficient of nanofluids. In the case of nanofluids which contain CuO nanoparticles, their results showed that the increase of nanofluid thermal conductivity in the ethylene glycol base was more than that of the distilled water base nanofluid. Also, the thermal conductivity of distilled water base nanofluid which contains MWCNTs increased at the rate of 11.3% in 0.01 volume fraction [6]. Totally improving the properties of a lubricant is a very important factor for saving total energy given to a mechanical system [7]. Nanoparticles also, due to their unique properties as a new kind of additive material for the purpose of improving the properties of lubricants, are very interesting and have attracted attention [8]. So far, the anti-wear and anti-friction properties of different nanolubricants [9-12] and also the heat transfer properties of fluids which contain different nanostructures [6,13,14] had been studied by many researchers. So, in the present research, we are surveying the effect of CuO nanoparticles on the other properties of engine oil such as viscosity as one of the most important factors which are used for correct selection of the required oil in a lubricating system and also flash point and pour point as the properties which are related to the quality of oil functionality in different work conditions.

Methods

Materials

Copper oxide nanoparticles were made on the basis of the presented method by Karthikeyan et al. [15]. The X-ray diffraction (XRD) pattern of the product obtained is identical to that of single-phase CuO with a monoclinic structure (Figure 1). Further comparison of XRD patterns with the JCPDS file (JCPDS 80–1268) data also confirms the same. By means of XRD, no obvious peaks of impurity were found.

The specifications of the oil used are shown in Table 1. The SAE 20W50 engine oil (Behran Pishtaz, Behran Oil Company, Tehran, Iran) was used as a base fluid. Behran Pishtaz oil is a distinguished engine oil which has been made upon the highest world standards and is suitable for most of today's light gasoline automobiles and heavy diesel vehicles.

Apparatus

A bath ultrasonic (P120 Elmasonic, Elma Hans Schmidbauer GmbH & Co. KGa, Singen, Germany), probe ultrasonic (Syclon Ultrasonic Homogenizer/JY92-IIN, Zhejiang, China), and planetary ball mill (PM100, Retsch, Haan, Germany) were used for opening agglomerated CuO nanoparticles from each other and dispersing them inside the base fluid. Also, we used the KD2 Pro (Decagon Devices, Pullman, WA, USA) for measuring the thermal conductivity of base oil and nanolubricants.

Preparation of nanolubricants

The concentration of nanoparticles has a great effect on the functionality of nanolubricants, with about 0.1 to 0.5 wt.% stated as the optimum concentration for nanoparticles in

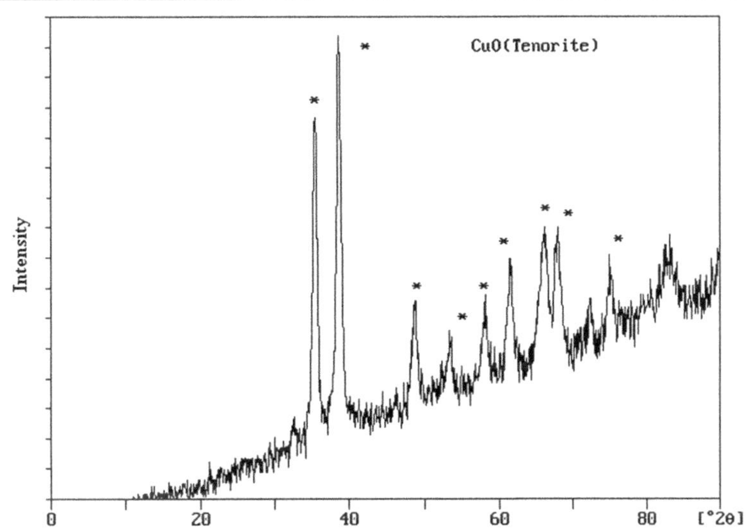

Figure 1 XRD pattern of CuO nanoparticles.

Table 1 Specification of SAE 20W50 engine oil

Specification	Value
Viscosity at 40°C (cSt)	172.11
Viscosity at 100°C (cSt)	19.8
Density at 15°C (kg/m^3)	889
Viscosity index	133
Pour point (°C)	−27
Flash point (°C)	224

most of the investigation works and studies [12,16]. Also, in the present research, for the purpose of evaluating the rate and manner of the effect of CuO concentration on the properties of base oil, nanolubricants were made at three different concentrations of 0.1, 0.2, and 0.5 wt.% by adding CuO nanoparticles to the base oil. Basically, nanofluids are solid–liquid mixtures which are earned from dispersing nanoparticles with sizes smaller than 100 nm inside a base fluid [13]. For this purpose, researchers have used conventional methods which are common in investigation works such as dispersing using a mechanical mixer, bath and probe ultrasonic, planetary ball mill, and surface activation materials. Because the viscosity of the base fluid used in the present research was very high, we used planetary ball mill method for dispersing nanoparticles inside the base fluid. Functional specifications of the planetary ball mill which is used for making nanolubricants are shown in Table 2.

Measuring the viscosity, pour point, and flash point

Kinematic viscosity, pour point, and flash point of the base oil and also nanolubricants, which were made at three different concentrations, were measured on the basis of American Society for Testing and Materials (ASTM) D-445, ASTM D-97, and ASTM D-92, respectively. The accuracy measurements of these parameters according to the ASTM standards are shown in Table 3.

Measuring thermal conductivity

Energy transfer from high-energy particles to particles with lower energy in a material is called conduction, which takes place by the interaction between the particles. Thermal conductivity (*K*) of fluids is measured as their ability to conduct heat. For this purpose, we used the KD2 Pro equipment which is completely a portable field or laboratory apparatus for analyzing the thermal properties and uses the transient hot wire system for measuring thermal conductivity [17]. Thermal conductivities of base oil and nanolubricant with 0.1 wt.% concentration were measured at 20°C temperature.

Table 2 Functional specifications of ball mill

Specification	Value
Speed (rpm)	300
Time (h)	3
Weight of balls (g)	200
Weight of sample (fluid + nanoparticles) (g)	30

Table 3 Accuracy measurement of ASTM standards

ASTM standard	Accuracy
ASTM D-445	0.26%
ASTM D-92	±8°C
ASTM D-97	±3°C

Results and discussion

Preparation of nanolubricants

One of the most effective factors of the nanofluid properties is the rate of dispersion and stability of nanoparticles inside the base fluid. When dispersion of particles inside the base fluid is not good, it is possible that agglomeration and precipitation of nanoparticles occur; in this case, macroparticles, like inclusions, develop which cause damage to frictional surfaces and also obstruct the lubricant vents. In the present research, to disperse nanoparticles inside the base oil, we used three mechanical methods including bath and probe ultrasonic and planetary ball mill (Table 2) methods. To achieve the best and the most stable state, a 0.1 wt.% sample of oil/CuO has been made using each of the three mentioned methods. All of the samples were maintained inside the completely transparent glassy containers in a completely stagnant condition for about 720 h for the purpose of evaluating their stability conditions. In this period of time, the stability condition of all the samples were periodically and visually inspected and recorded. Nanoparticles in each of the three samples had precipitated a considerable amount, although the rate of precipitation in the sample which was made by the ball mill method was very less than those of the two other samples. It seems that due to the high viscosity of base oil and also the agglomerated state of nanoparticles, for opening and dispersing nanoparticles inside the base oil, we need a great amount of energy to supply in the ball mill method at a greater rate with respect to the two ultrasonic methods.

Viscosity

The rate of oil resistance against flowing is called viscosity, which is one of the most important factors for selecting suitable oil for an apparatus. Because of the extra importance of viscosity for a lubricant, in the present research, we have focused our attention on surveying the produced changes in the engine oil's viscosity which is the result of adding CuO nanoparticles.

For this purpose, viscosity of the engine oil which did not contain nanoparticles considered as base fluid and also viscosity of the lubricants which contain CuO nanoparticles at three concentrations 0.1, 0.2, and 0.5

wt.% and also at two temperatures 40°C and 100°C were measured. The obtained results are shown in Figure 2. As it can be seen, the viscosity of nanolubricants at each of the two mentioned temperatures had increased by increasing the concentration of nanoparticles, although the rate of changes in viscosity at lower concentrations is much smaller than that at higher concentrations. The highest amount of increase in viscosity with respect to the base fluid is 5.7%, which is related to the nanolubricant with 0.5 wt.% concentration and at 40°C temperature. The interesting point related to the viscosity of nanolubricants with 0.1 wt.% concentration is that in each of the two mentioned temperatures (40°C and 100°C), the viscosity of the oil which did not contain CuO nanoparticles had a very little amount of decrease. When nanoparticles are added to the oil, they are placed between the oil layers and lead to ease of fluid layer movement on each other. As a result, viscosity will decrease slightly. As concentration increases, nanoparticles agglomerate and create larger and asymmetric particles, which prevent movement of oil layers on each other, so viscosity will increase.

So finally, it can be concluded that the viscosity of SAE 20W50 engine oil which contains CuO nanoparticles is a function of concentration of nanoparticles and at lower concentrations, it had no appreciable change with respect to the viscosity of the base oil.

Thermal conductivity

The results related to measuring the thermal conductivity coefficient of the base lubricant and the lubricant which contains CuO nanoparticles are shown in Figure 3. As it can be seen, the thermal conductivity of the engine oil which contains CuO nanoparticles had increased with respect to the base oil, although the rate of increase is only 3%. We must notice that in the case of different nanofluids, various parameters are effective in the rate of produced changes and also in their properties, including the kind of base fluid and kind of nanoparticles suspended inside the base fluid.

Flash point and pour point

The trend of changes of flash point as a function of CuO nanoparticle concentration is shown in Figure 4a. Flash point is the lowest temperature at which the oil's vapor in contact with air and as exposed to an ignition takes fire at a moment and then rapidly extinguishes. It can be seen that adding CuO nanoparticles to the base oil causes an increase in the flash point of the base oil. It can be concluded that the increase of thermal conductivity through adding of nanoparticles is attributed to the increase of oil resistance against ignition. Also, flash point has a direct relation with concentration of nanoparticles, although this relation is not linear and the intensity changes in lower concentrations are more than the changes in higher concentrations. The rate of increase in flash point of the nanofluid at 0.1 wt.% concentration with respect to the base fluid is 7.5%, and the highest amount of increase is related to the 0.5 wt.% sample, which is 13%.

The highest rate of wear in an engine is on the first moment the engine starts to work, which is produced because sufficient oil does not reach to parts, and to avoid this problem and reduce its effect, it is required that the oil be pumpable and sufficiently be flowable to easily, and with high speed, reach all parts of the engine. This property of oil is evaluated with a parameter called pour point. In fact, the oil's pour point is the boundary point of temperature in which the fluid cannot flow either. Because of the importance of this subject in the present research, the produced changes in the rate of pour point of base oil under the effect of adding CuO nanoparticles at different concentrations were studied. The trend of changes of pour point as a function of CuO nanoparticle concentration is shown in Figure 4b. As it is observed, the amount of nanolubricant pour

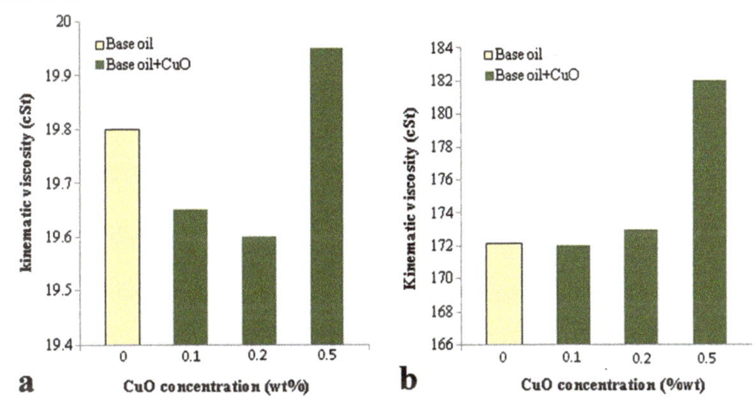

Figure 2 Kinematic viscosity of lubricants at 100°C (a) and at 40°C (b).

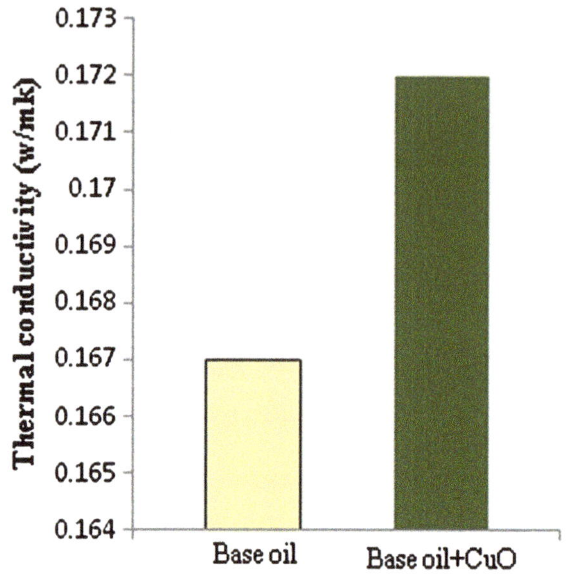

Figure 3 Thermal conductivity of nanolubricant and base lubricant.

point is seriously a function of the concentration of CuO nanoparticles such that in 0.1 and 0.5 wt.% concentrations, the pour point has decreased, but in the 0.2 wt.% concentration, there was 3.7% improvement with respect to the base oil.

Experimental

For synthesis of CuO nanoparticles, at first, Cu $(CH_3COO)_2{\cdot}2H_2O$, CH_3COOH, and solid NaOH are supplied. In our typical procedure, 600 ml of 0.2 M copper acetate aqueous solutions is mixed with 2 ml glacial acetic acid in a round-bottomed flask. Under vigorous stirring, 0.16 g of solid NaOH (pellets) is added rapidly to the boiling solution at 100°C, until the pH value of the mixture reached 6 to 7. At this stage, the color of the solutions changes from blue to black, and a black precipitate is produced simultaneously. After cooling to room temperature, the precipitate is centrifuged, washed twice with triply distilled water, ethanol, and acetone, and finally dried at room temperature.

The nanolubricants used in this study were produced by the following procedure. For example, for 0.1 wt.% sample, at first, 0.025 g of CuO nanoparticles was mixed with 25 g base oil inside the cup of the planetary ball mill, and then the ball mill worked based on the condition in Table 2.

Conclusions

Dispersing nanoparticles inside engine oil, due to the base oil's high viscosity, is a very difficult work. Upon the experiments which have been done in this research, using planetary ball mill for dispersing copper oxide nanoparticles inside the base fluid is the most suitable method because it supplies high energy preventing their agglomeration and precipitation. With increase in the concentration of nanoadditives, some of the properties of engine oil improved, but it must be noticed that an increase in concentration causes a reduction in lubricating properties of engine oil due to agglomeration and precipitation of nanoparticles and as a result causes damage to frictional surfaces. So determining the appropriate concentration to achieve better properties is a very important issue. In the present research, properties of CuO/oil nanofluids at three different concentrations were evaluated. Flash point and viscosity of engine oil had a direct relation with the concentration of nanoparticles. Thermal conductivity and flash point increased 3% and 7.5% in 0.1 wt.% concentration, respectively. The amount of pour point only at 0.2 wt.% concentration had 3.7% improvement with respect to the base fluid but decreased about 7.4% in other concentrations. Based on obtained results, it seems that the nanolubricant with 0.2 wt.% concentration is the best sample for CuO/oil because in this sample pour point and flash point are improved and viscosity has not changed much.

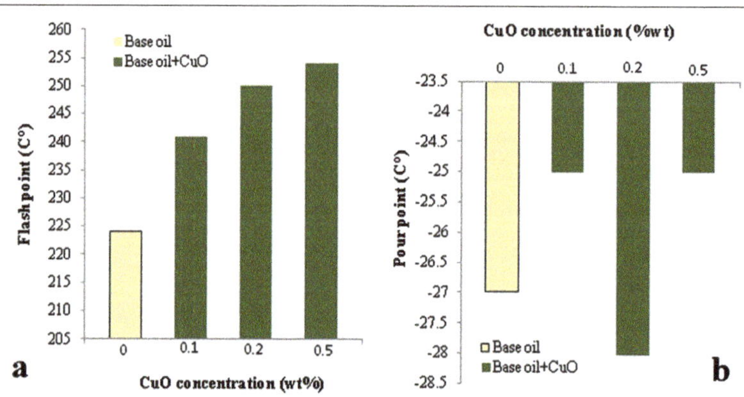

Figure 4 Flash point (a) and pour point (b) of nanolubricants.

Competing interests
The authors declare that they have no competing interests.

Authors' contributions
EE carried out all the stages of preparation and property measurement of nanolubricants. AR and MA carried out all the stages of CuO nanoparticle synthesis and analyzing them. HA and SSM participated in the interpretation of results, drafted the manuscript, and provided financial support. All authors read and approved the final manuscript.

Acknowledgments
The authors fully thank the CM Research Center and Laboratory for analyzing the oil and wear particles (Alborz Tadbirkaran) and especially the Research Institute of Petroleum Industry of Iran for providing the equipment in the direction of doing researches.

Author details
[1]Department of Mechanical Engineering of Agricultural Machinery, University of Tehran, Mesbah Cross, Karaj 4111, Iran. [2]Nanotechnology Research Center, Research Institute of Petroleum Industry (RIPI), West Blvd. Azadi Sport Complex, Tehran 14665-1998, Iran.

References
1. Bakunin VN, Suslov AY, Kuzmina GN, Parenago OP, Topchiev AV (2004) Synthesis and application of inorganic nanoparticles as lubricant components – a review. J Nanopart Res 6:273–284
2. Wu Y, Tsuia W, Liub T (2007) Experimental analysis of tribological properties of lubricating oils with nanoparticle additives. Wear 262:819–825
3. Thottackkad MV, Perikinalil RK, Kumarapillai PN (2012) Experimental evaluation on the tribological properties of coconut oil by the addition of CuO nanoparticles. Int J Precis Eng Manuf 13:111–116
4. Li X, Zhu D, Wang X (2009) Experimental investigation on viscosity of Cu-H_2O nanofluids. Journal of Wuhan University of Technology-Mater Sci Ed 24:48–52
5. Chen L, Xie H, Yu W, Li Y (2011) Rheological behaviors of nanofluids containing multi-walled carbon nanotube. J Dispersion Sci Technol 32:550–554
6. Hwang YJ, Ahn YC, Shin HS, Lee CG, Kim GT, Park HS, Lee JK (2006) Investigation on characteristics of thermal conductivity enhancement of nanofluids. Curr Appl Phys 6:1068–1071
7. Choi Y, Lee C, Hwang Y, Park M, Lee J, Choi C, Jung M (2009) Tribological behavior of copper nanoparticles as additives in oil. Curr Appl Phys 9:124–127
8. Lee CG, Hwang YJ, Choi YM, Lee JK, Choi C, Oh JM (2009) A study on the tribological characteristics of graphite nano lubricants. Int J Precis Eng Manuf 10:85–90
9. Ji X, Chen Y, Zhao G, Wang X, Liu W (2011) Tribological properties of $CaCO_3$ nanoparticles as an additive in lithium grease. Tribol Lett 41:113–119
10. Lee J, Cho S, Hwang Y, Lee C, Kim S (2007) Enhancement of lubrication properties of nano-oil by controlling the amount of fullerene nanoparticle additives. Tribol Lett 28:203–208
11. Ku BC, Han YC, Lee JE, Lee JK, Park SH, Hwang YJ (2010) Tribological effects of fullerene (C60) nanoparticles added in mineral lubricants according to its viscosity. Int J Precis Eng Manuf 11:607–611
12. Lee J, Cho S, Hwang Y, Cho HJ, Lee C, Choi Y, Ku BC, Lee H, Lee B, Kim D, Kim S (2009) Application of fullerene-added nano-oil for lubrication enhancement in friction surfaces. Tribol Int 42:440–447
13. Wang X, Mujumdar S (2007) Heat transfer characterization of nanofluid: a review. Int J Therm Sci 46:1–19
14. Kim D, Kwon Y, Cho Y, Li C, Cheong S, Hwang Y, Lee J, Hong D, Moon S (2009) Convective heat transfer characteristics of nanofluids under laminar and turbulent flow conditions. Curr Appl Phys 9:119–123
15. Karthikeyan NR, Philip J, Raj B (2008) Effect of clustering on the thermal conductivity of nanofluids. Mater Chem Phys 109:50–55
16. Li W, Zheng S, Cao B, Ma S (2011) Friction and wear properties of ZrO_2/SiO_2 composite nanoparticles. J Nanopart Res 13:2129–2137
17. Decagon Devices Company (2013) KD2 Pro. http://www.decagon.com/products/environmental-instruments/Thermal-Properties-Instruments/kd-2-pro/. Accessed 18 Aug 2012

Generation of PVP fibers by electrospinning in one-step process under high-pressure CO_2

Wahyudiono[1], Siti Machmudah[1,2], Kanako Murakami[3], Satoko Okubayashi[4] and Motonobu Goto[1*]

Abstract

Background: Electrospinning is a process of electrostatic fiber formation using electrical forces to produce polymer fibers from polymer solution in nano/micrometer scale diameters. Various polymers have been successfully electrospun into ultrafine particles and fibers in recent years, mostly in solvent solution and some in melt form. In this work, electrospinning was conducted under high-pressure carbon dioxide (CO_2) to reduce the viscosity of polymer solution. The experiments were conducted at 313 K and approximately 8.0 MPa. Polyvinylpyrrolidone in dichloromethane was used as a polymer solution with 4 wt.% of concentration. The applied voltage was 17 kV, and the distance of nozzle and collector was 8 cm. The morphology and structure of the fibers produced were observed by scanning electron microscopy.

Results: When the CO_2 pressure was 5 MPa, the resultant fibers had an average diameter of 2.28 ± 0.38 to 4.93 ± 1.02 μm. The ribbon-like morphology was formed with increasing pressure of CO_2 at 8 MPa with a tip 0.75-mm inside diameter.

Conclusions: The results show that the depressurization of CO_2 at the end of experiment assists the removal process of the polymer solvent and produces the porous nature of fibers without collapsing or foaming. These behaviors hold the potential to considerably improve devolatilization electrospinning processes.

Keywords: Electrospinning, High-pressure CO_2, Polyvinylpyrrolidone, Electrospun

Background

Electrospinning is an interesting process for producing nonwoven fibers from polymer with average diameters in the range of nano- to micrometers [1-5]. This process utilizes a high-voltage source to inject charge of a certain polarity into a polymer solution, which is then accelerated toward a collector of opposite polarity. The ease of electrospinning has proven to be a relatively simple and versatile method for forming nonwoven fibrous mats. However, there are a number of processing parameters that can greatly influence the properties of the generated fibers, such as viscoelastic force and surface tension, which have been found to depend on solution concentration [6-9], gravitational force which is dependent on solution density, and electrostatic force which has been found to depend on the applied electrostatic field and conductivity of the solution [8,9]. Additionally, temperature, humidity, and airflow in the electrospinning chamber also affect the result of the electrospinning process [10,11].

Supercritical fluid can be defined as a substance for which both pressure and temperature are above the critical point. The special combination of gas-like viscosity and liquid-like density of a supercritical fluid makes it an excellent solvent for various applications. Supercritical fluids have been used successfully as solvents, anti-solvents, or plasticizers in polymer processing, e.g., polymer modification, polymer composites, polymer blending, microcellular foaming, particle production, and in polymer synthesis [8,9,12-16]. Carbon dioxide (CO_2) is the most commonly used supercritical fluid because of its low critical temperature ($T_c = 304$ K) and pressure ($P_c = 7.38$ MPa), low toxicity, and high purity at a low cost. It was a good solvent for many nonpolar compounds and polymers. Its solvent power depends on temperature and pressure and also on weak interactions with the chain groups in the polymer. Furthermore, it is nonflammable, and its use does not contribute to the net global warming effect. Being a gas

* Correspondence: mgoto@nuce.nagoya-u.ac.jp
[1]Department of Chemical Engineering, Nagoya University, Furo-cho, Chikusa-ku, Nagoya 464-8603, Japan
Full list of author information is available at the end of the article

under ambient conditions favors its easy removal from polymeric products, thus saving costs on other secondary operations such as drying and solvent removal.

In this work, the application of high-pressure CO_2 as an electrospinning processing aid to produce fibers with various morphologies was conducted. Polyvinylpyrrolidone (PVP) has been chosen as a starting material because it was soluble in water and other polar solvents. In solution, it has excellent wetting properties and readily forms films. This makes it good as a coating or an additive to coatings. Recently, it was used for improving the dissolution rates of poorly water-soluble drugs in pharmaceutical technologies [15]. Shen et al. [17] introduced the application of CO_2 as an electrospinning processing aid at near-critical point to create fibers polymer. They used 6.5 wt.% PVP in dichloromethane (DCM) as a polymer solution feed with 20 kV of peak voltage applied. At 3.45, 4.96, and 5.10 MPa, the fibers spun which had porous internal structure with a coherent external skin have produced clearly. Similar experiments with 6.5 wt.% PVP in DCM in supercritical CO_2 resulted also in fibers spun with porous internal structure [18]. Liu et al. [18] concluded that the fibers created with a 2.5-cm distance from the nozzle to the collector had a much different morphology than the fibers obtained with three times longer nozzle-to-collector distance (8.5 cm) which housed in a nonconductive polyether ether ketone (PEEK) tube (6.35-cm inside diameter (ID), 7.62-cm outside diameter (OD), ±400-ml volume). They reported also that the polymer jet experiences longer flight times due to the longer nozzle-to-collector distance, which allows more time for the fibers to stretch and elongate before depositing on the collector. However, polymer solutions containing CO_2 exhibit extremely complex phase-equilibrium behaviors due to the large differences among the physical properties of the polymer, solvent, and CO_2; these induce complicated physicochemical interactions in polymer solutions containing CO_2 [16]. Therefore, it is not easy to predict the fiber formation mechanisms at these conditions.

Methods

Figure 1 shows our apparatus for electrospinning under high-pressure CO_2. The main apparatus consisted of a nonconductive PEEK (6.00-cm ID, 15.00-cm OD, ±565-ml volume) autoclave including cartridge heaters coupled with an electric fan, a high-voltage power supply, a high-pressure pump, a high-pressure syringe pump, a back-pressure regulator, and a stainless steel syringe with a volume of 8 ml. Before starting electrospinning experiments, the PEEK vessel was heated to around the desired temperature of 313 K (in fact, the temperature range was from 311 to 319 K). After the desired temperature was reached, CO_2 was pumped into the PEEK vessel through the PEEK capillary tube to desired pressures (5 and 8 MPa). The polymer solution was injected into the PEEK vessel when the desired conditions were reached. The high-pressure stainless steel syringe placed in the high-pressure syringe pump was used to inject the polymer solution via the PEEK capillary tube with a 0.5-mm ID; the polymer solution flow rate was 0.05 ml/min. At the

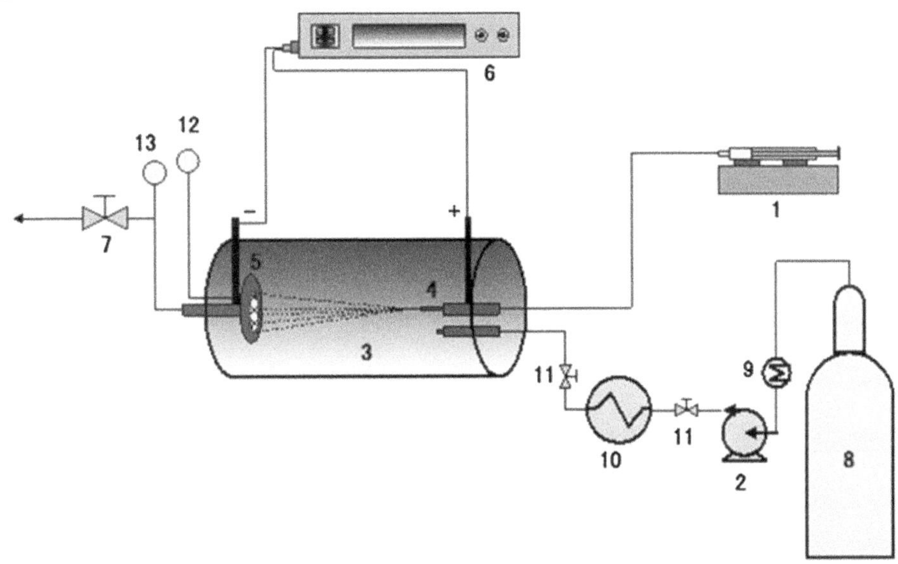

Figure 1 Schematic diagram of the electrospinning system under high-pressure CO_2. (**1**) High-pressure syringe pump, (**2**) high-pressure pump, (**3**) PEEK autoclave inclusive heater, (**4**) nozzle, **5**) fiber collector, (**6**) high-voltage power supply, (**7**) BPR, (**8**) CO_2 cylinder, (**9**). chiller, (**10**) heat exchanger, (**11**) needle valve, (**12**) temperature monitor, and (**13**) pressure monitor.

same time, a high-voltage power supply was applied to generate electrostatic force (17 kV). This apparatus transferred the polymer solution and CO_2 separately through the nozzles placed in the stainless steel flange (anode electrode), whose diameters were 0.25 to 0.75 mm and 3.175 mm, respectively. The morphologies of the electrospun fibers were observed using a scanning electron microscope (SEM) (JSM-6390LV, JEOL Ltd., Akishima, Tokyo, Japan), and the fiber diameter was measured from the SEM image using an image-analyzer software (Image J 1.42).

Results and discussion

Figure 2a,b shows pictures of the fibers obtained when electrospinning was conducted at room temperature and under pressurized CO_2, respectively. One of the challenges faced when applying electrospinning fiber formation to the surface of a collector is solvent evaporation. In this case, DCM as a polymer solvent was removed before the solution achieved the target within the short distance between the tip and collector. This is apparent in images of fibers produced by electrospinning without and with pressurized CO_2. At room temperature, the fibers have been generated clearly; however, the polymer solvent seems clear, resulting in the wet appearance of the fibers obtained. It indicates that the evaporation process of polymer solvent was slow. On the contrary, the fibers produced under pressurized CO_2 were dry, with no remaining apparent polymer solvent; this indicates that CO_2 may assist the evaporation process of the polymer solvent by depressurization at the end of the experiment, allowing the evaporation of the polymer solvent to occur more quickly. These results also showed that the use of supercritical CO_2 as a polymer synthesis solvent provides several important issues, such as solubility and drying. Because the solubility of a supercritical solvent, such as CO_2, can be tuned by controlling the temperature and pressure, it is possible to form fibers within a thermodynamic window where the polymer has

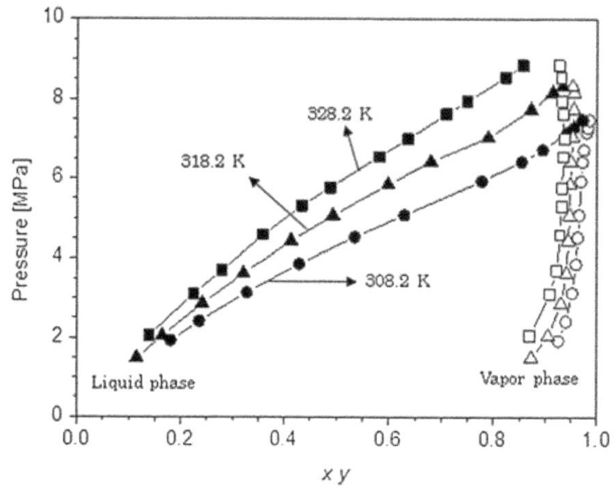

Figure 3 Phase compositions for the system CO_2 and DCM. Temperatures were at 308.2, 318.2, and 328.2 K (*x*, liquid fraction; *y*, vapor phase).

been softened, but not dissolved. As a result, CO_2 at supercritical conditions may reduce the polymer viscosity at much lower temperatures than are necessary when using melt processing [19]. As an advantage, the fibers' product can be isolated from the reaction media (supercritical CO_2) by simple depressurization, resulting in a dry polymer product. Figure 3 depicted the phase compositions for the system CO_2 and DCM at 308.2, 318.2, and 328.2 K that were plotted directly from Tsivintzelis' data [20]. From this figure, it could be seen that the CO_2 apparently had sufficient affinity to carry a portion of the DCM. As the pressure of the CO_2 was increased, the amount of displaced DCM also increased. This is, of course, a beneficial effect in terms of removing the solvent from the polymer.

Compared with another polymeric nanofiber technique, such as drawing, template synthesis, phase separation, and self-assembly, electrospinning is a unique process that is capable of producing fibers with diameters ranging over several orders of magnitude, from the micrometer range typical of conventional fibers to the

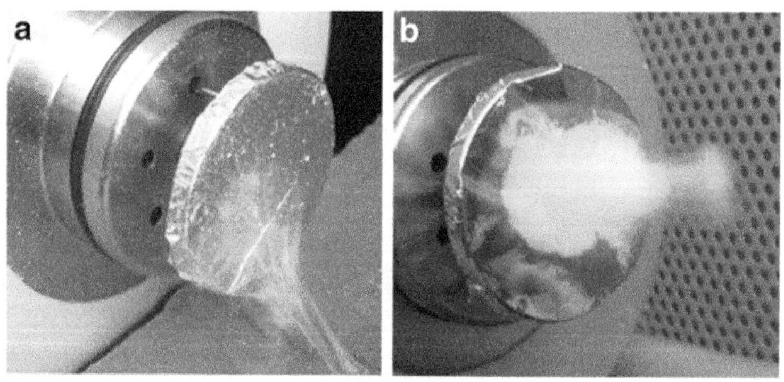

Figure 2 Pictures of fibers obtained (a) without and (b) with pressurized CO_2 at 5 MPa.

Figure 4 SEM images of fibers obtained under pressurized CO_2 at 5.0 MPa. It is with a power supply of 17 kV at various tip diameters. Tip diameter = (**a**) 0.25 mm, (**b**) 0.50 mm, and (**c**) 0.75 mm.

nanometer range. Despite the simplicity of electrospinning, the hydrostatic pressure in the capillary tube, the distance between the tip and the collector, the feed rate, and the size of the nozzle had high influence on the fiber structure and morphology of electrospun fibers associated with the electrospinning process. Figure 4 shows SEM images of PVP fibers electrospun from a polymer concentration of 4 wt.% with different IDs of the tip in a range from 0.25 to 0.75 mm. These images clearly depicted that there was no correlation between the tip diameter used and the average fiber diameter obtained in the solution electrospinning process, but a broader range of fiber diameter was obtained with a bigger needle diameter. These results indicate that the ID of the tip has a certain effect on the electrospinning process. Using the Image J 1.42 tool, from each image, at least 200 different fibers were randomly selected, and their diameters were measured to generate an average fiber diameter. Figure 5 shows the use of a tip with 0.75-mm ID yielded fibers with an average diameter of 4.11 ± 2.00 μm; a tip with 0.5-mm ID afforded an average diameter of 4.93 ± 1.02 μm, and a tip with a 0.25-mm ID resulted in an average diameter of 2.28 ± 0.38 μm. Like many other electrospinning parameters, the effect of the tip diameter on the fiber diameter is not absolute. Increase in the tip diameter was found to directly increase the fiber diameter, and fibers would easily split when they have big tip diameters [21,22]. However, Mo et al. [23] reported that a smaller internal tip diameter was found to reduce the clogging as well as the amount of beads on the electrospun fibers, and too big a tip size of internal diameter has been found to cause the blocking of nozzles and occurrence of beads. The reduction in the clogging could be due to less exposure of the solution to the CO_2 atmosphere during electrospinning. Therefore, the decrease in the internal diameter of the tip might cause a reduction in the diameter of the electrospun fibers. Finally, when the smaller tip diameter (0.25 mm) was used, good results were achieved. This means that a smaller tip diameter might decrease the polydispersity of the fiber diameters at these conditions.

Figure 6 shows the morphology of the electrospun fibers obtained when the electrospinning process was performed

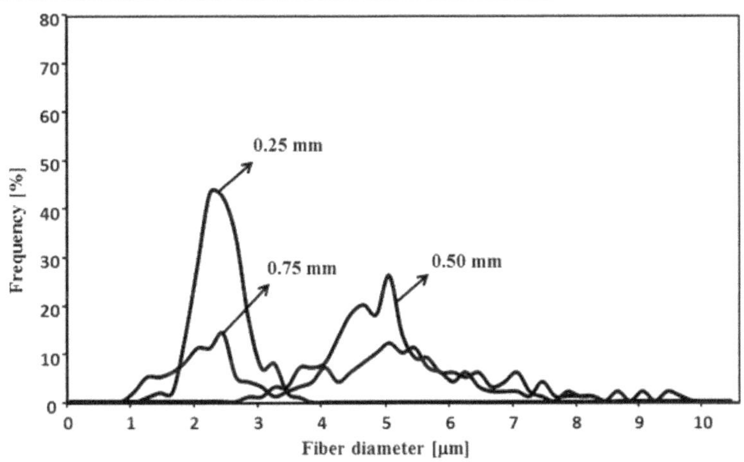

Figure 5 Diameter distribution of fibers. They are fabricated at 5 MPa and 17 kV of applied voltage with different tip diameters.

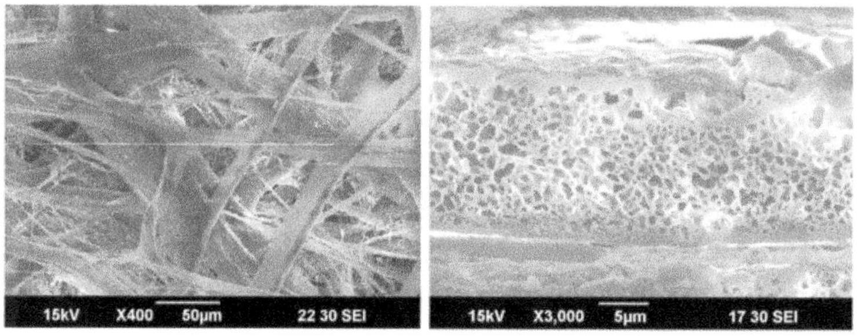

Figure 6 SEM images of PVP fibers obtained under supercritical CO_2 at 8 MPa and 320 K. Its tip diameter is 0.75 mm. Ribbon-like morphology of PVP fibers (left-side SEM) and a close-up of one of the same fibers cut open (right-side SEM).

at supercritical condition with a tip of 0.75 mm ID. Some of these fibers are cylindrical, while other fibers seem more flattened, with ribbon-like morphology. Ribbon morphology is most likely due to the collapse of rapidly solidified outer shell of electrospun jets [24]. Koombhongse et al. [24] explained that fibers in the form of ribbons with various cross sections that resulted from a thin skin formed by the rapid evaporation of the solvent. Remaining solvent escaped by diffusion through the skin. The skin had little influence during the early part of the jet path, where the jet diameter was much larger than the thickness of the skin. Another possible mechanism is non-axisymmetric jet instability originating from a perturbation of jet cross-sectional shape. Due to their irregular shape and their nonuniformity of the electrospun fibers, it is difficult to obtain a good focus on the fiber image under SEM to determine the fiber diameter. Therefore, the measurement for the size of the electrospun fibers went by their width. It should be noted that, for this particular case, the words 'diameter' and 'width' were used synonymously. Roughly, the diameter of fibers was estimated from 5 to 20 μm. When they were cut with a blade to investigate the internal fiber morphology, some of them have multiple pores. The pore sizes are estimated to be 1 to 2 μm. Porosity could possibly occur by occlusion of the solvent phase as the fibers agglomerate. Another more likely possibility is that the solvent phase forms porous regions by nucleation and growth in the polymer-rich phase as more CO_2 is transferred into the fibers. The depressurization ensures that the fibers retain their porous nature without collapsing or foaming [25]. Watkins and McCarthy [26] also reported that the presence of CO_2 at supercritical conditions could remove solvents from polymer blends by depressurization. They described that the infusion of CO_2 into a variety of semicrystalline and glassy polymer substrates and thermally initiated radical polymerizations within the swollen substrates to generate polystyrene-substrate polymer blends. As reported by Shin et al. [27], that type of CO_2 is not a good solvent to dissolve the PVP polymer. In their system (PVP + DCM + supercritical CO_2), PVP was not soluble in CO_2 at pressures as high as 290 MPa and at temperatures up to 480 K. On the other hand, the PVP was readily soluble in dichloromethane. In similar systems, Gokhale et al. [28] performed supercritical antisolvent for particle formation with CO_2 as a medium. With infrared spectroscopy, they suggested that no significant effect on the PVP particles formed. These observations supported the results obtained, i.e., electrospun fibers which were produced by electrospinning under pressurized CO_2 did not change their properties. Currently, the PEEK autoclave which attached in the electrospinning apparatus was being modified. This modification was expected to allow for the observation of the inside of the PEEK autoclave using a charge-coupled device camera and to stabilize the experiment temperature.

Experimental

Polyvinylpyrrolidone (MW 1,300,000) was purchased from Sigma-Aldrich (St. Louis, MO, USA) and used as received. As a solution solvent, DCM (99.0%) was obtained from Wako Pure Chemical Industries, Ltd. (Osaka, Japan) and was used without further purification. As a polymer solution, PVP was dissolved in DCM at a concentration of 4 wt. % at room temperature. This concentration was selected based on the previous researcher's reports [17,18]. After the PVP had been dissolved in DCM, it was poured into a stainless steel syringe.

Conclusions

The production of PVP fibers by electrospinning at 17 kV was studied at a temperature and pressure of 313 K and approximately 8.0 MPa, respectively. At room temperature, the fibers obtained seemed wet; conversely, the fibers produced were dry with no remaining polymer solvent when electrospinning was conducted under pressurized CO_2. At 5 MPa, when a smaller tip diameter (0.25-mm ID) was used, the low polydispersity of the fiber diameters was achieved with an average diameter of 2.28 ± 0.38 μm. The ribbon-like structures of PVP was formed with increasing pressure of CO_2 at 8 MPa with a tip of 0.75-mm ID. The results thus show that the depressurization of CO_2 at the end of experiment assists the removal process of the polymer solvent and produces the porous nature of fibers without collapsing or foaming.

Competing interests
The authors declare that they have no competing interests.

Authors' contributions
W and SM conducted the experiment and all data analyses. KM carried out the SEM analysis. SO participated in electrospinning development. MG supervised this work and provided all experimental and analytical equipment. All authors read and approved the final manuscript.

Acknowledgments
This work was supported by the Grants-in-Aid for Scientific Research by the Ministry of Education, Culture, Sports, Science and Technology, Japan.

Author details
[1]Department of Chemical Engineering, Nagoya University, Furo-cho, Chikusa-ku, Nagoya 464-8603, Japan. [2]Department of Chemical Engineering, Sepuluh November Institute of Technology, Kampus ITS Sukolilo, Surabaya 60111, Indonesia. [3]Department of Applied Chemistry and Biochemistry, Kumamoto University, Kumamoto 860-8555, Japan. [4]Department of Advanced Fibro-Science, Kyoto Institute of Technology, Kyoto 606-8585, Japan.

References
1. Huang M, Zhang Y, Kotaki M, Ramakrishna S (2003) Compos Sci Technol 63:2223
2. Ramakrishna S, Fujihara K, Teo WE, Lim TC, Ma Z (2005) In: An introduction to electrospinning and nanofibers. World Scientific Publishing, Singapore, p 90
3. Doshi J, Reneker DH (1995) J Electrostat 35:151
4. Casper CL, Stephens JS, Tassi NG, Bruce CD, Rabolt JF (2004) Macromolecules 37:573
5. Theron SA, Zussman E, Yarin AL (2004) Polymer 45:2017
6. Cloupeau M, Prunet-Foch B (1990) J Electrostatics 25:165
7. Demir MM, Yilgor I, Yilgor E, Erman B (2002) Polymer 43:3303
8. Baumgarten PK (1971) J Colloid and Interface Science 36:71
9. Fong H, Chun I, Reneker DH (1999) Polymer 40:4585
10. Lee JS, Choi KH, Ghim HD, Kim SS, Chun DH, Kim HY, Lyoo WS (2004) J Appl Polym Sci 93:1638
11. Wannatong L, Sirivat A, Supaphol P (2004) Polym Int 53:1851
12. Nalawade SP, Picchioni F, Marsman JH, Janssen LPBM (2006) J Supercritical Fluids 36:236
13. Wahyudiono, Murakami K, Machmudah S, Sasaki M, Goto M (2012) High Pressure Research An International Journal 32:54
14. Wahyudiono, Murakami K, Machmudah S, Sasaki M, Goto M (2012) Jpn J Appl Phys 51:08HF07
15. Ignatova M, Manolova N, Rashkov I (2007) Eur Polym J 43:1112
16. Bungert B, Sadowski G, Arlt W (1998) Ind Eng Chem Res 37:3208
17. Shen Z, Thompson BE, McHugh MA (2006) Macromolecules 39:8553
18. Liu J, Shen Z, Lee SH, Marquez M, McHugh MA (2010) J Supercritical Fluids 53:142
19. Natalia L, Gary T (2004) J. Supercritical Fluids 31:329
20. Tsivintzelis I, Missopolinou D, Kalogiannis K, Panayiotou C (2004) Fluid Phase Equilibria 224:89
21. Zhao SL, Wu XH, Wang LG, Huang Y (2004) J Appl Polym Sci 91:242
22. Zeng J, Chen X, Xu X, Liang Q, Bian X, Yang L, Jing X (2003) J Appl Polym Sci 89:1085
23. Mo XM, Xu CY, Kotaki M, Ramakrishna S (2004) Biomaterials 25:1883
24. Koombhongse S, Liu W, Reneker DH (2001) J Polymer Science Part B 39:2598
25. Dixon DJ, Bodmeier RA, Johnston KP (1993) AIChE J 39:127
26. Watkins JJ, McCarthy TJ (1994) Macromolecules 27:4845
27. Shin MS, Lee JH, Kim H (2008) Fluid Phase Equilibria 272:42
28. Gokhale A, Khusid B, Dave RN, Pfeffer R (2007) J Supercritical Fluids 43:341

Green synthesis of silver nanoparticles using carob leaf extract and its antibacterial activity

Akl M Awwad[1*], Nidá M Salem[2] and Amany O Abdeen[1]

Abstract

Background: This paper describes a rapid and eco-friendly method for green synthesis of silver nanoparticles from aqueous solution of silver nitrate using carob leaf extract (*Ceratonia siliqua*) in a single-pot process.

Results: Formation of stable silver nanoparticles at different concentrations of $AgNO_3$ gave mostly spherical particles with a diameter ranging from 5 to 40 nm. It was observed that the use of carob leaf extract makes a fast and convenient method for the synthesis of silver nanoparticles and can reduce silver ions into silver nanoparticles within 2 min of reaction time without using any severe conditions. Green synthesis of silver nanoparticles (AgNPs) was characterized by UV-visible (UV-vis) spectroscopy, scanning electron microscopy, atomic absorption spectroscopy, Fourier transform infrared spectroscopy, and X-ray diffraction (XRD). The UV–vis spectra gave surface plasmon resonance for synthesized silver nanoparticles at 420 nm. The XRD analysis showed that the AgNPs are crystalline in nature and have face-centered cubic geometry.

Conclusion: Further, the AgNPs showed an effective antibacterial activity toward *Escherichia coli* pathogen.

Keywords: Biosynthesis, Silver nanoparticles, Carob leaf extract, Characterization, Antibacterial activity

Background

Silver nanoparticles (AgNPs) have become the focus of intensive research owing to their wide range of applications in areas such as catalysis, optics, antimicrobials, and biomaterial production. Silver nanoparticles exhibit new or improved properties depending upon their size, morphology, and distribution. Various approaches using plant extract have been used for the synthesis of metal nanoparticles. These approaches have many advantages over chemical, physical, and microbial syntheses [1-10] because there is no need of the elaborated process of culturing and maintaining the cell, hazardous chemicals, high-energy requirements, and wasteful purifications. Many research papers reported the synthesis of silver nanoparticles using plant extracts such as *Helianthus annuus, Basella alba, Oryza sativa, Saccharum officinarum, Sorghum bicolor,* and *Zea mays* [11]; pine, persimmon, ginkgo, magnolia, and platanus leaves [12]; *Jatropha curcas* seeds [13]; *Acalypha indica* leaf [14]; banana peel [15]; *Chenopodium album* leaf [16]; *Rosa rugosa* [17]; *Trianthema decandra* roots [18]; *Ocimum sanctum* stems and roots [19]; *Sesuvium portulacastrum* leaves [20]; *Murraya koenigii* (curry) leaf [21]; *Macrotyloma uniflorum* seeds [22]; *Ocimum sanctum* (Tulsi) leaf [23]; *Garcinia mangostana* (mangosteen) leaf [24]; *Stevia rebaudiana* leaves [25]; *Nicotiana tobaccum* leaf [26]; *Ocimum tenuiflorum, Solanum trilobatum, Syzygium cumini, Centella asiatica, and Citrus sinensis* leaves [27]; *Arbutus unedo* leaf [28]; *Ficus benghalensis* leaf [29]; mulberry leaves [30]; and *Olea europaea* leaves [31].

In this study, we have synthesized silver nanoparticles using carob leaf extract for reduction of Ag^+ ions to Ag^0 nanoparticles from silver nitrate solution within 2 min of reaction time at ambient temperature. It was also shown that the average size of silver nanoparticles can be controlled to 5 to 40 nm by varying the concentration of silver nitrate and the volume of leaf extract. Further, biosynthesized silver nanoparticles are found to be highly effective against *Escherichia coli* bacteria.

Methods

Preparation of carob leaf extract

Carob leaves (Figure 1) were collected from carob trees at the campus of the Royal Scientific Society, Amman,

* Correspondence: akl.awwad@yahoo.com
[1]Royal Scientific Society, El Hassan Science City, Amman 11941, Jordan
Full list of author information is available at the end of the article

Figure 1 Picture of carob leaves.

Jordan. The leaves were washed several times with distilled water to remove the dust particles and then sun-dried to remove the residual moisture. The dried carob leaves were cut into small pieces and boiled in a 250-ml glass beaker along with 200 ml of sterile distilled water for 10 min. After boiling, the color of the aqueous solution changed from watery to yellow color. The aqueous extract was separated by filtration with Whatman No. 1 filter paper (Maidstone, UK) and then centrifuged at 1,200 rpm for 5 min to remove heavy biomaterials. The carob leaf extract was stored at room temperature to be used for biosynthesis of silver nanoparticles from silver nitrate.

Synthesis of AgNPs

In a typical reaction procedure, 5 ml of carob leaf extract was added to 100 ml of 1×10^{-3} M aqueous $AgNO_3$ solution, with stirring magnetically at room temperature. The yellow color of the mixture of silver nitrate and carob leaf extract at 0 min of reaction time changed very fast at room temperature after 2 min to a black suspended mixture. This indicated that carob leaf extract speeds up the biosynthesis of silver nanoparticles more than other plant leaves [12-18]. The concentrations of $AgNO_3$ solution and leaf extract were also varied at 1 to 4 mM and 5% to 10% by volume, respectively. UV-visible (UV–vis) spectra showed strong surface plasmon resonance (SPR) band at 420 nm and thus indicating the formation of silver nanoparticles. The AgNPs obtained by carob leaf extract were centrifuged at 15,000 rpm for 5 min and subsequently dispersed in sterile distilled water to get rid of any uncoordinated biological materials.

Characterization techniques

UV–vis absorption spectra were measured using a Shimadzu UV-1601 spectrophotometer (Kyoto, Japan). Crystalline metallic silver nanoparticles were examined using an X-ray diffractometer (Shimadzu, XRD-6000) equipped with Cu Kα radiation source using Ni as filter at a setting of 30 kV/30 mA. All X-ray diffraction (XRD) data were collected under the experimental conditions in the angular range $3° \leq 2\theta \leq 50°$. Fourier transform infrared (FT-IR) spectra for *Ceratonia siliqua* leaf extract powder and silver nanoparticles was obtained in the range 4,000 to 400 cm^{-1} with an IR-Prestige-21 Shimadzu FT-IR spectrophotometer, by KBr pellet method. Scanning electron microscopy (SEM) analysis of synthesized silver nanoparticles was done using a Hitachi S-4500 SEM machine (Chiyoda-ku, Japan). Concentration of silver ions was analyzed by atomic absorption spectroscopy (AAS; AA-6300, Shimadzu).

Results and discussion

FT-IR spectrum

The FT-IR spectrum obtained for carob leaf extract (Figure 2) displays a number of absorption peaks, reflecting its complex nature. Strong absorption peaks at 3,309 and 3,421 cm^{-1} result from stretching of the -NH band of amino groups or is indicative of bonded -OH hydroxyl group. The absorption peaks at about 2,897 cm^{-1} could be assigned to stretching vibrations of -CH_2 and CH_3 functional groups. The peaks at 1,712 and 1,581 cm^{-1} indicate the fingerprint region of CO, C-O, and O-H groups. The intense band at 1,068 cm^{-1} could be assigned to the C-N stretching vibrations of aliphatic amines. The FT-IR spectrum also shows bands at 1,581 and 1,446 cm^{-1} identified as amide I and amide II which arise due to carbonyl (C=O) and amine (–NH) stretching vibrations in the amide linkages of the proteins, respectively. The absorption band at 1,446 cm^{-1} could be attributed to methylene scissoring vibrations from the proteins. FT-IR study indicates that the carboxyl (–C=O), hydroxyl

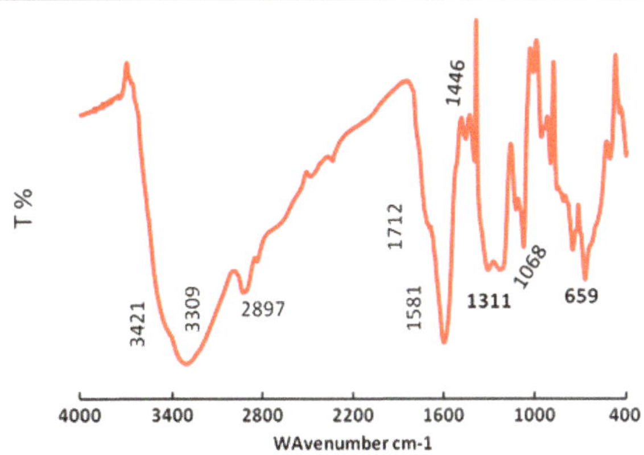

Figure 2 FT-IR spectrum of carob leaf extract.

(-OH), and amine (N-H) groups in carob leaf extract are mainly involved in reduction of Ag^+ ions to Ag^0 nanoparticles. The FT-IR spectroscopic study also confirmed that the protein present in carob leaf extract acts as a reducing agent and stabilizer for the silver nanoparticles and prevents agglomeration. The carbonyl group of amino acid residues has a strong binding ability with metal, suggesting the formation of a layer covering silver nanoparticles and acting as a stabilizing agent to prevent agglomeration in the aqueous medium.

UV–vis absorbance study

The addition of carob leaf extract to silver nitrate ($AgNO_3$) solution resulted in color change of the solution from transparent to brown due to the production of silver nanoparticles. The color changes arise from the excitation of surface plasmon vibrations with the silver nanoparticles. The SPR of silver nanoparticles produced a peak centered near 420 nm. UV–vis absorbance of the reaction mixture was taken from 0 till 2 min (Figure 3). It was observed that the absorbance peak was centered near 420 nm, indicating the reduction of silver nitrate into silver nanoparticles. It was also observed that the reduction of silver ions into silver nanoparticles started at the start of reaction and reduction was completed at almost 2 min at room temperature, indicating rapid biosynthesis of silver nanoparticles.

XRD studies

Analysis through X-ray diffraction was carried out to confirm the crystalline nature of the silver nanoparticles. The XRD pattern showed numbers of Bragg reflections that may be indexed on the basis of the face-centered cubic structure of silver. A comparison of our XRD spectrum with the standard confirmed that the silver particles formed in our experiments were in the form of nanocrystals, as evidenced by the peaks at 2θ values of 38. 28°, 44.04°, 64.34°, and 77.28° corresponding to (111), (200), (220), and (311) Bragg reflections, respectively, which may be indexed based on the face-centered cubic structure of silver. X-ray diffraction results clearly show that the silver nanoparticles formed by the reduction of Ag^+ ions by the carob leaf extract are crystalline in nature. The unassigned peaks at 2θ = 27.96°, 32.28°, and 46.18° denoted by (*) in Figure 4 are thought to be related to crystalline and amorphous organic phases. It was found that the average size from XRD data and using the Debye-Scherrer equation was approximately 18 nm. The presence of structural peaks in XRD patterns and the average crystalline size around 18 nm clearly illustrate that the AgNPs synthesized by our green method were nanocrystalline in nature. The average particle size of silver nanoparticles synthesized by the present green method can be calculated using the Debye-Scherrer equation [19,22]:

$$D = K\lambda/\beta\cos\theta,$$

where D is the crystallite size of AgNPs, λ is the wavelength of the X-ray source (0.1541 nm) used in XRD, β is the full width at half maximum of the diffraction peak, K is the Scherrer constant with a value from 0.9 to 1, and θ is the Bragg angle.

FT-IR spectra of biosynthesized silver nanoparticles

Results of the FT-IR study of biosynthesized AgNPs showed sharp absorption peaks located 3,410, 1,608, 1,558, and 1,408 cm^{-1} (Figure 5). The absorption peak at 1,608 cm^{-1} may be assigned to the amide I bond of proteins arising from carbonyl stretching in proteins, and the peak at 3,410 cm^{-1} is assigned to OH stretching in alcohols and phenolic compounds. The absorption peak at 1,608 cm^{-1} is close to that reported for native proteins, which suggests that proteins are interacting with biosynthesized silver nanoparticles and also their secondary structure was not affected during reaction

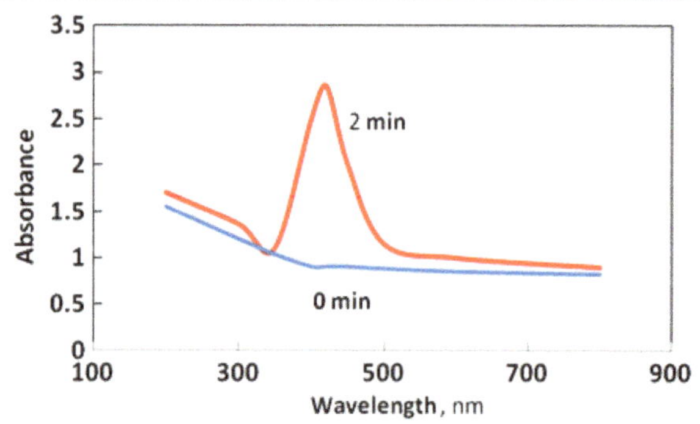

Figure 3 UV–vis spectrum. Showing absorption of 10^{-3} M aqueous solution of silver nitrate with carob leaf extract as a function of time.

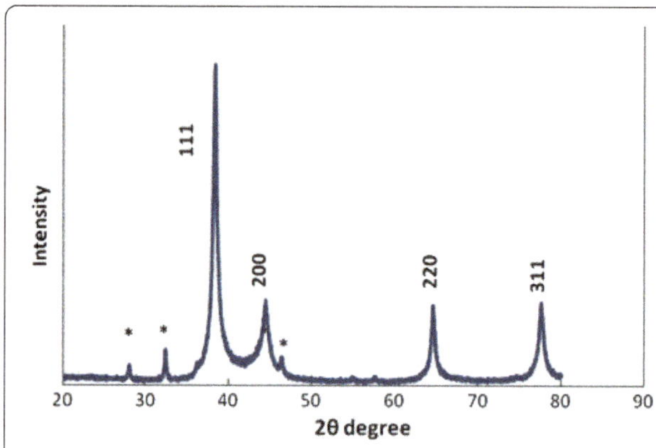

Figure 4 XRD pattern of silver nanoparticles biosynthesized by treating silver nitrate with carob leaf extract.

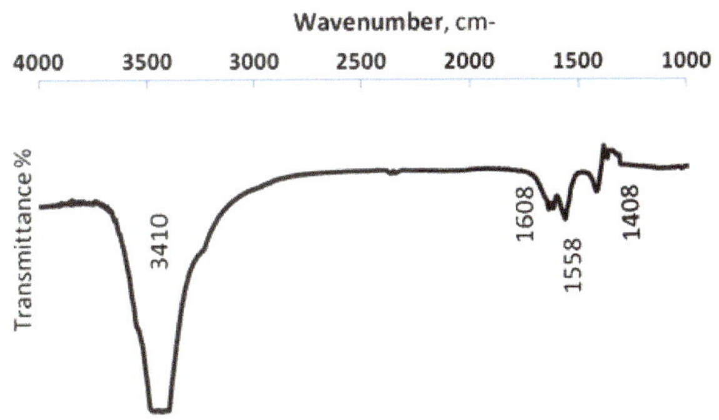

Figure 5 FT-IR of dried silver nanoparticles biosynthesized by carob leaf extract.

with Ag^+ ions or after binding with Ag^0 nanoparticles. This FT-IR spectroscopic study confirmed that the carbonyl group of amino acid residues has a strong binding ability with silver, suggesting the formation of a layer covering silver nanoparticles and acting as a capping agent to prevent agglomeration and provide stability to the medium. These results confirm the presence of possible proteins acting as reducing and stabilizing agents.

Atomic absorption spectroscopy analysis

Silver ion concentration was analyzed by AAS which showed the conversion of Ag^+ ions into Ag^0 nanoparticles. Initially, a standard solution of 10 ppm of $AgNO_3$ was prepared and analyzed with AAS at 0 min. The Ag^+ ion concentration in the reaction solution, after adding carob leaf extract, was monitored at different time intervals. The result showed a decrease in concentration of Ag^+ ions with reaction time, indicating the complete conversion of Ag^+ ions into Ag^0 nanoparticles at 2 min of reaction time (Figure 6).

SEM analysis of silver nanoparticles

SEM analysis shows uniformly distributed silver nanoparticles on the surfaces of the cells (Figure 7). The silver nanoparticles were spherical in shape with particle size range from 5 to 40 nm. The larger silver particles may be due to the aggregation of the smaller ones, due to the SEM measurements.

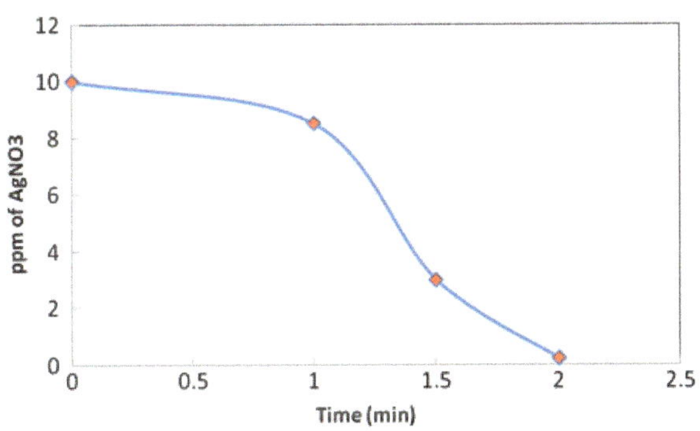

Figure 6 AAS analysis of silver nitrate (ppm) with reaction time (min) using carob leaf extract.

Figure 7 SEM image of silver nanoparticles synthesized by carob leaves extract.

Antibacterial assays

Biosynthesized silver nanoparticles were analyzed for their antimicrobial activity against *E. coli* by disk diffusion method (Figure 8). It was observed that microbial growth of *E. coli* was independent on AgNP concentration. The minimum inhibitory concentration (MIC) was determined as the lowest concentration of silver nanoparticles that inhibited the visible growth of *E. coli*. It was found that the MIC (μg/l) for silver nanoparticles is 0.5, for silver nitrate 1.8 mg, and for standard antibiotic 0.6 mg. The zone of inhibition ranged from 8 to 12 mm. These results indicated that the silver nanoparticles

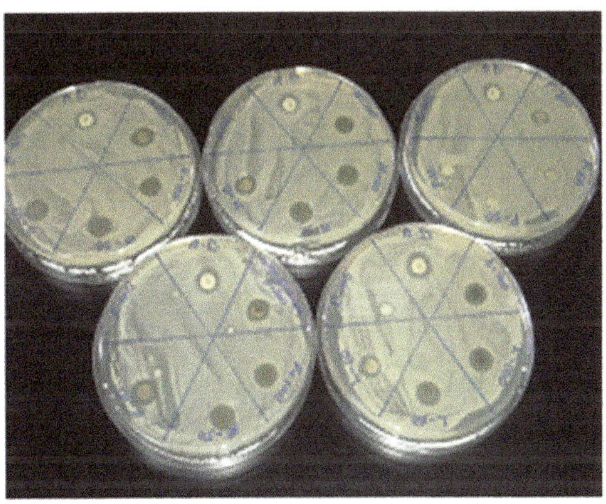

Figure 8 Antibacterial activity of silver nanoparticles against *E. coli*.

synthesized using carob leaf extract have stronger activity than standard antibiotic.

Experimental

For synthesis of silver nanoparticles, at first, carob leaf extract is prepared by boiling the dried small pieces of carob leaves in sterile distilled water. In a typical reaction procedure, 5 ml of carob leaf extract is added to 100 ml of 1×10^{-3} M aqueous solution of silver nitrate ($AgNO_3$), with stirring magnetically at room temperature. At this stage, the color of the reaction mixture changes from yellow to black and black suspended particles are produced, indicating the formation of silver nanoparticles. The black mixture is centrifuged, washed three times with distilled water and ethanol, and finally dried at 60°C. Synthesized silver nanoparticles were characterized by UV–vis spectroscopy, SEM, AAS, FT-IR, and XRD. Afterwards the antibacterial activity of silver nanoparticles was studied by diffusion method.

Conclusions

We have developed a fast, eco-friendly, and convenient green method for the synthesis of silver nanoparticles from silver nitrate using carob leaf extract at ambient temperature. Carob leaf extract is found suitable for the green synthesis of silver nanoparticles within 2 min at ambient conditions. Spherical, polydisperse AgNPs of particle sizes ranging from 5 to 40 nm with an average size of 18 nm are obtained. Color changes occur due to surface plasmon resonance during the reaction with the ingredients present in the carob leaf extract resulting in the formation of silver nanoparticles, which is confirmed by XRD, FT-IR, UV–vis spectroscopy, and SEM. FT-IR spectroscopic study confirmed that the carbonyl group of amino acid residues has a strong binding ability with silver, suggesting the formation of a layer covering silver nanoparticles and acting as a capping agent to prevent agglomeration and provide stability to the medium, yet further research is needed in this area to explore the possible biomolecule responsible for the bioreduction process. The antibacterial activity of biologically synthesized silver nanoparticles was evaluated against *E. coli* pathogen.

Competing interests

The authors declare that they have no competing interests.

Authors' contributions

AMA participated in the design of the study and performed the green synthesis method and characterization of synthesized nanoparticles. NMS participated in the design of the study and carried out the extraction of carob leaf extract and analysis of antibacterial activity results. AOA performed the antibacterial activity. All authors read and approved the final manuscript.

Acknowledgements

This work was supported by the SABEQ Program and Royal Scientific Society, Jordan.

Author details

[1]Royal Scientific Society, El Hassan Science City, Amman 11941, Jordan. [2]Plant Protection Department, Faculty of Agriculture, Jordan University, Amman 11942, Jordan.

References

1. Liu YC, Lin LH (2004) New pathway for the synthesis of ultrafine silver nanoparticles from bulk silver substrates in aqueous solution by sonoelectrochemical methods. Electrochem Commun 6:78–86
2. Vorobyova SA, Lesnikovich AI, Sobal NS (1999) Preparation of silver nanoparticles by interphase reduction. Colloid Surf A 152:375–379
3. Bae CH, Nam SH, Park SM (2002) Formation of silver nanoparticles by laser ablation of a silver target in NaCl solution. Appl Surf Sci 197:628–634
4. Mandal D, Blonder ME, Mukhopadhyay D, Sankar G, Mukherjea P (2000) The use of microorganisms for the formation of metal nanoparticles and their application. Appl Microbiol Biotechnol 69:485–492
5. Basavaraja S, Balaji DF, Lagashetty A, Rajasab AH, Venkataraman A (2008) Extracellular biosynthesis of silver nanoparticles using the fungus *Fusarium semitectum*. Mater Res Bull 43:1164–1170
6. Kowshik M, Ashtaputre S, Kharrazi SS, Vogel W, Urban J, Kulkarni SK, Paknikar M (2003) Extracellular synthesis of silver nanoparticles by a silver-tolerant yeast strain MKY3. Nanotechnology 14:95–100
7. Keki S, Torok J, Deak G (2000) Silver nanoparticles by PAMAM-assisted photochemical reduction of Ag^{+}. J Colloid Interface Sci 229:550–553
8. Jha AK, Prasad K (2010) Green synthesis of silver nanoparticles using *Cycas* leaf. Inter J Green Nanotechno: Phys Chem 1:110–117
9. Yu D-G (2007) Formation of colloidal silver nanoparticles stabilized by Na+ –poly(gamma-glutamic acid)-silver nitrate complex via chemical reduction process. Colloids Surf B 59:171–178
10. He S, Guo Z, Zhang Y, Zhang S, Wang J, Gu N (2007) Biosynthesis of gold nanoparticles using the bacteria *Rhodopseudomonas capsulate*. Mater Lett 61:3984–3987
11. Leela A, Vivekanandan M (2008) Tapping the unexploited plant resources for the synthesis of silver nanoparticles. Afr J Biotechnol 7:3162–5315
12. Song JY, Kim BS (2009) Rapid biological synthesis of silver nanoparticles using plant leaf extract. Bioprocess Biosyst Eng 32:79–84
13. Bar H, Bhui DK, Gobinda SP, Sarkar PM, Pyne S, Misra A (2009) Green synthesis of silver nanoparticles using seed extract of *Jatropha curcas*. Physicochem Eng Aspects 348:212–216
14. Krishnaraj C, Jagan EG, Rajasekar S, Selvakumar P, Kalaichelvan PT, Mohan N (2010) Synthesis of silver nanoparticles using *Acalypha indica* leaf extracts and its antimicrobial activity against water borne pathogens. Colloids Surf B Biointerfaces 76:50–56
15. Bankar A, Joshi B, Kumar AR, Zinjarde S (2009) Banana peel extract mediated novel route for synthesis of silver nanoparticles. Colloid Surf A Physicochem Eng Aspect 368:58–63
16. Dwivedi AD, Gopal K (2010) Biosynthesis of silver and gold nanoparticles using *Chenopodium album* leaf extract. Colloid Surf A Physicochem Eng Aspect 369:27–33
17. Dubey SP, Lahtinen M, Sillanpaa M (2010) Green synthesis and characterization of silver and gold nanoparticles using leaf extract of *Rosa rugosa*. Colloid Surf A Physicochem Eng Aspect 364:34–41
18. Geethalakshmi E, Sarada DV (2010) Synthesis of plant-mediated silver nanoparticles using *Trianthema decandra* extract and evaluation of their anti microbial activities. Int J Eng Sci Tech 2:970–975
19. Ahmad N, Sharma S, Alam MK, Singh VN, Shamsi SF, Mehta BR, Fatma A (2010) Rapid synthesis of silver nanoparticles using dried medicinal plant of basil. Colloids Surf B Biointerfaces 81(1):81–86
20. Nabikhan A, Kandasamy K, Raj A, Alikunhi NM (2010) Synthesis of antimicrobial silver nanoparticles by callus and leaf extracts from saltmarsh plant, *Sesuvium portulacastrum* L. Colloids Surf B Biointerfaces 79:488–493
21. Christensen L, Vivekanandhan S, Misra M, Mohanty AK (2011) Biosynthesis of silver nanoparticles using *murraya koenigii* (curry leaf): an investigation on the effect of broth concentration in reduction mechanism and particle size. Adv Mater Letters 2:429–434
22. Vidhu VK, Aromal A, Philip D (2011) Green synthesis of silver nanoparticles using *Macrotyloma uniflorum*. Spectrochim Acta A Mol Biomol Spectros 83:392–397

23. Singhal G, Bhavesh R, Kasariya K, Sharma AR, Singh RP (2011) Biosynthesis of silver nanoparticles using *Ocimum sanctum* (Tulsi) leaf extract and screening its antimicrobial activity. J Nanopart Res 13:2981–2988
24. Veerasamy R, Xin TZ, Gunasagaran S, Xiang TF, Yang EF, Jeyakumar N, Dhanaraj SA (2011) Biosynthesis of silver nanoparticles using mangosteen leaf extract and evaluation of their antimicrobial activities. J Saudi Chemical Society 15:113–120
25. Yilmaz M, Turkdemir H, Kilic MA, Bayram E, Cicek A, Mete A, Ulug B (2011) Biosynthesis of silver nanoparticles using leaves of *Stevia rebaudiana*. Mater Chem Phys 130:1195–1202
26. Parasad KS, Pathak D, Patel A, Dalwadi P, Prasad R, Patel P, Selvaraj K (2011) Biogenic synthesis of silver nanoparticles using *Nicotiana tobaccum* leaf extract and study of their antimicrobial effect. Afr J Biotechnol 10:8122–8130
27. Logeswari P, Silambarasan S, Abraham J (2012) Synthesis of silver nanoparticles using plants extract and analysis of their antimicrobial property. J Saudi Chem Soc. doi:10.1016/j.jscs.2012.04.007
28. Kouvaris P, Delimitis A, Zaspalis V, Papadopoulos D, Tsipas SA, Michalidis N (2012) Green synthesis and characterization of silver nanoparticles produced using *Arbutus Unedo* leaf extract. Mater Letters 76:18–20
29. Saxena A, Tripathi RM, Zafar F, Singh P (2012) Green synthesis of silver nanoparticles using aqueous solution of *Ficus benghalensis* leaf extract and characterization of their antimicrobial activity. Mater Letters 67:91–94
30. Awwad AM, Salem NM (2012) Green synthesis of silver nanoparticles by mulberry leaves extract. Nanosci Nanotechno 2:125–128
31. Awwad AM, Salem NM, Abdeen A (2012) Biosynthesis of silver nanoparticles using *Olea europaea* leaves extract and its antibacterial activity. Nanosci Nanotechno 2:164–170

Modeling and simulation of an industrial secondary reformer reactor in the fertilizer plants

Ali Ashour AL-Dhfeery* and Ala'a Abdulrazaq Jassem

Abstract

In this paper, the industrial secondary reformer reactor has been modeled and simulated at steady-state operation conditions. It aims to modify a complete mathematical model of the secondary reformer design. The secondary reformer is a part of a subprocess in a higher scale unit of ammonia synthesis. It is located after the primary reformer in the ammonia plant where it is used in the synthesis of ammonia. The reactions in the secondary reformer reactor are assumed to be carried inside two reactors in series. The first reactor, upper section, is the combustion zone, while the second reactor, bottom section, is the catalyst zone with nickel catalyst based on the alumina. In order to create a reactor design model for the secondary reformer in the industrial ammonia plant, combustion and catalyst zones have been studied. The temperature and compositions of gas in the combustion zone are predicted using the atomic molar balance and adiabatic flame temperature model. In the catalyst zone, the temperature and composition profiles along the axial distance are predicted using a one-dimensional heterogeneous catalytic reaction model with the assumption that the reactions are first order in methane partial pressure.

Most of the methane is converted in the part of one fourth to one half of the catalyst zone length from inlet to outlet. The temperature gradient between the gas and catalyst surface decreases along the axial distance, and both temperatures approach the same value at the bottom of the catalyst bed. Finally, the results of this simulation have been compared with the industrial data taken from the existing ammonia unit in the State Company of Fertilizers South Region in the Basra/Iraq, which show a relatively good compatibility.

Keywords: Ammonia, Autothermal reformer, Steam reforming, Reforming reactions, Hydrogen production

Background

The industrial secondary reformer reactor plays an important role in the ammonia plant. The aim of the secondary reformer reactor is to produce and adjust the amount of hydrogen and nitrogen gases. A schematic diagram of the industrial-scale secondary reformer reactor is shown in Figure 1. The process gas leaves the primary reformer through the transfer line; then, it will enter the secondary reformer at the top through the combustion chamber. Process air is introduced into this combustion chamber, in which the oxygen of the air is burnt, liberating heat and raising the temperature of the product gases from the primary reformer, where N_2 will be produced through this section. The partially oxidized gas passes through the catalyst zone to produce the hydrogen, and it is provided with a Ni/$MgAl_2O_4$ catalyst inside the reactor shell. Finally, the reformat gases leave the secondary reformer through the outlet nozzle at the bottom of the secondary reformer reactor.

The results of previous studies by various researchers mainly by Ebrahimi et al. [1], Khorsand and Deghan [2], Yu [3], Raghunandaqnan and Reddy [4], and Ravi et al. [5] have included the design and evaluation of the performance of a secondary reformer with different parameters.

Mathematical models which combine the predictions of the overall consumption of the reactants with a prediction of the product distribution are very scarce for the secondary reformer in literature. Only few research groups studied the combustion zone in the secondary reformer reactor explicitly. One reason might be the unavailability of the mechanism of methane and hydrogen combustion in the literature due to the wide distribution of detailed mechanisms for the combustion of

* Correspondence: chemicalengineer82@yahoo.com
Department of Chemical Engineering, Basra University, Basra, 61004, Iraq

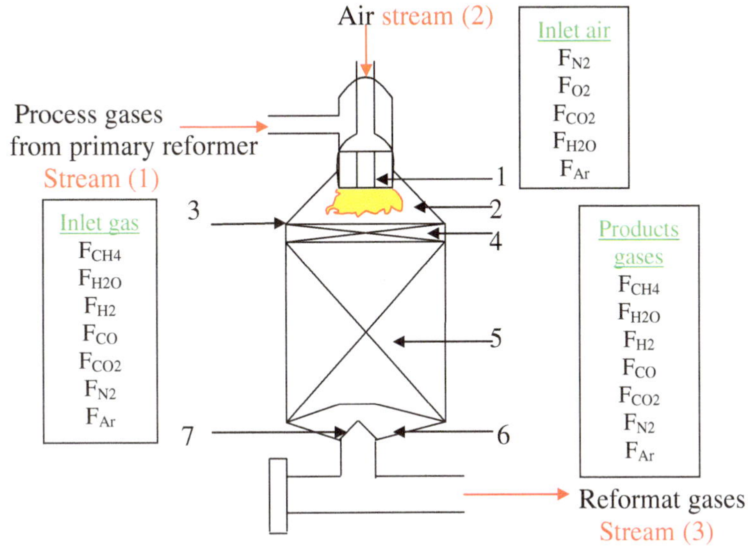

Figure 1 Schematic diagram of the secondary reformer. 1, burner; 2, combustion zone; 3, alumina bricks plate; 4, alumina balls (diameter = 40 mm); 5, catalyst zone; 6, alumina balls (diameter = 40 mm, 75 mm, and 120 mm); 7, cone.

natural gas only, which consists mainly of methane. Therefore, numerous researchers focused attention on the autothermal reformer which represented a stand-alone process wherein the entire natural gas conversion is carried out by an internal combustion with oxygen as presented by Amirshaghaghi et al. [6], Li et al. [7], Pina and Borio [8], Skukri et al. [9], and Hoang and Chan [10].

Conversional technology employs a single reactor and single nickel catalyst, but in this study, the reactor is divided into two reactors in series. The first reactor, upper section, is the combustion zone that is filled with inert gases from the primary reformer without catalyst; the combustion is carried out by adding an air stream, not oxygen as in the previous studies in autothermal reformer; in the second reactor, bottom section, which is the catalyst zone filled with nickel catalyst, the classic reforming reactions occur. The originality of the work lies in this proposal.

The results of the mathematical model such as mole fraction, temperature profile, and pressure drop of the combustion and catalyst zones will be compared relative to the industrial values taken from the secondary reformer reactor in the State Company of Fertilizers South Region (SCFSR) in the Basra/Iraq to show the degree of accuracy.

Methods

Mathematical model of combustion zone

The homogenous (non-catalytic), exothermic, irreversible reactions take place in the combustion zone. The methane is combusted through numerous radical reactions, but it can be represented as a methane combustion reaction. Also, H_2 will be converted to H_2O in the combustion zone. The overall reaction mechanism of methane and hydrogen with oxygen in the combustion zone can be considered as follows:

$$CH_4 + 3/2O_2 \rightarrow CO + 2H_2O \Delta H_r^\circ = -519.760 \text{kJ/mol} \quad (1)$$

$$H_2 + 1/2O_2 \rightarrow H_2O \Delta H_r^\circ = -242 \text{kJ/mol} \quad (2)$$

An input–output model of the combustion zone based on global mass and energy balances using a simple mathematical model consists of atomic molar balance and adiabatic flame temperature with the assumption that the feed and inlet air to the reactor are fully mixed along the flame, that complete oxygen consumption is considered, and that the conditions inside and outside of the combustion zone are equal.

For the combustion zone, the atomic molar balance as shown in Figure 1 can be derived as follows:

$$\text{For argon}: F_{IN,1} + F_{IN,2} = F_{OUT} \quad (3)$$

$$\text{For nitrogen}: F_{IN,1} \times 2 + F_{IN,2} \times 2 = F_{OUT} \times 2 \quad (4)$$

$$\text{For carbon}: F_{IN(CH_4),1} + F_{IN(CO),1} + F_{IN(CO_2),1} + F_{IN(CO_2),2} = F_{OUT(CH_4)} + F_{OUT(CO)} + F_{OUT(CO_2)} \quad (5)$$

$$\text{For hydrogen}: F_{IN(CH_4),1} \times 4 + F_{IN(H_2O),1} \times 2 + F_{IN(H_2),1} \times 2 + F_{IN(H_2O),2} \times 2 = F_{OUT(CH_4)} \times 4 + F_{OUT(H_2O)} \times 2 + F_{OUT(H_2)} \times 2 \quad (6)$$

$$\text{For oxygen}: F_{IN(H_2O),1} + F_{IN(CO),1} + F_{IN(CO_2),1} \times 2 + F_{IN(O_2),2} \times 2 + F_{IN(H_2O),2} + F_{IN(CO_2),2} \times 2 = F_{OUT(H_2O)} + F_{OUT(CO)} + F_{OUT(CO_2)} \times 2 \quad (7)$$

$$\text{Over all material balance}: F_{tot} = F_{OUT(CH_4)} + F_{OUT(H_2O)} + F_{OUT(H_2)} + F_{OUT(CO)} + F_{OUT(CO_2)} + F_{OUT(N_2)} + F_{OUT(Ar)} \quad (8)$$

where $F_{IN,1}$ is the molar flow rate of inlet process gases in the stream number.1 (kmol/h), $F_{IN,2}$ is the molar flow rate of inlet air in the stream number 2 (kmol/h), F_{OUT} is the molar flow rate of each gas from the combustion zone (kmol/h), F_{tot} is the total molar flow rate of outlet gases from the combustion zone (kmol/h).

The secondary reformer is operated close to adiabatically, and hence, the temperature is given by the adiabatic heat balance. Adiabatic flame temperature can be described by the following equation:

$$\Delta H = \sum_{OUT} F_i.H_i - \sum_{IN} F_i.H_i = 0.0 \quad (9)$$

where, ΔH is the enthalpy change (kJ/kmol), F is the molar flow rate (kmol/h), H_i is the specific enthalpy of the i-component (kJ/kmol).

Mathematical model of catalyst zone

In the catalyst zone, heterogeneous endothermic, reversible reactions take place. The product gas of the combustion zone is directed to the catalytic zone of the secondary reformer. The catalyst bed involves three reversible reactions. Anyway, the following equations can be written as follows:

$$CH_4 + H_2O \leftrightarrow CO + 3H_2 (\text{Steammethanereforming}) \Delta H_r^\circ = +206.100 \text{kJ/mol} \quad (10)$$

$$CO + H_2O \leftrightarrow CO_2 + H_2 \ (\text{Water} - \text{gasshift}) \Delta H_r^\circ = -41.150 \text{kJ/mol} \quad (11)$$

$$CH_4 + 2H_2O \leftrightarrow CO_2 + 4H_2 (\text{Carbondioxidereforming}) \Delta H_r^\circ = +165 \text{kJ/mol} \quad (12)$$

The reaction rate expressions of Froment and Xu described the kinetic rate of disappearance or production of species as a function of the partial pressures. The kinetic model of the reactions on a Ni/$MgAl_2O_4$ catalyst are based on a Langmuir-Hinshelwood reaction mechanism in which rate expressions can be described by the following equations [11]:

$$\text{For methane steam reforming reaction(SMR)}: R_1 = k_1 / \left(P_{H_2}^{2.5}.DEN^2\right) \left[P_{CH_4} * P_{H_2O} - \left(P_{H_2}{}^3 * P_{CO}/K_{e,1}\right)\right] \quad (13)$$

$$\text{Rateofreactionforwater} - \text{gasshiftreaction(WGS)}: R_2 = k_2 / (P_{H_2}.DEN^2) \left[P_{CO} * P_{H_2O} - \left(P_{H_2} * P_{CO_2}/K_{e,2}\right)\right] \quad (14)$$

$$\text{Reaction rate for carbon dioxide reforming reaction (CDR)}: R_3 = k_3 / \left(P_{H_2}^{3.5}.DEN^2\right) \left[P_{CH_4} * P_{H_2O}{}^2 - \left(P_{H_2}{}^4 * P_{CO_2}/K_{e,3}\right)\right] \quad (15)$$

where the denominator (DEN) is defined by the following equation:

$$DEN = 1 + K_{ad,CO} \times P_{CO} + K_{ad,H_2} \times P_{H_2} + K_{ad,CH_4} \times P_{CH_4} + K_{ad,H_2O} \times P_{H_2O}/P_{H_2} \quad (16)$$

where R_k is the rate of reaction for reaction k (kmol/kg_{cat}·h), k_k is the constant of reaction rate for reaction k (accordant to the Arrhenius constant), $K_{e,k}$ is the equilibrium constant for reaction k (bar^2), $K_{ad,i}$ is the adsorption coefficient of the i-component (accordant to the unit of pre-exponential factor for adsorption of i-component $A_{ad,i}$), DEN is the denominator in the expressions for the reaction rates (dimensionless unit), and P_i is the partial pressure of i-components (bar).

In the heterogeneous reaction, the actual reaction rate is affected by the molecular diffusion into the micropore inside the catalyst. Therefore, the effectiveness factor should be considered in order to apply the kinetic equations to the industrial reactor design. The effectiveness factor can be estimated using the following equation as noted by Froment and Bischoff [12]:

$$\eta_k = (1/\phi_k)\left((1/\tanh(3\phi_k)) - (1/3\phi_k)\right) \quad (17)$$

where η_k is the reaction effectiveness factor for reaction k (dimensionless unit) and the Φ Thiele modules (dimensionless unit).

The spherical catalyst pellet is assumed to find the effectiveness factor to apply on the design of the industrial catalytic reformer. If the catalyst pellet is spherical, the Thiele Modulus can be defined by the following equation with a characteristic length, $D_s/6$, as noted by previous researchers including Rostrup-Nielsen and Froment and Bischoff [12]:

$$\phi_k = \frac{D_s}{6}\sqrt{\frac{k_{vk}.\rho_b.(1+K_{e,k})}{K_{e,k}.D_{e,k}}} \tag{18}$$

where D_S is the equivalent pellet diameter (m). This is defined as the diameter of a sphere with the same external surface area per unit volume of the catalyst pellet, $D_s = 6.(1-\epsilon_B)/S_B$; where, the S_B is the specific surface area of bed (m^3/m^2) and, the S is the specific surface area of pellet (m^3/m^2), the A_p is surface area of pellet while, v_p is the volume of pellet ϵ_B is the porosity of packed bed (m^3_{void}/m^3_r). $k_{v,k}$ is the volumetric kinetic constant for reaction k (to convert the unit of constant of reaction rate) ($m^3/Kg_{cat}{\cdot}h$). ρ_b is the bulk density (kg/m^3).The effective diffusivity of the i-component for reaction k is defined as follows [3]:

$$De_{,k} = 1/\left[\left(1/D_{Kn,i}\right) + \left(1/D_{i,mix}\right)\right] \tag{19}$$

where $D_{e,k}$ is the effective diffusivity of reaction k (m^2/h). $D_{i,mix}$ is the molecular diffusivity of the i-component (m^2/h). $D_{Kn,I}$ is the Knudsen diffusivity (m^2/h).

In the secondary reformer reactor, the gas product from the combustion zone is directed to the catalyst bed, while the catalytic reactions as shown in Equations 10, 11, and 12 take place on the nickel catalyst. Therefore, the mass balance for i-component and heat balance in axial direction based on an adiabatic one-dimensional heterogeneous catalytic reaction model can be described by the following equation [3]:

$$(\partial F_i/\partial \ell) = \sum_{k=1}^{3} \upsilon_{i,k}.\eta_k.R_k.\rho_C(1-\epsilon_B).A \tag{20}$$

where ℓ corresponds to axial coordinates (m) and υ_i is the stoichiometric coefficient of the i- component in the chemical reaction equations (dimensionless unit). ρ_c is the density of catalyst (kgc/m^3_r), A is the cross section area (m^2). The change in the temperature throughout the reactor length has been given as follows [12]:

$$F_g.C_{p,g}.\rho_g.(\partial T_g/\partial \ell) = \sum_{k=1}^{3}\left(-\Delta H^{\circ}_{r,k}\right).\eta_k.R_k.\rho_C(1-\epsilon_B).A \tag{21}$$

where F_g is the molar flow rate of gases in the catalyst zone (kmol/h), C_P is the specific heat capacity ($kJ/kmol{\cdot}K$), T is the temperature of gases (K), $\Delta H^o_{r,k}$ is the standard heat of reaction k (kJ/mol), and ρ_g is the density of gases (kg/m^3).

The temperature of the catalyst surface is a very important factor in heterogeneous reactions. It was evaluated relative to the bulk gas temperature [3].

$$T_s = T_g +^{g} (((\sum(-\Delta H^{\circ}_{r,k}).\eta_k.R_k).\rho_C(1-\epsilon_B).D_S)/(6.h.(1-\epsilon_B)) \tag{22}$$

where T_s is the temperature surface of the catalyst (K) and h is the heat transfer coefficient ($kJ/m^2{\cdot}K{\cdot}h$) which is represented as $h = (Nu{\cdot}K/D_S)$.

The thermal conductivity could be estimated from the following equation [13]:

$$K = \mu_g\left(C_{P,g} + \left(10.4/M_{wt,i}\right)\right) \tag{23}$$

where K is the thermal conductivity of gases ($W/m{\cdot}K$), C_P is the specific heat capacity ($kJ/kg{\cdot}K$), M_{wt} is the molecular weight of gases (kg/kmol), and μ_g is the viscosity of gases (Pa·s).

For the packed bed, the Nusselt number equation could be formulated by the following equation [14]:

$$Nu = 2.0 + 1.3Pr^{0.15} + 0.66Pr^{0.31}Re_P^{0.5} \tag{24}$$

where Nu is the Nusselt number (dimensionless unit), Pr is Prandtl's number (dimensionless unit), and Re_p is the Reynolds number of a particle (dimensionless unit).

For the flow through ring packing often used in industrial packed columns, the Ergun equation is considered as a good semi-empirical correlation for predicting pressure drop as follows [14]:

$$\frac{\partial P}{\partial \ell} = -\left(\left(\left(150(1-\epsilon_B)^2.\mu_g.u\right)/(\epsilon_B{}^3.D_S^2)\right) + ((1.75.(1-\epsilon_B)\rho_b.u^2)/(\epsilon_B^3.D_S))\right) \tag{25}$$

where P is the pressure of gases (bar), u is the velocity of gases (m/h) $u = Q/A$,and Q is the volumetric flow rate (m^3/h).

Solution procedure

The solution starts with an assumption of the initial total molar flow rate of outlet gases (F_{tot}) of the combustion zone. Chemical compositions and temperature are calculated using the atomic molar balance and the adiabatic flame temperature equations. Consequently, the final calculation results will be used as input data to the catalytic zone. The reaction rate equations are used with mass and heat balance to determine the temperature and composition profile for each component as a function of reactor length in the catalyst zone. The set of differential equations will be solved using the numerical analysis, Euler method. The iteration calculation will apply and repeat until the molar fraction calculation and reaction temperature will agree with the industrial value. Otherwise, a new guessing value of the total molar flow rates of outlet gases from the combustion zone will be assumed.

Results and discussion

Combustion zone

The actual operating conditions such as feed molar flow rate, temperature, and pressure have been tabulated in Table 1. These data are used for modeling and evaluating the theoretical performance of the industrial secondary reformer reactor.

Table 1 Actual operations conditions of industrial secondary reformer in SCFSR in Basra/Iraq [15]

Parameters	Input feed process gas Stream 1	Input air Stream 2
Temperature (°C)	795	550
Pressure (bar)	32	32
Molar flow rates of the components (kmol/h)		
CH_4	426.040	-
H_2O	3,684.010	11.190
H_2	2,932.100	-
CO	403.460	-
CO_2	537.440	0.520
N_2	15.990	1,357.360
Ar	5.830	16.350
O_2	-	365.080

Table 2 Output data of the industrial secondary reformer and simulation [15]

Parameters	Output of combustion zone (simulation data)	Output of secondary reformer (simulation data)	Output of secondary reformer (plant data)
Temperature (°C)	1,096.644	1,009.630	969
Pressure (bar)	32	31.918	31.570
Mole flow rates (kmol/h) of the components			
CH_4	183.698	19.001	18.640
H_2O	4,182.983	3,985.640	3,992.350
H_2	2,929.004	3,455.730	3,449.710
CO	645.806	777.861	785.200
CO_2	537.947	570.589	563.600
N_2	1,373.380	1,373.380	1,373.360
Ar	22.180	22.180	22.180
O_2	-	-	-
Total molar flow rate (kmol/h)	9,875	10,204.400	10,205.504
H_2/N_2	2.132	2.516	2.511
$(CO+H_2)/N_2$	2.602	3.082	3.083

According to the simulation results of the mathematical model, the temperature of the combustion chamber is 1,096.644°C. The values of the components' molar flow rate have been given in Table 2, as there are no industrial data available in this zone; therefore, they are not comparable.

Catalyst zone

In this section, the influence of temperature has been discussed for describing the effect of these parameters on the design of the secondary reformer relative to Basra Fertilizer Plant.

Figure 2 depicts the profile of temperature gases and catalyst surface as a function of reactor length. The sharp decline along the reactor occurred due to the general behavior of methane steam reforming, and carbon dioxide reforming reactions are prevailing endothermic reforming reactions that control the reaction pathway, while the water-gas shift reaction is a slightly exothermic reaction. The profile of gas and catalyst temperatures is a very important parameter effect on the rates of chemical reactions.

In the region of the catalyst entrance, the rate of reactions for methane steam and carbon dioxide reforming reactions is very fast. This is due to the higher effect of operation conditions such as partial pressure of methane and the value of feed temperature. After the middle of the catalyst zone, the trend of reaction rates in the Figure 3 has declined slowly due to the effect of endothermic reactions relative to chemical reactions in Equations 10 and 12. The reforming reaction rates at the inlet catalyst zone are considerably higher than the water-gas shift reaction because the last reaction became more active under a low temperature range; then, in middle of the catalyst zone it begins to increase because the temperature becomes low [9]. The water-gas shift reaction is negative along the axial distance from the top of the bed due to inversion of the water-gas shift reaction inside the catalyst particles [8].

Figure 4 depicts the decline of diffusivities along the reactor length due to the diffusion phenomena that occurred from high to low concentration regions (concentration gradient). Also, there are other driving forces (besides concentration differences) such as temperature and pressure gradients [16]. The overall reaction rate is mainly controlled by the diffusion rate due to the rate of mass transfer (diffusivity) of the reactants to the catalyst pellet which is low; the concentration of reactants in the catalyst pellet will be low because it is consumed by the reaction as fast as it arrives [17] so that the conditions' mean reaction occurs before the reactant has diffused far into the pellet [12].

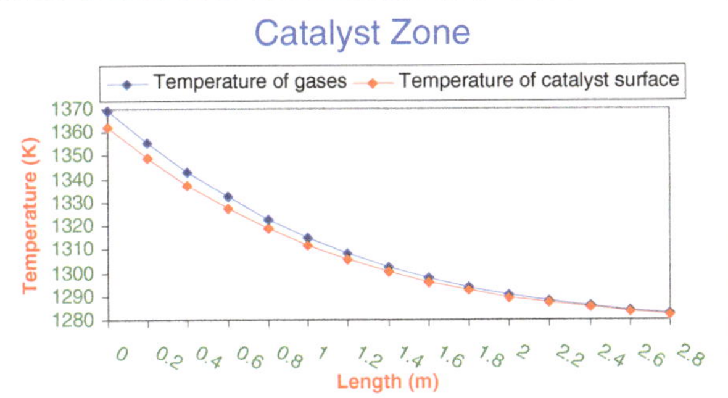

Figure 2 Temperature distribution along the axial distance of the catalyst bed relative to inlet operation conditions. $T = 1{,}369.644$ K and $P = 32$ bar.

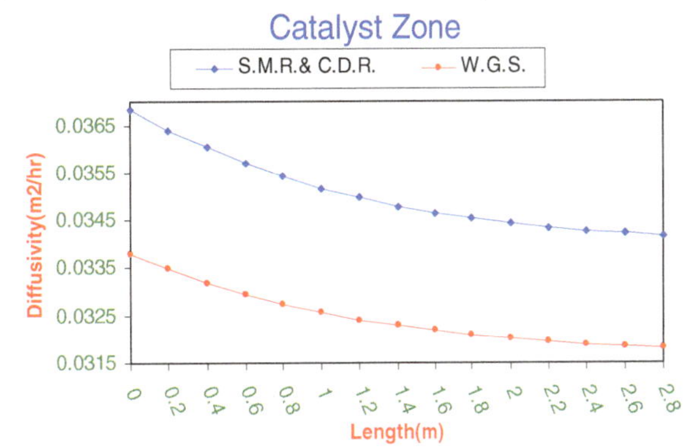

Figure 4 Diffusivity profile along the length of the catalyst bed relative to inlet operation conditions. $T = 1{,}369.644$ K and $P = 32$ bar.

The effectiveness factor is an important parameter to predict the temperature and composition profiles for realizing the actual reaction rate with pore diffusion.

In this study, the effectiveness factors show low values ranging from 0.006 to 0.001 along the axial distance of the catalyst bed as shown in Figure 5 due to many factors such as large pellets, low diffusivity, and high values of the constant rate of reactions. Note that the effectiveness factor $<<1$ means that only the surface near to the outer surrounding of the pellet is effective. In this case, the catalyst in the central portion of the pellet is not utilized [12].

The theoretical results' prediction from the mathematical model has been compared with the industrial data collected from the manual and daily log sheet of the Basra Fertilizer Plant as summarized in Table 2. The results depicted high accuracy between simulation results and plant data.

Figure 6 depicts the molar flow rate distribution of each component along the axial distance of the catalyst bed. Methane is almost completely consumed at inlet conditions of the catalyst bed due to the high temperature of gases coming from the combustion zone. High temperature accelerates the reaction rates of steam methane reforming and carbon dioxide reforming. Most of the methane conversion had been achieved in the half of the catalyst bed. Then, the rate of reactions of the methane steam reforming and carbon dioxide reforming has been decreased as shown in Figure 3 because the temperature and partial pressure of methane had been decreased.

The sharp increase in the H_2 content is mainly explained by SMR and CDR as shown in Figure 3 along the catalyst bed. Thus, the hydrogen production is dominated by chemical reactions in Equations 10 and 12.

The sharp increase in CO along the reactor length occurred due to the methane steam reforming and water-gas shift reactions as shown in Figure 3 relative to the chemical reactions in Equations 10 and 11.

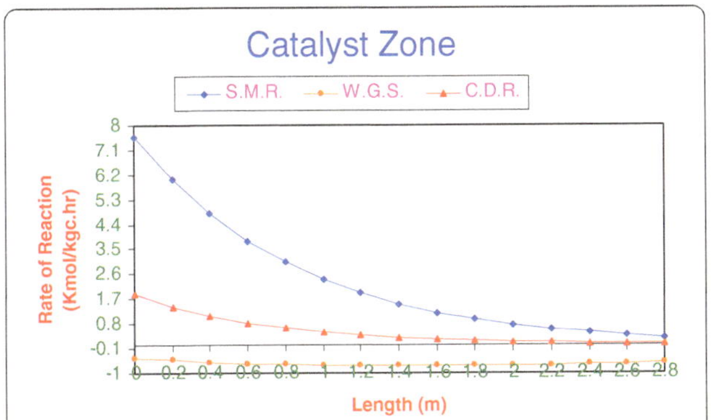

Figure 3 Rates of reaction profile along the length of the catalyst bed relative to inlet operation conditions. $T = 1{,}369.644$ K and $P = 32$ bar.

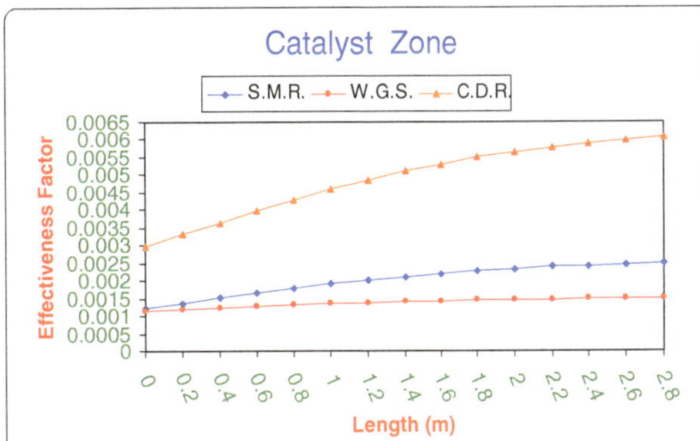

Figure 5 Effectiveness factor profile along the length of the catalyst bed relative to inlet operation conditions. $T = 1{,}369.644$ K and $P = 32$ bar.

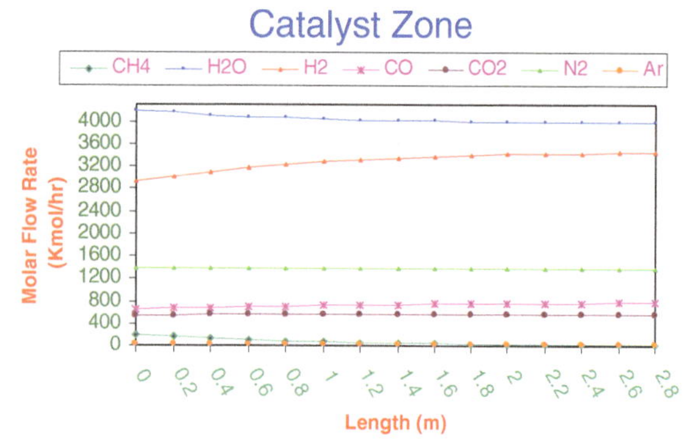

Figure 6 **Molar flow rate distribution of each component along axial distance of catalyst bed relative to inlet operation conditions.** $T = 1{,}369.644$ K and $P = 32$ bar.

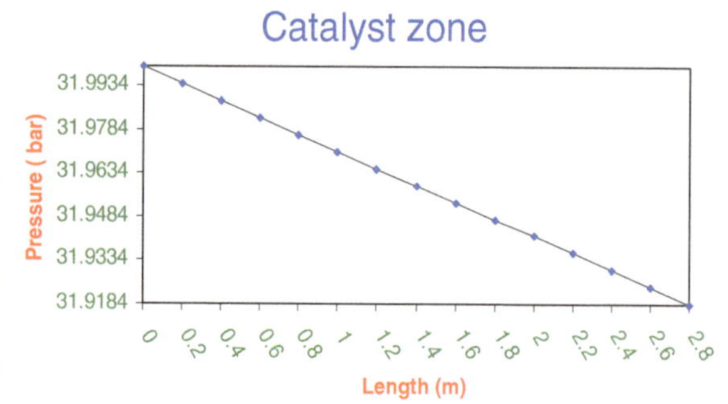

Figure 7 **Pressure distribution along the axial distance of the catalyst bed relative to inlet operation conditions.** $T = 1{,}369.644$ K and $P = 32$ bar.

The CO_2 generation is described by the reactions of water-gas shift and carbon dioxide reforming as mentioned in Equations 11 and 12. The increase in CO_2 production occurred due to carbon dioxide reforming as shown in Figure 3. Then, the decline occurred due to CO_2 which was consumed by the water-gas shift reaction because of the inversion of the water-gas shift reaction inside the catalyst particles. (It's negative along the axial distance from the top of the bed). Nitrogen and argon are constant along the reactor length because neither nitrogen nor argon reacted with any other components.

In the gas phase of the packed bed catalytic reaction system, pressure drop is one of the important parameters. From a practical point of view, the value of the pressure drop gives a good indication about catalyst performance inside the reaction shell. There are many reasons for the increase of pressure drop, for example, with a long time of operation; the catalyst may be thermal cracking due to the high operation temperature, so the value of pressure drop will increase.

The decline of pressure along the reactor occurred due to the friction between particles of gases, particles of gases with pellets of the catalyst, and particles of gases with a wall of reactor as shown in Figure 7.

Experimental

In this section, the analysis of inlet and outlet synthesis process gases for the industrial secondary reformer reactor in Basra Fertilizer Plant has been carried out. The experimental part will investigate, specify, and tabulate the actual operation conditions of the industrial secondary reformer as summarized in Table 1.

Currently, describing the industrial secondary reformer reactor at Basra Fertilizer Plant, the industrial catalyst and the operation conditions must be understood with more details.

The wall of the secondary reformer is fabricated from carbon steel with three refractory layers that are lined to protect the wall and to reduce the heat loss. The inside diameter of the secondary reformer is 3.450 m with a lining thickness of 440 mm. The alumina brick plate is located above the catalyst bed in order to distribute the combusted gas evenly to the catalyst bed. The alumina balls of 40 mm in diameter are packed below the alumina brick plate with a depth of 300 mm and with a volume of 2.500 m^3 to protect the catalyst. The nickel catalysts, RKS-2, with one hole are packed with a volume of 26 m^3; the height of the catalyst bed is 2.8 m as depicted in Table 3. The catalyst is supported by a ceramic ball and cone made of high alumina bricks which are comprised of bricks with slot holes that provide a highly stable support for the catalyst which bears the catalyst weight. The catalyst support is covered with alumina balls of 40 mm with heights of 300 mm, 75 mm, and 120 mm and with volumes of 2.800, 4.100, and 4.100 m^3, respectively.

Conclusion

The secondary reformer for hydrogen and nitrogen production is mathematically investigated by a series of

Table 3 Physical characteristic of the catalyst [15]

Catalyst pellet	
Model name	RKS-2
Shape	Rings
Nominal Size: OD × ID × H	19 × 9 × 19 mm
Chemical composition	
NiO wt.%	9%
$MgAl_2O_4$ wt.%	Balance
Filling density	1.070 Kg/L

OD, outer diameter; ID, inner diameter; H, height.

simulation of operation conditions which have been collected from the documents of Basra Fertilizer Plant.

A mathematical model of the industrial secondary reformer in Basra Fertilizer Plant with all assumptions made had been completed. The theoretical results obtained have been compared with the industrial secondary reformer reactor details. The results show a good compatibility as shown in Table 2. The model predicts the temperature and composition profiles of the gas that leaves the combustion chamber and investigates the temperature and concentration gradient inside the catalyst particle. The value of pressure drop depicts low values along the axial distance of the catalyst bed; therefore, pressure drop can be assumed negligible. The overall reaction rate is mainly controlled by the diffusion rate. In the catalyst zone entrance, methane steam reforming and carbon dioxide reforming reaction rates at the catalyst zone inlet were considerably higher than the water-gas shift reaction. The diffusion into the pellet is relatively slow so that these conditions' mean reaction occurs before the reactant has diffused far into the pellet. The effectiveness factor $<<1$ means that only the surface near from the outer surrounding of the pellet is effective; therefore, the diffusion and effectiveness factor gave a good indication about the shape of the catalyst that is used in the catalyst zone as a ring without a central portion because the chemical reaction takes place at the outer surface. Finally, this simulation model depicts a high reliability for the designing and testing of an industrial secondary reformer, not for Basra Fertilizer Plant, but it is extended to any secondary reformer reactor in fertilizer plants.

Abbreviations

CDR: carbon dioxide reforming reaction; SCFSR: State Company of Fertilizers South Region; SMR: methane steam reforming reaction; WGS: water-gas shift reaction.

Acknowledgement

We acknowledge the full cooperation of the State Company Fertilizer Plant South Region in Basra/Iraq to introduce the industrial data for this investigation.

References

1. Ebrahimi H, Behroozsarand A, Zamaniyan A (2010) Arrangement of primary and secondary reformers for synthesis gas production. Chemical Engineering Research and Design **88:**1342–1350
2. Khorsand K, Deghan K (2007) Modeling and simulation of reformer autothermal reactor in ammonia unit. Petroleum and Coal J 49:64–71
3. Yu YH (2002) Simulation of secondary reformer in industrial ammonia plant. Chem Eng Tech J **25:**307–314
4. Raghunandanan KS, Reddy KV (1994) Development of a mathematical model for a secondary reformer. Chem Eng. Tech J 17:273–279
5. Ravi K, Dhingra SC, Guha BK, Joshi YK (1989) Simulation of primary and secondary reformers for improved energy performance of an ammonia plant. Chem Eng Tech J **12:**358–364
6. Amirshaghaghi H, Zamaniyan A, Ebrahimi H, Zarkesh M (2010) Numerical simulation of methane partial oxidation in the burner and combustion chamber of autothermal reformer. Applied Mathematical Modeling **34:**2312–2322
7. Li Y, Wang Y, Zhang X, Mi Z (2008) Thermodynamic analysis of autothermal steam and CO_2 reforming of methane. International Journal of Hydrogen Energy **33:**2507–2514
8. Pina J, Borio DO (2006) Modeling and simulation of an autothermal reformer. Latin American Applied Research J **36:**289–294
9. Shukri MI, Songip AR, Ahmad A, Nasri NS (2004) Simulation study of methane autothermal reforming for hydrogen production. In: Proceedings of Advances in Malaysian fuel cell research and development., pp p149–p158
10. Hoang DL, Chan SH (2004) Modeling of a catalytic autothermal methane reformer for fuel cell applications. Applied Catalysis J 268:207–216
11. Xu J, Froment GF (1989) Methane steam reforming: II. Diffusional limitations and reactor simulation, AIChE J. 35
12. Smith JM (1970) Chemical engineering kinetics. Chemical engineering series., McGraw-Hill; New York
13. Sinnott RK (1999) Unit operations. In Related Butterworth-Heinemann titles in the chemical engineering series, vol.6, 3rd edition., Butterworth-Heinemann, Oxford
14. Coulson JM, Richardson JF, Bucharust JR, Harker JH (2002) Unit operations. In Related Butterworth-Heinemann titles in the chemical engineering series, vol 2, 5th edition., Butterworth-Heinemann, Oxford
15. Documents of State Company of Fertilizer Plant South Region
16. Geankoplis CJ (1983) Transport processes and unit operation. Prentice Hall Professional Technical Reference, New Jersey
17. Luyben W (1996) Process modeling, simulation, and control for chemical engineers, 2nd edition., Mc Graw Hill; New York

An experimental investigation on the fabrication of W-Cu composite through hot-press

Mostafa Roosta[1*], Hamidreza Baharvandi[1] and Hossein Abdizade[2]

Abstract

In this paper the feasibility of fabricating W-Cu composite by hot-press has been studied and the best parameters for hot-pressing were acquired. W/20 vol%Cu composite was successfully prepared by this process, in which at first the powder of tungsten and copper, containing 20% vol copper were mixed in a ball mill for 3 h at rotation speed of 200 rpm. Then the mixed powders were hot-pressed for 3 h at a compact pressure of 30 MPa and temperatures of 1,250°C, 1,350°C, and 1,450°C. The composites made have been investigated and revealed the making W-Cu composite with good properties included density, hardness, modules of elasticity, flexural strength, and microstructure. Also, the sample made at 1,450°C possesses better properties, and its microstructure showed the tungsten matrix and the copper reinforcement separately; the X-ray diffraction patterns showed that no compound has been formed between W and Cu.

Keywords: Tungsten-copper, Composite materials, Hot-press

Background

Composite materials are engineered or naturally occurring materials made from two or more constituent materials with significantly different physical or chemical properties which remain separate and distinct at the macroscopic or microscopic scale within the finished structure. W-Cu composites are made by mixing a refractory phase called tungsten, which possess high strength and low coefficient of thermal expansion, and a Cu-phase having high thermal and electrical conductivity [1]. The combination of these elements optimizes some properties such as ductility, strength, and corrosion and wear resistance. In the recent years, W-Cu composites have gained great importance in automotive, electrical, and military industry because of their high thermal conductivity, low thermal expansions, high wear resistance and excellent electrical conductivity. They are used as electrical contacts, resistance welding electrodes, heat sinks, electro-forging dies, packaging materials, and so on [2,3].

The fabrication of a full density W-Cu composite is very difficult. Because of the big difference between melting points of tungsten (approximately 3,400°C) and copper (approximately 1,083°C), there is no overlap of sintering temperature ranges. Also, tungsten-copper system has no mutual solubility which means poor sinter ability [1]. The infiltration of a porous-sintered tungsten skeleton by liquid copper is one of the most common methods for producing W-Cu composite. However, in this method defects like pores, copper lakes, and tungsten agglomerates form easily, so this technique results in a poor quality [2-4]. Other methods to fabricate W-Cu composites include the following: thermal-mechanical process [2], metal powder injection molding [5], hot extrusion [6], liquid sintering and hot-hydrostatic extrusion [7-9], microwave sintering [10], resistance sintering [11], vacuum plasma spray [12,13], mechanical alloying, and other new methods [14-18].

Although one of the most important methods in producing composites is hot-press, in which the mixed powders will be subjected to high pressure and high temperature at the same time, this method has not been used as sole way for making W-Cu composite. Of all the advantages that this method contains, increasing the properties of row and sintered material such as density, strength, and fatigue properties is the main one. In this experiment, we use hot-press to produce W-Cu composite because of its simplicity and low cost.

* Correspondence: mostafa_rosta@yahoo.com
[1]Malek-Ashtar University of Technology, P.O. Box 13336–95711, Tehran, Iran
Full list of author information is available at the end of the article

Results and discussion

Physical properties

The bulk density of pieces made at 30 MPa and temperatures of 1,250°C, 1,350°C and 1,450°C was measured about 15.21, 15.76, and 16.25 g/cm^3 respectively. Also, the porosity of specimens can be measured through this method. For example, the apparent porosity which does not include the volume of the sealed pores was evaluated about 0.78% for composite made at 1,450°C.

Since in Brinell hardness test method, the diameter of ball indenter is big and pressure is applied to the bulk of sample, this method was used to calculate the hardness of the work pieces. According to the results, the hardnesses of the samples made at 30 MPa and temperatures of 1,250°C, 1,350°C and 1,450°C were equal to 131, 149, and 163 HB respectively.

For calculating modulus of elasticity in accordance with ASTM C769 test standard (ASTM International, West Conshohocken, PA, USA), the speed of sound in the sample must be calculated first. In this way, a 2.5 × 2.5 cm sample is made and then an ultrasonic pulse with frequency of 4 MHz will be sent to the surface of samples and then will be received. By recording the return time of sound and sample thickness, the speed of sound (v) passing the sample was obtained; by using the following relations, modulus of elasticity (E) in GPa was obtained:

$$E = \rho v^2 \qquad (1)$$

Thus according to calculated density for the composite, the modulus of elasticity are 312.5, 325.4, and 341.8 GPa for samples made at 1,250°C, 1,350°C and 1,450°C, respectively.

Flexural strength of samples is obtained based on the ASTM C1161 test standard. According to this test standard, specimens with dimensions of 3 × 4 × 45 mm are made and the surfaces are prepared by polishing. Then the load was applied by flexural strength machine with speed of 0.5 mm/min and distance (l) of 40 mm between two bases of applying load. By drawing the force-strain diagram, the maximum fracture force (F) is recorded. Using the following relation, flexural strength or modulus of rupture (MOR) is calculated in MPa,

$$MOR = \frac{3Fl}{2bd^2} \qquad (2)$$

where F is fracture force (N); B is sample width (mm); D is thickness of sample (mm); and L is the distance between two bases of applying load (mm).

Correspondingly, the flexural strength of composite made at 1,250°, 1350° and 1450°C is 381.2, 350.7 and 343.8 MPa.

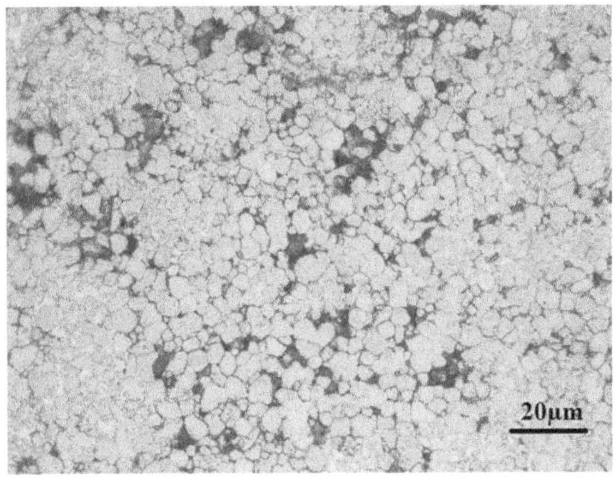

Figure 1 Optical micrographs of W/20 vol%Cu prepared by hot-press with 30 MPa pressure at 1,400°C.

The study of microstructure

Since it is clear that the best result was obtained from the sample made at 30 MPa pressure and temperatures of 1,450°C, we perform the microstructure study just for this sample. After preparation and polishing, the parts as a result were seen under optical microscopy and scanning electron microscopy. The etching of work pieces is done according to ASTM 98 C standard, in which 10 g of ferrocyanide potassium ($K_3Fe(CN)_6$) and 10 g of NaOH is added to 10 ml distilled water. As shown in Figure 1, the optical micrographs of W-Cu composite made in 1,450°C and 30 MPa copper and tungsten phases are visible. The porosity in the composite is also seen, one of the main defects that exists in hot-press.

Also by secondary scanning electron microscopy (SEM) and back scattered electron (BSE) the morphology of samples and their particle size and shape are studied. Figure 2 shows the SEM, and Figure 3 shows the BSE micrographs of the W/20 vol%Cu composite prepared by hot-press at 1,450°C and 30 MPa.

In Figure 4, the elemental analysis map of composite can be seen. It is clear in the picture that the composite is made of tungsten, copper, and nickel.

On the other hand, as it is clear in the back-scattered electron images, the composite is made of dark phase and bright phase. According to the feature of back-scattered electron images, the bright phase is related to heavier elements because heavier elements reflect more light. However, to prove this issue, energy dispersive X-ray spectroscopy (EDX)-line scan and elemental analysis of both phases can be used. As shown, the dominant bright phase is about 80% volume and the dark phase of copper can also be seen in the image. To prove that

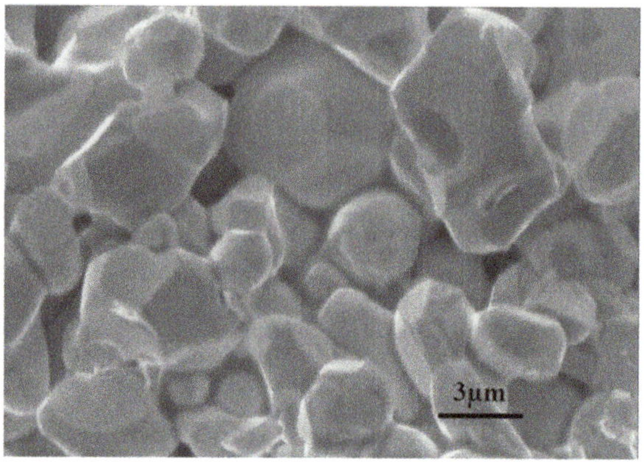

Figure 2 SEM micrographs of W/20 vol%Cu prepared by hot-press with 30 MPa pressure at 1,400°C.

bright phase belongs to tungsten and dark phase belongs to copper, an EDX-line scan from a bright grain and EDX-line scan from the dark grains are taken and shown in Figures 5 and 6.

Also, the fact that the bright phase belongs to tungsten and dark phase belongs to copper can be shown by elemental analysis of the bright and dark phases. Figure 7 shows the elemental analysis of bright and dark phases.

Figure 8 shows the XRD patterns of starting tungsten and copper powders before production of composite. Also the XRD result of W/20%volCu is present in the image. As clearly seen in the pictures for the pure powders before making composite, only peaks related to copper and tungsten are seen. However, in the XRD patterns of composite made from tungsten and copper, there are only peaks related to these two elements, which show that no compound between these two elements has been formed. Furthermore, because of the low amount of copper in the composite, about 10 weight percent, peaks related to copper have a low intensity in comparison with peaks related to tungsten.

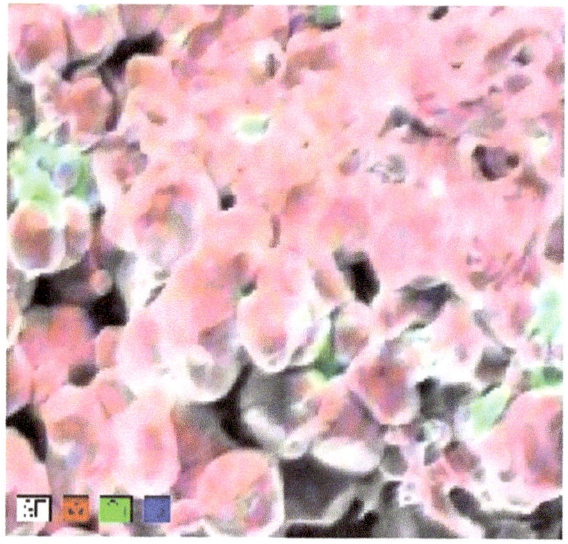

Figure 4 Map of composite W/20 vol%Cu; red parts represent W and green parts represent Cu.

Figure 3 BSE micrographs of W/20 vol%Cu prepared by hot-press with 30 MPa pressure at 1,400°C.

Methods

The elemental tungsten powder and copper powder were used as raw materials. In order to make W/20vol%Cu composite, weight percent of powders must be calculated first. For this purpose, the weight percent of copper and tungsten in the desired composite by Equation 3 is calculated, according to density and volume percent of Cu and W, into percent $Wt_{Cu} = 10.34$ and percent $Wt_W = 89.66$.

$$\%Wt_{Cu} = \frac{W1}{W1 + W2} \times 100 = \frac{\rho 1 V1}{\rho 1 V1 + \rho 2 V2} \times 100 \quad (3)$$

On the other hand, the theoretical density of composite is easily calculated by adding up the mass of each component:

$$\begin{aligned} m &= m_f + m_m \\ m &= \rho V \\ \rho V &= \rho_f V_f + \rho_m V_m \\ \rho V &= \rho_f f V + \rho_m (1-f) V \\ \rho &= f \rho_f + (1-f) \rho_m \end{aligned} \quad (4)$$

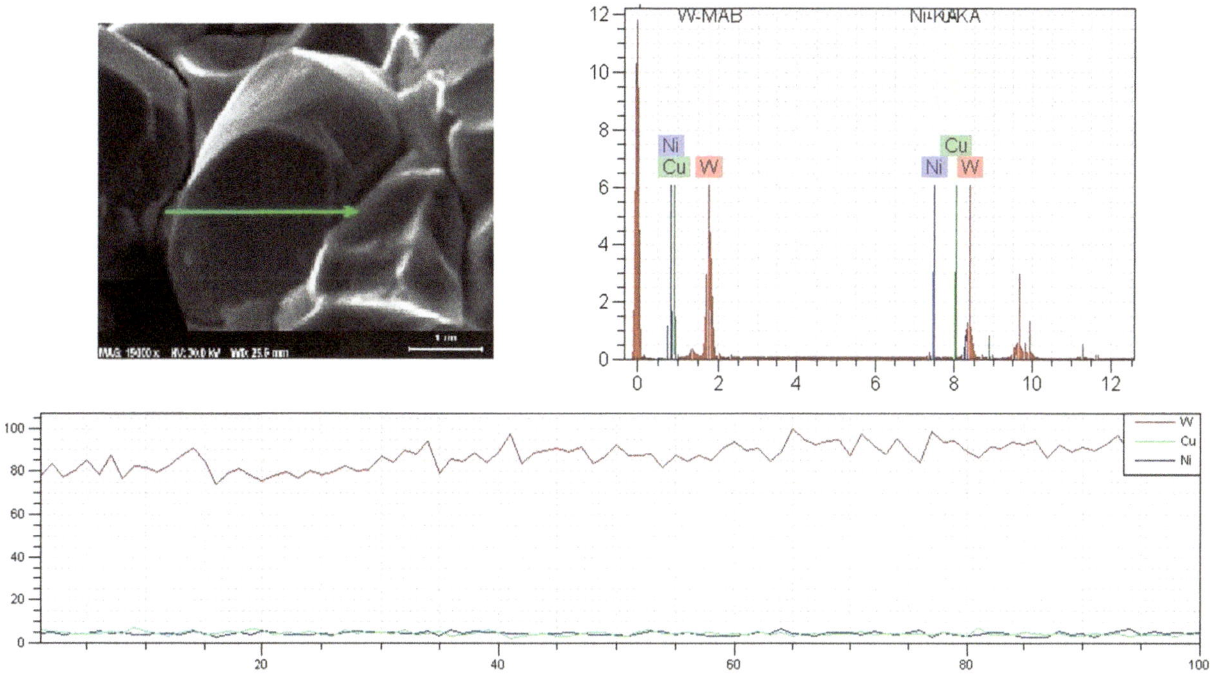

Figure 5 **EDX-line scan of bright grains, showing that the bright grains belong to tungsten.**

Where f is the volume fraction of fibers or the second phase, ρ is density, m is mass, V is volume, m_f is mass of second phase and m_m is mass of matrix. According to these equations, the theoretical density of composite is equal to $\rho = 17.22$ g/cm^3.

The required amount of powders is calculated considering the density of composite and size of the piece. In this experiment, we attempt to produce a piece with dimensions of $12 \times 12 \times 0.4$ cm, so the volume will be 57.6 cm^3 and $m = 991.87$ g. Hence, we need 991.872 g of tungsten-copper

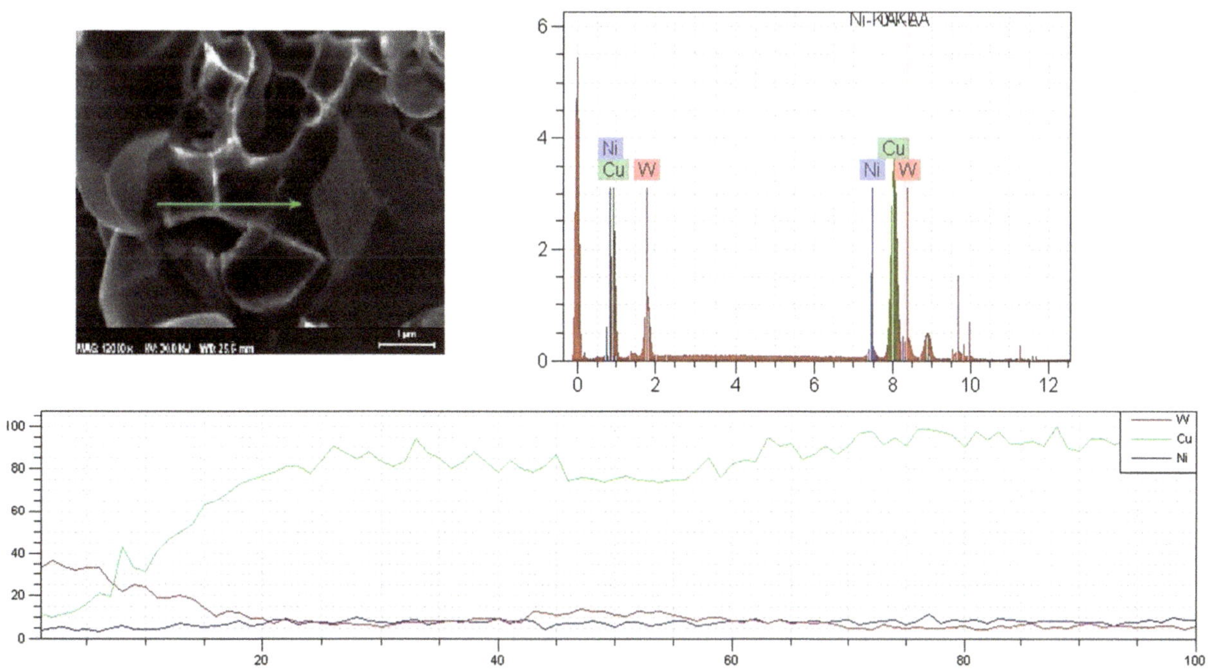

Figure 6 **EDX-line scan of dark grains, showing that the dark grains belong to copper.**

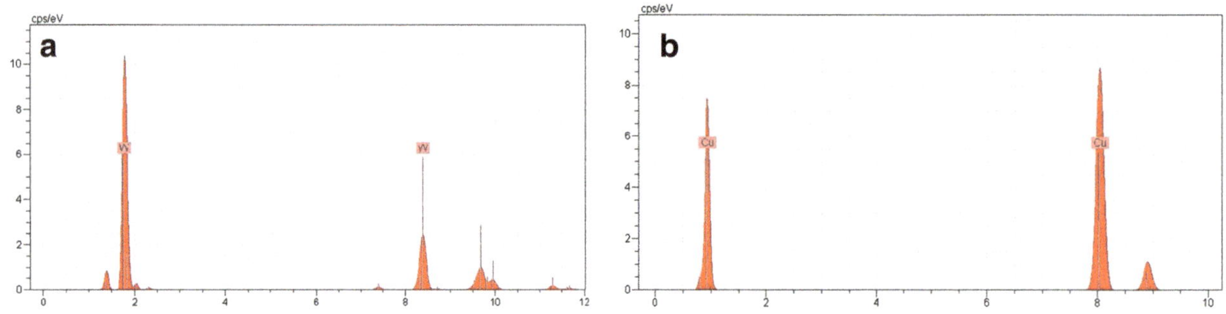

Figure 7 Elemental analysis. a) bright phase **b**) dark phase.

powder to make the piece, and the value of each of these two components is calculated as follows:

$$W_{Cu} = 10.34/100 \times 991.87 = 102.56g$$

$$W_W = 991.872–102.56 = 889.3g$$

Thus to fabricate a piece with aforementioned dimensions from W/20% volCu composite, it must be seen how much tungsten and copper powder is required. Therefore at first stage, the required amount of each powder was weighed and then mixed for 3 h in a ball mill, with rotation speed of 200 rpm. A 2%wt of nickel powder was added to the composite as well as to reduce the composite density, considering the low density of nickel (approximately 8.9 g/cm^3), in comparison with the composite (approximately 17.22 g/cm^3). The mixed powders, after being dried, were performed at room temperature under 80 MPa pressure and then hot-pressed for 3 h with the compaction pressure of 30 MPa at three different temperatures of 1,250°C; 1,350°C; and 1,450°C to obtain the optimum temperature. The heating rate was chosen at 10°C/min. The sintering process is shown in Figure 9. The hot-pressing was also done under high-purity argon (99.99%) atmosphere to avoid the powder from being oxidized.

The density of composite was measured through Archimedes test, according to ASTM B311 standard. This test standard is mostly used to determine the density of powder metallurgy parts and is based on the water displacement method. The hardness was evaluated in accordance with Brinell hardness test method ASTM E10. In addition, the bending strength of the composite has been measured by bending test according to ASTM C1161, and the modulus of elasticity was measured as well by ASTM C769 test standard method, in which the modulus of elasticity is evaluated by the help of sound velocity in the work piece. Lastly, the microstructure of made specimens was investigated by optical microscopy and scanning electron microscopy.

Conclusion

In this study, the feasibility of fabricating W/Cu composite via hot-press was studied. The W/20 vol%Cu composites were made at 30 MPa pressure and temperatures of

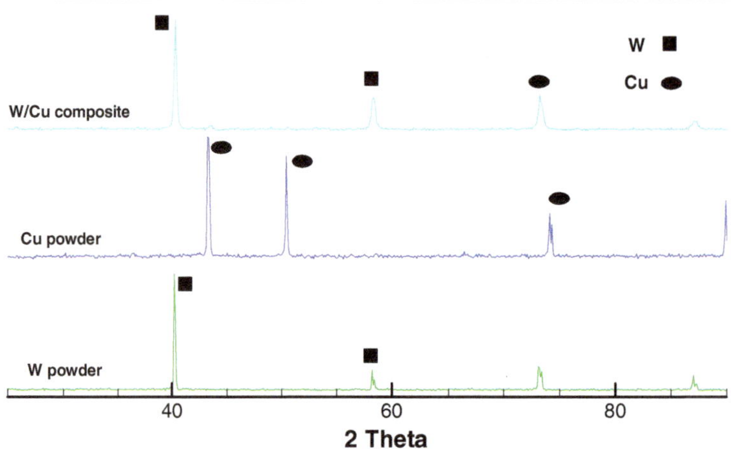

Figure 8 XRD patterns of starting powders and the produced composite after being hot-pressed.

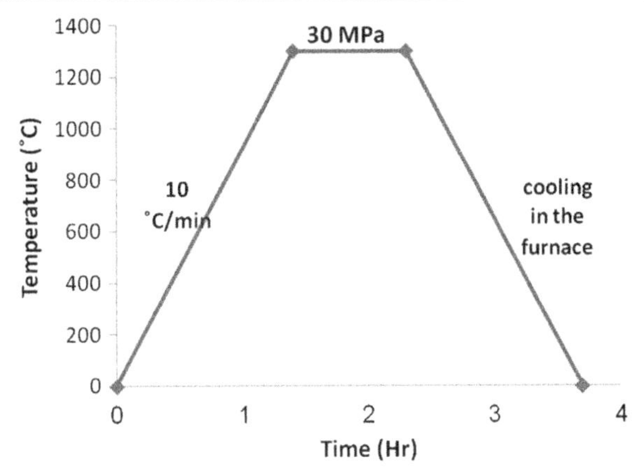

Figure 9 The sintering process.

1,250°C, 1,350°C and 1,450°C showed good properties, including density, hardness, modulus of elasticity, and flexural strength. The microstructure showed the tungsten matrix and the copper reinforcement distinctly, and the XRD patterns showed that no compound has been formed between W and Cu. The best results were obtained for the sample made at temperatures of 1,450°C.

Competing interests
The authors declare that they have no competing interests.

Acknowledgment
The authors would like to express their gratitude to Dr. Ebrahim Afshari for providing language help and useful suggestion in writing the article.

Author details
[1]Malek-Ashtar University of Technology, P.O. Box 13336-95711, Tehran, Iran. [2]School of Metallurgy and Materials Engineering, University of Tehran I. R., Tehran, Iran.

Authors' contributions
MR, the first author, carried out making the composite and testing it. HB, participated in the design of tests and coordination . HA, conceived of the study, and helped to draft the manuscript. All authors read and approved the final manuscript.

References
1. Dongdong G, Shen Y (2008) Mater Sci Engin 489:169–177
2. Ho PW, Li QF, Fuh JYH (2008) Mater Sci Engin 485:657–663
3. Amirjan M, Zangeneh-Madar K, Parvin N (2009). Int J Refractory Metals Hard Mater (in press)
4. Yunping L, Yu S (2008) Int J Refract Metals Hard Mater 26:540–548
5. Ibrahim A, Abdallah M, Mostafa SF (2009) Abousree Hegazy A. Mater Design 30:1398–1403
6. Li D, Liu Z, Yu Y, Wang E (2008) J Alloys Comp 462:94–98
7. Li D, Liu Z, Yu Y, Wang E (2008) Int J Refractory Metals Hard Mater 26:286–289
8. Liu BB, Xie JX, Qu XH (2008) Compos Sci Technol 68:1539–1547
9. Luo S, Yi J, Guo Y, Peng Y, Li L (2009) J Ran J J Alloys Comp 473:L5–L9
10. Maneshian MH, Simchi A (2008) J Alloys Comp 463:153–159
11. Junbo W, Wen'ge C, Bingjun D (2006). J Wuhan Univ Technol-Mater Sci Ed 21 (4):32-35.
12. Li DR, Liu ZY, Yu Y, Wang ED (2009) Mater Sci Engin 499:118–122
13. Zhou Z, Du J, Song S, Zhong Z, Ge C (2007) J Alloys Comp 428:146–150
14. Lee J (2008) Int J Refractory Metals Hard Mater 26:290–294
15. Pintsuk IG, Smid J, Doring E, Hohenauer W, Linke J (2007) J Mater Sci 42:30–39
16. Kim S, Kim Y, Suk M (2007) Metals Mater Int 13(5):391–394
17. Kang H, Kang SB (2004) J ThermalSprayTechnology,13(2):223-228.
18. Hong SH, Kim BK (2003) Mater Lett 57:2761–2767

Combination of nanophotocatalysis with electro-Fenton-like process in the removal of phenol from aqueous solution: GC analysis and response surface approach

Mahmoud Zarei[1*], Alireza Khataee[1], Mehrangiz Fathinia[1], Fardin Seyyednajafi[2] and Heidar Ranjbar[2]

Abstract

Background: This study focuses on the application of photoelectro-Fenton (PEF) with Mn^{2+} as catalyst combined with a photocatalytic process using supported TiO_2 nanoparticles for the removal of phenol from aqueous solutions. TiO_2 nanoparticles (Degussa P-25) immobilized on glass plates were used as photocatalyst, and carbon nanotube-polytetrafluoroethylene electrode was employed as cathode. Response surface methodology (RSM) was performed to evaluate the effects of the four key factors (initial amount of Mn^{2+}, initial concentration of phenol, reaction time, and applied current) on the PEF/TiO_2 process.

Results: The application of RSM after analysis of variance showed a high coefficient of determination value (R^2 = 0.918) and thus indicated a high correlation between the observed and predicted values. The optimum operation conditions were located by graphical response surface and contour plots. The optimum initial amount of Mn^{2+}, initial phenol concentration, applied current, and reaction time were found to be 0.3 mM, 15 mg/L, 300 mA, and 150 min, respectively. The observed experimental response for removal efficiency was 69.36 in optimum conditions of variables. The total organic carbon measurements showed 95% mineralization of 20 mg/L phenol at 15 h using the PEF/TiO_2 process. The degradation product evolution was verified by gas chromatography analysis.

Conclusions: An empirical relationship between the response and independent variables was attained and expressed by the second-order polynomial equation. Effect of experimental parameters on the removal efficiency of phenol was established by the response surface and contour plots of the model-predicted responses.

Keywords: Carbon nanotube, TiO_2 nanoparticles, Response surface design, Photodegradation, Electrochemical treatment

Background

Nowadays, one of the major problems facing industrialized nations is contamination of the environment by hazardous chemicals. Among the various wastes, phenol is a characteristic pollutant found in wastewaters and industrial effluents. This pollutant has been a typical molecule present in most aromatic products, highlighting its relevance. Phenol is lethal to most aquatic organisms exposed to concentrations of 10 to 100 ppm [1], being considered toxic to fish at a 1- to 2-ppm concentration level [2], and presents acute toxicity to man [3]. Due to its toxic properties, the efficient removal of this compound from wastewater is an environmental concern. In recent years, various technologies have been developed for environmental pollutant remediation [4-9]. The advanced oxidation processes (AOPs), which can produce oxidizing hydroxyl radicals ($OH^{\bullet}$), have been proven to be an efficient method for the degradation of organic

* Correspondence: zarei90211@yahoo.com
[1]Department of Applied Chemistry, Faculty of Chemistry, University of Tabriz, P.O. Box, 51666-16471, Tabriz, Iran
Full list of author information is available at the end of the article

pollutants [8,10,11]. Among AOPs, Fenton-based processes (reaction (1)) have been encountering a notable development due to their favorable results in combination with easy handling [12]:

$$Fe^{2+} + H_2O_2 + H^+ \rightarrow Fe^{3+} + OH^\bullet + H_2O \quad (1)$$

Ferrous ions and hydrogen peroxide are simultaneously produced in an aqueous medium by ferric ions, initially introduced at a catalytic concentration, (reaction (2)) and a bi-electronic reduction of the dissolved molecular oxygen (reaction (3)) [13,14]:

$$Fe^{3+} + e^- \rightarrow Fe^{2+} \quad (2)$$

$$O_2 + 2H^+ + 2e^- \rightarrow H_2O_2 \quad (3)$$

A Pt or a boron-doped diamond anode in an undivided cell is utilized in the electro-Fenton (EF) technique, while Fe^{2+} or Fe^{3+} is added to the solution to permit the degradation of pollutants by $OH^\bullet$ generated from reaction (1) in the medium. In the photoelectro-Fenton (PEF) method, the solution is treated under EF conditions, and ultraviolet light is simultaneously irradiated to accelerate the mineralization rate of organics. Main effects of the photocatalytic action of this irradiation can be related to (a) the photolysis of $Fe(OH)^{2+}$, which is the predominant Fe^{3+} species at pH 3.0, regenerating a greater amount of Fe^{2+} and producing more quantity of $OH^\bullet$ via reaction (4) [8,15,16]:

$$[Fe(OH)]^{2+} + h\upsilon \rightarrow Fe^{2+} + OH^\bullet \quad (4)$$

and (b) the photodecomposition of Fe^{3+} complexes with generated carboxylic acids, e.g., oxalic acid. Oxalic acids behave as photo-active complexes in the presence of ferric ions at acidic pH, which undergo photo-decarboxylation reaction (Equation 5) [17,18]:

$$Fe(III)(RCO_2)^{2+} + h\upsilon \rightarrow Fe^{2+} + CO_2 + R^\bullet \quad (5)$$

One of the forms of the Fe^{3+}/Fe^{2+} redox couple (E_0 = 0.77 V/NHE) is utilized in the classical electro-Fenton process. Some homogeneous redox couple $M(n+1)^+/Mn^+$ can also undergo Fenton-type reaction and lead to the formation of hydroxyl radical according to reaction (6) [19].

$$M^{n+} + H_2O_2 \rightarrow M^{(n+1)+} + OH^- + OH^\bullet \quad (6)$$

This process can be called as 'electro-Fenton-like' process. Several metal cations such as Mn(II), Fe(III), Ag(I), Cu(II), or Co(II) have been tested in common Fenton or photo-Fenton reactions [19,20]. The electro-Fenton-like process efficiency is affected by the standard potential of the $M(n+1)^+/Mn^+$ redox couple used and by the scavenging catalyst effect. Therefore, it is very important to gain insight in this electro-Fenton-like process activated by transition metals, which would certainly help widen the application of this promising technology for wastewater treatment.

On the other hand, heterogeneous photocatalysis through UV illumination on semiconductor surface is an attractive advanced oxidation process. TiO_2 is a wide bandgap semiconductor, and when it is illuminated with the light of λ < 390 nm, electrons are promoted from the valence band to the conduction band to give electron–hole pairs through reaction (7) [6,21,22]:

$$TiO_2 + h\upsilon(\lambda < 390\ nm) \rightarrow e^- + h^+ \quad (7)$$

The valence band (h^+) potential is positive enough to generate hydroxyl radicals at the surface of TiO_2, and the conduction band (e^-) potential is negative enough to reduce molecular oxygen, as shown in Equations 8 to 12:

$$e^- + O_{2(ads)} \rightarrow O^\bullet{}_{2(ads)}{}^- \quad (8)$$

$$e^- + H_{(ads)}{}^+ \rightarrow H^\bullet{}_{(ads)} \quad (9)$$

$$O^\bullet{}_{2(ads)}{}^- + H^+ \rightarrow HOO^\bullet \quad (10)$$

$$2HOO^\bullet \rightarrow H_2O_2 + O_2 \quad (11)$$

$$H_2O_2 + O^\bullet{}_{2(ads)}{}^- \rightarrow OH^\bullet + OH^- + O_2 \quad (12)$$

Moreover, the hydroxyl radicals formed are strong oxidants and attack organic pollutants present at or near the surface of TiO_2 (Equation 13). It causes photodegradation of organic matters (OM) [6,23].

$$OH^\bullet{}_{(ads)} + OM \rightarrow degradation\ of\ the\ OM \quad (13)$$

A serious inconvenience appears when treating iron-chelating organic compounds because an iron catalyst needed to carry out Fenton's reaction (Equation 1) is no longer available, and the process becomes ineffective. Phenol is able to form stable complexes with ferric ions [24]. So, in this work, we have studied the removal of phenol as a model pollutant from aqueous solutions using the process of PEF/supported TiO_2 nanoparticles with carbon nanotube-polytetrafluoroethylene (CNT-PTFE) electrode as the cathode and Mn^{2+} as the catalyst. The reaction of Mn^{2+} with H_2O_2 and the regeneration of Mn^{2+} are shown in Equations 14 and 15, respectively [25]:

$$Mn^{2+} + H_2O_2 + H^+ \rightarrow Mn^{3+} + H_2O + OH^\bullet \quad (14)$$

$$Mn^{3+} + e^- \rightarrow Mn^{2+} \quad (15)$$

As it has been reported previously [4-9], the efficiency of phenol removal by PEF is dependent on various parameters such as the initial amount of catalyst, initial concentration of phenol, reaction time, and applied current. To determine the influence of these parameters, experiments were carried out, varying systematically one of the parameters and keeping constant the others in conventional methods. This should be repeated for all of the influencing parameters which results in an unreliable number of experiments. Central composite design (CCD) was employed to optimize the effective parameters with the minimum number of experiments. In this case, it would be possible to estimate linear interaction and quadratic effects of the factors and to provide a prediction model for the response. So in this work, we have used response surface methodology (RSM) to study the influence of experimental parameters on the phenol removal efficiency of the PEF/TiO_2 process. Phenol removal efficiency was selected as the response for optimization, and the functional relationship between the response and the most significant independent variables (factors) was established by means of an experimental design. All the parameters are simultaneously applied in order to calculate their relative effect [26,27].

Methods

Chemicals

Phenol (λ_{max} = 270 nm, M_w = 94.11 g/mol) was purchased from Merck Co. (Dramstadt, Germany). Sulfuric acid, sodium sulfate, and manganese (II) sulfate were obtained from Merck Co. The other chemicals used in this study such as methanol, methyltrimethoxysilane, and HCl were obtained from Merck Co. and were used as received. Multi-walled carbon nanotubes were produced from Cheap Tubes Inc. (Brattleboro, VT, USA). Their specific surface area, inside diameters, and outer diameters were 233 m^2/g, 3 to 5 nm, and 8 to 15 nm, respectively. TiO_2 nanoparticles (Degussa P-25, Düsseldorf, Germany) having 80% anatase and 20% rutile were immobilized on glass plates. Their specific surface area (BET) and mean crystalline size were 49.7 m^2/g and 21 nm, respectively. The scanning electron microscopy image of used TiO_2 was reported previously [18].

Instruments

The solution pH was measured with a Metrohm 654 pH-meter (Herisau, Switzerland). The experiments were performed using a DC power supply (ADAK PS 808, Tehran, Iran). The removal of phenol from the aqueous solution was followed by using a UV-visible (UV–vis) spectrophotometer (Light wave S2000, Nottingham, England). Irradiation was carried out with a 6-W UV-C lamp (T5 fluorescent cool light, Guangdong, China), which was placed in a quartz tube inside a 2,000-mL batch cell. Total organic carbon (TOC) measurements were carried out using a Shimadzu TOC–VCSH (Columbia, MD, USA) TOC analyzer. Nitrogen sorption analyses were obtained with a Micromeritics Tristar sorptometer (Norcross, GA, USA) using standard continuous procedures at 77.15 K on calcined samples that had been degassed at 363 K for 1 h and then at 403 K under high vacuum for at least 10 h. The surface area was calculated according to the Brunauer-Emmett-Teller (BET) model [28] over a relative pressure range of 0.05 to 0.30. Pore diameter distribution was calculated according to the Barret-Joyne-Halenda method [28], modified by the Halsey thickness curve correction [29] on the desorption branch. The BET surface area, total pore volume, and pore size of TiO_2 nanoparticles were 49.7 m^2/g, 0.26 cm^3/g, and 24 nm, respectively. For gas chromatography (GC) analysis, a Shimadzu GC-2010 gas chromatograph (North America) with a 25-m to 0.22-mm CBP 1 capillary column was used.

Combined photoelectro-Fenton and photocatalytic system

The treatment of phenol was performed at constant current and room temperature (25°C) in an undivided, cubic tank with a capacity of about 2,000 mL. The CNT-PTFE electrode was selected as the cathode, and Pt sheets of 11.5 cm^2 area were used as the anode. The distance between electrodes was about 1 cm. The fabrication and schematic diagram of the CNT-PTFE electrode was explained previously [30]. The cathode, anode, and UV lamp (placed in a quartz tube) were centered in the cell, surrounded by the TiO_2 nanoparticles immobilized on glass plates, which covered the four sides of the inner wall of the cell. The schematic diagram of the used electrolytic-photocatalytic system was reported previously [18]. In all experiments, the solutions were vigorously stirred with a magnetic bar to ensure suitable transport of reactants towards/from the electrodes. Pure O_2 gas at 140 mL/min was fed into the diffusion cathode for the production of H_2O_2 from reaction (3). Solutions of phenol (2,000 mL) containing 0.05M Na_2SO_4 as background electrolyte were degraded in all cases at pH 4.0. The removal efficiency (R (%)) was defined by the following expression:

$$\text{Removal efficiency } (R(\%)) = \left(1 - \frac{A}{A_0}\right) \times 100 \quad (16)$$

where A_0 and A are the phenol aqueous solution absorbance at time 0 and t at λ_{max} = 270 nm, respectively.

Results and discussion

Comparative removal and mineralization of phenol

Phenol degradation was carried out by using EF, PEF, UV/TiO_2, and PEF/TiO_2 processes. In order to compare the above processes, 20-mg/L phenol solutions at pH 4 have been used as the sample in all experiments. Figure 1 shows the removal efficiency for the 20-mg/L phenol solution under electrolysis at 300 mA and an initial Mn^{2+} concentration of 0.3 mM after 180 min at room temperature (25°C). It can be observed that the highest

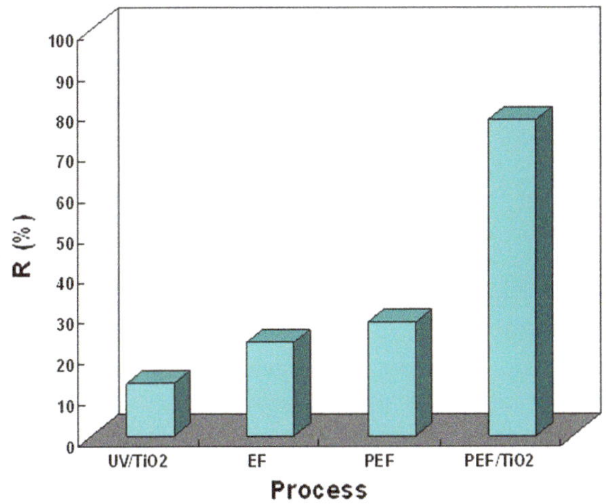

Figure 1 Removal efficiency (*R* (%)) for a 20-mg/L phenol solution at room temperature. I = 300 mA, pH = 4.0, $[Na_2SO_4]$ = 0.05M, $[Mn^{2+}]_o$ = 0.3 mM, reaction time = 180 min.

removal efficiency was obtained using the PEF/TiO_2 process. The results also showed that phenol removal after 180 min of reaction time follows a decreasing order: PEF/TiO_2 > PEF > EF > UV/TiO_2. Furthermore, the pollutant was destroyed during 180 min of PEF/TiO_2 process, yielding 78% of removal efficiency, whereas at the same reaction time, PEF, EF, and UV/TiO_2 processes led to 28%, 23%, and 13% removal efficiency, respectively. These results indicate that combining the PEF method with the photocatalytic process can effectively accelerate the removal efficiency. Figure 2 shows representative UV–vis spectral changes of the phenol solution as a function of reaction time through the PEF/TiO_2 process. The absorption peak at λ = 270 nm diminished and finally disappeared under 180 min of PEF/TiO_2 process. The efficiency of the PEF/TiO_2 process to mineralize the phenol solution was evaluated by TOC decay. This study was carried out with a solution of pH 4.0 containing 20 mg/L of phenol by applying 300-mA current for 15 h. A TOC abatement was observed attaining 95% of mineralization after 15 h of PEF/TiO_2 process. TOC decay shows that the PEF/TiO_2 process can effectively degrade the pollutant solution containing phenol.

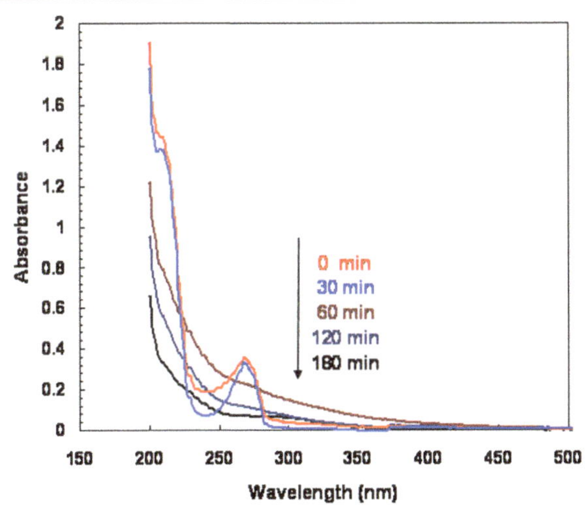

Figure 2 UV–vis spectral changes with electrolysis time for 20 mg/L phenol solution by PEF/TiO_2 process. pH = 4.0, I = 300 mA, $[Na_2SO_4]$ = 0.05 M.

Central composite design model

The four-factor CCD matrix and experimental results obtained in the PEF/TiO_2 process are presented in Table 1. The second-order polynomial response equation (Equation 17) was employed to attain the interaction between the dependent and independent variables.

$$Y = b_0 + \sum b_i x_i + \sum b_{ii} x_i^2 + \sum b_{ik} x_i x_k \qquad (17)$$

where Y is a response variable of removal efficiency. The b_i is the regression coefficients for linear effects; b_{ii} is the regression coefficients for squared effects; b_{ik} is the regression coefficients for interaction effects, and x_i is the coded experimental levels of the variables.

Based on these results, an empirical relationship between the response Y and the four significant independent variables (x_1, x_2, x_3, and x_4) was attained as shown below:

$$\begin{aligned} Y = {} & 25.0000 - 9.8964x_1 - 6.8971x_2 - 0.0011x_3 \\ & + 8.8131x_4 + 1.0521x_1x_2 + 3.6771x_1x_3 \\ & + 1.9271x_2x_3 + 1.0625x_2x_4 - 0.0002x_3x_4 \\ & - 4.5625x_1^2 - 0.0004x_2^2 - 1.0625x_3^2 + 0.0002x_4^2 \end{aligned} \qquad (18)$$

From Equation 18, it can be seen that the initial Mn^{2+} concentration, initial phenol concentration, and applied current have negative effect on the response compared to the reaction time, which has a positive effect on the removal of phenol. The phenol removal efficiencies (R (%)) have been predicted by Equation 18 and presented in Table 1. These results indicated good agreements between the experimental and predicted values of removal efficiency. The square of determination coefficient (R^2) quantitatively evaluates the correlation between the experimental data and the predicted responses. The experimental results and the predicted values obtained from the model (Equation 18) were compared. It was found that the predicted values matched the experimental values reasonably well with R^2 = 0.918. This implies that 91.8% of the variations for percent removal is explained by the independent variables, and this also means that the model does not explain only about 8.2% of variation.

Table 1 The four-factor central composite design matrix and the value of response function (*R* (%))

Run	[Mn^{2+}] (mM)	[Phenol] (mg/L)	*I* (mA)	Time (min)	Removal efficiency (%)	
					Experimental	Predicted
1	0	+2	0	0	10.50	23.29
2	0	0	−2	0	20.02	22.79
3	−1	−1	+1	−1	40.50	39.58
4	−2	0	0	0	39.00	50.12
5	0	0	0	0	25.10	25.00
6	0	0	+2	0	24.02	31.62
7	−1	−1	−1	−1	35.06	34.29
8	−1	−1	+1	+1	75.00	69.95
9	+1	+1	−1	+1	20.19	16.58
10	+1	+1	−1	−1	10.06	9.45
11	0	0	0	0	25.00	25.00
12	0	0	0	+2	46.00	51.45
13	+1	+1	+1	−1	15.20	12.25
14	0	0	0	0	23.50	25.00
15	+1	+1	+1	+1	25.00	20.12
16	+1	−1	−1	−1	20.12	20.75
17	0	0	0	0	25.00	25.00
18	+2	0	0	0	10.40	8.29
19	0	0	0	−2	10.00	13.95
20	+1	−1	+1	−1	25.00	25.29
21	−1	+1	+1	+1	53.00	48.41
22	+1	−1	−1	+1	30.00	32.12
23	0	0	0	0	25.00	25.00
24	0	0	0	0	24.50	25.00
25	0	−2	0	0	60.00	56.12
26	−1	−1	−1	+1	65.00	63.91
27	+1	−1	+1	+1	35.40	37.41
28	−1	+1	−1	+1	50.00	44.12
29	−1	+1	−1	−1	25.00	18.75
30	0	0	0	0	25.20	25.00
31	−1	+1	+1	−1	30.00	22.29

Adjusted R^2 (Adj-R^2) is also a measure of the goodness of a fit, but it is a useful tool for comparing the explanatory power of models with different numbers of independent variables. The adjusted R^2 will increase only if the new term improves the model more than would be expected by chance, and it will decrease when a predictor improves the model less than expected by chance. It corrects the R^2 value for the sample size and the number of terms in the model by using degrees of freedom in its computations. If there are many terms in a model and the sample size is not very large, Adj-R^2 may be visibly smaller than R^2 [31]. Here, the Adj-R^2 value (0.847) was very close to the corresponding R^2 value (see Table 2).

Table 2 summarizes the results of the quadratic response surface model fitting in the form of analysis of variance (ANOVA). ANOVA confirms the adequacy of the quadratic model [30,32]. ANOVA divides the sum of squares of each of the two sources of variation, the model and the experimental error, showing whether the variation from the model is significant or not when compared with the ones associated with residual error [30,31,33]. This comparison is performed by degrees of freedom (*F* value). If the model predicts the experimental results well, the *F* value should be greater than the tabulated value of *F* distribution for a certain number of degrees of freedom in the model at a level of significance α. The *F* value obtained, 12.82, is clearly greater than the tabulated *F* (2.352 at 95% significance), confirming the adequacy of the model fits.

Table 3 gives the Student's *t* distribution and the corresponding values, together with the parameter estimate. The *P* values show the significance of each of the coefficients. The larger the magnitude of the Student's *t* test and the smaller the *P* value, the more significant is the corresponding coefficient [30,32].

The Pareto analysis calculates the percentage effect of each factor on the response and gives more significant information to interpret the results. The following relation was used to perform this analysis [26,34]:

$$P_i = \left(\frac{b_i^2}{\sum b_i^2}\right) \times 100 \; (i \neq 0) \quad (19)$$

Figure 3 shows the Pareto graphic analysis. As can be seen in this figure, among the variables, the initial Mn^{2+} concentration (b_1, 37.01%), reaction time (b_4, 29.35%), and initial phenol concentration (b_2, 17.98%) produce the main effect on the removal efficiency.

Effect of variables as response surface and contour plots

The response surface and contour plots of the model-predicted responses were obtained by the Minitab software while two variables were kept constant and the others were varied within the experimental ranges. These plots were utilized to assess the interactive relationships

Table 2 Analysis of variance for the fit of removal efficiency from the central composite design

Source of variations	Sum of squares	Degree of freedom	Adjusted mean square	*F* value
Regression	7,304.47	14	521.748	12.82
Residuals	651.08	16	40.693	
Total	7,955.55			

R^2 = 0.918, Adj-R^2 = 0.847.

Table 3 Estimated regression coefficients and corresponding *t* and *P* values from the data of central composite design experiments

Coefficient	Parameter estimate	Standard error	*t* value	*P* value
b_0	25.0000	10.4757	2.228	0.041
b_1	−9.8964	4.9537	−1.998	0.063
b_2	−6.8971	4.9537	−1.392	0.183
b_3	−0.0011	0.0073	−0.151	0.881
b_4	8.8131	4.9537	1.779	0.094
b_{12}	1.0521	1.1929	0.882	0.391
b_{13}	3.6771	1.1929	3.082	0.007
b_{14}	0.0000	0.0000	0.463	0.650
b_{23}	1.9271	1.1929	1.615	0.126
b_{24}	1.0625	1.5948	0.666	0.515
b_{34}	−0.0002	0.0016	−0.118	0.908
b_{11}	−4.5625	1.5948	−2.861	0.011
b_{22}	−0.0004	0.0016	−0.274	0.787
b_{33}	−1.0625	1.5948	−0.666	0.515
b_{44}	0.0002	0.0016	0.118	0.908

between the process variables and treatment outputs for the removal of phenol.

In Figure 4, the response surface and contour plots were developed as a function of the initial Mn^{2+} concentration and applied current while the initial phenol concentration and reaction time were kept constant at 15 mg/L and 90 min, respectively, being the central levels. As can be seen in Figure 4, phenol removal efficiency decreased with increasing initial Mn^{2+} concentration. The catalyst concentration (e.g., Fe^{2+}, Fe^{3+}, Mn^{2+}, ...) is

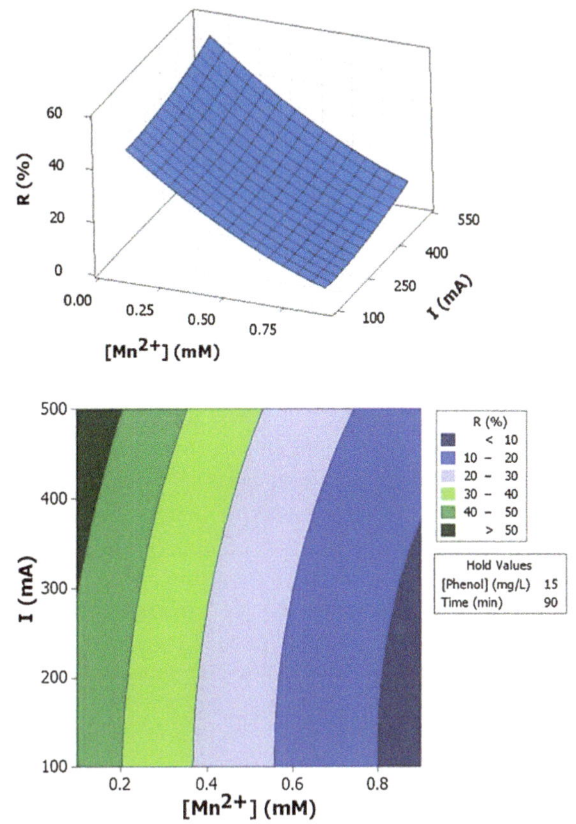

Figure 4 Removal efficiency as a function of initial Mn^{2+} concentration and applied current. The response surface plot and contour plot of the removal efficiency (*R* (%)) as a function of initial Mn^{2+} concentration (mM) and applied current (mA).

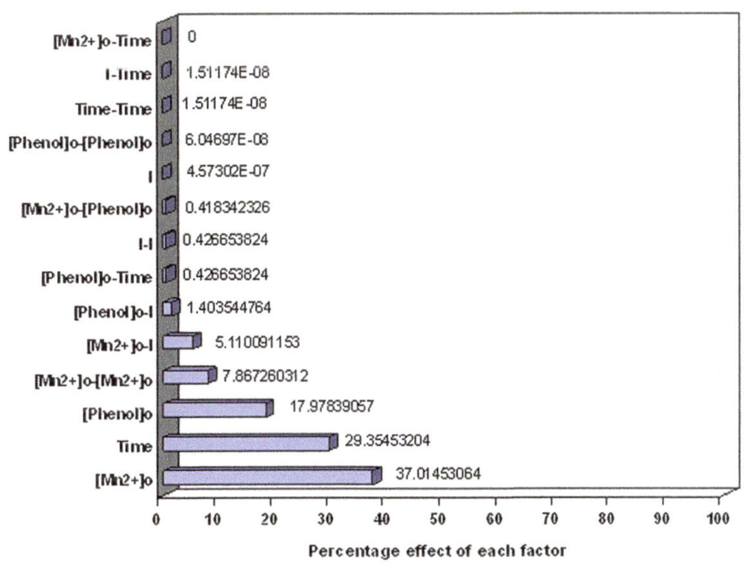

Figure 3 Pareto graphic analysis.

a key factor in EF processes and is a function of the cathode employed. For example, concentrations of 0.5 to 1.0 mM Fe^{2+} are optimum for carbon-PTFE GDE cathodes because of their lower ability for Fe^{2+} regeneration [8,35]. At high Mn^{2+} concentrations, the phenol removal rate decreased, which can be explained by metal deposition that leads to decreasing hydroxyl radical production according to reaction (21). Anodic deposition observed during the experiments has been reported by several studies [19,36,37] on acidic solutions according to reaction (20):

$$Mn^{2+} + 2H_2O \rightarrow MnO_{2(S)} + 4H^+ + 2e^- \quad (20)$$

On the other hand, the phenol removal efficiency increased with increasing applied current values from 100 to 500 mA. The enhancement in phenol removal efficiency with the applied current can be related to a high production of H_2O_2 through reaction (3).

The response surface and contour plots of the removal efficiency as a function of initial phenol concentration and applied current have been shown in Figure 5. As can be seen, an increase in initial phenol concentration from 5 to 25 mg/L decreases removal efficiency. This behavior is one of the properties of AOPs. In the constant conditions of the system (e.g., constant current, pH, etc.), a certain amount of hydroxyl radicals is produced while the phenol concentration is increased. Therefore, the amount of hydroxyl radicals is not enough to remove the high concentrations of the pollutant; removal efficiency then decreases [38].

Figure 6 displays the 2D and 3D plots of phenol removal efficiency as a function of applied current and reaction time at fixed initial Mn^{2+} and phenol concentrations of 0.5 mM and 15 mg/L, respectively. As can be seen, increasing reaction time increased the phenol removal efficiency at all applied currents.

Determination of optimal conditions and evolution of degradation

The model obtained using experimental data was used to verify the optimum values of PEF/TiO_2 process variables. The optimum values of the PEF/TiO_2 process variables to reach the maximum removal efficiency are shown in

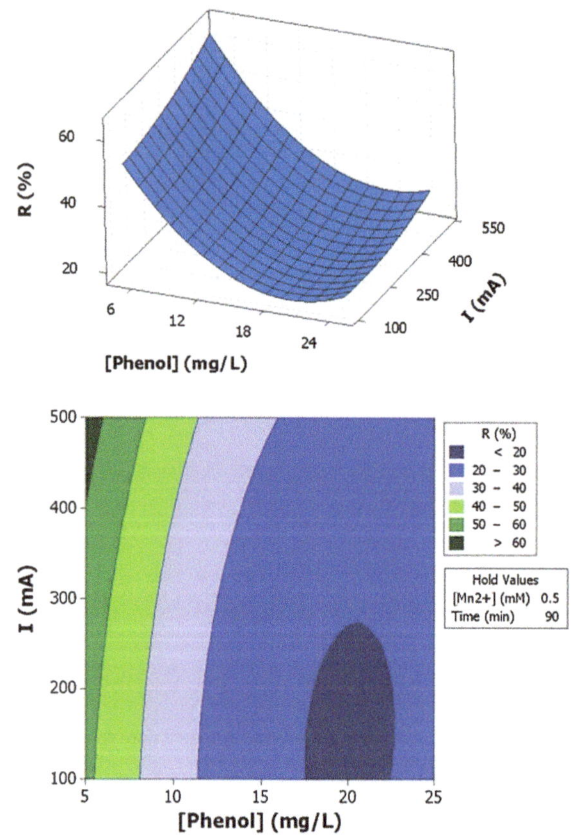

Figure 5 Removal efficiency as a function of initial phenol concentration and applied current. The response surface plot and contour plot of the removal efficiency (*R* (%)) as a function of initial phenol concentration (mg/L) and applied current (mA).

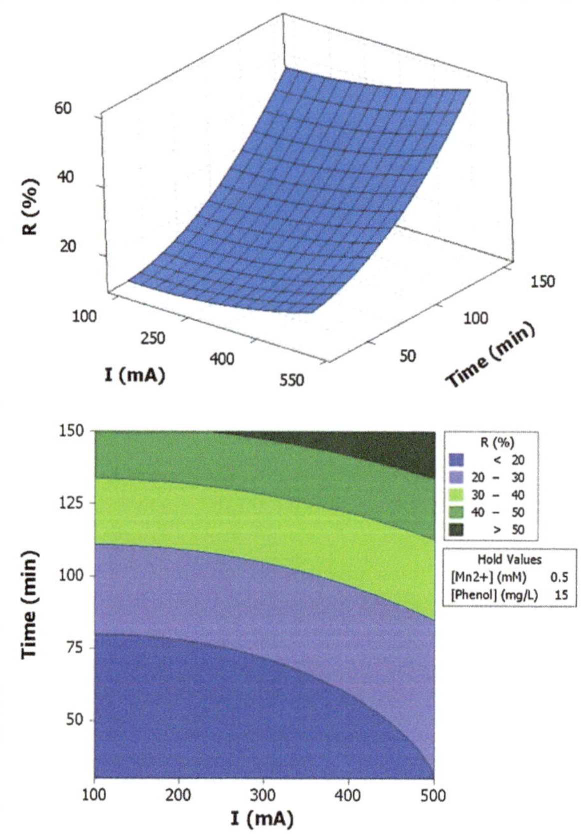

Figure 6 Removal efficiency (*R* (%)) as a function of applied current and reaction time. The response surface plot and contour plot of the removal efficiency (*R* (%)) as a function of applied current (mA) and reaction time (min).

Table 4. The predicted values were verified by further experimental test, and the result indicated that the maximal removal efficiency was obtained when the values of each operational parameter were justified at the optimum values. It implies that the strategy to optimize the removal conditions and to obtain the maximal removal efficiency by CCD for phenol removal by electrochemical and photocatalytic processes is successful. The theoretical response proposed by Minitab software for the removal of phenol was 70.35%. To confirm the used model adequacy and the validity of the resulted optimization conditions, additional experiment was performed under optimal operating conditions. The value of the experimental response was 69.36% and was found to be very close to the value proposed by the model (70.35%).

GC was used to follow the time evolution of stable aromatic intermediates generated during phenol mineralization. The samples were analyzed by GC with the following temperature program: 70°C for 3 min, 10°C/min up to 90°C and hold time 2 min, 10°C/min up to 200°C and hold time 5 min. The temperatures of the inlet and detector were 200 and 250°C, respectively. The GC chromatographs (Figure 7) of the samples collected at various time points throughout the PEF/TiO_2 process of a 20-mg/L phenol solution at pH 4, electrolyzed at 300 mA showed the disappearance of phenol peak and the formation of a new peak which elutes after the parent compound. This observed new peak disappeared after approximately 3 h. This behavior is consistent with the fast destruction of phenol and proves that the generated oxidation products can further react with $OH^{\bullet}$ to form various breakdown products and eventually mineralize.

Electrical energy consumption during the PEF/TiO_2 process

The removal of aqueous organic pollutants by photoelectrolytic processes is an electric energy-intensive process. So, the electric energy can represent a major fraction of the operating costs, and simple figures of merit based on electric energy consumption can be very useful and informative. In the case of low concentrations of a pollutant, which applies here, the appropriate figure of merit is the electrical energy per order (E_{EO}). E_{EO} is defined as the number of kilowatt hour of electrical energy required to reduce the concentration of a pollutant by one order of magnitude (90%) in 1 m^3 of contaminated water. The E_{EO} can be calculated according to the proposal of the Photochemistry Commission of the International Union of Pure and Applied Chemistry (see Equation 21) [18].

$$E_{EO} = \frac{P \times t \times 1{,}000}{V_s \times 60 \times \log \left(C_i / C_f \right)} \tag{21}$$

where P is the lamp power (kW) of the photochemical system, V_s is the solution volume (L), C_i and C_f are the initial and final pollutant concentrations, and t is the reaction time (min). So, the electrical energy (kWh/m^3) required to photochemically remove 20 mg/L of phenol from 2,000 mL of phenol solution in the optimized conditions (I = 0.3 A, pH = 4, $[Mn^{2+}]_0$ = 0.3 mM) is 15 kWh/m^3. The electrical energy consumption in the electrochemical process was calculated using the commonly used Equation 22 [18]:

$$E = \frac{U \times I \times t}{V_s \times 3.6 \times 10^3} \tag{22}$$

where E is the electrical energy (kWh/m^3), U is the cell voltage in volt (V), I is the current in ampere (A), t is the time of process per second, and V_s is the solution volume (L). Using Equation 22, the electrical energy (kWh/m^3) required to electrochemically remove 20 mg/L of phenol from 2,000 mL of its solution in the optimized conditions is 9 kWh/m^3. In this preliminary economic study, energy costs in the electrochemical and photochemical processes are taken into account as major cost items in the calculation of the operating cost as kilowatt hour per volume (m^3) of the water in the reactor (Equation 23).

$$\text{Operating cost} = a \left(E_{\text{Photochemical process}} + E_{\text{Electrochemical process}} \right) \tag{23}$$

where $E_{\text{Photochemical process}}$ (E_{EO}) and $E_{\text{Electrochemical process}}$ (E) are 15 and 9 kWh/m^3, respectively. 'a' is the electrical energy price in Iranian market in September 2012, which is equal US$ 0.036/kWh. So, the operating cost is equal to US$ 0.8640/m^3, which is very low in comparison with different AOP processes in different papers [39-41]. As an example, Brillas et al. [42] reported the results obtained from the different treatments of a 1,000-mg/L aniline solution at selected electrolysis times where most TOC removal was achieved. It was reported that anodic oxidation in the presence of electro-generated

Table 4 Optimum operating conditions of the process variables

Variable	Optimum value
	Phenol
$[Mn^{2+}]$ (X_1)	0.3 mM
[Phenol] (X_2)	15 mg/L
I (X_3)	300 mA
Time (X_4)	150 min
Experimental R	69.36%
Predicted R	70.35%

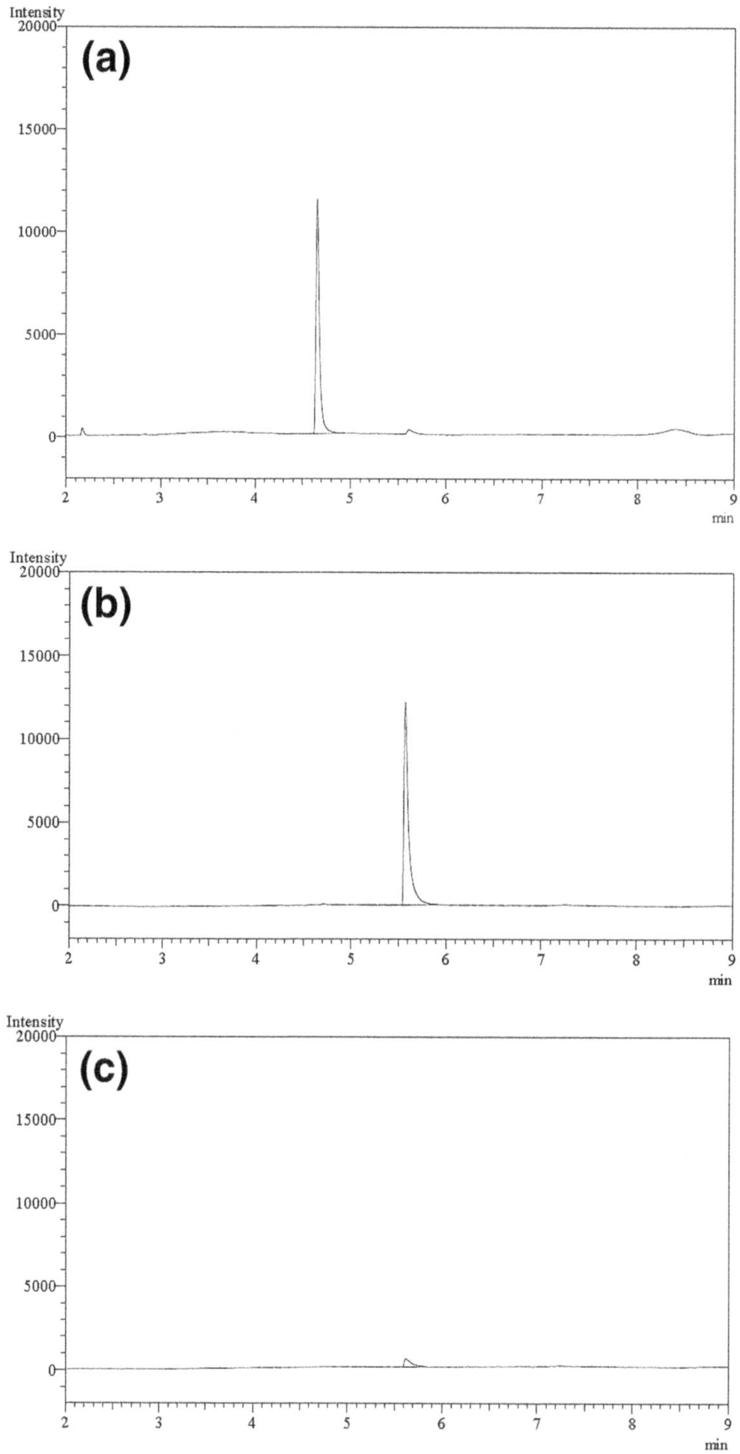

Figure 7 Gas chromatography analysis. Evolution of GC chromatographs during the PEF/TiO_2 process treatment of a 20-mg/L phenol aqueous solution at pH 4.0, I = 300 mA, [Na_2SO_4] = 0.05M. (**a**) Reaction time = 0 h; (**b**) reaction time = 6 h; (**c**) reaction time = 9 h.

H_2O_2 at 20 A has very low oxidizing power and very high energy cost (18% and 502 kWh/m^3 at 6 h, respectively) that it is not useful in practice. According to their reports, the EF process presented much more degradation ability with moderate consumption, lower than 45 kWh/m^3. For both processes, the increase in current leads to higher energy cost. On the other hand, higher degradations and energy costs were always obtained when the solution volume was reduced from 30 to 10 L.

Hanafi et al. [43] reported the treatment of olive mill wastewater containing polyphenols, considering modifying chemical oxygen demand (COD), dark color removal, and pH. Through an optimum time of 15 min, 2 mg/L of Cl_2 concentration, pH of 4.2, and density of 250 A/m^2, polyphenols were reduced by 70%, and energy consumption was 2.63 kWh/kg COD removed. Agladze et al. [39] reported that 87% to 90% of the initial COD of solution containing phenol was removed in 1 h with a 5-A current in the membrane cell and 40 kWh/m^3 of energy consumption, whereas the same process took 30 min of electro-Fenton treatment in an undivided cell with 22 kWh/m^3 of energy consumption. On the other hand, only 10 min of electrolysis with a 5-kWh/m^3 power cost was needed when the electrolyzed solution was treated with Fe(II)/H_2O_2 in a 1:1 molar ratio addition outside. They reported that the significant economy of specific energy consumption (29.4 kWh/m^3 for 92.6% COD removal against 39 kWh/m^3 for 91% TOC removed) was achieved in the present study for the ex-cell treatment.

Experimental

In this study, to find the optimum conditions for phenol degradation by photoelectro-Fenton combined with photocatalytic process, CCD was employed. CCD is a widely used form of RSM. A five-level CCD consisting of 31 experiments was employed in this work, including 16 cube points, seven replications at the center point, and eight axial points. Initial Mn^{2+} concentration (X_1), initial phenol concentration (X_2), applied current (X_3), and reaction time (X_4) were chosen as the main factors in order to evaluate the influence of operating parameters on phenol removal efficiency. Experimental data were analyzed using the Minitab 15 software (State College, PA, USA). For statistical calculations, the variables X_i were coded as x_i according to the following relationship:

$$x_i = \frac{X_i - X_o}{\delta X} \quad (24)$$

where X_o is the value of X_i at the center point and δX presents the step change [30,44,45]. The experimental ranges and the levels of the independent variables for phenol removal are given in Table 5.

Table 5 Experimental ranges and levels of the independent test variables

Variables	Ranges and levels				
	−2	−1	0	+1	+2
[Mn^{2+}] (mM) (X_1)	0.1	0.3	0.5	0.7	0.9
[Phenol] (mg/L) (X_2)	5	10	15	20	25
I (mA) (X_3)	100	200	300	400	500
Time (min) (X_4)	30	60	90	120	150

I, applied current.

Conclusions

The removal of phenol from aqueous solutions by a photocatalytic process using immobilized TiO_2 nanoparticles combined with photoelectro-Fenton process with CNT-PTFE electrode as cathode and Mn^{2+} ions as catalyst has been studied. A comparison of EF, PEF, UV/TiO_2, and PEF/TiO_2 processes for the removal of phenol solution was performed. The pollutant was destroyed during 180 min of the PEF/TiO_2 process, yielding 78% of removal efficiency, whereas at the same time, the PEF, EF, and UV/TiO_2 processes led to 28%, 23%, and 13% removal efficiency, respectively. Based on experimental results, an empirical relationship between the response and independent variables was attained and expressed by the second-order polynomial equation. Effect of experimental parameters on the removal efficiency of phenol was established by the response surface and contour plots of the model-predicted responses. Analysis of variance showed a high coefficient of determination value (R^2 = 0.918), thus ensuring a satisfactory adjustment of the second-order regression model with the experimental data. To follow the time evolution of stable aromatic intermediates generated during phenol mineralization, samples were analyzed by means of GC.

Competing interests

The authors declare that they have no competing interests.

Authors' contributions

MZ carried out the RSM studies and GC analysis, participated in the sequence alignment, and drafted the manuscript. AK conceived of the study, participated in its design and coordination and in the sequence alignment. MF carried out the experiments and participated in the statistical analysis. FS participated in the design of the study as an industrial advisor. HR participated in the study design and coordination as an industrial advisor. All authors read and approved the final manuscript.

Acknowledgements

This paper reports a part of the results of the project funded by Tabriz Petrochemical Complex. The authors wish to thank the Tabriz Petrochemical Complex and University of Tabriz for all the supports provided.

Author details

[1]Department of Applied Chemistry, Faculty of Chemistry, University of Tabriz, P.O. Box, 51666-16471, Tabriz, Iran. [2]R&D Center of Tabriz Petrochemical Complex, Tabriz, Iran.

References

1. (1983) In: Tatken RL, Lewis RJ (ed) Registry of toxic effects of chemical substances, 1981–2, vol. 1, Publication No. 83-107th edition. National Institute for Occupational Safety and Health (NIOSH), Cincinnati
2. Arana J, Tello Rendon E, Dona Rodriguez JM, Herrera Melian JA, Gonzalez Diaz O, Pérez Pena J (2001) Chemosphere 44:1017
3. Dreisbach RH (1983) Handbook of poisoning: prevention, diagnosis and treatment. Lang Medical Publications, Los Altos, p 401
4. Brillas E, Calpe JC, Casado J (2000) Water Res 34:2253

5. Kusvuran E, Gulnaz O, Irmak S, Atanur OM, Ibrahim Yavuz H, Erbatur O (2004) J Hazard Mater 109:85
6. Khataee AR, Vatanpour V, Amani Ghadim AR (2009) J Hazard Mater 161:1225
7. Fathinia M, Khataee AR, Zarei M, Aber S (2010) J Mol Catal A Chem 333:73
8. Brillas E, Sirés I, Oturan MA (2009) Chem Rev 109:6570
9. Wu W, Xiao X, Zhang S, Ren F, Jiang C (2011) Nanoscale Res Lett 6:533
10. Khataee AR, Zarei M (2011) Desalination 273:453
11. Khataee AR, Zarei M, Fathinia M, Jafari MK (2011) Desalination 268:126
12. Pignatello JJ, Oliveros E, MacKay A (2006) Crit Rev Environ Sci Technol 36:1
13. Almeida LC, Garcia-Segura S, Bocchi N, Brillas E (2011) Appl Catal B 103:21
14. Oturan MA, Pinson J, Bizot J, Deprez D, Terlain B (1992) J Electroanal Chem 334:103
15. Skoumal M, Rodriguez RM, Cabot PL, Centellas F, Garrido JA, Arias C, Brillas E (2009) Electrochim Acta 54:2077
16. Sun YF, Pignatello JJ (1993) Environ Sci Technol 27:304
17. Irmak S, Yavuz HI, Erbatur O (2006) Appl Catal B Environ 63:243
18. Khataee AR, Zarei M, Ordikhani-Seyedlar R (2011) J Mol Catal A Chem 338:84
19. Pimentel M, Oturan N, Dezotti M, Oturan MA (2008) Appl Catal B Environ 83:140
20. Türk H, Cimen Y (2005) J Mol Catal A Chem 234:19
21. Thompson TL, Yates JT (2006) Chem Rev 106:4428
22. Khataee AR, Kasiri MB (2010) J Mol Catal A Chem 328:8
23. Daneshvar N, Salari D, Niaei A, Khataee AR (2006) J Environ Sci Heal B 41:1273
24. Paiva-Martins F, Gordon MH (2005) J Agric Food Chem 53:2704
25. Balci B, Oturan MA, Oturan N, Sires I (2009) J Agric Food Chem 57:4888
26. Abdessalem AK, Oturan N, Bellakhal N, Dachraoui M, Oturan MA (2008) Appl Catal B Environ 78:334
27. Farhat-Khemakhem A, Farhat MB, Boukhris I, Bejar W, Bouchaala K, Kammoun R, Maguin E, Bejar S, Chouayekh H (2012) AMB Express 2:10
28. Brunauer S, Emmett PH, Teller E (1938) J Am Chem Soc 60:309
29. Halsey G (1948) J Chem Phys 16:931
30. Zarei M, Niaei A, Salari D, Khataee A (2010) J Hazard Mater 173:544
31. Santos SCR, Boaventura RAR (2008) Appl Clay Sci 42:137
32. Liu HL, Chiou YR (2005) Chem Eng J 112:173
33. Aleboyeh A, Daneshvar N, Kasiri MB (2008) Chem Eng Process 47:827
34. Haaland DP (1989) Experimental design in biotechnology. Marcel Dekker, New York
35. Sires I, Garrido JA, Rodriguez RM, Cabot Pl, Centellas F, Arias C, Brillas E (2006) J Electrochem Soc 153:D1
36. Wu MS, Chiang PCJ (2006) Electrochem Commun 8:383
37. Ghaemi M, Binder L (2002) J Power Sources 111:248
38. Daneshvar N, Aber S, Vatanpour V, Rasoulifard MH (2008) J Electroanal Chem 615:165
39. Agladze GR, Tsurtsumia GS, Jung BI, Kim JS, Gorelishvili G (2007) J Appl Electrochem 37:385
40. Esplugas S, Gimenez J, Contreras S, Pascual E, Rodriguez M (2002) Water Res 36:1034
41. Suty H, De Traversay C, Cost M (2004) Water Sci Technol 49:227–233
42. Brillas E, Casado J (2002) Chemosphere 47:241
43. Hanafi F, Assobhei O, Mountadar M (2010) J Hazard Mater 174:807
44. Kasiri MB, Aleboyeh H, Aleboyeh A (2008) Environ Sci Technol 42:7970
45. Khataee AR, Zarei M, Asl SK (2010) J Electroanal Chem 648:143

Permissions

All chapters in this book were first published in IJIC, by Springer; hereby published with permission under the Creative Commons Attribution License or equivalent. Every chapter published in this book has been scrutinized by our experts. Their significance has been extensively debated. The topics covered herein carry significant findings which will fuel the growth of the discipline. They may even be implemented as practical applications or may be referred to as a beginning point for another development.

The contributors of this book come from diverse backgrounds, making this book a truly international effort. This book will bring forth new frontiers with its revolutionizing research information and detailed analysis of the nascent developments around the world.

We would like to thank all the contributing authors for lending their expertise to make the book truly unique. They have played a crucial role in the development of this book. Without their invaluable contributions this book wouldn't have been possible. They have made vital efforts to compile up to date information on the varied aspects of this subject to make this book a valuable addition to the collection of many professionals and students.

This book was conceptualized with the vision of imparting up-to-date information and advanced data in this field. To ensure the same, a matchless editorial board was set up. Every individual on the board went through rigorous rounds of assessment to prove their worth. After which they invested a large part of their time researching and compiling the most relevant data for our readers.

The editorial board has been involved in producing this book since its inception. They have spent rigorous hours researching and exploring the diverse topics which have resulted in the successful publishing of this book. They have passed on their knowledge of decades through this book. To expedite this challenging task, the publisher supported the team at every step. A small team of assistant editors was also appointed to further simplify the editing procedure and attain best results for the readers.

Apart from the editorial board, the designing team has also invested a significant amount of their time in understanding the subject and creating the most relevant covers. They scrutinized every image to scout for the most suitable representation of the subject and create an appropriate cover for the book.

The publishing team has been an ardent support to the editorial, designing and production team. Their endless efforts to recruit the best for this project, has resulted in the accomplishment of this book. They are a veteran in the field of academics and their pool of knowledge is as vast as their experience in printing. Their expertise and guidance has proved useful at every step. Their uncompromising quality standards have made this book an exceptional effort. Their encouragement from time to time has been an inspiration for everyone.

The publisher and the editorial board hope that this book will prove to be a valuable piece of knowledge for researchers, students, practitioners and scholars across the globe.

List of Contributors

T Mazoochi
Anatomical Sciences Research Center, Kashan University of Medical Sciences, Kashan 87159-88141, Iran

M Hamadanian
Institute of Nanoscience and Nanotechnology, University of Kashan, Kashan 87317-51167, Iran
Department of Physical Chemistry, Faculty of Chemistry, University of Kashan, Kashan 87317-51167, Iran

M Ahmadi
Department of Chemistry, Faculty of Sciences, Shahid Beheshti University, Tehran 19835-389, Iran

V Jabbari
Anatomical Sciences Research Center, Kashan University of Medical Sciences, Kashan 87159-88141, Iran
Institute of Nanoscience and Nanotechnology, University of Kashan, Kashan 87317-51167, Iran

B Sammaiah
Department of Chemistry, Osmania University, Koti, Hyderabad 500195, India

D Sumalatha
Department of Chemistry, Osmania University, Koti, Hyderabad 500195, India

GS Satyanarayana Reddy
Department of Chemistry, Osmania University, Koti, Hyderabad 500195, India

M Rajeswari
Department of Chemistry, Osmania University, Koti, Hyderabad 500195, India

LN Sharada
Department of Chemistry, Osmania University, Koti, Hyderabad 500195, India

HR Pouretedal
Faculty of Applied Chemistry, Malek-ashtar University of Technology, Shahin-Shahr, 1387763681 Iran

M Kazemi
Department of Chemistry, Islamic Azad University, Shahreza Branch, Shahreza and Fadak Group, Isfahan 37541-13115, Iran

Javad Fadaee Kakhki
Department of Applied Chemistry, Quchan Branch, Islamic Azad University, Quchan, Iran

Mohammad Reza Abedi
Department of Applied Chemistry, Quchan Branch, Islamic Azad University, Quchan, Iran

Upendra Singh Yadav
Department of Petroleum Engineering, Indian School of Mines, Dhanbad Jharkhand 826004, India

Vikas Mahto
Department of Petroleum Engineering, Indian School of Mines, Dhanbad Jharkhand 826004, India

Emad Yousif
Department of Chemistry, College of Science, Al-Nahrain University, Baghdad, 10072, Iraq

Mustafa Abdallh
Department of Chemistry, College of Science, Al-Nahrain University, Baghdad, 10072, Iraq

Hassan Hashim
Department of Physics, College of Science, Al-Nahrain University, Baghdad 10072, Iraq

Nadia Salih
School of Chemical Sciences and Food Technology, Faculty of Science and Technology, Universiti Kebangsaan Malaysia, Bangi, Selangor 43600, Malaysia

Jumat Salimon
School of Chemical Sciences and Food Technology, Faculty of Science and Technology, Universiti Kebangsaan Malaysia, Bangi, Selangor 43600, Malaysia
Bashar Mudhaffar Abdullah
School of Chemical Sciences and Food Technology, Faculty of Science and Technology, Universiti Kebangsaan Malaysia, Bangi, Selangor 43600, Malaysia

Yip-Foo Win
Department of Chemical Science, Faculty of Science, Universiti Tunku Abdul Rahman, Perak Campus, Jalan Universiti, Bandar Barat, Kampar, Perak 31900, Malaysia

Riyas Sharafudeen
Department of Chemical and Process Engineering Technology, Jubail Industrial College, P. O. Box 10099, Jubail Industrial City 31961, Kingdom of Saudi Arabia

Sanny Verma
Chemical Sciences Division, Indian Institute of Petroleum (Council of Industrial and Scientific Research) Mohkampur, Dehradun, 248005, India

Suman L Jain
Chemical Sciences Division, Indian Institute of Petroleum (Council of Industrial and Scientific Research) Mohkampur, Dehradun, 248005, India

Bir Sain
Chemical Sciences Division, Indian Institute of Petroleum (Council of Industrial and Scientific Research) Mohkampur, Dehradun, 248005, India

Mohammad Hossein Nasirtabrizi
Department of Applied Chemistry, Islamic Azad University, Ardabil Branch, Ardabil 5614633167, Iran
Zeinab Mohammadpoor Ziaei
Department of Applied Chemistry, Islamic Azad University, Ardabil Branch, Ardabil 5614633167, Iran

Aiyoub Parchehbaf Jadid
Department of Applied Chemistry, Islamic Azad University, Ardabil Branch, Ardabil 5614633167, Iran

Leila Zare Fatin
Department of Applied Chemistry, Islamic Azad University, Ardabil Branch, Ardabil 5614633167, Iran

Morteza Hasanzadeh Kafshgari
Department of Chemical Engineering, University of Sistan and Baluchestan, PO Box 98164-161, Zahedan, Iran

Mohsen Mansouri
Department of Chemical Engineering, University of Sistan and Baluchestan, PO Box 98164-161, Zahedan, Iran

Mohammad Khorram
School of Chemical, Petroleum and Gas Engineering, Shiraz University, Shiraz, Iran

Shahab Rahimi Kashani
Materials Science Centre, School of Materials, University of Manchester, Grosvenor Street, Manchester M1 7HS, UK

Hossein Sid kalal
NFCRS, Nuclear Science and Technology Research Institute, Atomic Energy Organization of Iran, P.O. Box 11365-3486 Kargar Shomali street, Tehran, Iran

Ali Reza Khanchi
NFCRS, Nuclear Science and Technology Research Institute, Atomic Energy Organization of Iran, P.O. Box 11365-3486 Kargar Shomali street, Tehran, Iran

Elahe Konoz
Department of Chemistry, Islamic Azad University, Central Tehran Branch, Tehran, Iran

Sara Khaki
Department of Chemistry, Islamic Azad University, Central Tehran Branch, Tehran, Iran

Mohammad Taghiof
NFCRS, Nuclear Science and Technology Research Institute, Atomic Energy Organization of Iran, P.O. Box 11365-3486 Kargar Shomali street, Tehran, Iran

Majid Baghdadi
Graduate Faculty of Environment, University of Tehran, Tehran, Iran

Homayon Ahmad Panahi
Department of Chemistry, Islamic Azad University, Central Tehran Branch, Tehran, Iran

Ahmad Mousa
Department of Materials and Metallurgical Engineering, Al-Balqa Applied University, Salt 19117, Jordan

Noor Asnida Asli
Centre of Nanoscience and Nanotechnology (NANO-SciTech Centre), Institute of Science, Universiti Teknologi MARA, 40450 Shah Alam, Selangor D.E., Malaysia
School of Physics and Material Studies, Faculty of Applied Sciences, Universiti Teknologi MARA, 40450 Shah Alam, Selangor D.E., Malaysia

Muhammad Salleh Shamsudin
Centre of Nanoscience and Nanotechnology (NANO-SciTech Centre), Institute of Science, Universiti Teknologi MARA, 40450 Shah Alam, Selangor D.E., Malaysia
School of Physics and Material Studies, Faculty of Applied Sciences, Universiti Teknologi MARA, 40450 Shah Alam, Selangor D.E., Malaysia
University of Southampton Malaysia Campus (USMC), No. 3, Persiaran Canselor 1, Kota Ilmu Educity@Iskandar, 79200 Nusajaya, Johor D.T., Malaysia

Suriani Abu Bakar
Department of Physics, Faculty of Science and Mathematics, Universiti Pendidikan Sultan Idris, 35900 Tanjung Malim, Perak D.R., Malaysia

Mohamad Rusop Mahmood
Centre of Nanoscience and Nanotechnology (NANO-SciTech Centre), Institute of Science, Universiti Teknologi MARA, 40450 Shah Alam, Selangor D.E., Malaysia
NANO-Electronic Centre, Faculty of Electrical Engineering, Universiti Teknologi MARA, 40450 Shah Alam, Selangor D.E., Malaysia

Saifollah Abdullah
Centre of Nanoscience and Nanotechnology (NANO-SciTech Centre), Institute of Science, Universiti Teknologi MARA, 40450 Shah Alam, Selangor D.E., Malaysia
School of Physics and Material Studies, Faculty of Applied Sciences, Universiti Teknologi MARA, 40450 Shah Alam, Selangor D.E., Malaysia

Ahmad Tavasoli
School of Chemistry, College of Science, University of Tehran, Tehran 1417614411, Iran

Saba Karimi
School of Chemistry, College of Science, University of Tehran, Tehran 1417614411, Iran

Morteza Shoja
School of Chemistry, College of Science, University of Tehran, Tehran 1417614411, Iran

Ahlam M Farhan
Department of Chemistry, Faculty of Science for Women, Baghdad University, Baghdad, Iraq

Ammar H Al-Dujaili
Department of Chemistry, College of Education, Ibn Al-Haitham e, Baghdad University, Baghdad, Iraq

Akl M Awwad
Royal Scientific Society, El Hassan Science City, Amman, Jordan

Nasser Saghatoleslami
Department of Chemical Engineering, Ferdowsi University of Mashhad, Mashhad, 9177948974, Iran

Mohammad Amiri
Department of Chemical Engineering, Ferdowsi University of Mashhad, Mashhad, 9177948974, Iran

Jamal Rohi Golkhatmi
Department of Chemical Engineering, Ferdowsi University of Mashhad, Mashhad, 9177948974, Iran

Hakim Faraji
Department of Chemistry, Varamin-Pishva Branch, Islamic Azad University, Varamin, 33817-74895, Iran

Ramin Mohammad Ali Tehrani
Department of Chemistry, Shahre Rey Branch, Islamic Azad University, Tehran, 1800155144, Iran

Esmaiel Nouri
Department of Materials and Metallurgical Engineering, Iran University of Science and Technology (IUST), Narmak, Tehran 16846-13114, Iran

Mohammad Shahmiri
Department of Materials and Metallurgical Engineering, Iran University of Science and Technology (IUST), Narmak, Tehran 16846-13114, Iran

Hamid Reza Rezaie
Department of Materials and Metallurgical Engineering, Iran University of Science and Technology (IUST), Narmak, Tehran 16846-13114, Iran

Fatemeh Talayian
Department of Materials Engineering, Sahand University of Technology, Tabriz 11111-53317, Iran

Abdul Razak Mohamed Sikkander
Department of Chemistry, Periyar University, Salem, Tamil Nadu 636011, India
Department of Chemistry, Velammal Engineering College, Chennai, Tamil Nadu 600066, India

Chinnappiyan Vedhi
Department of Chemistry, V.O. Chidambaram College, Tuticorin, Tamil Nadu 628008, India

Paramasivam Manisankar
Department of Chemistry, Periyar University, Salem, Tamil Nadu 636011, India
Department of Industrial Chemistry, Alagappa University, Karaikudi, Tamil Nadu 630003, India

Mohammad Hossein Rasoulifard
Water and Wastewater Treatment Research Laboratory, Department of Chemistry, Faculty of Science, University of Zanjan, Zanjan, 45371-38791, Iran

Hamed Majidzadeh
Water and Wastewater Treatment Research Laboratory, Department of Chemistry, Faculty of Science, University of Zanjan, Zanjan, 45371-38791, Iran

Fatemeh Torkian Demneh
Water and Wastewater Treatment Research Laboratory, Department of Chemistry, Faculty of Science, University of Zanjan, Zanjan, 45371-38791, Iran

Elaheh Babaei
Water and Wastewater Treatment Research Laboratory, Department of Chemistry, Faculty of Science, University of Zanjan, Zanjan, 45371-38791, Iran

Mohammad Hassan Rasoulifard
Department of Vetinerary, Islamic Azad University of Tabriz, Tabriz Branch, Tabriz, 1655, Iran

Senthil Kumar Ponnusamy
Department of Chemical Engineering, SSN College of Engineering, Chennai 603 110, India

Ramalingam Subramaniam
Department of Chemical Engineering, University of Louisiana at Lafayette, Lafayette, LA 70504, USA

Vijender Singh
Department of Chemistry, KSKV Kachchh University, Bhuj, Gujarat 370001, India
Department of Chemistry, B.B.K. D.A.V. College for Women, Amritsar, Punjab 143005, India

Poonam Khullar
Department of Chemistry, B.B.K. D.A.V. College for Women, Amritsar, Punjab 143005, India

Pragnesh N Dave
Department of Chemistry, KSKV Kachchh University, Bhuj, Gujarat 370001, India

Navjot Kaur
Department of Chemistry, Lyallpur Khalsa College, Jallandhar, (Punjab) 144001, India

David Ebuka Arthur
Department of Chemistry, Ahmadu Bello University, Zaria, Kaduna State PMB 1044, Nigeria

Achika Jonathan
Department of Chemistry, Ahmadu Bello University, Zaria, Kaduna State PMB 1044, Nigeria

Paul Ocheje Ameh
Department of Chemistry, Ahmadu Bello University, Zaria, Kaduna State PMB 1044, Nigeria

Crystal Anya
Department of Chemistry, Federal University of Technology, Minna, Niger State PMB 65, Nigeria

Mohsen Mansouri
Department of Chemical Engineering, University of Sistan and Baluchestan, Zahedan 98164-161, Iran

Hossein Atashi
Department of Chemical Engineering, University of Sistan and Baluchestan, Zahedan 98164-161, Iran

Ali Akbar Mirzaei
Department of Chemistry, University of Sistan and Baluchestan, Zahedan 98135-674, Iran

Reza Jangi
Department of Chemical Engineering, University of Sistan and Baluchestan, Zahedan 98164-161, Iran

Hassan Farag
Chemical Engineering Department, Faculty of Engineering, Alexandria University, Alexandria, Egypt

Mona Ossman
Petrochemical Engineering Department, Faculty of Engineering, Pharos University, Alexandria, Egypt
City for Scientific Research CSAT, Borg Elarab, Alexandria, Egypt

Moustapha Mansour
Chemical Engineering Department, Faculty of Engineering, Alexandria University, Alexandria, Egypt

Yousra Farid
Chemical Engineering Department, Faculty of Engineering, Alexandria University, Alexandria, Egypt

Govindaraju Muthuraman
Department of Chemistry, Presidency College, University of Madras, Chennai, India

Mohamedjakkaria Mohamed Ibrahim
Department of Chemistry, Presidency College, University of Madras, Chennai, India

Nilesh Shinde
Department of Polymer and Surface Engineering, Institute of Chemical Technology, N M Parekh Marg, Matunga (East), Mumbai 400019, India

Mukesh Kathalewar
Department of Polymer and Surface Engineering, Institute of Chemical Technology, N M Parekh Marg, Matunga (East), Mumbai 400019, India

Anagha Sabnis
Department of Polymer and Surface Engineering, Institute of Chemical Technology, N M Parekh Marg, Matunga (East), Mumbai 400019, India

Nagavani Sunkaraneni
Department of Biochemistry, Kakatiya University, Warangal, Andhra Pradesh 506 009, India

Chandra Mouleswar Rao Jillepalli
Department of Biochemistry, Kakatiya University, Warangal, Andhra Pradesh 506 009, India

Madhukar Jeripothula
Department of Biochemistry, Kakatiya University, Warangal, Andhra Pradesh 506 009, India

Ehsan-o-llah Ettefaghi
Department of Mechanical Engineering of Agricultural Machinery, University of Tehran, Mesbah Cross, Karaj 4111, Iran

Hojjat Ahmadi
Department of Mechanical Engineering of Agricultural Machinery, University of Tehran, Mesbah Cross, Karaj 4111, Iran

Alimorad Rashidi
Nanotechnology Research Center, Research Institute of Petroleum Industry (RIPI), West Blvd. Azadi Sport Complex, Tehran 14665-1998, Iran

Seyed Saeid Mohtasebi
Department of Mechanical Engineering of Agricultural Machinery, University of Tehran, Mesbah Cross, Karaj 4111, Iran

Mahshad Alaei
Nanotechnology Research Center, Research Institute of Petroleum Industry (RIPI), West Blvd. Azadi Sport Complex, Tehran 14665-1998, Iran

Wahyudiono
Department of Chemical Engineering, Nagoya University, Furo-cho, Chikusa-ku, Nagoya 464-8603, Japan

Siti Machmudah
Department of Chemical Engineering, Nagoya University, Furo-cho, Chikusa-ku, Nagoya 464-8603, Japan
Department of Chemical Engineering, Sepuluh November Institute of Technology, Kampus ITS Sukolilo, Surabaya 60111, Indonesia

Kanako Murakami
Department of Applied Chemistry and Biochemistry, Kumamoto University, Kumamoto 860-8555, Japan

Satoko Okubayashi
Department of Advanced Fibro-Science, Kyoto Institute of Technology, Kyoto 606-8585, Japan

Motonobu Goto
Department of Chemical Engineering, Nagoya University, Furo-cho, Chikusa-ku, Nagoya 464-8603, Japan

Akl M Awwad
Royal Scientific Society, El Hassan Science City, Amman 11941, Jordan

Nidá M Salem
Plant Protection Department, Faculty of Agriculture, Jordan University, Amman 11942, Jordan

Amany O Abdeen
Royal Scientific Society, El Hassan Science City, Amman 11941, Jordan

Ali Ashour AL-Dhfeery
Department of Chemical Engineering, Basra University, Basra, 61004, Iraq

Ala'a Abdulrazaq Jassem
Department of Chemical Engineering, Basra University, Basra, 61004, Iraq

Mostafa Roosta
Malek-Ashtar University of Technology, P.O. Box 13336–95711, Tehran, Iran

Hamidreza Baharvandi
Malek-Ashtar University of Technology, P.O. Box 13336–95711, Tehran, Iran

Hossein Abdizade
School of Metallurgy and Materials Engineering, University of Tehran I. R., Tehran, Iran

Mahmoud Zarei
Department of Applied Chemistry, Faculty of Chemistry, University of Tabriz, P.O. Box, 51666–16471, Tabriz, Iran

Alireza Khataee
Department of Applied Chemistry, Faculty of Chemistry, University of Tabriz, P.O. Box, 51666–16471, Tabriz, Iran

Mehrangiz Fathinia
Department of Applied Chemistry, Faculty of Chemistry, University of Tabriz, P.O. Box, 51666–16471, Tabriz, Iran

Fardin Seyyednajafi
R&D Center of Tabriz Petrochemical Complex, Tabriz, Iran

Heidar Ranjbar
R&D Center of Tabriz Petrochemical Complex, Tabriz, Iran